Fine Sediment in Open Water

From Fundamentals to Modeling

ADVANCED SERIES ON OCEAN ENGINEERING

ISSN 1793-074X

Published

Vol. 44 Satellite SAR Detection of Sub-Mesoscale Ocean Dynamic Processes
by Quanan Zheng

Vol. 45 Ocean Surface Waves: Their Physics and Prediction (3rd Edition)
by Stanisław Ryszard Massel

Vol. 46 Tsunami: To Survive from Tsunami (2nd Edition)
by Tomotsuka Takayama, Kazumasa Katoh, Fumihiko Imamura, Yoshiaki Kawata, Susumu Murata and Shigeo Takahashi

Vol. 47 Coastal Engineering: Theory and Practice
by Vallam Sundar and Sannasi A. Sannasiraj

Vol. 48 Introduction to Coastal Engineering and Management (3rd Edition)
by J. William Kamphuis

Vol. 49 Design of Coastal Hazard Mitigation Alternatives for Rising Seas
by David Basco

Vol. 50 Tsunami: Engineering Perspective for Mitigation, Protection and Modelling
by V. Sundar, S. A. Sannasiraj, K. Murali and V. Sriram

Vol. 51 Turbulence in Coastal and Civil Engineering
by B. Mutlu Sumer and David R. Fuhrman

Vol. 52 Ocean Wave Dynamics for Coastal and Marine Structures
by V. Sundar

Vol. 53 Headland-Bay Beaches: Static Equilibrium Concept for Shoreline Management
by John Hsu, Jung L. Lee, Antonio H. F. Klein, Mauricio González and Raúl Medina

Vol. 54 Numerical Modeling of Tsunami Waves
by Juan Horrillo and William Knight, Zygmunt Kowalik

Vol. 55 Fine Sediment in Open Water: From Fundamentals to Modeling
by Johan C. Winterwerp, Thijs van Kessel, Dirk S. van Maren and Bram C. van Prooijen

*For the complete list of titles in this series, go to https://www.worldscientific.com/series/asoe

Advanced Series on Ocean Engineering — Volume 55

Fine Sediment in Open Water

From Fundamentals to Modeling

Johan C Winterwerp
Deltares, The Netherlands & Delft University of Technology, The Netherlands

Thijs van Kessel
Deltares, The Netherlands

Dirk S van Maren
Deltares, The Netherlands

Bram C van Prooijen
Delft University of Technology, The Netherlands

World Scientific

NEW JERSEY • LONDON • SINGAPORE • BEIJING • SHANGHAI • HONG KONG • TAIPEI • CHENNAI • TOKYO

Published by

World Scientific Publishing Co. Pte. Ltd.
5 Toh Tuck Link, Singapore 596224
USA office: 27 Warren Street, Suite 401-402, Hackensack, NJ 07601
UK office: 57 Shelton Street, Covent Garden, London WC2H 9HE

Library of Congress Control Number: 2021042904

British Library Cataloguing-in-Publication Data
A catalogue record for this book is available from the British Library.

Advanced Series on Ocean Engineering — Vol. 55
FINE SEDIMENT IN OPEN WATER
From Fundamentals to Modeling

ISBN 978-981-124-361-5 (hardcover)
ISBN 978-981-124-363-9 (ebook for institutions)
ISBN 978-981-124-364-6 (ebook for individuals)

For any available supplementary material, please visit
https://www.worldscientific.com/worldscibooks/10.1142/12473#t=suppl

Desk Editors: Vishnu Mohan/Michael Beale/Shi Ying Koe

Typeset by Stallion Press
Email: enquiries@stallionpress.com

To Walther van Kesteren, our dear and inspiring friend who regretfully could not contribute to this book.

E qual maggior sciochezza si puèo immaginar di quella che chiama cose preziose le gemme, l'argento e l'oro, e villissime la terra e il fango?

Galileo Galileï, 1632: Dialogo sopra i due massimi sistemi del mondo.

And can one imagine a greater delusion than calling gems, silver and gold precious, while calling soil and mud vile materials?

Preface

The current book continues the work of Winterwerp and Van Kesteren (2004) and Mehta (2013). Our main objective is to provide guidelines for good modeling practice to quantify the transport and fate of fine sediment in open water, with emphasis on the marine environment. We emphasize that a (numerical) model should never be an objective by itself. It is always a tool to help answering managerial or engineering questions, or to enhance our understanding of nature. Yet, in this book we do not elaborate on how to use a model as a tool, as this is too unique an issue.

We argue that the single most important requirement for good modeling is a thorough understanding of the physico-biological system to be studied — we refer to *system understanding* (Chapter 13). Such system understanding is imperative for developing a conceptual model of the aquatic system at hand, which then forms the basis for any further (numerical) modeling and setup of field surveys. This includes understanding of all relevant processes at various temporal and spatial scales, which are discussed in the first ten chapters of this book. A numerical model is therefore to be regarded as a mathematical formalization of a conceptual model. We elaborate extensively on the hydrodynamic, sedimentological and soil mechanic (and sometimes biological) time scales as these govern the dynamics in estuarine and coastal systems. We repeat some of the most important concepts presented earlier in Winterwerp and Van Kesteren (2004) to contain all relevant information in one volume. Moreover, hydrodynamic processes are discussed in detail, as it is water movement that transports fine sediment in open waters (Chapters 9 and 10). In Chapter 10, we also touch upon large-scale driving forces, governing coastal processes, such as the effects of earth rotation.

The numerical modeling addressed in the second part of this book also requires input of sediment properties, such as the parameters for erosion and deposition formulas. However, by its nature, the properties of fine sediment vary widely, and are therefore site-specific, thus (still) have to be determined experimentally. Appendix A summarizes methods for measuring these properties in the laboratory, while we emphasize the importance of protocols for sampling, sample storage and sample treatment.

In Chapters 14–18, we illustrate our reasoning with a number of examples from Dutch coastal waters. Though we applied our approach in numerous other (commercial) studies across the world, we limit our elaborated examples to studies in Dutch aquatic systems because of confidentiality and copyright reasons.

This book is mainly written for professional engineers working in estuaries and coastal systems, using more or less advanced numerical models. However, the first part of the book also contains material for graduate students, including those following the course on Sediment Dynamics at Delft University of Technology.

The authors would like to acknowledge gratefully the following contributions to this book: Sec. 2.1 was written by Dr. C. Chassagne, Sec. 7.6 was written by Dr. O.J. Kirichek, who also reviewed Appendix A, Chapter 8 is based on the class notes by Prof. P.M.J. Herman, who also reviewed this chapter in detail, Sec. 10.2.2 is largely based on the PhD-thesis by Dr. G.J. de Boer, who also reviewed this section. Mr. J. van Gils provided the information for Sec. 3.5, and reviewed this section.

Chapters 1–12 have been reviewed in detail by Prof. A.J. Mehta, Prof. Z.B. Wang reviewed Chapter 9, Dr. P. Vardon reviewed Chapters 2 and 5, and Mr. J.A.T.M. van Kester reviewed Chapter 11.

We received graphs and drawings from Dr. M. Barciela Rial, Mr. E. Hendriks, Dr. W. Jacobs, Dr. M.A. de Lucas Pardo, Mr. R. Matthew, Dr. S. te Slaa. Mrs. C. van der Boog prepared a number of drawings and Mr. Jianwei Sun helped with the list of references. The contributions by Thijs van Kessel and Bas van Maren were financially supported through Deltares Strategic Research Funds "Understanding System Dynamics" and "Resilient Ecosystems".

Han Winterwerp, Bas van Maren,
Thijs van Kessel, Bram van Prooijen

Contents

Chapter 1

Introduction

In the mid-1950s, the U.S. Army Corps of Engineers, San Francisco District asked Prof. Hans Albert Einstein for advice to resolve navigational problems in San Francisco Bay (Ettema and Mutel, 2014). Navigation was seriously hampered by rapid shoaling and bar formation with fine, muddy sediment in the channels, occurring in particular during summer, when the sediment load in the rivers was low. After some preliminary studies, it became evident that the processes of erosion and deposition of this fine sediment were not understood. Therefore, Einstein set up an extensive research program to study the behavior of fine cohesive sediment, including experimental work in the Hydraulic Laboratory of the University of California, Berkley, USA. This research was carried out with Ray B. Krone and Emmanuel Partheniades, his two PhD-students who pioneered our current understanding on the behavior of cohesive sediment.

Ray Krone carried out ground-breaking experiments in two straight flumes at Berkeley, revealing the processes of flocculation and floc break-up as a function of the hydraulic stresses on the clay particles. This research contributed greatly to the understanding of the transport and fate of fine sediment in San Francisco Bay. Emmanuel Partheniades focused more on the water-bed exchange processes, and carried out erosion and deposition experiments in a rotating annular flume at MIT. This highly innovative circular flume, in which the flow is driven by a rotating lid, whereas secondary currents are minimized by counter-rotating the flume itself, became one of the major workhorses for experimental research on cohesive sediment in the next two to three decades across the world.

In the early 1970s, the results from the erosion and deposition experiments by Krone and Partheniades were parameterized into a linear model, generally referred to as the Krone-Partheniades law for erosion and deposition. This law is currently being used as the bed-boundary condition in many (most) numerical models describing the transport and fate of fine sediment in open water systems. Though there is some dispute about who first formulated this linear model, it is certain that two of Prof. Krone's students were involved, i.e. Kandiah (1974; see also Kandiah and Arulanandan, 1974) and Ariathurai (1974; see also Ariathurai and Arulanandan, 1978). Ariathurai was probably the first to develop a numerical model for fine sediment transport (within San Francisco Bay) using this linear erosion/ deposition law.

While numerical modeling of the transport and fate of fine sediment took a great leap, Prof. Partheniades continued his research on the physics governing the behavior of these sediment until his retirement from the University of Florida in 1997. He summarized the results of his work in 2009 (Partheniades, 2009), part of which was carried out with Ashish J. Mehta. Prof. Mehta would become Partheniades' most prominent PhD-student continuing on the work started by Einstein, Krone and Partheniades, through his research with numerous MSc- and PhD-students, his teaching and the Intercoh conference series which he initiated. In 2013, Prof. Mehta (Mehta, 2013) summarized the developments on the science and application of cohesive sediment over half a century, in conjunction with an excellent introduction to the subject at graduate and undergraduate level.

While the work of the American school is disseminated and being followed globally, the pioneering work in the 1960s by Claude Migniot in France is less known, likely because it was all published in French. The main results of his work can be found in three publications in "La Houille Blanche" (Migniot, 1968, 1989a,b). Prof. Migniot taught at various universities in France and contributed to almost all subjects on cohesive sediment, such as settling and flocculation, consolidation and erosion processes. He initiated experimental work on mixed sediment and built up a considerable database on consolidation. His data on the settling velocity of flocculated and non-flocculated sediment is still relevant. Last but not least, Mignot was likely the first to address the non-Newtonian behavior of cohesive sediment (slurries) in great detail.

The initial research in the UK (at the Hydraulic Research Station Wallingford: M. Thorn, M. Owen and N. Odd and at IOS Taunton: R. Parker and R.B. Kirby) was mainly carried out *in-situ* through dedicated field surveys and measurements. The so-called Owen-tube for instance, became widely used for *in-situ* measuring settling velocities of fine sediment. Systematic research on the formation and behavior of fluid mud in harbor basins and navigation channels was first carried out in the UK.

Kirby and Parker (1974), Parker *et al.* (1994), Whitehouse *et al.* (2000) summarize the experience in the UK and the engineering rules developed in a handy booklet, which is still useful today.

Systematic research on cohesive sediment processes in The Netherlands started only in the mid-1980s by Van Leussen (1994) following on the more applied studies by Allersma (1971) and Augustinus (1988, 1978) in (British) Guyana and Suriname. The knowledge built up at Delft Hydraulics (now Deltares) and Delft University of Technology was summarized by Winterwerp and Van Kesteren (2004), in which the interaction between hydrodynamics, fine sediment and soil mechanical processes was emphasized.

Focus on near-shore marine environments

The focus of this book is on fine sediment in the marine environment, though we cover the entire land-ocean interaction of catchments, i.e. (1) rivers, (2) reservoirs and lakes, (3) tidal rivers, (4) estuaries, (5) regions of fresh water influence in the coastal zone (ROFI), and (6) the coastal zone. Tidal basins, lagoons, sea-arms, etc. fit in one of these categories. Figure 1.1 suggests that these systems can be huge, up to 1000s of km, though much smaller systems are more common. On the other hand, as the physical processes of fine sediments are dominated by molecular forces between the clay particles, we deal with a large range in scales, both temporal and spatial.

A proper system understanding starts with appreciating these scales. Sediment composition generally varies widely, from boulders to gravel, to sand, and finally to fine clay particles. We will not treat the behavior of sediment larger than a few 100 μm, in fact we focus on muddy systems or systems with (some) fine sand (up to around 200 μm). Throughout, we refer

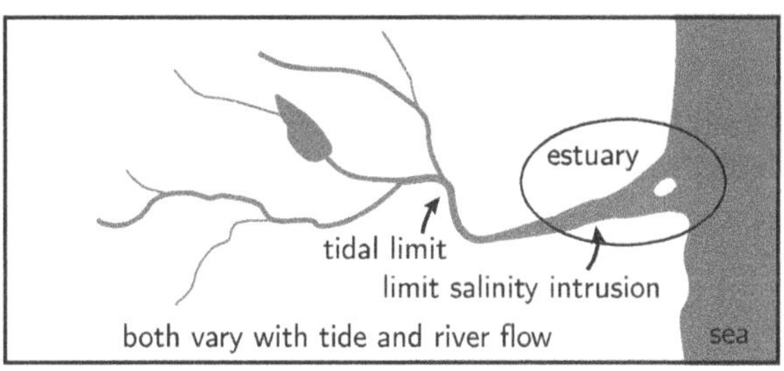

Fig. 1.1. Though focus is on the marine environment, this book treats basically the entire land–ocean interaction of catchments — note that both the tidal limit and salinity intrusion vary with tide and river flow.

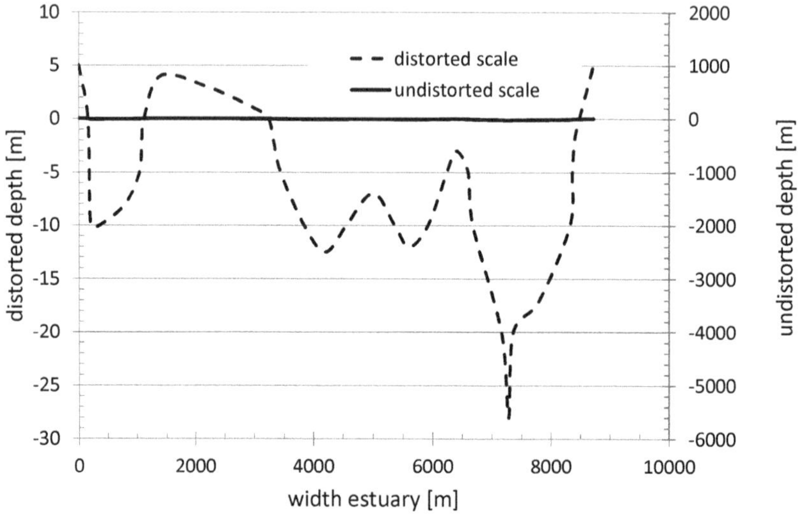

Fig. 1.2. Distorted and undistorted (i.e. vertical and horizontal scale are identical) profile of an arbitrary cross section in the Western Scheldt, The Netherlands.

to fines c.q. fine sediment and cohesive sediment, where cohesive sediment forms a specific class owing to their inherent cohesion. However, fines and cohesive sediment have in common that settling velocities are small, whereas silt (particle size between 2 and 64 μm) may exhibit pseudo-cohesion.

Only shallow, open water systems are treated. This implies that the cross section of such systems can be regarded as very thin laminae (typical depths of a few 10 m or smaller and widths of many to 10s of kilometers), e.g. Fig. 1.2. As a result, on short time scales, SPM (suspended particulate matter) concentrations are largely governed by water-bed exchange processes. These exchange processes therefore form an important part of this book, i.e. Chapters 2–7. However, in spite of their shallowness, three dimensional processes (e.g. gradients over the water depth) do play an important role, as discussed in next chapters, and particular in Chapters 9 and 10. On longer time scales, import/export of SPM becomes important, as these define the availability of fine sediment in the aquatic system. Processes contributing to this long-term availability of fine sediment are discussed in Chapters 9 and 10.

In Chapter 8, we discuss the important role of biological processes, not only on the sediment properties themselves, but also with respect to the overall hydrodynamics and interaction with the physical system.

In marine waters, pH generally measures around 8, thus buffering many chemical processes. These can therefore often be accounted for implicitly in the various parameters of physical processes, and chemical processes are touched upon only occasionally. However, in fresh water systems, variability in the chemical environment can be much larger, and care should be taken in applying the formulas and models presented in this book.

Transport modes

Obviously, the hydrodynamics in the aquatic system of Fig. 1.1 vary significantly, traveling from mountainous streams via wide meandering or breaching mainland rivers and estuaries to the coastal waters. Moreover, the bed composition changes from rocky substrates to sandy substrates, and to more muddy beds. These variations imply that the mode of sediment transport and the governing physical processes vary along these systems. Sediment transport (sediment load) measured at a cross-section in one of the river branches of Fig. 1.1, or at a cross section in the coastal zone is traditionally sub-divided into bed-material load and wash load. Bed-material load consists of sediment found in a sample taken from the bed. Sediment not found in the bed, but present in the water column, is referred to as wash load — see Fig. 1.3 for this traditional scheme.

Bed-material load may consist of bed load and/or suspended load. Bed load occurs as saltation of grains, i.e. small jumps of sediment particles close to the bed. In suspended load, sediment particles are mixed over part of or over the entire water column by turbulent motions in the flow. For sandy sediment, many algebraic transport formulas have been derived for bed load and suspended load as functions of flow velocity, water depth and grain size, e.g. Einstein (1950), Vanoni (1940), Raudkivi and Witte (1990) and Van Rijn (1993).

All textbooks on sediment transport address these modes of transport. However, owing to the large variations in hydrodynamics, sediment classified as wash load may become suspended load, or even bed load elsewhere (generally downstream) in the system. Also, at one location, sediment may

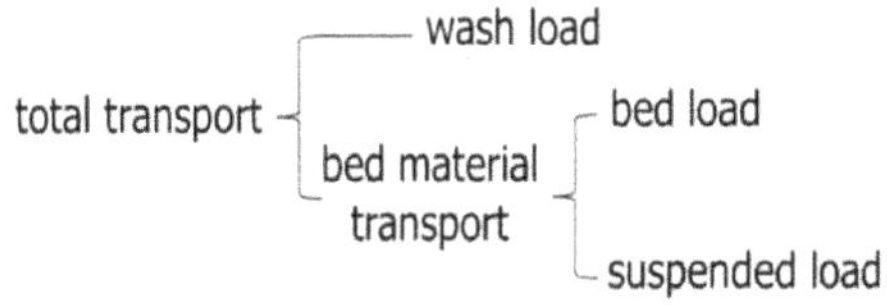

Fig. 1.3. Traditional diagram with the various transport modes.

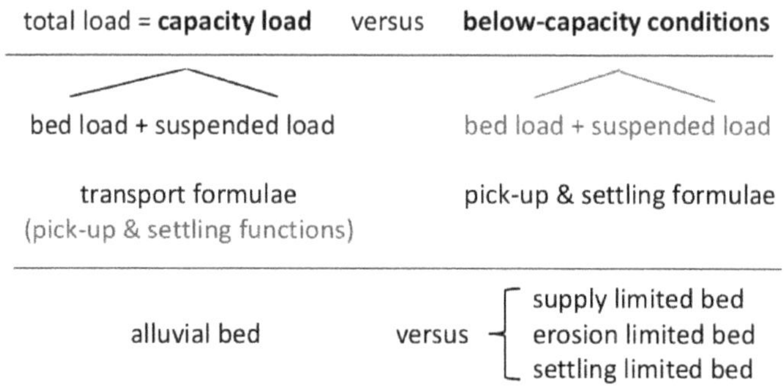

Fig. 1.4. Alternative diagram with transport modes, including beds with/of fine sediment.

behave as bed load at low river flows, becoming suspended at higher river flows.

Let us next analyze a river whose bed consists of rock or stiff boulder clay. In this case, transport of sand, which would otherwise be defined as bed-material transport, would be identified as wash load. On the other hand, even under the most dynamic flow regimes, e.g. mountain streams, sands and clays are found in some reaches of the river bed, for instance at the lee-side of boulders. Furthermore, in muddy systems, the availability of fines is often limited, and its load cannot easily be classified as suspended load nor as wash load. In other words, the traditional scheme of Fig. 1.3 does not cover all possibilities. Therefore, we propose an alternative scheme, distinguishing between capacity and below-capacity conditions, e.g. Fig. 1.4.

Capacity and below-capacity conditions

Equilibrium load, as commonly referred to in literature, is governed by the transport capacity of the flow (see also Sec. 7.1). Therefore, we refer to capacity conditions, carrying capacity or capacity load, instead of the commonly used expression of "equilibrium load". Moreover, we use the "capacity" concentration C_{cap} to characterize these capacity conditions instead of the commonly used "equilibrium" concentration C_{eq} to prevent confusion with the many equilibrium and off-equilibrium concepts treated in this book. This capacity is determined by turbulent motions (vertical mixing) of the flow and the mass (size/buoyancy) of the sediment particles. At microscale this implies that sedimentation and erosion balance — the number of particles picked up from the bed by erosion equals the number of particles that settle on the bed by sedimentation (but are generally different particles!). Moreover, it is implicitly assumed that the vertical profile of SPM is in equilibrium (balance between settling and vertical mixing throughout the water column), see Sec. 3.4. If the flow velocity were to increase, more sediment would be picked up and brought into the water column, the carrying capacity of the flow increases, and the sediment load increases till a new equilibrium between deposition and erosion is attained. For coarse sediment (sand), this occurs rapidly, for bed load even instantaneously. Of course this requires that all sediment to reach transport capacity conditions is available and can be picked up from the bed. We refer to an alluvial bed. Capacity load may consist of bed load and/or suspended load, and can be quantified with the algebraic transport formulas mentioned above. Under specific conditions, to be discussed in Chapter 3, SPM is distributed over the water column in the form of a Rouse profile. However, actual SPM profiles may deviate considerably from this profile, which is one reason of the wide spread in these transport formulas (Sec. 3.4.3).

Below-capacity conditions occur when the bed cannot supply sufficient sediment to reach capacity. They can occur under so-called supply-limited, erosion-limited and/or settling-limited conditions. Supply-limited conditions, for example, are found in sandy beds with some fines, as sketched in Fig. 1.5(b); these conditions are further discussed in Sec. 6.6. Erosion-limited conditions are found for cohesive sediment beds with a vertical strength distribution. Only a small part of the bed surface can be eroded where strengths are below the critical shear strength of erosion, e.g. Fig. 1.5(c), and Secs. 6.3 and 6.5. Settling-limited (or better settling-affected) conditions occur when no equilibrium (Rouse) profile of SPM is formed (often under tidal conditions), because the low settling velocity of SPM implies long settling times, hence limiting the amount of sediment settling on the bed during slack water, as explained in Sec. 3.4.

Wash load is a form of below-capacity transport. This load is mainly governed by upstream conditions, i.e. the sediment supply. For below-capacity conditions, no (algebraic) transport equations can be derived. Sediment exchange with the bed has to be described by pick-up formulas. Note that such settling and pick-up functions can also be used for assessing capacity transport rates, in which case the settling/pick-up functions and the total transport formulas should be mutually consistent (e.g. Van Rijn, 1993).

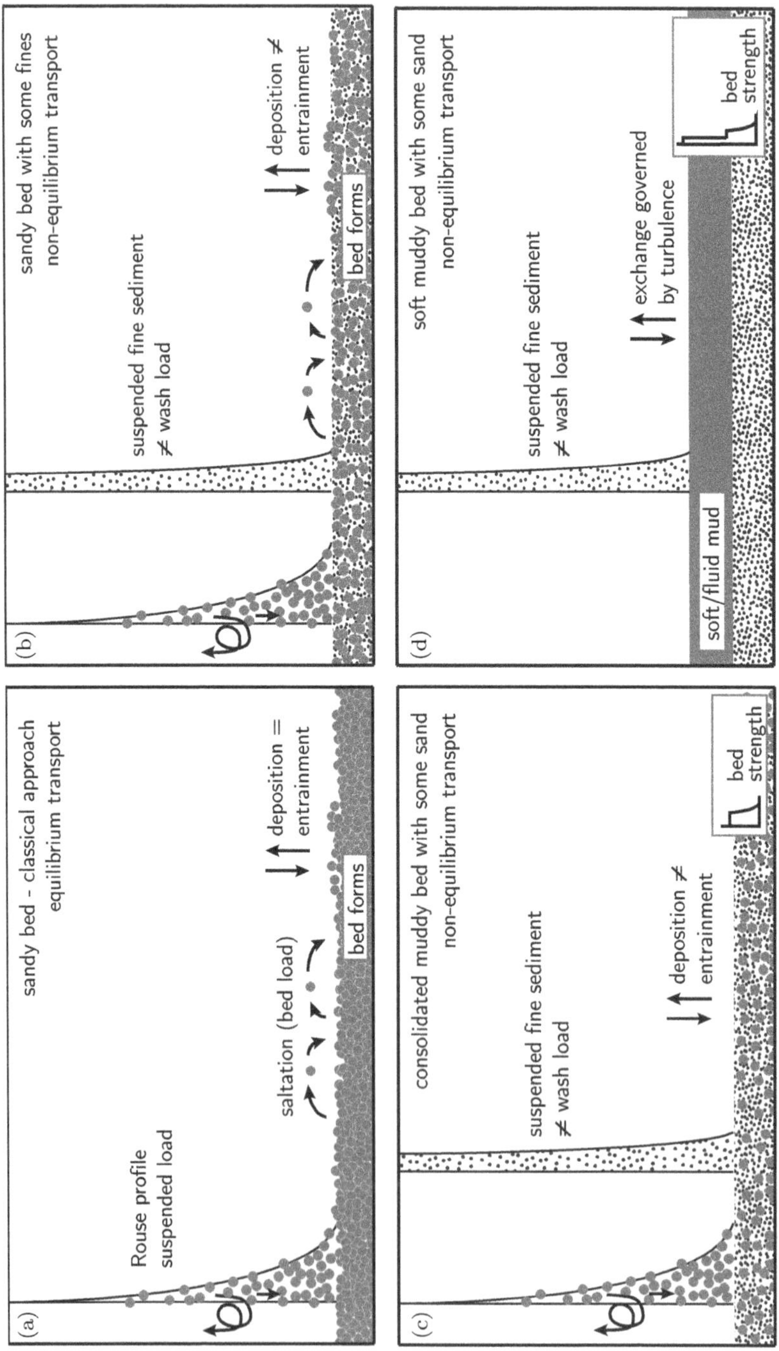

Fig. 1.5. Modes of transport: (a) sandy systems, (b) sand–mud mixtures, (c) mud–sand mixtures, (d) fluid mud conditions.

The distinction between capacity and below-capacity load is particularly relevant for describing the transport of cohesive sediment, and of sediment mixtures (sand–mud mixtures). Moreover, under tidal conditions, the time scales of the various processes to reach capacity should be related to the tidal time scale, as discussed in Sec. 3.4, and Chapter 9.

This distinction is visualized in Fig. 1.5. Figure 1.5(a) reflects capacity transport (capacity conditions) for sandy systems, as subject of many textbooks on sediment transport. Part (or all) of the sand is transported as bed load, which induces bed forms. If some fines were found in the bed, they would be washed from the bed when the sandy substrate erodes, see Fig. 1.5(b), or buried when the bed forms migrate (Sec. 6.6). The amount of fines in the water column is therefore governed by the erosion rate of the sandy substrate. Such a system generally exhibits capacity conditions for the sandy fractions, whereas the fines in the water column would not be at capacity. When the fraction of cohesive sediment in the bed increases, the bed becomes cohesive (Sec. 2.2), and the water-bed exchange of sediment is governed by the muddy substrate of the bed, as sketched in Fig. 1.5(c). In this case, the transport of non-cohesive and cohesive sediment is generally below capacity. Then the bed exhibits starved-bed conditions (i.e erosion and supply limited) for both the sand and mud fractions. This behavior is largely governed by the soil mechanical properties of the bed. Chapter 5 discusses the strengths of and stresses within a bed, and addresses the important role of over-consolidation and pseudo-cohesion. A special case occurs when the bed consists of large amounts of very soft cohesive sediment, which may have fluid-like characteristics (referred to as fluid mud, e.g. Fig. 1.5(d)). Water-bed exchange is then largely governed by hydrodynamic processes (mixing and entrainment, see Secs. 3.2.2 and 7.5.1), and SPM in the water column may, or may not be at capacity conditions. We touch upon this special situation in Chapter 7.

In the above, capacity is discussed at one particular location. However, the driving forces and type of sediment and its properties vary (largely) over an aquatic system, both in space and time. This induces a "memory" in the system: local SPM-concentrations and bed composition are affected by conditions elsewhere and previously occurring. Sediment transport has to therefore be regarded from a system-wide perspective as well. Because of the system's memory, one can only analyze dynamic conditions, i.e. the parameter(s) to be studied exhibit(s) time-varying behavior (sometimes cyclic, sometimes more erratic) within a specific time window. For fine sediment studies under morphostatic conditions, this window amounts to one or a few years, accounting for seasonal effects and year-to-year variations therein. The bathymetry does not change (the morphos (= form) is static) within this window, but sediment composition within the bed and SPM-concentrations in the water column can exhibit more or less cyclical behavior, governed by the hydrodynamic, sedimentological and soil mechanic time scales, which form therefore a key issue in this book.

A morphostatic approach is feasible if bed level changes over that time window are small. If not, a morphodynamic approach is required, accounting for bathymetrical changes in response to hydrodynamics and gradients in sediment transport. Also in this approach a dynamic equilibrium is possible, implying that the bathymetry (in particular channel-shoal complexes) attains a particular pattern — one refers to a morphodynamic equilibrium. Time scales to reach morphodynamic equilibrium are often long, and scale with the spatial dimensions of the system at hand (De Vriend, 1991a). Discussions on morphodynamics and morphodynamic equilibrium are subject of other textbooks and are beyond the scope of the present book.

Setup of this book

This book treats the behavior and modeling of fine sediment mainly in the marine environments, with emphasis on cohesive sediment. Cohesive sediment, when forming a soil, is often referred to as mud. Mud is a mixture of water, minerals (clay, silt and fine sand), organic material and sometimes gas. Cohesion can be understood from a microscopic and a macroscopic perspective, as discussed in detail in Chapter 2. In that chapter, we also introduce other basic sediment properties necessary to explain and understand the full sediment cycle, i.e. floc formation, settling, consolidation (bed formation) and erosion. Our discussion on these sediment properties begins in the muddy bed, being relevant to the water column as well. Furthermore, our understanding can only be complete by including a thorough analysis of the driving hydrodynamic forces, which are discussed in Chapter 3. Chapters 9 and 10 explain how asymmetries in these driving forces determine the large-scale horizontal transport of fine sediment in estuaries and in the coastal zone.

In Appendix A, we discuss protocols for measuring the various relevant properties of fine sediment in the laboratory. Some of these measurements are described in detail in national/international standards, such as in the AMST (American), NEN (Dutch), BS (British) norms. However, not all of these norms are appropriate for determining the properties of mud in the natural environment. In that case, we propose alternatives.

The first ten chapters, in conjunction with Appendix A are meant to provide the information for carrying out engineering studies through numerical modeling, or otherwise, of muddy aquatic systems. Yet, we maintain that "modeling remains an art" in spite of progress made in hardware and software, and that the modeling of muddy systems is full of pitfalls. The relevant equations are given in Chapter 11, whereas Chapter 12 describes some model schematizations. We discuss best modeling practice, both at an abstract level in Chapter 13, and in the form of simple examples (see Chapters 14 through 18). Appendix B summarizes some of the most important terms used.

Chapter 2

Cohesive Sediment Properties

Mud is a mixture of minerals, organic matter, (a lot of) water, and sometimes gas, as sketched in Fig. 2.1. The minerals consist of clay, silt and fine sand. Water sits in the pores of the sediment, or is bound to the clay. The relevant organic matter consists, amongst others, of organic polymers, such as polysaccharides. In this book, we treat saturated condition only, hence the gas content is zero.

In this chapter, we discuss the properties of cohesive sediment and granular material, to establish whether a soil exhibits cohesive behavior. Understanding cohesion requires analysis of the mutual interactions between individual clay particles, and the role of pore water chemistry and organic matter (organic polymers). This is discussed in Sec. 2.1.

Understanding the mechanical response at a larger scale requires analysis of the soil mechanical properties and behavior of a soil, as presented in Sec. 2.2. Later in this book, the processes of consolidation, erosion and the behavior of fluid mud are treated, based on a conceptual soil mechanical model. This model is based on a slight modification of the well-known Critical State Model (CSM), explained in Sec. 2.3. This Critical State Model is used to explain observed behavior in a qualitative way, not quantitatively.

2.1 Microscopic cohesive sediment properties[1]

Some (macroscopic) behavior of mud, such as swelling,[2] cohesion, flocculation and its mechanical properties (rheology, when in a (quasi-)fluid state),

[1]This section was written by Dr. C. Chassagne.

[2]In colloidal science, swelling refers to the uptake of water by clay particles (microscopic swelling), whereas in soil mechanics, swelling describes the expansion of a soil by the to-flow of water (macroscopic swelling).

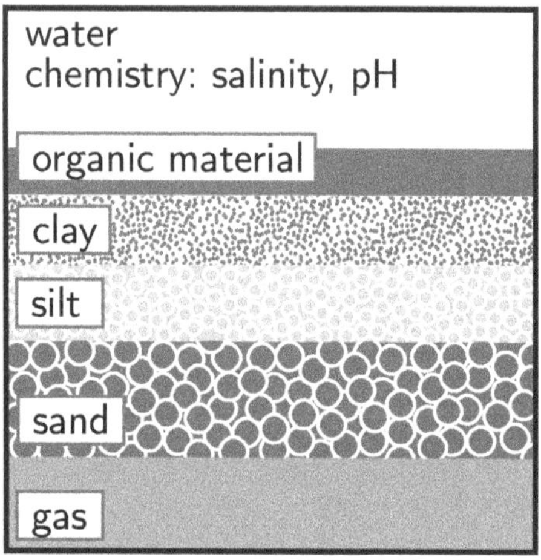

Fig. 2.1. Mud is a mixture of minerals (clay, silt and fine sand), organic material, a lot of water, and sometimes gas. The composition of mud generally changes over time (e.g. tide, season) and space (e.g. channel, intertidal).

can be understood only when studied at microscopic scale, i.e. at the scale of clay particles. Clay particles measure in the μm range or smaller; this class of particles is called colloidal. The behavior of colloidal suspensions has been studied extensively, and is a discipline by itself (Verwey and Overbeek, 1999; Wall, 2010). The reader is referred to, for instance, the Handbook of Clay Science (Bergaya and Lagaly, 2013) for more in-depth information.

The most relevant clays in open water

Most soils are formed by the breakdown of rock, and are therefore referred to as weathering products, such as feldspar and mica. As weathering proceeds, these minerals may be broken down further, forming clays.

Feldspar and mica, though, are non-clay minerals thus non-cohesive, yet can have a size in the clay range, i.e. of the order of a micron, see Sec. 2.2. These minerals belong to the family of tecto-silicates with a three-dimensional framework of silicate tetrahedra with SiO_2 at a 1:2 ratio. These minerals comprise nearly 75% of the earth crust. Tecto-silicates are alumino-silicates, i.e. minerals composed of aluminum, silicon, oxygen, and counter-cations Mitchell (1976). An exception is quartz with its crystalline structure composed of SiO_2 only.

Table 2.1. Clay mineral groups, layer type, swelling type and chemical formula.

Group	Layer type	Swelling index I_s	Chemical formula
Kaolinite	1:1	0.08–0.07	$Al_2Si_2O_5(OH)_4$
Illite[1]	2:1	0.65–0.28	$(K,H_3O)(Al,Mg,Fe)_2(Si,Al)_4O_{10}[(OH)_2,(H_2O)]$
Smectite[2]	2:1	3.60–0.26	$(Na,Ca)_{0.33}(Al,Mg)_2(Si_4O_{10})(OH)_2{\cdot}nH_2O$

Notes: [1]clay, mica, [2]montmorillonite, bentonite.

Clay minerals are hydrous alumino-silicates which form flat hexagonal sheets. These exhibit intrinsic cohesive behavior because of the silica tetrahedron SiO_4^{4-} and the octahedral complexes, which comprise a metal cation M^{m+} and six anions X^{b-}. The cation is usually Al^{3+} or Mg^{2+} in octahedral coordination with oxygen or hydroxyl groups. These combine in the form of sheets, which are the basic structures of clay minerals (Sposito, 2008). Three of the most common groups of clay minerals are given in Table 2.1.

Swelling behavior

The swelling index I_s is defined as:

$$I_s = -\frac{de}{d\left(\log \sigma_{zz}^{sk}\right)} \tag{2.1}$$

where e is the void ratio and σ_{zz}^{sk} the effective stress, see also Chapters 4 and 5. The swelling index is measured with a strain–stress test. Its definition is similar to the compression index, measuring the soil response to compression. The swelling index may vary considerably, as it depends strongly on the pore water chemistry (Mitchell, 1976).

Bonding between tetrahedral and octahedral sheets is of the primary valence type, and therefore very strong. A stack of such tetrahedral and octahedral sheets is called a unit layer. A variety of bonds can be formed between such unit layers, which may be so weak that the physical and chemical behavior of clays is affected by environmental conditions. Typical interlayers are shown in Fig. 2.2. Three types of interlayer bonding are common, as summarized in Table 2.2.

Isomorphous substitution in clay minerals (such as montmorillonite) induces net negative charges in the interlayer space to compensate for the presence of cations to achieve electro-neutrality. Many of these cations are exchangeable so they can be replaced by other cations. The amount of exchangeable cations is referred to as CEC (cation exchange capacity).

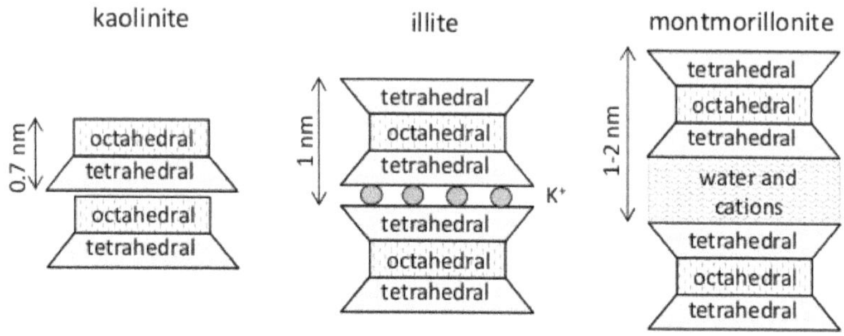

Fig. 2.2. Schematic of the principal clay minerals in terms of layers; octahedral refers to gibbsite (Al) or brucite (Mg) layers, tetrahedral refers to silica layers.

Table 2.2. Interlayer forces.

Interlayer bonding	Characteristics
Van der Waals forces	Bonding is weak; minerals cleave in parallel planes
Hydrogen bonding	Hydrogen bonds remain stable in the presence of water
Interlayer cation	Depending on the cation, bonding can be strong or weak

The swelling index is closely related to the interlayer bonding. In kaolinite, layers are kept together by Van der Waals forces and hydrogen bonds. This bonding is strong and interlayers therefore do not swell in contact with water. The large swelling rate of smectite is still not fully understood. It is thought that cations in the interlayer hydrate in contact with water, and that the hydration energy overcomes attractive forces between layers, leading to a delamination of the clay.

Surface charge

The surface of Si-sheets in a 1:1 clay mineral contains oxygen atoms, while the surface of Al- and Mg-sheets contains hydroxyl groups. In a 2:1 clay mineral, either side of the surface of Si-sheets consists of oxygen atoms. At the edges of both 1:1 and 2:1 clay minerals, oxygen atoms and cations (Si^{4+}, Al^{3+} and Mg^{2+}) are exposed, which may react with molecules in the ambient (pore) water. Depending on the pH of the aqueous environment, the edges consist of different amounts of $-O^-$, $-OH$ and $-OH_2^+$ groups. They can have positive, neutral or negative charges, depending on the pH. When positively charged (at pH $< 4 - 6$ for most clays), edge to face aggregation can occur, leading to so-called card-house structures (see also Sec. 4.3). The faces consist of hydroxyl and siloxane groups, which can dissociate depending on pH, leading to an overall negative surface charge for most clays at pH > 7.

When the overall surface charge of clays is close to zero, which happens beyond a threshold salinity, rapid flocculation is expected, in line with the DLVO theory, see below. This can be assessed from electrophoretic measurements, as discussed in the following paragraphs.

The surface charges on clay particles, together with the environmental conditions (pH, salinity) influence particle–particle interactions, which in turn affect the rheological behavior of clays, see also Chapter 5 and Sec. 7.3.

Electrostatic forces and pore water chemistry

As the surface of clay particles is charged, electrostatic (Coulombic) forces between particles are the fundamental forces determining the physical properties of clays. The overall surface charge of clay particles is generally negatively charged, as sketched in Fig. 2.3, while its edges can be either negatively or positively charged (see above). Hence, clay particles in low-salinity environments at neutral pH would repel each other. This Coulombic repulsive force is proportional to the overall charge of the particles, while decreasing exponentially with distance d from the particle.

However, these repulsive electrostatic forces can be compensated by ions dissolved in the ambient (pore) water. Such solutions are called electrolytes. Positively charged cations cluster around negatively charged clay particles, neutralizing this charge at the border of the so-called diffusive double layer (Fig. 2.3). If this diffusive double layer is thin enough, mutual

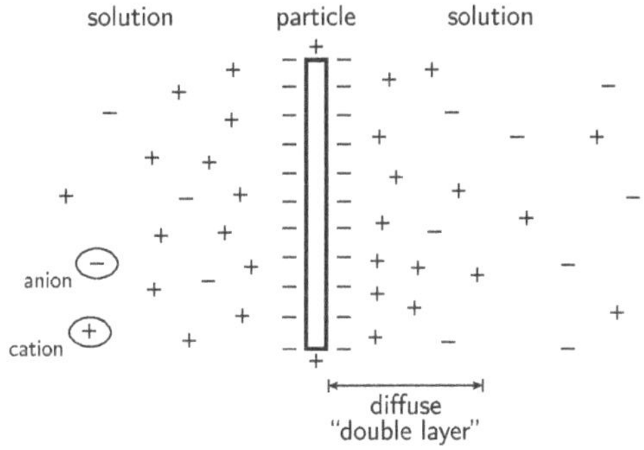

Fig. 2.3. Schematic of clay plate with electric charge in a solution with cations (electrolyte). At some distance, the particles electrical charge is neutralized. A good estimate for this distance is given by the so-called Debye length, which is a measure for the diffuse double layer thickness.

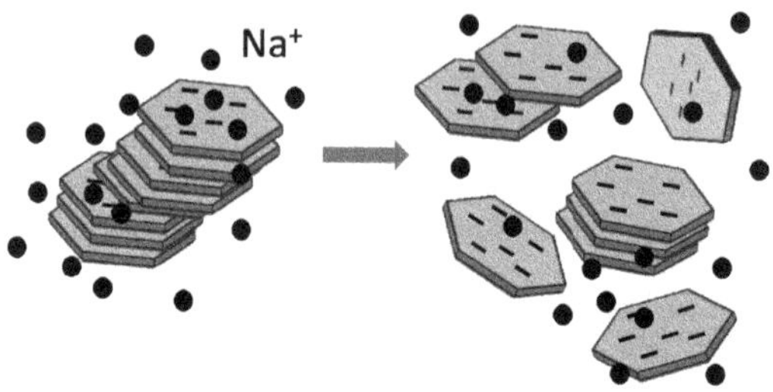

Fig. 2.4. Schematic of stack formation (left) of alkali smectite clay platelets in an aqueous dispersion with sodium cations and subsequent delamination (right) of such a stack, redrawn from Lagaly and Jasmund (1993).

particle attraction by molecular forces (the Van der Waals forces) becomes dominant, and stacks of clay particles may be formed, e.g. Fig. 2.4, left panel. The Van der Waals force depends on the composition of the particles and the surrounding medium only, and decreases with d^{-3}–d^{-7}, depending on the particle shape. The interaction between attractive and repulsive forces forms the basis of the so-called DLVO-theory (i.e. Derjauin–Landau–Verwey–Overbeek; see Verwey and Overbeek, 1999), which quantifies the interactions of colloids in an electrolyte.

These stacks form the so-called primary particles (or flocculi, e.g. Krone, 1986), i.e. the smallest clay aggregates found in natural open waters. Through aggregation, these primary particles can form larger structures, called aggregates or flocs. These primary particles are therefore the building blocks of flocs. The dynamics of floc formation are discussed in Sec. 4.3. One important parameter in the flocculation process is the aggregation coefficient (efficiency). This parameter describes the likelihood of two colliding particles to stick together. The aggregation coefficient is usually found empirically, by fitting data, though efforts have been made to relate it to the microscopic properties of the particles.

The aggregation coefficient is largely determined by the surface charge of clay particles. As a rule of thumb, particles with small surface charge are more prone to aggregation, whereas particles with large surface charge are less likely to form aggregates ("small" and "large" are defined more precisely below). This rule applies when colloidal clay particles are dispersed in an electrolyte, i.e. a solution with ions. However, this picture becomes more complex with other types of particles in a suspension, such as polymeric substances — their role is discussed in the rest of this section.

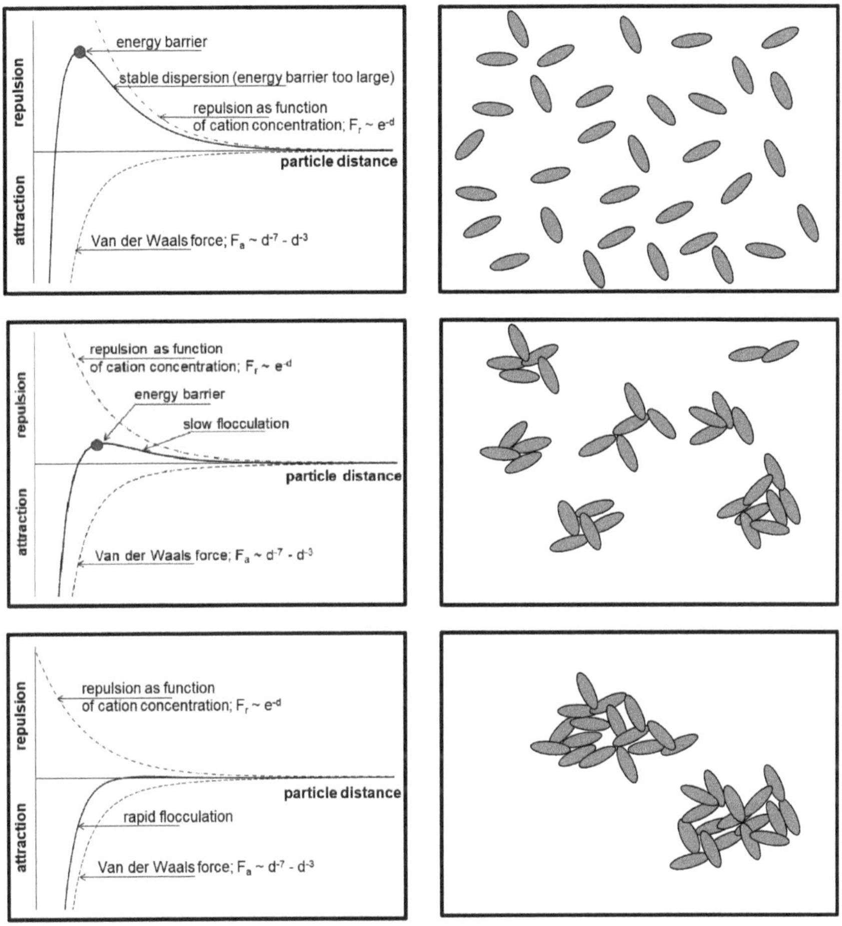

Fig. 2.5. Interaction energy between colloidal particles immersed in various electrolyte solutions as function of the distance between particles. The sum of the Coulombic repulsion and Van der Waals attraction determines the so-called energy barrier, which has to be overcome for particle aggregation. The upper row reflects a stable suspension, the middle row slow aggregation and the low row rapid flocculation.

Energy barrier

The DLVO-potential is schematically displayed in Fig. 2.5, showing the sum of repulsive and attractive forces between particles as a function of electrolyte concentration. The net force may have a peak, beyond which it reduces rapidly. This peak is known as the energy barrier for flocculation/aggregation and occurs close to the particles. Its energy is measured in kT_K units, where k is the Boltzmann constant ($k = 138 \cdot 10^{-23}$ J/K) and T_K is temperature (in degrees K(elvin)). The energy kT_K is associated with Brownian motion, as the molecular diffusion coefficient D_s for small particles with diameter D amounts to $D_s = kT_K/6\pi\mu D$, where μ is the dynamic viscosity of the ambient fluid. In colloidal suspensions, where particles undergo Brownian motion, aggregation is possible when the energy barrier is smaller than about $10kT_K$.

It is not necessary to calculate the DLVO potentials of clay suspensions to determine whether or not the energy barrier is smaller than $10kT_K$. For spherical particles with a homogeneous surface charge, it can be demonstrated that the energy barrier corresponds to an electric surface potential of about $kT_K/e = 25$ mV, where e is the electron unit charge. The electric surface charge can be measured by titration, using Gauss's law.

In practice, electrophoresis experiments are more practical, as these are much easier to perform than titration experiments. These measurements yield the electric potential at the shear plane around particles moving in an electric field. In ideal situations, this shear plane is located at the surface of a particle, in which case the surface electric potential and the electric potential of the plane of shear are identical. The International Union of Pure and Applied Chemistry (IUPAC) refers to the "zeta potential" (ZP), reflecting the electric potential at the shear plane (Delgado *et al.*, 2005). Consequently, when the absolute value of the zeta potential is smaller than about 25 mV, a suspension of spherical particles would be unstable, implying aggregation (flocculation). On the other hand, if the absolute value of the zeta potential is larger than about 25 mV, a suspension would be stable, and particles do not aggregate.

Note that suspensions of particles with identical surface charge can still have different zeta potentials, as the value of the potential (linked to the Coulomb potential) strongly depends on the conductivity of the ambient fluid as well, as sketched in Fig. 2.3. This explains why particles can remain dispersed in fresh water, where they have a high zeta potential, while they may aggregate in brackish or saline waters. This is illustrated in Fig. 2.5, where clay particles are dispersed in fresh water, mildly flocculate in a KCl solution, and strongly flocculate in a $AlCl_3$ solution. The strong flocculation in the latter case is explained by the higher conductivity induced by trivalent ions (Al_3^+) than by monovalent ions (K^+), implying that the zeta potential is lower with trivalent than with monovalent cations.

pH

H^+-ions are the smallest cations that exist. Furthermore, H^+-ions may interact chemically with a clay particle, contrary to most other cations. The surface charge of particles can therefore be pH-dependent. Thus, also the zeta potential is pH-dependent, and within a certain pH range, the zeta potential of these particles can become close or equal to zero, inducing aggregation. Because of this chemical interaction very stable aggregates (stacks) are formed and aggregation through H^+-ions is then often irreversible.

However, the majority of (cat)ions do not bind chemically to clay particles — these ions are called indifferent, as they only interact electrostatically with clay particles. As discussed above, indifferent ions decrease the Coulombic repulsion between particles, and therefore indifferent ions in a suspension induce aggregation (the DLVO-theory).

Models for spherical particles show large errors in predicting the zeta potential of clay suspensions; therefore models for non-spherical particles should be applied (Chassagne and Bedeaux, 2008; Chassagne *et al.*, 2009). Nonetheless, zero surface charge yields zero electrophoretic mobility and zero zeta potential, regardless of the theory used. This implies that suspensions are certainly unstable at very small zeta potentials (in practice, for clays, typically below $|5|$–$|10|$ mV). Therefore, many clays suspended in sea water (containing both monovalent and multivalent ions), aggregate at salinities above about 2 ppt (Mehta, 2013). A small amount of salt appears sufficient to induce flocculation.

As clay particles are not homogeneously charged, the relation between zeta potential and surface charge is not always representative for the aggregation ability of suspensions. Below a pH of 4–6, some clays (kaolinite, illite) have positively charged edges. Aggregation between these positive edges and the negatively charged faces is then possible, see Fig. 2.6, even though the overall zeta potential may be negative. This face-edge flocculation process yields so-called card-house structures, which are quite fragile however, see Fig. 2.7.

Stacks of clay platelets (Fig. 2.4) can delaminate under the influence of electrostatic forces between the clay particles, leading to swelling, i.e. the uptake of water. Smectite clays, for instance, are renown for their ability to swell (montmorillonite belongs to the smectite family). Stacks of platelets of other clays, such as kaolinite, do not or hardly delaminate or swell, as the space between these platelets is closed by hydrogen bonds, largely stabilizing these stacks.

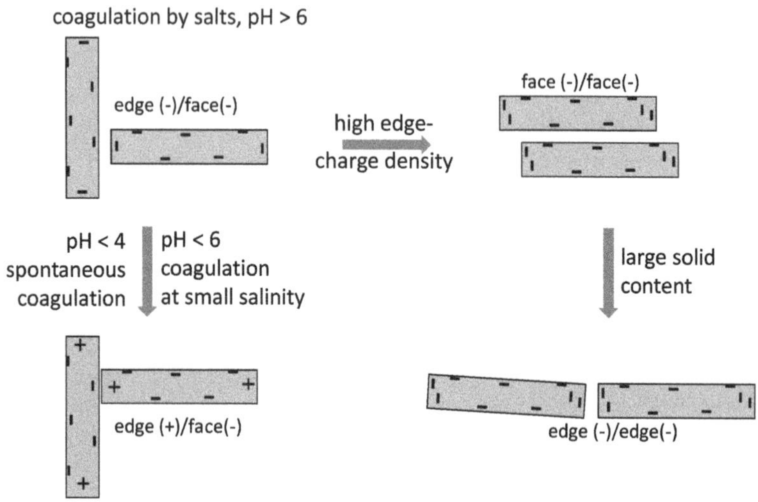

Fig. 2.6. Aggregation (coagulation) of clay platelets in saline solutions (electrolytes), further to Bergaya and Lagaly (2013). Note that in nature, instead of individual platelets, the smallest particles often consist of stacks of platelets (flocculi), as sketched in Fig. 2.4. The aggregation process of stacks of platelets is basically similar to the aggregation of individual platelets.

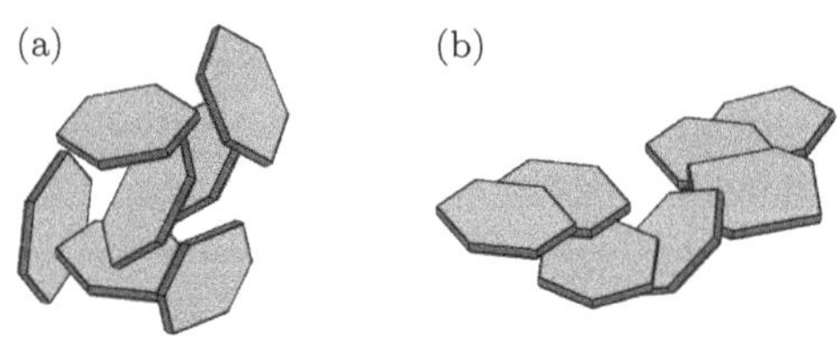

Fig. 2.7. Depending on pH and salinity, aggregation between clay particles can lead to different structures: (a) schematic of a fragile card-house structure, where edges and faces of the particles are in contact and (b) a band-type network with face-face aggregation (redrawn from Bergaya and Lagaly, 2013). See also Fig. 2.6.

The role of organic matter and organic polymers

Next, the role of organic matter in the aggregation/flocculation processes is discussed. Polyelectrolytes and polymers are commonly found in natural waters. Polyelectrolytes are polymers bearing an electrolyte group. These groups dissociate in water, so that polyelectrolytes become charged. Polyelectrolytes and polymers stem from the decomposition of organic matter and/or are excreted by living organisms (see also Chapter 8). Their structure, size and charge can vary greatly. Yet, to give an idea, a typical polyelectrolyte can be depicted as a long flexible carbon chain with some grafted surface charges, located anywhere along the chain. Due to their flexibility, the behavior of polyelectrolytes in natural waters differs largely from that of solid particles. The shape of polyelectrolytes is affected by environmental conditions: polymers can coil or uncoil, depending on the water chemistry and hydrodynamic conditions. Like solid particles, polyelectrolytes can aggregate mutually, but also with clay particles. Depending on the surface charge of polyelectrolytes, they can aggregate in different ways with clay particles.

Cationic polyelectrolytes

In the following, we discuss the aggregation/flocculation behavior of anionic (negatively charged) and cationic (positively charged) polyelectrolytes. Both types can aggregate with clay particles, but in a different way. Note that cationic polyelectrolytes are generally abundant in natural waters. Typical examples are polysaccharides, such as EPS, and proteins composed of peptides and amino acids. However, also anionic polyelectrolytes are encountered in natural open waters, as much biological matter is negatively charged. Moreover, anionic polyelectrolytes are often used in (waste) water treatment and other industrial processes.

Aggregation is enhanced by flocculants and coagulants. Following the definitions in sanitary engineering, flocculants are molecules, whereas coagulants are ions. Both may interact with clay particles.

Adding cationic flocculant to a clay suspension results in extremely rapid flocculation, as the negatively charged clay particles experience Coulombic attraction with the positively charged polyelectrolyte. Though cationic polyelectrolytes are abundant in natural waters (see above), synthetic use of these flocculants is strongly regulated, as their strong, irreversible bonding to the particles may be harmful to the environment. Therefore, less effective anionic flocculants are often used in industry.

At the low shear rates occurring in open waters, i.e. up to a few 10 s^{-1}, e.g. Fig. 3.4, flocs of clays and cationic polyelectrolytes generally have an open structure, as sketched in Fig. 2.8. At larger shear rates, flocs become more compact because the polyelectrolytes collapse on the clay particle, while the aggregate itself densifies over time, as more polyelectrolyte-clay links are formed. This is expected to occur under the influence of external stresses as well, such as during compaction in the bed, which may explain

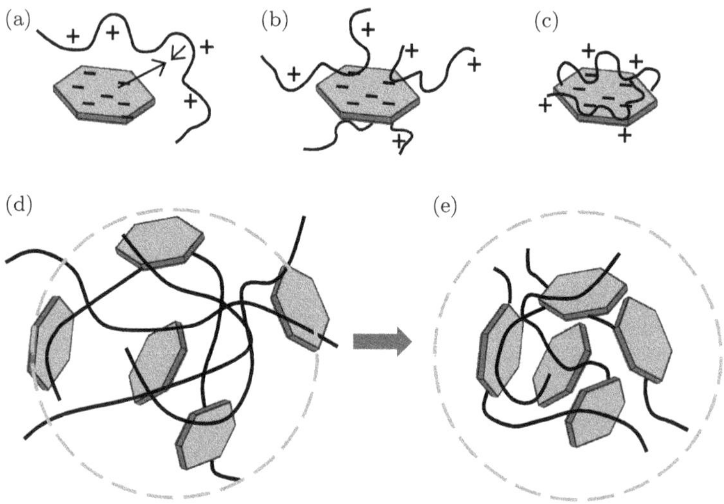

Fig. 2.8. Schematic of the interaction between a clay particle and a cationic polyelectrolyte. As a result of Coulombic attraction (a) floc formation is fast. At low shear, particles coated with cationic polyelectrolytes can exhibit a "hairy" structure (b). Under the influence of shear, these hairs can collapse onto the clay platelet and aggregates become denser (c), and dense flocs are formed through irreversible aggregation ((d) and (e)).

its denser structure (large fractal dimension $n_f = 2.6$–2.8) compared to flocs in the water column ($n_f = 1.9$–2.2), see Secs. 4.3 and 5.1.

Anionic polyelectrolytes

Even though anionic polyelectrolytes are negatively charged, aggregation with negatively charged clay particles is possible, provided the ambient water contains sufficient cations. These cations form bridges between anionic polyelectrolytes and negatively charged clay particles. Note that fresh water (even tap water!) generally contains so many cations that flocculation between clay particles and anionic polyelectrolytes is common. However, the flocculation process itself is quite different from the case with cationic polyelectrolytes, as sketched in Fig. 2.9.

This alternate flocculation process can best be understood from a flocculation experiment at low shear rate in which anionic polyelectrolyte is added to a clay suspension with a small amount of cations. After a few minutes, clouds of particles are formed, which eventually turn into flocs. Aggregation is slow as the cations need to be dispersed in between the clay particles and anionic polyelectrolytes. More salt in the suspension enhances the flocculation rate.

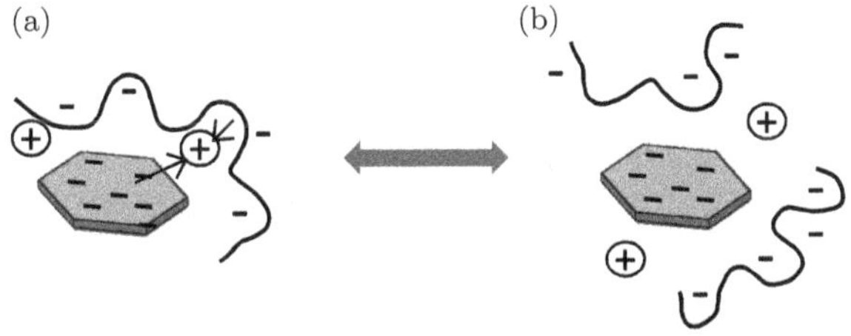

Fig. 2.9. Schematic of the interaction between a clay particle and an anionic polyelectrolyte. At low shear, aggregates can be formed because of cations (indicated by an ⊕) in the ambient water bridging the negative charges (a). At higher shear rates, the polyelectrolytes detach from the clay particles (b), but the flocs reform again at lower shear — this flocculation process is reversible.

In the first part of this section, we introduced electrophoretic mobility measurements (measuring the zeta potential (ZP)) as an instrument for assessing flocculation rates of clay particles in a solution of cations (the coagulants). However, with other components in a suspension, such as polyelectrolytes, our previous reasoning on the correlation between low zeta potential and flocculation ability should be refined, as the electrophoretic mobility of a mixture of suspended particles provides only information on its average mobility. For instance, if a large fraction of particles would have a small zeta potential and the remainder a high one, the measured average zeta potential does not contain information on the aggregation ability of the individual fractions. In other words, the average zeta potential may be beyond the critical value below which aggregation is possible, while the fraction of particles with low ZP may still form aggregates.

Let us further analyze mixtures of clays and polyelectrolytes. In general, the mass ratio between clay particles and polyelectrolytes in such mixtures is of the order of 10^2–10^3 (the concentration of clay is of the order of 10s to 1000s mg/L, while the concentration of polyelectrolytes never exceeds a few mg/L). As clay and polyelectrolyte particles are both in the μm range, such mixtures contain many more clay particles than polyelectrolytes. This implies that electrophoretic mobility measurements of such (non-aggregated) mixtures will provide proper information on the clay surface charge only. Hence, electrophoretic mobility measurements are useful (Ibanez Sanz, 2018) for such mixtures. Note that when aggregates are formed, electrophoretic mobility measurements provide information on the ZP of the flocs, and not of the individual clay particles or polyelectrolytes.

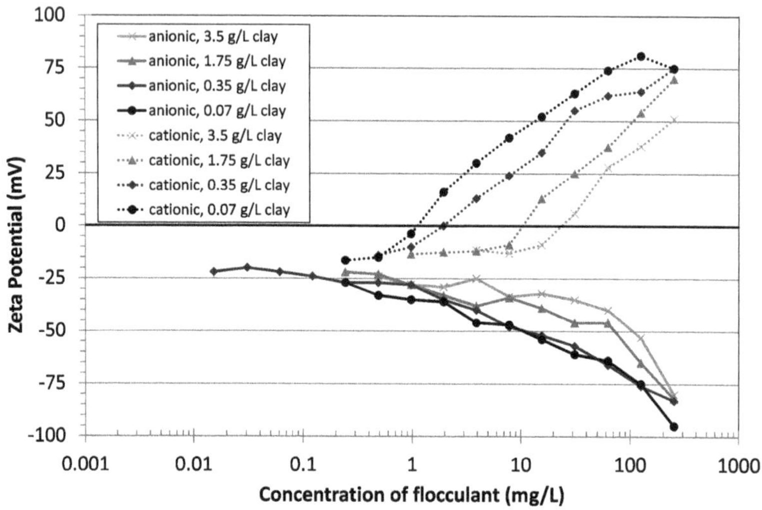

Fig. 2.10. Zeta potential as function of the concentration of cationic (upper series of data points) and anionic (lower series of data points) polyelectrolytes at different clay concentrations (redrawn from Ibanez Sanz, 2018).

Ibanez Sanz (2018) carried out experiments on river clay from The Netherlands in demi-water, measuring the ZP of clay-polyelectrolyte mixtures as a function of the concentration of polyelectrolytes at different clay concentrations. The results are presented in Fig. 2.10, showing two different classes of behavior.

First, the effects of cationic polyelectrolytes on clay suspensions at various concentrations are discussed. At low polyelectrolyte concentrations, the electrophoretic mobility (ZP) is negative, decreasing in absolute value with increasing polyelectrolyte concentration (Fig. 2.10). The electrophoretic mobility becomes zero when the clay/polyelectrolyte ratio becomes about 100–200, and optimal flocculation is expected. Accompanying settling and consolidation experiments also showed very rapid formation of large flocs, and clear supernatant water.

At larger polyelectrolyte concentration, the electrophoretic mobility becomes positive. Yet, large flocs are still being formed by a small fraction of clay particles, though at a smaller rate. However, a large amount of polyelectrolytes remains unbound in the suspension, while the suspension's viscosity increases, which reduces the flocculation rate further. The positive electrophoretic mobility reflects positively charged flocs. The large electrophoretic mobility thus becomes misleading in assessing the flocculation potential of the mixture.

The ZP at low anionic polyelectrolyte concentrations is also negative, but increases in absolute value with increasing polyelectrolyte concentration, as the amount of negatively charged particles increases. Thus no optimum concentration for anionic polyelectrolytes exists.

As these experiments were done in ultra-pure water with an electrical conductivity of less than a few μS/m, the amount of cations in the system becomes the limiting factor for flocculation at the largest clay concentrations.

The discussion above on the role of anionic and cationic polyelectrolytes is based on laboratory experiments, using well-defined clays and polyelectrolytes under controlled conditions. In natural systems, the processes are even more complex, because of the presence of organic matter. Organic matter is classified in three categories: living and dead organic matter and humus.

Living organic matter encompasses all microbiological organism, plant roots, fungi and bacteria. Dead organic matter (detritus) is composed of carbohydrates (from sugars to cellulose), fats, proteins and detritus. Humus contains organic matter that does not decompose quickly and is therefore stable. Humus is formed from fulvic and humic acids, lignins, oils, fats and waxes. Humus is a colloidal substance binding to clay minerals, but generally delays or even prevents flocculation. The role of living organisms on fine sediment dynamics is discussed in Chapter 8.

2.2 Macroscopic cohesive sediment properties

As mentioned above, mud (soil with cohesive characteristics) consists of a mixture of minerals, organic material, a lot of water, and sometimes gas (which is neglected here). This composition generally varies over space and changes over time. Clay content is generally higher under mild hydrodynamic conditions, such as at the head of bays and along the banks of estuaries. Tidal channels are generally sandier, owing to the large flow velocities. Navigation channels form an important exception, as their depth is often beyond their morphodynamic equilibrium, attracting fines — the processes in and around navigation channels are discussed in Secs. 7.6 and 10.5. Also over time, the sediment composition may change because of milder hydrodynamic conditions during slack water and/or neap tide. Last but not least, sediment composition often varies significantly over a year, owing to seasonal cycles in supply from marine and/or riverine sources, biological activity (e.g. Chapter 8) and wave forcing, in particular in temperate climates.

We use various definitions for quantifying the concentration of sediment in open water and in the bed, as summarized in Table 2.3. In general, we work with mass and/or volumetric concentrations. However, we also borrow from soil mechanics, adopting its parameters. Note that the (gravimetric) water content W and void ratio e can become larger than unity. We use the superscripts $\cdot^{cl}$, $\cdot^{si}$, $\cdot^{sa}$ and $\cdot^{m}$ to refer to the sediment type, i.e. the clay, silt, sand and mud (= clay + silt (+ water)) fraction, respectively, and the subscripts $\cdot_s$, $\cdot_w$ and $\cdot_f$ to refer to the sediment (i.e. solid) phase, the water (i.e. fluid) phase and the flocculated phase.

Particle size distribution

An important parameter for classifying sediment is the grain size distribution. Figure 2.11 presents an example of a more muddy and a more sandy sample. The sandy sample exhibits a bi-modal distribution, and the distribution of the muddier sample is also not fully regular. Figure 2.11 also illustrates the definitions of clay ($D < 2\,\mu$m), silt ($2 < D < 64\,\mu$m) and sand ($D > 64\,\mu$m), according to Dutch standards (NEN 6104). Note that other standards (e.g. ASTM D422, BS 1377) use slightly different definitions, whereas the American standards also distinguish between clay and colloids. Furthermore, the grain size distribution is a geometric measure, and does not represent the fraction of (clay) minerals. Weaver (1989), for instance, analyzed samples from the Mississippi River, showing that clay minerals can have sizes up to $60\,\mu$m, whereas non-clays (quartz and feldspar) can have sizes below $2\,\mu$m, as was mentioned in Sec. 2.1.

The distinction between silt and sand is important, as silt can exhibit pseudo-cohesive behavior at moderate strain rates, as explained below in the subsection on pseudo-cohesion.

The particle size distribution in Fig. 2.11 is obtained on a so-called deflocculated sample (see Appendix A.3). Deflocculation can be done following various protocols, which break up cohesive sediment flocs to a smaller or larger degree. For flocculation studies, one would be interested in the primary particles (building stones, flocculi) of these flocs, which are considerably larger than the smallest clay particles in a sample. For soil mechanical analyses, flocs have to be broken down to the clay (colloid) fraction, whereas studies on the settling velocity may use a fully flocculated sample. Note that

Table 2.3. Definitions of concentrations and solid contents.

Symbol	Parameter	Definition	Unit	Comments
c	Mass concentration	Mass solids/total wet volume	kg/m^3	
c_{gel}	Gelling concentration	Mass solids/total wet volume	kg/m^3	Space filling - "fluid mud"
e	Void ratio	Volume pores/volume solids	–	$e = (1 - \phi_s)/\phi_s$; can be > 1
n	Porosity	Volume pores/total wet volume	–	$n = 1 - \phi_s$
W	(Gravimetric) water content	Mass water/mass solids	%	Can be $> 100\%$
ϕ_f	Volume concentration flocs	Volume flocs/total wet volume	–	$\phi_f = c/c_{gel}$, ≤ 1
ϕ_s	Volume concentration solids	Volume solids/total wet volume	–	$\phi_s = c/\rho_s$
ρ_b	Bulk density	Mass wet sediment/total wet volume	kg/m^3	
ρ_{dry}	Dry bed density	Mass solids/total wet volume	kg/m^3	$\rho_{\text{dry}} \equiv c$
ρ_f	Floc density	Mass wet flocs/total wet volume	kg/m^3	
ρ_s	Specific density solids	Mass solids/solids volume	kg/m^3	
ρ_w	Specific density water	Mass water/water volume	kg/m^3	
ξ^i	Solids contents fraction i	Mass solid fraction i/total mass of solids	–	

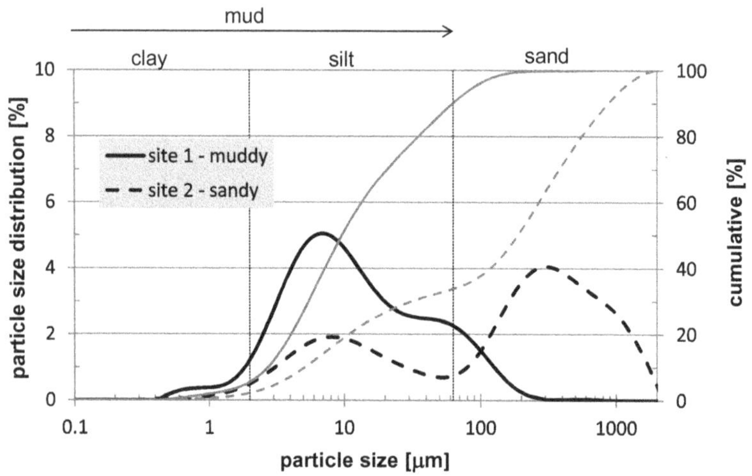

Fig. 2.11. Particle size distribution of a muddy and of a sandy sample, as measured with a Malvern particle sizer. The sample was deflocculated (see Appendix A.3). Different standards (e.g. American — ASTM; English — BS and Dutch — NEN, this graph) use different definitions for clay, silt and sand as elaborated in e.g. Winterwerp and Van Kesteren (2004).

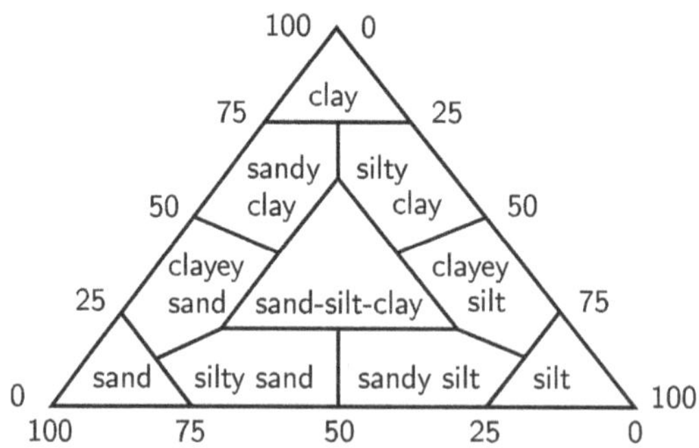

Fig. 2.12. Traditional sediment triangle (ternary diagram) for sediment classification (Shepard, 1954).

the particle size of the finer fraction is difficult to measure, and inaccuracies in the clay fraction may be large (e.g. Appendix A.3).

Traditionally, the particle size distribution can be visualized in a so-called ternary diagram (sediment triangle). Sedimentologists attribute properties to sediment/soil, based on the sand-silt-clay ratio, as shown in Fig. 2.12 (Flemming, 2000). The rationale for this distinction lies in the

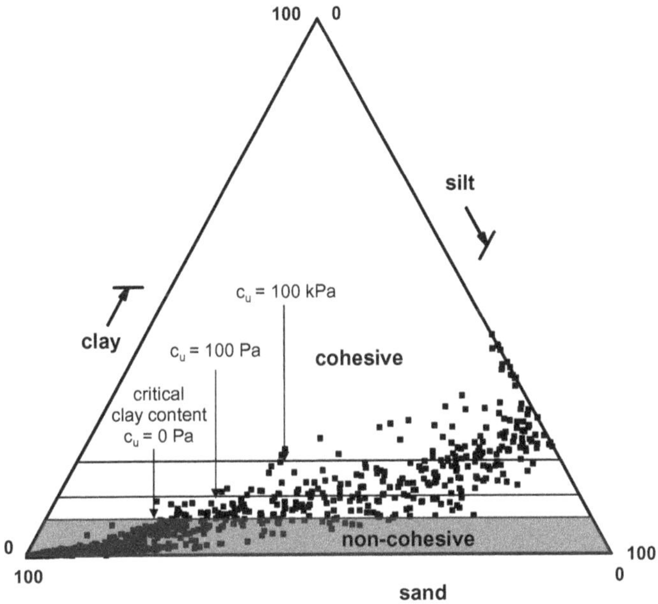

Fig. 2.13. Sediment triangle distinguishing between cohesive and non-cohesive behavior; c_u is undrained shear strength, see below.

different physical behavior of these components: clay is cohesive, sand is inert, whereas silt may exhibit pseudo-cohesive properties (see below).

The empirical sediment triangle can be quantified (e.g. Fig. 2.13), distinguishing between cohesive and non-cohesive behavior, accounting for the development of a granular skeleton (formed by clay, silt or sand) in the soil. The sediment-phase diagram thus obtained is described in Van Ledden (2003). In this book, we focus on the cohesive/non-cohesive transition of the diagram only.

One of the keystones of the sediment-phase diagram is the observation by Flemming (2000), that the silt-clay ratio in many estuarine and coastal waters is more or less constant. Apparently clays and silt follow similar transport paths, whether as part of the same flocs, or through spatial sorting. As this ratio differs across the world, it can be used as a fingerprint of the source of sediment. In the Western Scheldt, for instance, the silt–clay ratio of marine sediment is about 4, whereas further up-estuary in the Sea-Scheldt, the silt–clay ratio of riverine sediment decreases to 3 (Van Ledden, 2003).

Flemming's observation has important practical consequences, as the entire sediment composition can be established from the sand fraction only, when the silt–clay ratio would be known. Thus the clay content in a particular aquatic system has to be determined thoroughly only once through accurate psd-measurements. Following this, the easy-measurable sand content (wet sieving, see Appendix A.3) suffices for assessing the entire sediment composition, see Fig. 2.13.

Mechanical behavior

At the microscopic scale, i.e. the scale of the clay particles (μm's) cohesion is explained from the physico-chemical properties of clay particles and ambient/pore water, as elaborated in Sec. 2.1. At the macroscopic scale, cohesion can be understood from e.g. the mud's mechanical response to deformation in e.g. a triaxial test. A mud sample is placed in an impermeable cylindrical (rubber) membrane, with stiff porous disks at both ends. It is placed in a cell (cylindrical container) that can be pressurized. An additional axial load can be added via the top disk. After preparation and placement, the sample is allowed to consolidate to equilibrium (dissipation of excess pore pressures — see Chapter 5), or not. A triaxial test can be executed under drained or undrained conditions. Thus, three types of tests exist (e.g. Lambe and Whitman, 1979):

(1) unconsolidated undrained (UU),
(2) consolidated undrained (CU),
(3) consolidated drained (CD).

The left part of Fig. 2.14 sketches the loads on a sample, where x, y and z represent principal stress directions (see Sec. 2.3), while $\sigma_x = \sigma_y$, and where x and y are coordinates in the horizontal plane. Prior to consolidation, generally $\sigma_x = \sigma_y = \sigma_z = 0$, as water can drain through the porous disks.

First the soil response to consolidated drained (CD) conditions is discussed. After consolidation, the cell pressure is increased ($\sigma_x = \sigma_y = \sigma_z$), and the excess pore pressures generated are allowed to dissipate. Then an additional load $\Delta\sigma$ is placed on top of the cylinder, slowly increasing, allowing excess pore pressures to dissipate. This load is increased until the sample yields (fails). The response of the sample, in terms of effective stresses (Sec. 5.1), can be visualized in the Mohr–Coulomb diagram of Fig. 2.14, where τ represents shear stress. This test is repeated at a variety of cell pressures ($\sigma_{x,y}$) and normal loads $\Delta\sigma$, of course on another, undisturbed sample, obtaining a family of Mohr circles. The envelope of the various

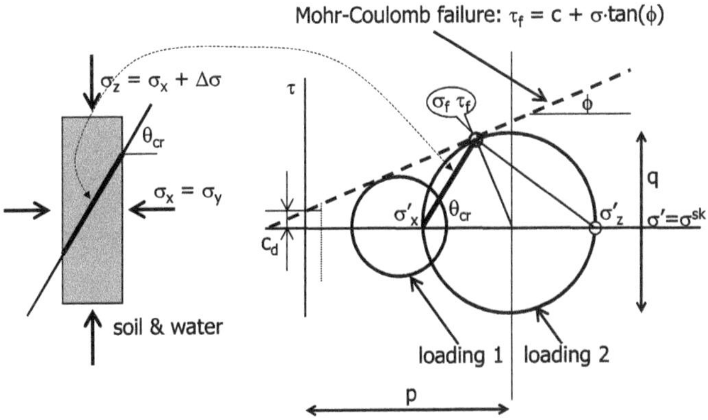

Fig. 2.14. Schematic of a triaxial test with the Mohr–Coulomb failure envelope in a normal stress–shear stress diagram (Mohr's Circle). This test is carried out under pre-consolidated drained (CD) conditions, and the Mohr diagram reflects effective stresses only. Also indicated are the isotropic and deviatoric stresses p and q (see Fig. 2.25).

Mohr circles is known as the Mohr–Coulomb failure envelope:

$$\tau_f = c_d + \sigma_f \tan\phi_d \tag{2.2}$$

where subscript $\cdot_d$ refers to drained conditions, and subscript $\cdot_f$ to failure. For non-cohesive sediment, this envelope is a straight line crossing the origin of the Mohr diagram. Its slope is referred to as the internal friction angle ϕ (also known as the angle of repose, or the "equilibrium slope of a pile of sand"). However, for non-cohesive sediment, the envelope crosses the vertical τ-axis at a value c_d, which is known as the drained strength of the mud sample, also referred to as the mud true cohesion. For cohesive sediment, c_d can be considered as a material parameter, depending on clay type, pore water chemistry and organic matter, though classical soil mechanics argues that c_d depends also on the stresses considered (Roscoe *et al.*, 1958). Moreover, the failure envelope for cohesive sediment is generally not a straight line (Schofield and Wroth, 1968), but we ignore these subtleties here.

For cohesive sediment, the friction angle ϕ_d for drained conditions in Fig. 2.14, Eq. (2.2) amounts to 10–30°. Because of the generation of under-pressure, ϕ_d decreases rapidly to zero under undrained conditions. The (apparent) build-up of strength under such undrained conditions is known as pseudo-cohesion (see also Sec. 2.3), and vertical slopes in the soil become possible — see below.

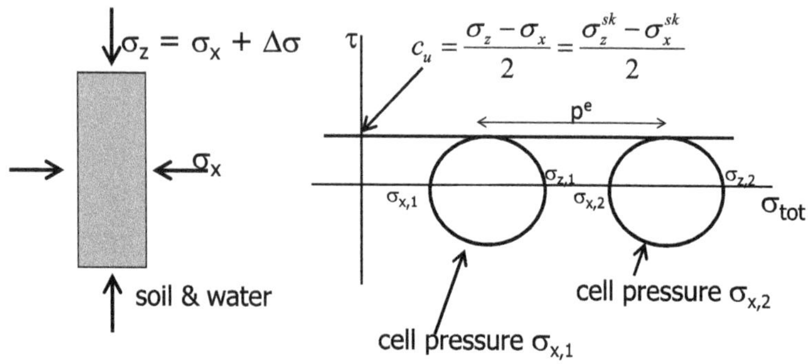

Fig. 2.15. Schematic of a triaxial test with the Mohr–Coulomb failure envelope in a normal stress–shear stress diagram (Mohr's Circle). This test is carried out under undrained conditions, thus representing the undrained strength — note that Mohr's Circle reflects effective stresses only (e.g. Eq. (5.1)).

Consolidated undrained (CU) triaxial tests will yield the same Mohr diagram, though the stress path towards the failure envelope differs a bit (Lambe and Whitman, 1979).

Though technically possible, drained triaxial tests are highly cumbersome because the mud discussed in this book is generally of low strength (see also below). However, triaxial tests can also be carried out under unconsolidated undrained (UU) conditions. A sample is placed in the triaxial cell, and a test is started immediately, while pore water cannot escape during loading. The Mohr diagram for such UU triaxial tests is sketched in Fig. 2.15. As water cannot escape, the difference between total stress and water pressure (i.e. the effective stress, see Sec. 5.1) remains constant, and all Mohr circles at yield (collapse) are identical, but move to the right over the σ_{tot}-axis with increasing cell pressure, thus pore water excess pressure. In terms of the classical Mohr circle, $\phi = 0$ in Fig. 2.15 (or almost zero, when compression of water is accounted for). The Mohr–Coulomb failure envelope now runs (almost) horizontal, and reflects the apparent or undrained cohesion of the soil c_u, which almost equals the undrained shear strength s_u ($c_u \approx s_u$), as $c_d \ll s_u$. In the following, we refer to the undrained shear strength c_u reflecting the cohesive nature of the soil.

According to Schofield and Wroth (1968), the ratio between the undrained shear strength and the effective stress in the bed σ_{zz}^{sk} (see Sec. 5.1) is constant (i.e. $c_u \approx 0.25\sigma_{zz}^{sk}$), whereas Skempton (1961) found $c_u \approx (0.11 + 0.37\text{PI})\,\sigma_{zz}^{sk}$ — see next paragraphs for the definition of PI.

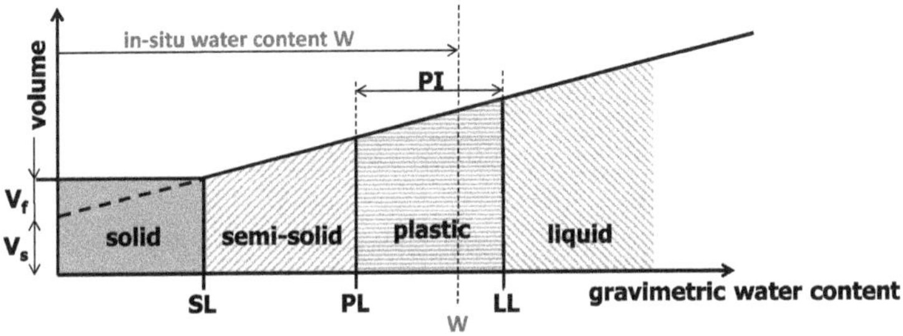

Fig. 2.16. The Atterberg limits and the plastic limit (PL), distinguishing between solid and plastic behavior, and the liquid limit (LL), distinguishing between plastic and fluid behavior. Note that at the LL, c_u = 1–2 kPa, whereas at the PL, c_u = 100–200 kPa.

Table 2.4. The Atterberg limits based on the gravimetric water content W.

Symbol	Parameter	Definition/comments
SL	Shrinkage limit	Minimal water content
PL	Plastic limit	Transition between solid and plastic behavior
LL	Liquid limit	Transition between plastic and fluid behavior
PI	Plasticity index	PI = LL − PL
LI	Liquidity index	Normalized water content: $\mathrm{LI} = \frac{W - \mathrm{PL}}{\mathrm{LL} - \mathrm{PL}}$
A	Activity	$A = \frac{\mathrm{PI}}{\xi_{cl} - \xi_0}$

Atterberg limits

The mechanical properties of cohesive sediment (soils) can be correlated with the so-called Atterberg limits. These are based on the (gravimetric) water content of a soil W (see Table 2.3), characterizing a soil–water mixture as a solid, a plastic material or a fluid, as shown in Fig. 2.16. The transition between solid and plastic, and plastic and fluid behavior is indicated by the liquid limit (LL) and the plastic limit (PL). The difference between LL and PL is known as the plasticity index PI, whereas the liquidity index LI refers to the normalized actual water content, see Table 2.4. The shrinkage limit SL reflects a minimum in the water content of a soil.

Procedures for determining PL and LL applicable for soft, fine sediment are described in Appendix A.4. Though the determination of the Atterberg limits is highly empirical and a bit arbitrary, a huge database exists, relating the mechanical properties of soils to these Atterberg limits. Hence, the Atterberg limits can give a first-order estimate of the mechanical properties of muddy soil (Locat and Demers, 1988). For instance, the undrained shear strength of a soil sample taken near the bed surface (thus at low isotropic stress) at the liquid limit amounts to about $c_u = 1$–$2\,\text{kPa}$, whereas at the plastic limit, $c_u = 100$–$200\,\text{kPa}$. Winterwerp and Van Kesteren (2004) and Jacobs (2011) plotted the strength of soil samples against the relative water content (W/PI), yielding power law relations with good correlations for soil strength.

The effects of organic matter (e.g. polysaccharides, etc.) are implicitly included in the Atterberg limits, as sample treatment (drying) should be done with dry (not too hot) air only (see Appendix A.4). The effect of organic matter on the mechanical properties of mud can be considerable, as shown in Sec. 5.2.

The Atterberg limits are traditionally plotted in a so-called plasticity chart, e.g. Fig. 2.17. The correlation between PI and PL yields information on the degree of plasticity. Data on natural soils typically fall between the A-line and the B-line in Fig. 2.17. Montmorillonite clays fall on the A-line, whereas the B-line lies in between the values for illite and kaolinite clays. Organic-rich soils fall below the B-line. Hence, the plasticity chart also gives a first indication on the clay minerals (mixture). Moreover, from the plasticity chart one can assess whether the Atterberg Limits have been determined accurately, i.e. are more or less in line with international standards.

Note that the data in Fig. 2.17 represent bed sediment from only one aquatic system (in this case the Dutch Port of IJmuiden), with varying sand

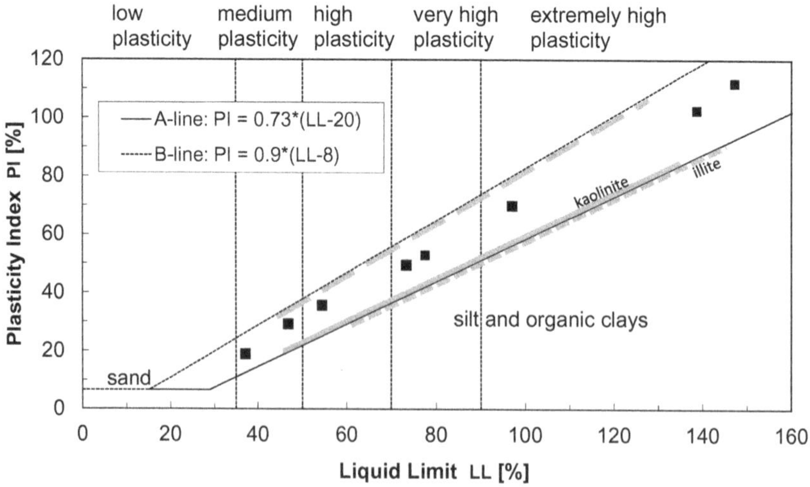

Fig. 2.17. The plasticity chart relates the plasticity index PI to the liquid limit LL with data from IJmuiden (The Netherlands). For about PI $<$ 7%, the soil is non-cohesive.

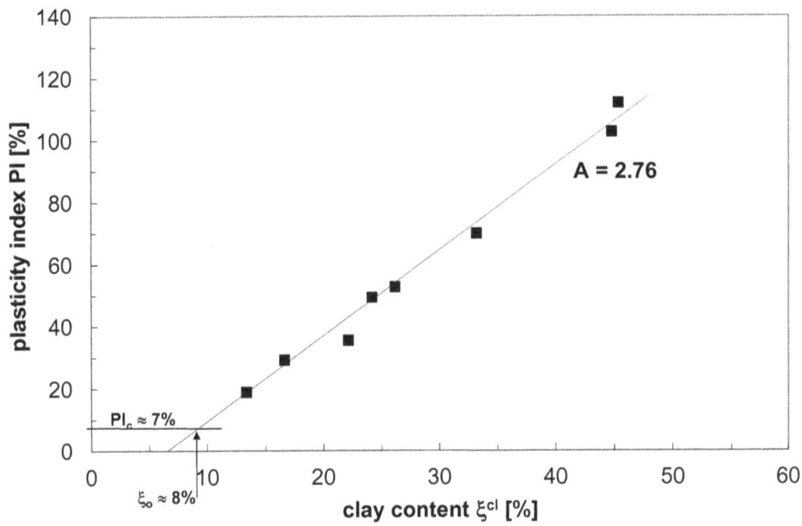

Fig. 2.18. The activity plot relates the plasticity index to the clay content, assessing the critical clay content beyond which cohesive behavior starts — see Winterwerp and Van Kesteren (2004) for the data.

content. The higher the sand content, the lower LL and PI. Experimental data (e.g. Mitchell, 1976; Skempton, 1985) reveal that for PI $<$ PI$_c$ $\sim$ 7%, the soil no longer exhibits cohesive behavior.

The activity plot relates PI as a function of the clay content ξ^{cl} of a sample. An example is given in Fig. 2.18. In soil mechanics literature, the clay's activity is defined as A = PI$/\xi^{cl}$ (Skempton, 1961; Lambe and Witman, 1979).In this book the clay content ξ^{cl} is normalized with its critical value ξ_0, accounting for the offset at PI$_c$ $\sim$ 7% in Fig. 2.17, defining the transition between non-cohesive and cohesive behavior (see also Winterwerp and Van Kesteren (2004) and Sec. 5.2).

Using the empirical information of Fig. 2.17, i.e. that cohesive behavior starts at PI$_c$ $>\sim$ 7%, the activity plot gives the critical clay content ξ_0 for cohesive behavior. This transitional clay content is drawn in the sediment triangle of Fig. 2.13, further classifying the physical characteristics of a sediment.

Typical values for ξ_0 range from 5% to 10%, hence cohesive behavior starts at fairly small clay contents (see Sec. 5.2). For a clay-silt ratio of 1:4, this would imply a critical mud content of 25–50%, beyond which a bed would exhibit cohesive behavior.

The slope of the activity plot represents the activity A of the clay, which is a function of the clay minerals, e.g. Eq. (2.3):

$$\text{PI} = A\left(\xi^{cl} - \xi_0\right) \tag{2.3}$$

Typical values for various clay minerals are given in Lambe and Witman (1979) for kaolinite: $A = 0.38$; illite: $A = 0.9$; and sodium montmorillonite: $A = 7.2$. Hence, the activity in Fig. 2.18 is indicative for a soil which contains a significant amount of montmorillonite. Again, lower PI-values are obtained at higher sand content.

Further, PI can be determined directly from LL and PL, or by using Eq. (2.3) on the basis of the clays activity and the clay content. In the latter case, we use PI* in stead of PI (Jacobs, 2011).

Rheological behavior

As discussed above, the drained and undrained strength/cohesion, c_d and c_u, can be obtained from triaxial tests. However, such tests are not easy for soft sediment (often below the liquid limit, see below), hence expensive. Therefore, we advocate vane tests for determining the undrained shear strength. Figure 2.19 shows a typical vane curve; the procedure for carrying out such tests is outlined in Appendix A.6. The dimensions of the vane are chosen in relation to the strength of the soil. With increasing deformation (rotation angle γ), the stress (torque on the vane axis) increases till a peak value, and then relaxes to a shoulder, which reflects the undrained shear strength c_u. The peak strength is indicative for the local pore water underpressures induced by deforming the soil, the breakage of bonds between the flocs, and local readjustment of the particles to their critical state (see below). The rotational speed of a vane should be large enough to

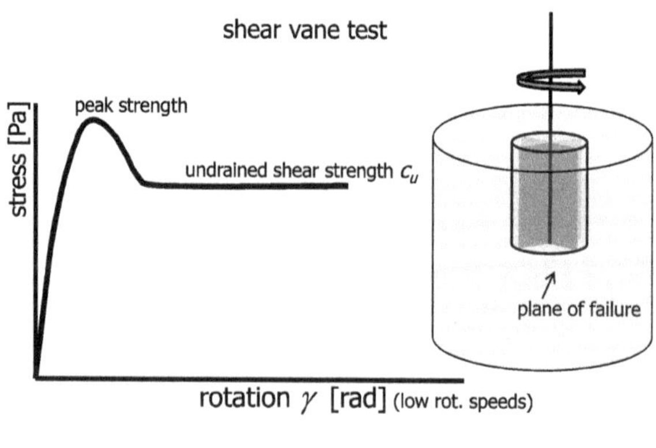

Fig. 2.19. Measuring undrained shear strength with shear vane test.

maintain undrained conditions. If not, a more gradual curve is measured, with erroneous (too low) values for c_u. If the rotational speed is too large (very large $\mathrm{Pe_w}$, see below), large pore water under-pressure occurs, and the plane of failure is likely to become irregular, while c_u again becomes less accurate. As a rule of the thumb, the peak strength should not exceed c_u by 20–50%.

The rheological properties of cohesive sediment are generally determined with a roto-viscometer, in which a soil sample is completely sheared. This can be done at a sequence of constant shear rates, $\dot{\gamma}$, until equilibrium, at constant acceleration of the deformation, in which case a non-equilibrium flow curve is attained, or in oscillatory mode — see also Sec. 7.3. The resulting flow curve is often continued up to high shear rates (up to several 100 s^{-1}), which is well beyond shear rates found in natural open waters.

Figure 2.20 shows a sketch of a flow curve (for equilibrium conditions), where viscosity decreases with shear rate, referred to as shear-thinning behavior, typical for most natural cohesive sediment (Sec. 7.3). The crossing with the y-axis defines the yield strength τ_y of soil. Note that the actual value of τ_y is a function of the lowest shear rate that can be set, or of its acceleration. In modern roto-viscometers, the shear stress (torque) can be set as well, i.e. these instruments are stress-controlled. But whatever instrument is used, τ_y is difficult to determine accurately, and literature reports much scatter in τ_y-values.

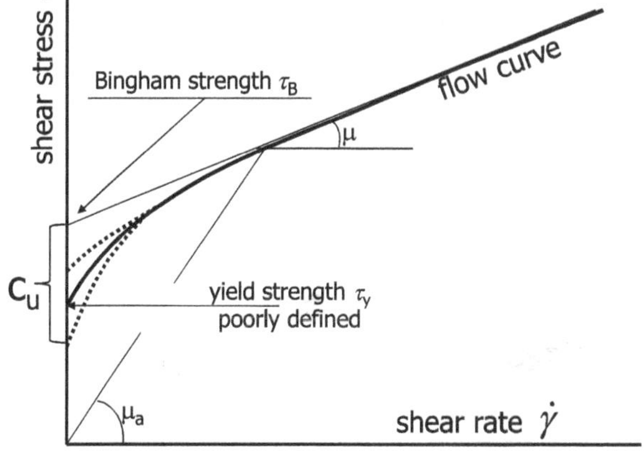

Fig. 2.20. Comparison of yield strength, Bingham strength and undrained strength in flow diagram and definition of viscosity and apparent viscosity.

The tangent to the flow curve defines the viscosity of a soil at a specific shear rate. For the shear-thinning material of Fig. 2.20, the viscosity decreases with increasing shear rate $\dot{\gamma}$. This tangent at higher shear rates, in conjunction with its crossing of the ordinate, defines the so-called Bingham model:

$$\tau = \tau_B + \mu\dot{\gamma} \tag{2.4}$$

as shown in Fig. 2.20, where τ_B reflects the so-called Bingham strength and μ the dynamic viscosity of the material. The apparent viscosity μ_a, which has to be applied in numerical models (Sec. 7.3), is defined as

$$\mu_a = \tau/\dot{\gamma} \tag{2.5}$$

Commonly, τ_B, τ_y and c_u are of the same order of magnitude, with τ_B generally the larger, provided tests have been carried out at similar isotropic stresses. Figure 2.21 shows the results of roto-viscometer tests on a number of cohesive soils from surface waters in The Netherlands, showing strengths well below the liquid limit. Indeed these soils are very soft, and standard soil mechanical tests (triaxial tests) are difficult to carry out.

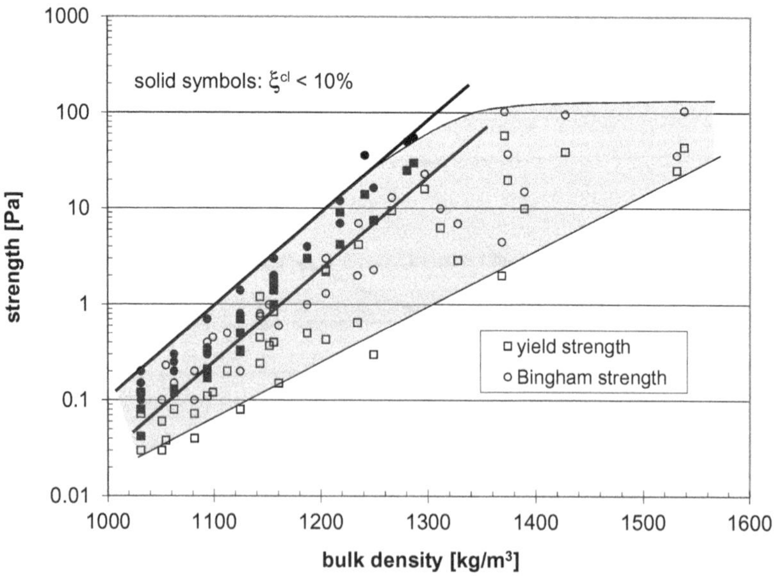

Fig. 2.21. Yield strength measured for a series of mud samples from The Netherlands, as a function of the mud's bulk density.

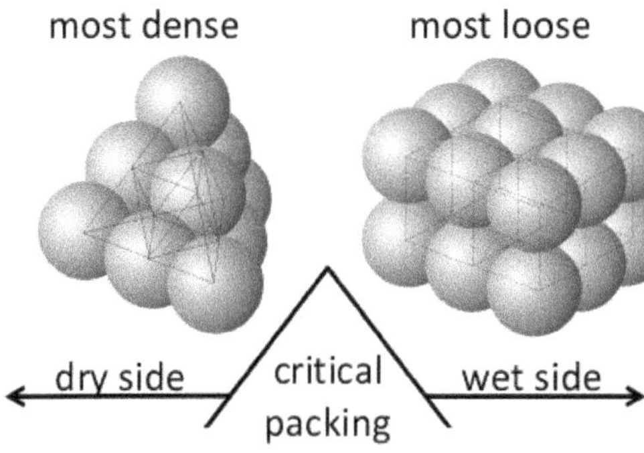

Fig. 2.22. Particle arrangement at most dense and most loose packing. To the left of the critical packing, we find the dry side, whereas to the right, the wet side occurs.

Section 7.3 contains some further discussions on the rheological behavior of cohesive sediment (see also Mehta, 2013).

Soil skeleton

Soils of granular material (sand) can be characterized by their degree of packing. Figure 2.22 presents a diagram of a pile of spheres at their most dense and at their most loose packing, whereas the pores are filled with water, representing fully saturated conditions only. Deformation of the most densely packed pile requires expansion (dilatation) and ingress of water. If the volume of the soil increases, we refer to the dry side of material response, as discussed in Sec. 2.3. On the other hand, deformation of the most loosely packed soil induces compaction, and water is squeezed out (egress) of the soil — the volume of this soil reduces — we refer to the wet side. Obviously, there should be a packing state in between, at which deformation of the pile of spheres does not induce changes in volume. Deformation of such a soil configuration induces some local rearrangement of the spheres (particles) only, but at larger scale, no water flows in or out of the soil. This state of packing is known as the critical packing, and the soil is at its so-called critical state. Energy required to deform a soil at this critical state is minimal, compared to other states of packing.

This critical state is important for fine grained soils in particular, as pore size scales with particle size. The hydraulic conductivity of such soils is therefore small, of the order of 10^{-10}–10^{-14} m/s. Thus substantial pore water flow requires large pressure gradients. Rapid deformations therefore take place under undrained conditions.

The critical state is also locally attained at the plane of failure around a vane, during a vane test (Fig. 2.19). Locally, particles rearrange while being deformed, until the critical state, so that water does not have to flow through the entire sample. This explains why vane tests at too low rotational speeds give too low strengths.

Pseudo-cohesion

Deformation of a soil can also be analyzed from the viewpoint of continuity. Displacing a particle in a water–soil matrix creates a void, which has to be filled instantaneously by (pore) water — if not, cavitation occurs. As a result, deformation induces pore water excess pressure (see also Sec. 5.1; Eq. (5.1)). The dissipation rate is governed by Terzaghi's equation (5.15); the solution of which can be written as

$$\frac{p_e}{p_{e,undr}} \sim \exp\left\{\frac{\sqrt{\mathrm{Pe_w}}}{2}\right\} \tag{2.6}$$

where p_e and $p_{e,undr}$ are the actual pore water excess pressure, and the pore water excess pressure under fully undrained conditions, respectively. Using Eq. (2.6), the Péclet number $\mathrm{Pe_w}$ can be written as

$$\mathrm{Pe_w} \equiv \frac{V\ell}{c_v} \approx \frac{z^2}{c_v t} \tag{2.7}$$

where V is a characteristic deformation rate, ℓ is a typical length scale, and c_v is the consolation coefficient. The consolidation coefficient describes the rate at which pore water pressure gradients are dissipated (Sec. 5.1).

Equation (2.6) is plotted in Fig. 2.23, exhibiting the buildup of excess pore pressure as a function of $\mathrm{Pe_w}$, i.e. rate of deformation. Soil deformations at large $\mathrm{Pe_w}$ imply undrained conditions: pore water cannot escape from or flow into the soil fast enough, and local pore water excess pressure (positive or negative) will be induced. Particles displaced at high $\mathrm{Pe_w}$

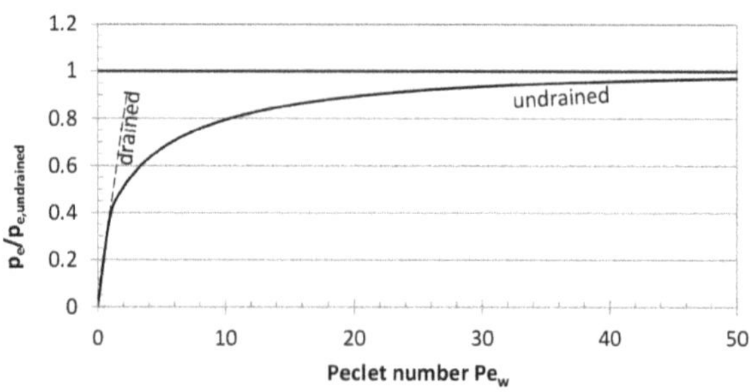

Fig. 2.23. Pore water excess pressure as function of Péclet number, i.e. Eq. (2.6).

Table 2.5. Cohesion and pseudo-cohesion at stress–strain conditions common in open water.

	Clay	Silt	Sand
Cohesion	√	—	—
Pseudo-cohesion	√	√	—

experience suction, requiring large stresses to deform. This phenomenon is referred to as apparent (or pseudo) cohesion; we refer to c_{ps}, which is thus a function of $\mathrm{Pe_w}$. c_{ps} can be regarded as an under-pressure induced critical stress: soil deformation can only take place at stresses beyond c_{ps}. The role of pseudo-cohesion is further discussed in Sec. 5.2 and Chapter 6. Breaching of sand along steep slopes is a well-known manifestation of this apparent cohesion (Mastbergen and Van den Berg, 2003). Quicksand is another manifestation of undrained behavior, the soil becoming fluid upon deformation.

Table 2.5 presents a first-order classification of the behavior of clay, silt and sand under stress-strain conditions common in estuaries and open seas. However, when deformations rates are small, i.e. small $\mathrm{Pe_w}$ drained conditions prevail, and pore water excess pressures are not generated. We refer to e.g. Winterwerp and Van Kesteren (2004) for further details.

The effects of pseudo-cohesion play an important role when soil is over-consolidated, such as on intertidal mud flats (e.g. Sec. 5.2 and 6.3.1).

2.3 The critical state model

The Mohr-diagram of Fig. 2.14 is highly convenient for visualizing the stability and yielding (failure) of a soil in response to external stress (the Mohr–Coulomb failure envelope). However, the stresses in the Mohr-diagram are generally not invariant: only under specific conditions, such as in a triaxial test, the main directions of loading are in the principal stress direction. Therefore, the invariant $p-q$ diagram is introduced in this section, elaborated upon further in various sections below. Here p represents the isotropic stress and q the deviatoric stress (Roscoe *et al.*, 1958; Schofield and Wroth, 1968), defined as

$$p = \frac{\sigma_1^{sk} + \sigma_2^{sk} + \sigma_3^{sk}}{3} \tag{2.8}$$

$$q = \sqrt{\frac{\left(\sigma_1^{sk} - \sigma_2^{sk}\right)^2 + \left(\sigma_2^{sk} - \sigma_3^{sk}\right)^2 + \left(\sigma_3^{sk} - \sigma_1^{sk}\right)^2}{2}} \tag{2.9}$$

where σ_i^{sk} represent effective stresses (e.g. Eq. (5.1)), and $\cdot_1$, $\cdot_2$ and $\cdot_3$ are the principal stress directions. We presume that consolidating/consolidated soil is isotropic in horizontal direction, i.e. $\sigma_1^{sk} = \sigma_2^{sk}$. This is also the condition in the triaxial tests of Figs. 2.14 and 2.15, thus $\sigma_1 = \sigma_2 = \sigma_x = \sigma_y$ and $\sigma_3 = \sigma_z$, where z is in vertical direction. In that case, the stress tensors p and q reduce to:

$$p = \frac{2\sigma_x^{sk} + \sigma_z^{sk}}{3} \tag{2.10}$$

and

$$q = \sigma_z^{sk} - \sigma_x^{sk} \tag{2.11}$$

This implies that p represents the average normal (principal) stress in the Mohr-diagram (Fig. 2.14) and q the diameter of the Mohr circles. Thus from Fig. 2.15, $q = 2c_u$.

The critical state model (CSM) by (Schofield and Wroth, 1968), which integrates the Granta-gravel and the modified Cam-Clay models, is assumed to apply in a qualitative sense to describe soft soil response to external loading. This model treats soil behavior in an idealized way, and therefore the coefficients derived may vary for natural soils. The CSM describes p and q as functions of the soil specific volume. Here, we prefer to introduce the void ratio e instead, which is acceptable because of the linear relation between specific volume and e (see also Jacobs, 2011 and Winterwerp and Van Kesteren, 2004). Though we do not use the CSM in a quantitative way, the model can help in understanding the stress development at the water-bed interface, where abrupt or smooth transition from fluid to solid behavior occurs.

In the original CSM, exponential relations between p and e are assumed. However, Winterwerp and Van Kesteren (2004) found power law relations from Seepage Induced Consolidation tests on very soft sediment. Section 5.2 elaborates further on this power law behavior. We therefore modify the CSM by using power law instead of exponential relations. Note that such power-law behavior is consistent with the self-similar approach introduced in Sec. 4.3 and 5.1.

The CSM describes the state of a soil in a three-dimensional p–q–e diagram as conceptually shown in Fig. 2.24. In this diagram, yield (failure) curves (effective stresses only) are drawn as the envelopes around the gray planes. These yield curves are equivalent to the Mohr–Coulomb envelopes in the Mohr-diagram. These gray planes include all possible stable states of a soil at a specific void ratio. Soil states beyond these yield planes are

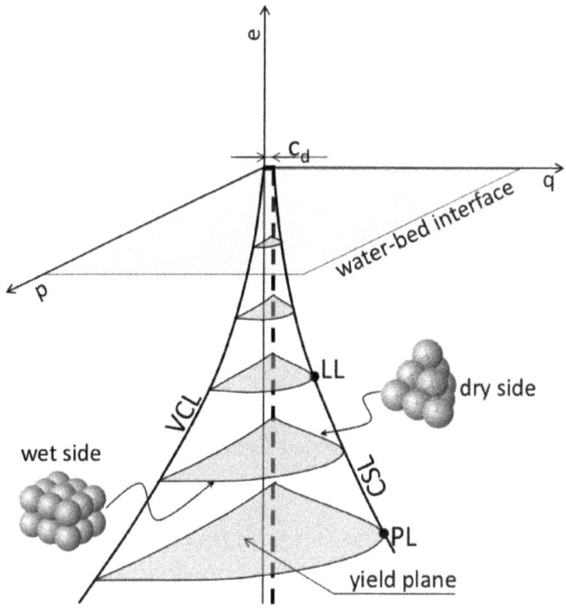

Fig. 2.24. Three-dimensional sketch of Cam-Clay model in p–q–e diagram (only effective stresses) with critical state line (CSL) and virgin compression line (VCL), and liquid limit (LL) and plastic limit (PL). Yield surfaces are represented as gray planes (almost) perpendicular to the e-axis. The part of the diagram left of the CSL (i.e. large p), is known as the wet side, whereas to the right of the CSL, the dry side is found (diagram not to scale).

impossible: when soil fails, its state moves along the e-axis towards smaller void ratio. The top of these yield curves connect, forming the so-called critical state line (CSL), i.e. packing conditions at which a soil yields (fails) without volume changes, as discussed above. Also indicated are the liquid limit (LL) and plastic limit (PL), which, by definition, are located on the CSL. Note that the water content of soft soils is well below the LL, thus at large e and typically at the wet side of the diagram, as is the case with all data in Fig. 2.21.

Figure 2.24 also reflects that for smaller e-values, the soil is stronger, indicated by larger yield planes. Note that cohesive soils exhibit true cohesion, which would imply a residual q in Fig. 2.24 at $p = 0$, as indicated by c_d. True cohesion allows for negative (i.e. tensile) isotropic stresses (p).

As discussed above, soil deformation may occur at stresses below or above the critical state, requiring the ingress (dry side) or egress (wet side) of water upon deformation.

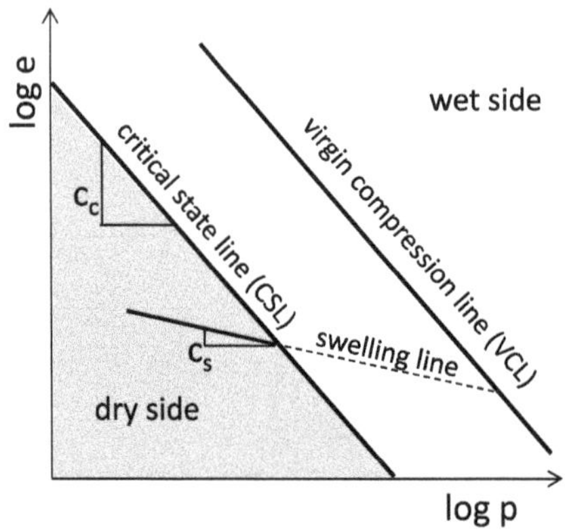

Fig. 2.25. Two-dimensional sketch of the modified CSM in p–e diagram, depicting parallel CSL- and VCL-lines, and dry and wet side.

Figure 2.24 also shows the so-called virgin compression line (VCL), which represents the maximum isotropic stress possible at zero deviatoric stress ($q = 0$). For soils, this is a hypothetical stress condition, as VCL requires frictionless conditions, which can occur in fluids only, e.g. Sec. 5.1 and Chapter 7.

For granular sediment, the yield curves in Fig. 2.24 are parallel to the p–q plane, i.e. deformations prior to yield do not induce volume changes. This is however not the case for cohesive sediment, where the yield planes are (slightly) curved, as cohesive soils exhibit plastic deformation — see Sec. 5.2. This is best illustrated in a two dimensional p–e diagram, e.g. Fig. 2.25. Contrary to the original CSM, we use a double logarithmic diagram, further to the assumed power-law behavior (Chapter 5).

In the modified CSM, the CSL and VCL run straight and parallel on double logarithmic axes (see also Secs. 5.1 and 5.2). The slope of these lines defines the compression coefficient C_c. Note that compression of cohesive soils contains an irreversible, plastic and a reversible, elastic deformation.

Upon unloading, negative pore water pressure may be induced, attracting water, thus causing the cohesive sediment sample to swell (Sec. 5.2). Such swelling occurs along the so-called swelling line (SL), indicated in Fig. 2.25, which runs at a slope C_s, known as the swelling coefficient. For dense soils at large isotropic stresses, C_s is a factor 1/7 to 1/3 smaller than C_c (e.g. Lambe and Whitman, 1979). Note that the compression and swelling coefficient differ from the consolidation coefficient defined in Sec. 5.1, as the latter includes the effects of hydraulic conductivity.

The CSM was set up for well-consolidated soils, and not for the soft soils in this book. The original CSM assumes elliptical yield surfaces, while the CSL runs through the top of these yield surfaces. These assumptions are incompatible with a constant drained shear strength, i.e. when c_d is a material property. However, there is ample evidence in literature (Tsiampousi *et al.*, 2013) that the shape of the yield surface deviates from elliptic. Therefore, we refrain from defining the shape of this yield surface in further detail. Instead, we assume a yield surface shape which allows for the CSL to run through the top of the yield surfaces, while the yield surfaces cross the q–e plane at the constant c_d. The VCL and CSL are also assumed to follow power-law relations, bearing in mind that the CSM is used only in a qualitative way.

Chapter 3

The Benthic Boundary Layer

3.1 Introduction

Historic perspective

In the 19th century there was a big dispute between scientists studying fluid dynamics. On the one hand, great advances were made with the development of potential flow theory, a non-viscous approach (see, e.g. Lamb, 1932). The developments in mathematics stimulated flow computations around complex bodies and in complex systems. *Vice versa*, fluid dynamics also initiated significant developments in mathematics itself. Potential flow theory was very successful enabling spectacular progress in many engineering fields, such as aeronautical engineering. For instance, potential flow theory has been used up to the 1990s for designing airfoils for advanced airplanes.

On the other hand, experimentalists argued that the theoretical approach could not be complete, as all experiments and daily experience showed that flowing fluid (water, air, ...) exerts drag on bodies in that flowing medium. James Froude was one of the most renowned scientists from this camp, making large steps in nautical engineering, while developing experimental resistance laws.

Around the turn of the 20th century, Prandtl resolved this paradox by realizing that a thin boundary layer develops close to a body moving through a fluid. Within this boundary layer, large velocity gradients develop, responsible for the measured drag. Outside this boundary layer, the flow is basically non-viscous, and potential flow theory still holds. Within the boundary layer, the flow can be either laminar or turbulent (though also a turbulent boundary layer may contain a laminar part). As the hydraulic drag induced within a laminar boundary layer is much smaller than in a turbulent one, the design of structures (airplane wings!) with a laminar boundary layer has been a holy grail in many engineering disciplines.

Prandtl and his colleagues experimentally showed that a major part of the boundary layer exhibits a logarithmic velocity profile. Since then, many scientists tried to derive this logarithmic velocity profile from first principles. This, however, appeared only possible after some important assumptions.

Later, this logarithmic velocity profile was also applied in open water flow, assuming this is true throughout the water column. This chapter therefore treats the flow in shallow water systems as boundary layer flow.

Turbulence and the boundary layer

We treat open waters, such as rivers and (coastal) seas, assuming that bed-level variations (morphological changes) occur at a much larger time scale than variations in the hydrodynamics. The bed can then be treated as a solid wall, forming a solid boundary to the flow. Therefore, the flow over such a bed is referred to as wall flow or boundary layer flow. In this chapter, we discuss the behavior of turbulent boundary flow within a gradually changing mean flow. The stream-wise velocity varies over depth, often following a logarithmic profile. We first consider the properties and characteristics of turbulent flow; secondly, the logarithmic velocity profile is derived; thirdly, the effects of a non-uniform density is considered. Subsequently, the effects of waves and density-driven stratification on the mixing are considered. Finally, we elaborate on the distribution of particles over the water column.

3.2 Boundary layer flow

3.2.1 *Turbulence*

The flow in open channels is generally turbulent, indicated by the Reynolds number *Re*, which represents the ratio of inertia and viscous forces within a flow. For open water flow, the overall Reynolds number is defined as $\mathrm{Re} \equiv Uh/\nu$, where U is the characteristic flow velocity, h the characteristic water depth, and ν the kinematic viscosity (see below). It is generally assumed that for $Re \gtrsim 2400$, open channel flow is turbulent. Note that many other Reynolds numbers are used, such as the turbulent Reynolds number, the wave Reynolds number, etc. (see further in this chapter).

This section does not aim at providing a monograph on turbulence — the reader is referred to one of the many excellent text books on turbulence, e.g. classics such as Hinze (1975) and Tennekes *et al.* (1972), or more recent ones, e.g. Pope (2001). However, it is useful to elaborate a bit on the most fundamental aspects of turbulence, necessary for understanding the transport of fine sediment.

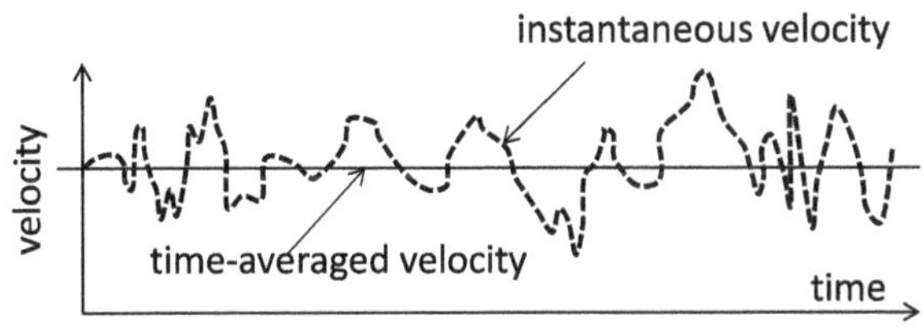

Fig. 3.1. Instantaneous velocity signal measured in a stationary turbulent flow.

Figure 3.1 shows an example of a velocity measurement in a turbulent flow. The instantaneous velocity signal is very irregular, almost erratic within certain bounds. However, if averaged over a sufficiently long time $\tau_{\rm tur}$, a constant signal is retrieved. The period $\tau_{\rm tur}$ depends on the time scale of the turbulent eddies and is generally longer in deeper water than in shallow water. Typical periods for $\tau_{\rm tur}$ range from 5 to 10 minutes. Velocity variations with longer time scales are to be interpreted as modifications of the mean flow.

For a further understanding of turbulent flow properties, it is illustrative to decompose the instantaneous velocity u into its turbulence mean and its fluctuating part: this procedure is known as the Reynolds decomposition:

$$u\left(t\right) = \bar{u} + u'\left(t\right) \tag{3.1}$$

where the overbar indicates averaging over the turbulence time scale τ_{tur}, and a dash refers to the turbulent part of the velocity. Then, by definition:

$$\int_{\tau_{tur}} u \,\mathrm{d}t = \bar{u} \quad \text{and} \quad \int_{\tau_{tur}} u' \,\mathrm{d}t = 0 \tag{3.2}$$

The magnitude of the fluctuations generally scales with the shear in the mean flow. For wall-bounded flow, the shear scales with the shear near the wall. Nezu and Nakagawa (1993) presented approximate relations for the magnitude of the turbulent velocity fluctuations, measured in turbulent shear flow (i.e. not free turbulence):

$$\frac{u'_{\rm rms}}{u_*} = 2.30 \exp\left(-\frac{z}{h}\right), \quad \frac{w'_{\rm rms}}{u_*} = 1.27 \exp\left(-\frac{z}{h}\right) \tag{3.3}$$

where u_* is the shear velocity, see Eq. (3.18). As an example, the vertical distribution of the rms-value of u' for a mean velocity of 1 m/s in 5 m deep open channel flow is depicted in Fig. 3.2. The mean flow is based on a logarithmic flow distribution over the vertical, see also further in this

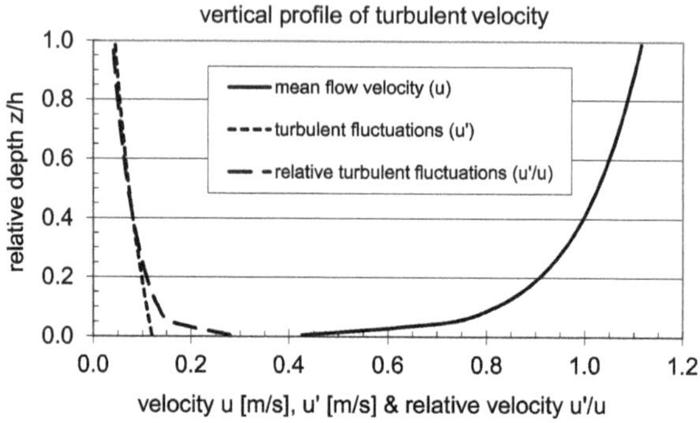

Fig. 3.2. Vertical distribution of turbulent horizontal velocity according to Nezu and Nakagawa (1993).

chapter. The rms-value of u' exceeds 0.1 m/s near the bed, and relative values increase up to 10–20% of the mean flow velocity.

The total kinetic energy K of a flow consists of a contribution by the mean flow E_k and a contribution by the fluctuating, turbulent velocities k_T (which is generally denoted by k, but which may be confused with the (turbulent) wave number $k(= 2\pi/\lambda$, λ is turbulent wave length)):

$$\begin{aligned} K &= 0.5(u^2 + v^2 + w^2) = 0.5((\bar{u} + u')^2 + (\bar{v} + v')^2 + (\bar{w} + w')^2) \\ &= 0.5(\bar{u}^2 + \bar{v}^2 + \bar{w}^2) + 0.5(\overline{u'^2} + \overline{v'^2} + \overline{w'^2}) = E_k + k_T \end{aligned} \tag{3.4}$$

where u, v and w are velocities in x, y and z-direction (z is vertical direction, see also Chapter 11).

Nezu and Nakagawa (1993) give an approximation also of the vertical distribution over the water column of the mean turbulent kinetic energy in a wall-bounded shear flow:

$$\frac{k_T}{u_*^2} = 4.78 \exp\left(-\frac{z}{h}\right) \tag{3.5}$$

Although the kinetic energy is presented as a single parameter, it reflects the sum of all turbulent motions at all (turbulent) time/length scales. This is exhibited by a power density spectrum for the turbulent kinetic energy k_T, the higher frequencies of which follow the well-known relationship $k_T \propto k^{-5/3}$ (where k is the turbulent wave number), see e.g. Fig. 3.3 (Tennekes *et al.*, 1972). Thus, turbulence quantities are not fully stochastic. Here the two ends of the turbulence spectrum are of relevance. The lowest frequency is defined by the water depth h and a characteristic horizontal velocity U, thus $f_{t,\min} = U/h$ — we do not treat the horizontal

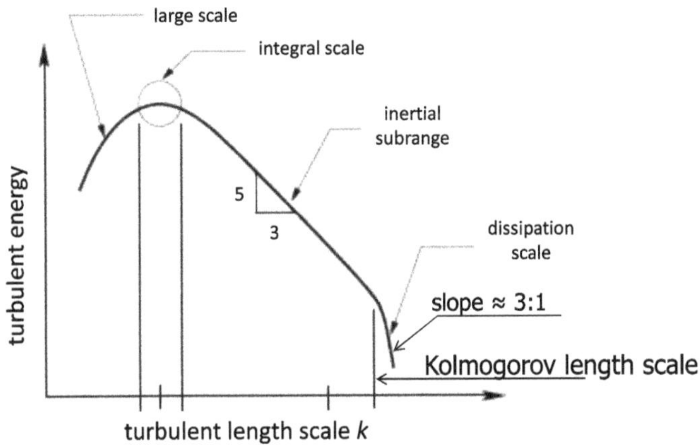

Fig. 3.3. Turbulent energy spectrum with Kolmogorov scale.

components of turbulence in this book. The largest frequency occurs at the Kolmogorov length scale λ_K below which turbulence is dissipated into heat: $f_{t,\max} \equiv f_{\lambda_K} = \sqrt{\varepsilon_T/\nu}$, where ε_T is the turbulence dissipation rate per unit mass and λ_K is the Kolmogorov length scale (see also Sec. 4.3).

The dissipation rate is defined as (Tennekes *et al.*, 1972):

$$\varepsilon_T \equiv 2\nu\overline{s_{ij}s_{ij}} \quad \text{with} \quad s_{ij} = \frac{1}{2}\left(\frac{\partial u_i'}{\partial x_j} + \frac{\partial u_j'}{\partial x_i}\right) \tag{3.6}$$

where ν is the kinematic viscosity (see below) and s_{ij} the turbulence strain rate tensor. It can be shown (e.g. Tennekes *et al.*, 1972) that in isotropic turbulence ε_T and k_T are related:

$$\varepsilon_T \sim \frac{k^{3/2}}{l} \sim \frac{u_*}{l} \tag{3.7}$$

where ℓ is a characteristic length scale. The vertical distribution of the dissipation rate in open channel shear flow is approximated by Nezu and Nakagawa (1993):

$$\frac{\varepsilon_T h}{u_*^3} = \frac{1 - z/h}{\kappa z/h} \tag{3.8}$$

The turbulent Reynolds number is unity at the Kolmogorov scale λ_k, owing to its definition $Re_T\,(= u_\lambda \lambda/\nu)$. This implies $u_{\lambda_K} \approx \nu/\lambda_K$, and the velocity gradient G (shear rate) at the Kolmogorov scale becomes:

$$G \equiv \left.\frac{\mathrm{d}u}{\mathrm{d}z}\right|_{\lambda_K} = \frac{\nu}{\lambda_K^2} \tag{3.9}$$

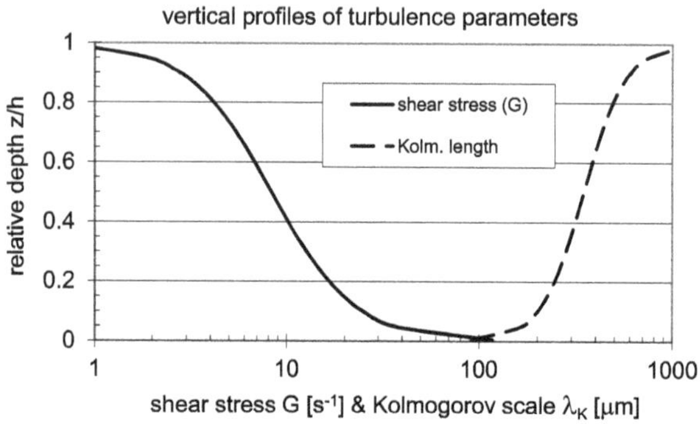

Fig. 3.4. Vertical distribution of Kolmogorov scale λ_K and shear rate in 5 m deep open channel shear flow, $U = 1$ m/s, assuming a logarithmic velocity profile, using Eqs. (3.9) and (3.10).

Or, substituting from Eq. (3.8):

$$\lambda_K = \left(\frac{\kappa z \nu^3}{u_*^3 (1 - z/h)} \right)^{1/4} \tag{3.10}$$

Both parameters, i.e. G and λ_k are important in assessing the aggregation and break-up of flocs (see Sec. 4.3). As an example, the vertical distribution of G and λ_k for a mean velocity of 1 m/s in 5 m deep open channel shear flow is presented in Fig. 3.4, showing typical G-values of the order of $10\,\mathrm{s}^{-1}$.

Averaging over the turbulence time scale τ_{tur} becomes more complicated in case of non-stationary flow, as for example found in tidal systems. Figure 3.5 shows a mean tidal signal with turbulent velocity fluctuations. In general, however, the time scale T of the tidal is much larger than the turbulent time scale: $T \gg \tau_{tur}$ so that a running average over τ_{tur} is sufficient.

Determination of the turbulent properties in the presence of waves can be more difficult, as the time scales of waves and turbulence are relatively close. For example, determination of the turbulent kinetic energy from a measured velocity signal of a few minutes will result in contamination by the wave orbital motions. In such cases, a spectral analysis is needed, e.g. Zhu *et al.* (2016).

Thus the flow velocity consists of a time-averaged component and a fluctuating part. The fluctuating part, the turbulence velocity, determines the turbulent kinetic energy. The turbulent motion plays a crucial role in the mixing of momentum and sediment. In the following section, turbulence is further elaborated for the specific case of boundary layer flow.

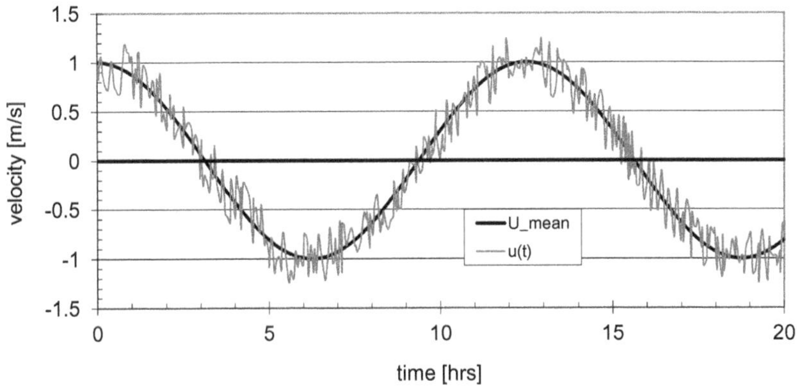

Fig. 3.5. Velocity in a tidal flow. Averaging out the turbulent motion requires a window that is much larger than the turbulence time scale, but much smaller than the tidal time scale.

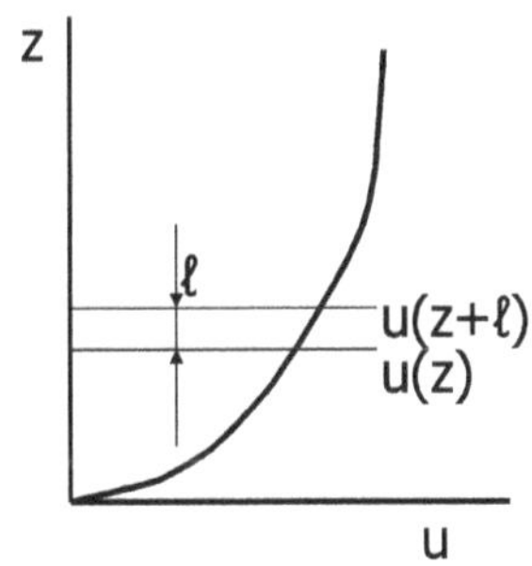

Fig. 3.6. Diagram depicting the turbulence mixing length, as proposed by Prandtl.

3.2.2 *Turbulent boundary layer in steady flow*

In this section, the turbulent flow parameters introduced above are coupled to the mean flow characteristics with a so-called a turbulence model. The first "turbulence model" was developed by Prandtl. He related turbulent velocity fluctuations in a boundary layer to the mean vertical velocity gradient, as shown in Fig. 3.6. The idea is that u' at level z is determined by the mean velocities at distance ℓ from that level. Thus temporal variations in velocity are converted into spatial variations over the length scale ℓ ("frozen" turbulence). This length scale is known as the turbulence mixing length. As sketched in Fig. 3.6, Prandtl proposed:

$$u' = u(z + \ell) - u(z) \tag{3.11}$$

Thus:

$$\lim_{\ell \to 0} \{u'\} = \ell \lim_{\ell \to 0} \left\{ \frac{u(z+\ell) - u(z)}{\ell} \right\} = \ell \frac{\mathrm{d}u}{\mathrm{d}z} \tag{3.12}$$

Further to the definition of the turbulent energy, the cross-correlations in the flow velocity are elaborated upon. We present the vertical x–z plane only — in three dimensions the three-dimensional turbulence stress tensor would be obtained. Averaging the cross-correlations between u and w over the turbulence time scale τ_{tur} yields (see Eq. (3.1)):

$$\overline{uw} = \overline{(\bar{u} + u')(\bar{w} + w')} = \bar{u}\bar{w} + \overline{\bar{u}w'} + \overline{u'\bar{w}} + \overline{u'w'} = \bar{u}\bar{w} + \overline{u'w'} \tag{3.13}$$

in which $\bar{u}\bar{w}$ reflects shear by the main flow, and $\overline{u'w'}$ turbulent shear, the so-called Reynolds stress τ_{xz}:

$$\tau_{xz} = -\overline{\rho u'w'} \tag{3.14}$$

Because of the fluctuating velocities, also the (bed) shear stress fluctuates, which can be characterized by a probability distribution. This is further elaborated in Sec. 6.3.3.

Substitution from Eq. (3.12) and assuming $|w'| \approx |u'|$ and $\rho' \ll \rho$ (which is known as the Boussinesq approximation), one obtains:

$$\tau_{xz} = -\overline{\rho u'w'} \approx -\rho\overline{u'w'} = -\rho\ell^2 \left|\frac{\mathrm{d}u}{\mathrm{d}z}\right| \frac{\mathrm{d}u}{\mathrm{d}z} \equiv \rho\nu_T \frac{\mathrm{d}u}{\mathrm{d}z} \tag{3.15}$$

Here the eddy viscosity $\nu_T \left(= \ell^2 \mathrm{d}u/\mathrm{d}z \approx \ell u'\right)$ is introduced, assuming that turbulent diffusion can be modeled in a similar way as Fickian (molecular) diffusion. This approach is known as the Boussinesq concept. Note that we generally use the kinematic viscosity [m^2/s], while for molecular viscosity the dynamic (absolute) viscosity $\mu\,(\equiv \rho\nu)$ [Pa·s] is commonly used.

For completeness, we give the full three-dimensional turbulence stress tensor $\sigma_{T,ij}$ as well:

$$\sigma_{T,ij} = \rho\nu_T \left(\frac{\partial u_i}{\partial x_j} + \frac{\partial u_j}{\partial x_i} \right) - \frac{2}{3}\delta_{ij}\rho k = 2\mu_T S_{ij} - \frac{2}{3}\delta_{ij}\rho k \tag{3.16}$$

in which S_{ij} is the turbulence strain tensor and δ_{ij} is the Kronecker delta ($\delta = 1$ for $i = j$ and $\delta = 0$ otherwise). Generally it is assumed that ν_T is identical in all directions, which implies isotropic turbulence. However, close to the bed, turbulence is not isotropic, as indicated by Eq. (3.3) and Fig. 3.2 (see Nezu and Nakagawa, 1993). This anisotropy is accounted for in more sophisticated turbulence models (see, e.g. Pope, 2001). In the following, we continue in 2DV-mode (x–z plane) for didactic reasons.

Prandtl and his followers found experimental evidence that the velocity profile within a turbulent boundary layer has a logarithmic shape. It is not possible to derive this shape theoretically from first principles, but it can be made plausible from physical arguments. For this, the stress balance in open channel flow is elaborated upon:

$$\underbrace{\frac{\partial u}{\partial t}}_{\text{I}} + \underbrace{\frac{1}{\rho}\frac{\partial p}{\partial x}}_{\text{II}} - \underbrace{\frac{1}{\rho}\frac{\partial \tau}{\partial z}}_{\text{III}} = 0 \tag{3.17}$$

Term I and II (thus neglecting term III) describe non-viscous (potential) flow, i.e. outside the boundary layer — this becomes relevant in Sec. 3.2.4. Term I and III describe the water movement in the wave boundary layer (Sec. 3.2.4) and term II and III together describe stationary viscous flow. The latter is depicted by the solid line in Fig. 3.7. In the following paragraphs, we assume such stationary flow. Note that term II and III (i.e. $\partial p/\partial x = \partial \tau/\partial z$) also describe slowly varying flow, thus applicable for the major part of a tidal cycle, except around slack water.

For homogeneous conditions (i.e. only fresh or saline water, no density gradients by salinity, etc.), $\partial p/\partial x$ is independent of z. Next, it is assumed that close to the bed, the turbulent mixing length Eq. (3.11) scales with z: $\ell = \kappa z$, where κ is the Von Kármàn constant (≈0.41). A characteristic velocity close to the bed follows from dimensional analysis:

$$u_* \equiv \sqrt{\tau_b/\rho} \tag{3.18}$$

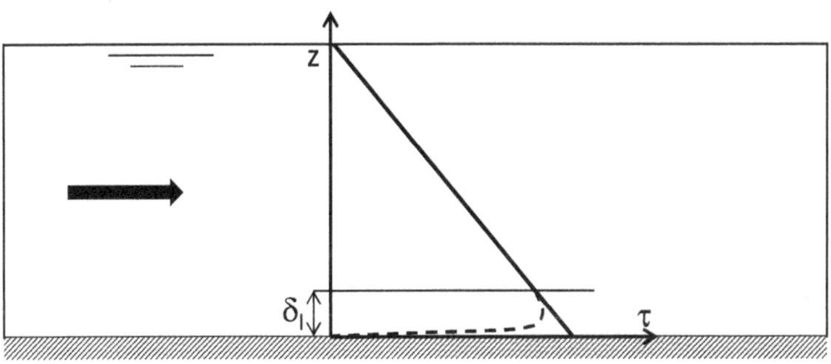

Fig. 3.7. Linear increase of the shear stress from surface to bottom for stationary flow. Near the bed the flow is laminar and the shear stress is dominated by viscosity. The thickness δ_ℓ reflects this laminar sublayer; note that it is much thinner than indicated in the figure.

where τ_b is the bed shear stress (i.e. $\tau_{z=0}$), and u_* is known as the shear velocity. Then, close to the bed, one finds:

$$\frac{\partial u}{\partial z} \approx \frac{u_*}{\ell} = \frac{u_*}{\kappa z} \tag{3.19}$$

Integration over z gives gives the logarithmic velocity profile:

$$\frac{u}{u_*} = \frac{1}{\kappa} \ln z + \text{const} = \frac{1}{\kappa} \ln \frac{z}{z_0} \tag{3.20}$$

in which z_0 is an integration constant, known as the roughness length or height (see Sec. 3.3). Integration over the water depth h yields:

$$\frac{u(z)}{\bar{u}} = \frac{\ln(z/z_0)}{\ln(\kappa h/z_0)} \tag{3.21}$$

Further to Eq. (3.14), and substituting from Eq. (3.19), a parabolic viscosity profile is found:

$$\tau = \tau_b \left(1 - \frac{z}{h}\right) = \rho u_*^2 \left(1 - \frac{z}{h}\right) = \rho u_* \kappa z \left(1 - \frac{z}{h}\right) \frac{\partial u}{\partial z} = \rho \nu_T \frac{\partial u}{\partial z} \tag{3.22}$$

Thus, a logarithmic velocity profile and a parabolic eddy viscosity profile ($\nu_T = u_* \kappa z(1 - z/h)$) are two sides of the same coin.

In practice, also more empirical velocity profiles are used (e.g. Nezu and Nakagawa, 1993), such as:

$$\frac{u(z)}{\bar{u}} = \left(\frac{z}{h}\right)^n \tag{3.23}$$

with $1/7 < n < 1/5$. Note that Eq. (3.23) is not consistent with a parabolic viscosity profile.

Over (very) smooth walls, however, Prandtl and his followers found that close to that wall, the flow in the boundary layer is laminar, as shown in Fig. 3.8. Within this laminar part, the so-called viscous sublayer, the stress profile deviates from its linear form (p becomes constant), and the velocity profile reads: $u(z)/u_* = zu_*/\nu$. The thickness of this laminar layer measures about $\delta_l \approx 11.6\,\nu/u_*$. At a mean flow velocity of 1 m/s, a typical value for δ_l would be 0.2 mm.

Further away from the wall, the velocity profile in open water flow also deviates from its logarithmic shape, which is accounted for by a so-called wake function. Thus, outside the laminar sublayer and the buffer layer,

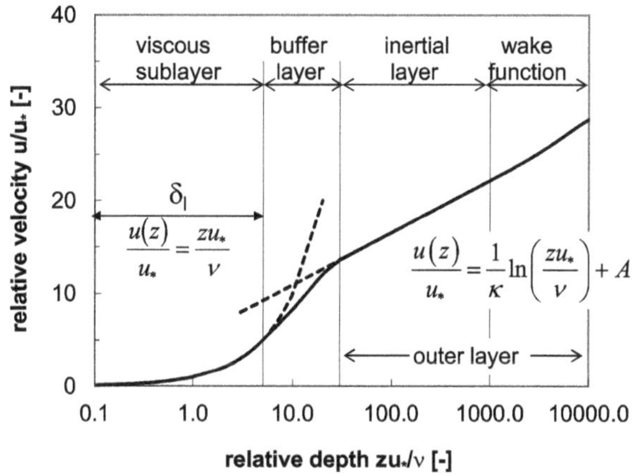

Fig. 3.8. Laminar/logarithmic velocity profile over a smooth wall and wake function.

which reflects the transition between laminar and turbulent behavior, the velocity profile is described by

$$\frac{u(z)}{u_*} = \frac{1}{\kappa} \ln\left(\frac{zu_*}{\nu}\right) + A \tag{3.24}$$

with $A \approx 5.6$, see e.g. Pope (2001) for further details, whereas Coles (1956) proposed:

$$\frac{u(h) - u(z)}{u_*} = -\frac{1}{\kappa} \ln\left(\frac{zu_*}{\nu}\right) + \frac{2\Pi}{\kappa} \cos^2\left(\frac{\pi z}{2h}\right) \tag{3.25}$$

with $\Pi \approx 0.2$.

3.2.3 *Effect of stratification on the turbulent boundary layer*

A constant and uniform density was assumed in the previous derivation of the logarithmic velocity profile. In many natural systems, the density is not uniform over the water depth. This is often a consequence of variations in salinity and/or temperature. However, also sediment concentration gradients can be of importance (Chapter 7). Sediment particles can follow turbulent motions when the particle settling velocity is much smaller than the (vertical) turbulent motions, i.e. $w_s \ll w'$. The settling velocity of fine sediment is generally at most a few mm/s, often much smaller (see Sec. 4.1), whereas the turbulent fluctuations measure many cm/s. Thus fine

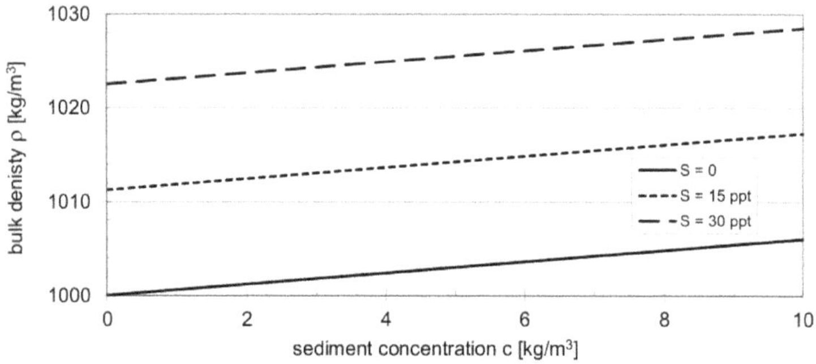

Fig. 3.9. Bulk density as function of salinity and SPM-concentration.

sediment suspensions can generally be treated as a one-phase fluid, which is our approach in this book.

Because of this one-phase assumption, the bulk density of water is not only affected by temperature and salinity, but by SPM values as well. An approximate relation for the fluids bulk density ρ_b reads:

$$\rho_b[\text{kg/m}^3] \approx \rho_0(T) + 0.75S[\text{o/oo}] + 0.6c[\text{kg/m}^3] \tag{3.26}$$

in which T is the temperature [°C], S the salinity (mass salt/mass sea water) and c the sediment concentration. Equation (3.26) is known as the equation of state and is visualized in Fig. 3.9. A more accurate equation of state was published by Millero and Poisson (1981). It is to be understood that relatively small gradients in SPM (and in salinity and temperature) can have large effects on the overall hydrodynamics and net transport of (fine) sediments. This is further elaborated in Sec. 9.1 on gravitational circulation.

Here, the effects of vertical density gradients in the water column are further discussed. Figure 3.10 presents a cartoon of a gradually stratified water column. If, by vertical mixing, two parcels of water were to be exchanged over the vertical, the parcel with denser water would end up in a less dense environment, and *vice versa* the less dense parcel would end up in the denser near-bottom water. Gravity opposes this mixing: work has to be done to overcome these buoyancy effects, which can only be supplied through the turbulent energy of the flow. In other words, stratified fluids dampen turbulent motion by consuming turbulence energy.

This reduction in vertical mixing has been studied extensively for temperature- and salinity-induced stratification (e.g. Turner, 1973), resulting in a large number of so-called damping functions. One of the

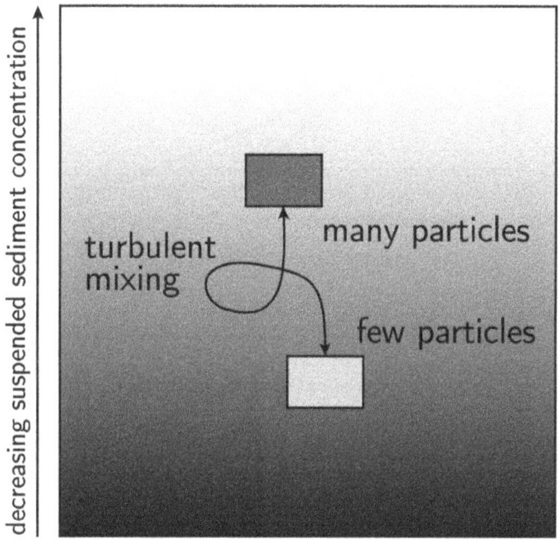

Fig. 3.10. Cartoon of buoyancy effects on vertical mixing in a gently stratified water column.

better-known functions is by Munk and Anderson (1948):

$$\nu_{T,\text{eff}} = \nu_{T,0}(1 + 10\text{Ri})^{-0.5} \tag{3.27}$$

$$\varepsilon_{T,\text{eff}} = \varepsilon_{T,0}(1 + 3.33\text{Ri})^{-1.5} \tag{3.28}$$

in which subscript $\cdot_0$ refers to non-stratified conditions, and the gradient Richardson number Ri is defined as

$$Ri \equiv -\frac{g\partial\rho/\partial z}{\rho(\partial u/\partial z)^2} \tag{3.29}$$

Ri is a measure for the stability of a stratified flow (see e.g. Hinze, 1975; Turner, 1973). Note that damping of matter and momentum are different for reasons explained in Sec. 3.4. Figure 3.11 shows large reductions in vertical exchange of momentum and matter at already moderate Richardson numbers.

Self-stratification by suspended (fine) sediments can induce high near-bed SPM-concentrations, which may affect the entire flow field and sediment transport dramatically, e.g. Sec. 4.4, Fig. 4.16 and Sec. 7.1.

The literature contains many other damping functions. However, today, buoyancy effects on vertical mixing are accounted for in the commonly used turbulence models. As an example, the k-equation of the $k - \varepsilon$ turbulence

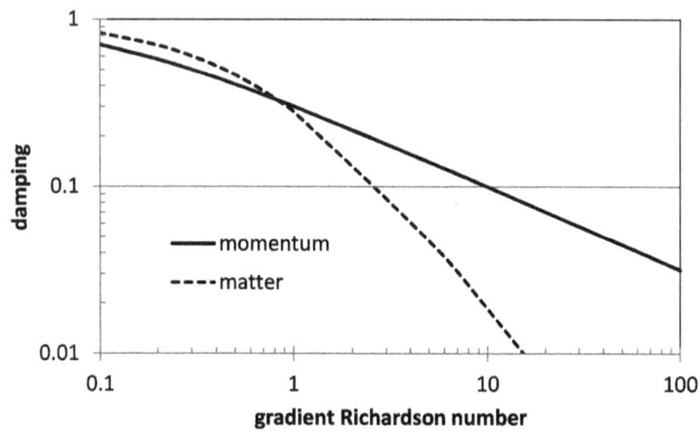

Fig. 3.11. Munk–Anderson damping functions for momentum and matter in stratified flow.

model is given (see also Chapter 11):

$$\frac{\mathrm{D}k_T}{\mathrm{D}t} = \underbrace{\frac{\partial}{\partial z}\left\{(\nu + \nu_T)\frac{\partial k}{\partial z}\right\}}_{\text{I}} + \underbrace{\nu_T\left(\frac{\partial u}{\partial z}\right)^2}_{\text{II}} + \underbrace{\varepsilon_T}_{\text{III}} - \underbrace{\frac{g}{\rho}\frac{\nu_T}{\sigma_T}\frac{\partial \rho}{\partial z}}_{\text{IV}} \tag{3.30}$$

in which k_T is the kinetic turbulent energy (Eq. (3.4)), and term I represents diffusion of turbulent energy, term II turbulence production, term III turbulent dissipation and term IV buoyancy destruction (i.e. damping of vertical mixing). The original $k - \varepsilon$ model for homogeneous conditions does not contain term IV. From Eq. (3.30) the time scale for turbulence decay T_T can be defined:

$$T_T = \frac{\ell}{\sqrt{k}} = \frac{h}{u_*} \tag{3.31}$$

where $\ell = h$ in open channel flow.

Note that also the ε-equation is to be modified — (see Rodi, 1984) and Sec. 11.1.

3.2.4 *Wave boundary layer*

The subject of waves is a discipline by itself, and many excellent text books have been published (Holthuijsen, 2010). Also the role of waves on the transport of coarser sediments (sand), in particular in the coastal zone, has been studied extensively (see e.g. Fredsøe and Deigaard, 1992; Soulsby *et al.*, 1993; Van Rijn, 1993).

The role of waves in relation to fine cohesive sediment can potentially be threefold:

(1) waves augment bed shear stresses, hence enhance erosion (this is discussed in Sec. 3.3);
(2) waves may mix some sediment into the water column (this contribution is small, as explained below);
(3) the presence of soft near-bed layers of fine sediment (fluid mud) dampens waves and thereby propel the mud in wave direction (streaming, Secs. 7.4 and 7.5.2).

Muddy foreshores are characterized by their shallowness and gentle slopes. Bars and bed forms are rare, as fines are generally not transported as bed load. Also the wave climate itself is generally fairly mild — if not, fine sediments would not accumulate on such foreshores. Hence, wave-driven (longshore) transport, so important in sandy systems, hardly plays a role along muddy coasts. Moreover, waves generally do not break on muddy foreshores, but dissipate by friction within the mud layer (e.g. Sec. 7.4). Hence, the role of breaking waves is not discussed in this book.

In the following, the wave number k and frequency ω are frequently used, defined as:

$$k = 2\pi/L \quad \text{and} \quad \omega = 2\pi/T \tag{3.32}$$

where L is the wavelength (we use L to distinguish from the turbulent length scale λ) and T the wave period. The dispersion relation for deep water (short) waves is implicit. However, two very good approximations can be found in literature, e.g. by Eckart (1948):

$$k = k_0/\tanh\{k_0 h\} \tag{3.33}$$

where k_0 is the deep-water wave number; and by Guo (2002):

$$k = \frac{\omega^2}{g}\left(1 - \exp\left\{-\left(\omega\sqrt{h/g}\right)^{5/2}\right\}\right)^{-2/5} \tag{3.34}$$

in which h is the local water depth and g the acceleration of gravity. In particular Guo's approximation is accurate within 1% over a large range of wave numbers and is to be recommended in numerical models, as the full implicit dispersion equation does not have to be resolved (iteratively).

To show the limited influence of waves on mixing in the water column, it is illustrative to elaborate on the profile of the horizontal (orbital) velocity under waves, which is obtained from solving Eq. (3.17), omitting the

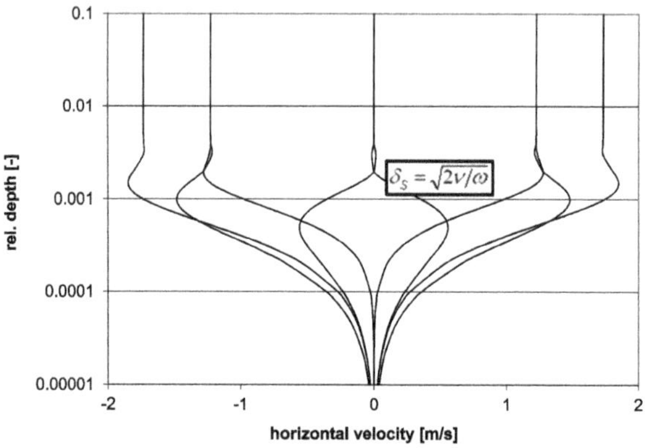

Fig. 3.12. Vertical profile of horizontal orbital motion — solution to Eq. (3.35); $T = 5\,\mathrm{s}$, $H_s = 1\,\mathrm{m}$ and $h = 2\,\mathrm{m}$.

pressure gradient term II, and assuming harmonic boundary conditions:

$$u(z,t) = a_b[\cos\{\omega t\} - e^{-z/\delta_S}\cos\{\omega t - z/\delta_S\}] \tag{3.35}$$

in which δ_S is the Stokes' boundary layer thickness (see below). The orbital excursion at the bed a_b is given by $a_b = H_s\omega/\sinh\{kh\}$ with the significant wave height H_s. This solution was first presented by Longuet-Higgins (1953). For the full solution, i.e. including the vertical orbital motion, the reader is referred to e.g. Holthuijsen (2010). As an example, the solution to Eq. (3.35) is presented in Fig. 3.12 (note that 1 m wave height in 2 m water depth is not realistic, and certainly violates linear assumptions, but is used for didactic reasons, exaggerating the effects), showing a boundary layer with large vertical velocity gradients. This layer amounts to about 0.1–0.5% of the water depth (for 2 m water depth this implies 2–10 mm), and is known as the Stokes' wave boundary layer δ_S:

$$\delta_S = \sqrt{2\nu/\omega} \tag{3.36}$$

where ν is the kinematic viscosity of the water. Note that δ_S is not the thickness of the turbulent boundary layer under waves — see below.

Above the Stokes' boundary layer, the flow is basically laminar (hence no vertical mixing — see above). The boundary layer remains so thin, because the flow does not get time to develop — while growing, the flow reverses again. Thus velocity gradients near the bed are large, and so are the wave-induced bed shear stress and turbulence production. The difference with the near-bed conditions under tides is summarized in Table 3.1 — note that "large" and "small" are relative adjectives.

Table 3.1. Time scales for boundary layer development.

Tidal wave	Wind wave (sea and swell)
Long wave: ω small	Short wave: ω large
Thick boundary layer (water depth)	Thin boundary layer (mm's–cm's)
Entire water column turbulent	Above boundary layer: laminar flow
Small near-bed velocity gradient	Large near-bed velocity gradient
Small bed shear stress	Large bed shear stress
Small near-bed turbulence production	Large near-bed turbulence production

The Stokes boundary layer for sea and swell is generally very thin. However, within this boundary layer, turbulence production is large, driven by vertical velocity gradients. Moreover, turbulence is advected (orbital motions) and diffused away from its source higher into the water column, and the turbulent boundary layer δ_T is therefore (much) thicker than δ_S. The exact value of δ_T is not well established, and various values are found in literature, i.e. $\delta_T = 15\delta_s$ (Fredsøe and Deigaard, 1992); $\delta_T = (2-4)\kappa\,|u_{*fw}|/\omega$ (Grant and Madsen, 1979); while Van Kesteren and Bakker (1985) propose $\delta_T = 2\pi\kappa\,|u_{*fw}|/\omega$, in which u_{*w} and u_{*fw} are the shear velocity by waves, and by the combined effect of currents and waves, respectively. Note that δ_T generally amounts to a few cm–dm, hence substantially larger than δ_S, but still much smaller than the water depth, except in very shallow conditions, such as over almost drying intertidal areas.

Current-wave interaction

It is illustrative elaborating on the effect of interacting waves and currents on vertical turbulent mixing, following Grant and Madsen (1979), assuming flow and waves are in the same direction. The basic idea is that the effective roughness for the flow increases (representing the large turbulence production near the bed), as sketched in Fig. 3.13. Grant and Madsen introduce a wave–current shear velocity u_{*fw}:

$$|u_{*fw}| = \sqrt{u_{*f}^2 + u_{*w}^2} \tag{3.37}$$

Within the turbulent wave-boundary layer, the (mean) eddy viscosity ν_T scales linearly with depth and reads:

$$z_0 \leq z \leq \delta_w: \quad \nu_T = \kappa\,|u_{*fw}|\,z = \kappa\sqrt{u_{*f}^2 + u_{*w}^2}\,z \tag{3.38}$$

The near-bed shear stress $\tau_b = \rho_w u_{*f}^2$ is (assumed) constant throughout the wave-boundary, but as yet unknown, and yields a logarithmic velocity

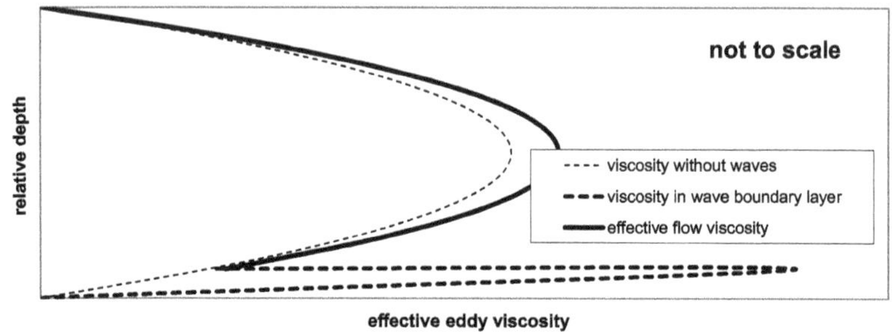

Fig. 3.13. Cartoon of wave-current interaction on vertical mixing further to Grant and Madsen (1979) — not to scale.

profile based on $|u_{*fw}|$. Above the wave-boundary layer, wave-induced turbulence is not notable and Grant and Madsen assume a parabolic viscosity profile, which yields the usual logarithmic velocity profile. The two profiles are matched at δ_T. The wave-induced turbulence contributes to the mean flow above the wave-boundary through an increase in effective roughness length z_{fw}:

$$\delta_T \geq z_0: \quad \frac{z_{fw}}{z_0} = \left(\frac{\delta_w}{z_0}\right)^\beta; \quad \beta = 1 - \left(1 + \left(\frac{u_{*w}}{u_{*f}}\right)^2\right)^{-1/2} \tag{3.39}$$

The effective bed shear stress τ_e and shear velocity u_*, as applied in the preceding paragraphs, follow from applying the logarithmic law of the wall, using z_{fw} instead of z_0.

If currents and waves are not co-directional, the reader is referred to the original paper by Grant and Madsen (1979) or Soulsby and Clarke (2005).

3.3 Bed shear stresses

The velocity profiles discussed in the previous sections are all related to the bed shear stress τ_b, as the bed is the main producer of turbulence in shallow water flows. Not only the velocity profile, but also the erosion of (fine) sediment is a function of that bed shear stress. The bed shear stress is therefore a crucial parameter in fine sediment studies, which needs ample attention. Unfortunately, the bed shear stress cannot be measured directly and the determination of the bed shear stress in numerical models still depends on various assumptions. In this section, we discuss the bed shear stress for uniform (or slowly varying) flow, the bed shear stress under waves and its interaction with uniform flow, and the effects of stratification.

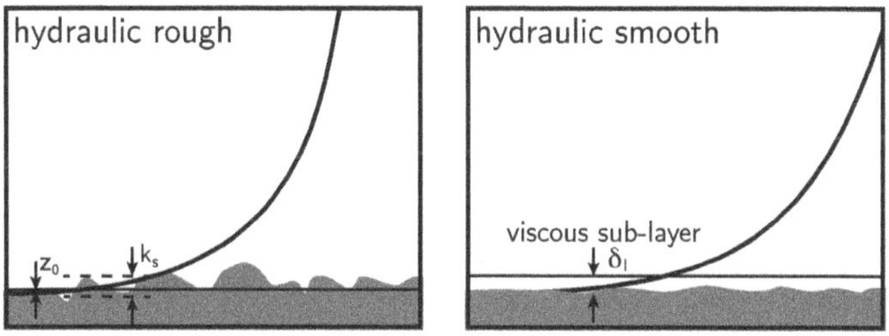

Fig. 3.14. Cartoon of hydraulically rough and smooth conditions in relation to the viscous (laminar) sublayer.

Table 3.2. Roughness height for hydraulically smooth and rough conditions.

Hydraulically smooth	$\mathrm{Re}_* < 5$	$z_0 = 0.11\nu/u_*$
Transitional	$5 < \mathrm{Re}_* < 70$	$z_0 = 0.11\nu/u_* + k_s/30$
Hydraulically rough	$\mathrm{Re}_* > 70$	$z_0 = k_s/30$

3.3.1 *Bed shear stress in steady and tidal flow*

As explained in Sec. 3.2.2, part of the boundary layer is laminar if the wall is (very) smooth and if the flow velocity is small (the thickness of the laminar sublayer decreases with increasing flow velocity). We refer to hydraulically smooth conditions. In case of a rougher wall, without a laminar sublayer, we refer to hydraulically rough conditions. This distinction is sketched in Fig. 3.14, depicting that for hydraulically rough conditions, the viscous sublayer drowns in between protruding bed elements. These bed elements are characterized by the Nikuradse roughness height k_s. The transition from hydraulically smooth to rough conditions is given by the roughness Reynolds number, defined as $Re_* = u_* k_s/\nu$, as shown in Table 3.2. Table 3.2 also presents values of the roughness length z_0 in the various regimes. Note that in the smooth regime, the roughness height depends on u_*, thus on the bed shear stress. This implicit relation has therefore to be solved iteratively.

In open channel flow, hydraulically rough conditions are generally encountered. Finding k_s then becomes the challenge. There are three methods to determine the bed shear stress in open water:

(1) **From measured vertical velocity profiles:** If one assumes a logarithmic velocity profile, velocity data can be plotted in a $u-\log z$ graph, while fitting Eq. (3.19) through the data points. Next to the assumption of a logarithmic profile (which does not hold in accelerating or

decelerating flow, nor under stratified conditions), this procedure gives the local bed shear stress only. This approach suffers from inaccuracies, as it implies an extrapolation towards the very near bed position.

(2) **From turbulence measurements:** Under non-stratified conditions, the vertical shear stress distribution is (almost) linear (see Fig. 3.7). Hence, from measuring the turbulent stresses $\overline{u'v'}$, $\overline{u'w'}$ and $\overline{v'w'}$ at two heights, the bed shear stress can be obtained from extrapolation, assuming isotropy. This requires high resolution measurements, capturing the turbulence time scales.

(3) **From measured water levels:** In stationary flow, the (energy) slope in water level directly yields the total energy loss, hence the mean, effective bed shear stress over the reach considered (term II and III of Eq. (3.17)). Under non-stationary conditions, a (numerical) hydrodynamic model can be deployed, tuning the roughness coefficient(s) to reproduce the measured water levels (and phases under tidal conditions). Again, a mean, effective value is obtained for the reach considered. Local bed shear stresses cannot be determined, as water level differences can only be determined accurately over a sufficiently long reach.

In practice, a variety of roughness coefficients is used, summarized in Table 3.3, with their mutual relation. Equation (3.40) presents the relation between the bed shear stress with these coefficients as a function of a characteristic (depth-mean) velocity U:

$$\tau_b = c_f U^2 \rho = \frac{g}{C^2}\rho U^2 = \frac{gn^2}{h^{1/3}}\rho U^2 = \frac{f}{8}\rho U^2 \tag{3.40}$$

From which we find:

$$\frac{u_*}{U} = c_f = \frac{\sqrt{g}}{C} = \frac{gn^2}{h^{1/3}} = \frac{f}{8} \tag{3.41}$$

Above, we discuss the effective bed shear stress, which reflects the total momentum losses in the flow. We refer to the total (effective) hydraulic drag. This total hydraulic drag in open water (river, estuary, coastal) is determined by skin friction and various energy losses, such as flow separation over bed forms, bank irregularities, etc. These latter losses are lumped in a parameter known as form drag. This drag determines the velocity profile and therefore needs to be established for flow computations. Note that for modeling, the model resolution also affects the effective

Table 3.3. Friction coefficients with their range and (typical values) in open channel flow.

			Typical values	Unit
z_0	Roughness length		0.003–7 (0.2)	mm
k_s	Nikuradse roughness height	$k_s = 30z_0$	0.001–0.2 (0.005)	m
C	Chézy coefficient	$C = 18\log(12h/k_s)$	50–110 (60)	$m^{1/2}/s$
f	Darcy–Weisbach	$f = 8g/C^2$	0.007–0.03 (0.02)	—
	White Colebrook	$\frac{1}{\sqrt{f}} = -2\log\left\{\frac{k_s}{12R_h} + \frac{2.51}{\mathrm{Re}\sqrt{f}}\right\}$		
n	Manning	$n = \sqrt[6]{h}/C$	0.011–0.027 (0.023)	$s/m^{1/3}$
S	Strickler	$S = 1/n$	37–83 (43)	$m^{1/3}/s$

form drag — this is how one accounts for sub-grid effects. Thus:

$$\text{total drag} = \text{form drag} + \text{skin friction} \tag{3.42}$$

For sediment transport, the processes in a thin layer near the bed determine water-bed exchange processes. Thus, the water-bed exchange processes are merely governed by the skin friction, the drag force that acts on the sediment particles. The distinction between skin friction, form drag and total drag, in conjunction with a variety of drag predictors, has been treated in numerous papers and textbooks, such as Van Rijn (1993). This distinction is very relevant for sandy systems, such as rivers, and many coastal seas. For muddy conditions, this distinction is less important, because muddy beds are generally smooth, while bed forms are generally not formed, owing to the transport of fine sediments in suspension, rather than as bed load. Yet, a distinction has to be made, as mass erosion (Sec. 6.4) and biota (Chapter 8) may induce irregularities in the bed surface. Figure 3.15 presents a cartoon of the near-bed processes relevant for water-bed exchange processes.

For sandy beds, Van Rijn (1993) proposed $k_s = 3D_{90}$ for the skin friction under hydraulically rough conditions, where D_{90} is the 90-percentile of the sediment grain size. For muddy systems, the primary particles are not a direct measure for the roughness height, as they are generally aggregated to flocs of larger size (Sec. 4.3). Also the flocs themselves do not provide any information on the skin friction, however, as the near-bed conditions are often formed by a fluff layer (Sec. 6.5).

Note that in numerical models, distinction between total drag and skin friction is often not made. Correct modeling of the erosion rates then may

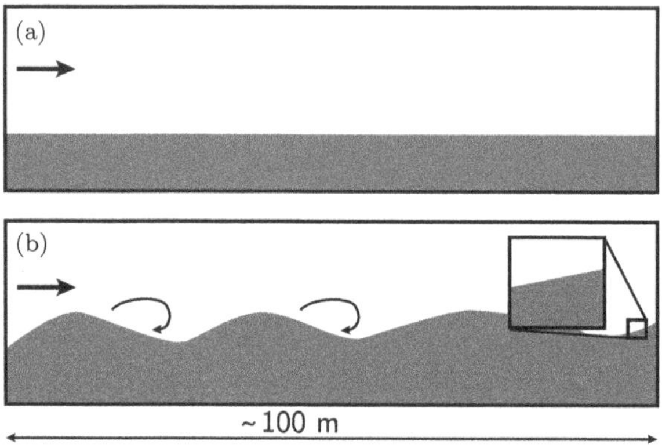

Fig. 3.15. (a) For a smooth horizontal bed, the total drag force equals skin friction, as there are no bed forms. (b) In case of bed forms, the total drag consists of drag by bed forms and skin friction. For sediment transport, one generally accounts for skin friction only as a sediment particle does not "see" the bed form. The flow in the wake of bed forms does however induce additional turbulence. Therefore, a local variation in bed shear stresses between crests and troughs is likely.

require tuning the critical shear stress for erosion to unrealistically large values. This pragmatic engineering approach can be efficient, provided that bed properties do not change too much.

Bed shear stress distribution

Owing to the stochastic nature of turbulence, turbulent properties, such as turbulent velocities and bed shear stresses, are characterized by a distribution around a mean value. Let us assume that the turbulent velocities have a Gaussian distribution. The bed shear stress distribution then becomes skewed, as the bed shear stress τ_b scales with u^2 (e.g. Hofland and Battjes, 2006). However, the turbulent flow velocities deviate from such a Gaussian distribution. Furthermore, turbulence is non-isotropic, especially close to the wall. Therefore, the distribution proposed by Hofland and Battjes (2006) is not accurate, as shown by Van Prooijen and Winterwerp (2010).

However, the bed shear stress distribution is generally skewed, e.g. as measured by Obi *et al.* (1996). His distribution (though for fairly low Re ≈ 6600) is presented in Fig. 3.16. The significance of this distribution for fine sediment dynamics is elaborated upon in Sec. 6.3.3.

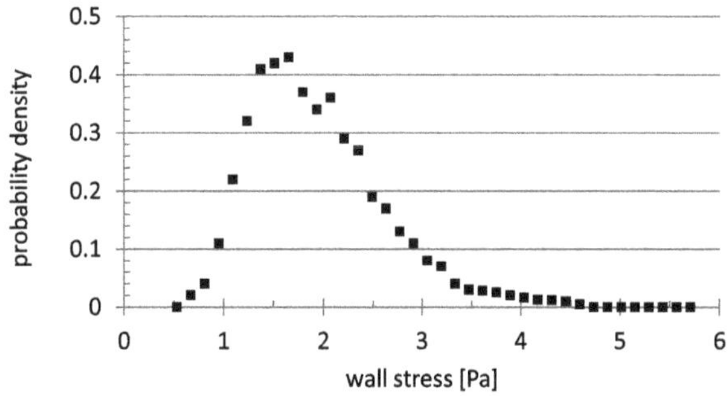

Fig. 3.16. Probability density distribution of bed shear stress as measured by Obi *et al.* (1996).

3.3.2 *Bed shear stress for waves*

Also under waves, the bed shear stress is a function of the bed's friction coefficient. Here, we present the formulas by Swart (1977) and Soulsby (see Whitehouse *et al.*, 2000). The friction coefficient f_w is defined by

$$\tau_{b,w} = 0.5 f_w \rho U_w^2 \tag{3.43}$$

where $\tau_{b,w}$ is the wave-induced bed shear stress and U_w is the orbital (wave) velocity amplitude at the bed. The widely-used friction formula by Swart (1977) reads:

$$\begin{aligned} &\text{for } \frac{a_b}{k_s} > \frac{\pi}{2}\text{:} \quad f_w = 0.00251\exp\left\{5.21\left(\frac{a_b}{k_s}\right)^{-0.19}\right\}, \quad \text{and} \\ &\text{for } \frac{a_b}{k_s} \leq \frac{\pi}{2}\text{:} \quad f_w = 0.31 \end{aligned} \tag{3.44}$$

where k_s is the Nikuradse roughness height (see Sec. 3.3.1). The orbital excursion a_b is obtained from linear wave theory:

$$a_b = \frac{U_w}{\omega} = \frac{\omega H_s}{\sinh\{kh\}} \tag{3.45}$$

Equation (3.44) is plotted in Fig. 3.17. In comparison with the Darcy–Weisbach friction coefficient (Table 3.3), with typical values between 0.007 and 0.03, the wave friction coefficient f_w is substantially larger for most wave conditions, reflecting the large turbulence production near the bed.

The friction formula by Soulsby is based on the wave Reynolds number defined as $\mathrm{Re_w} \equiv U_w a_b/\nu$, as shown in Table 3.4. The distinction between

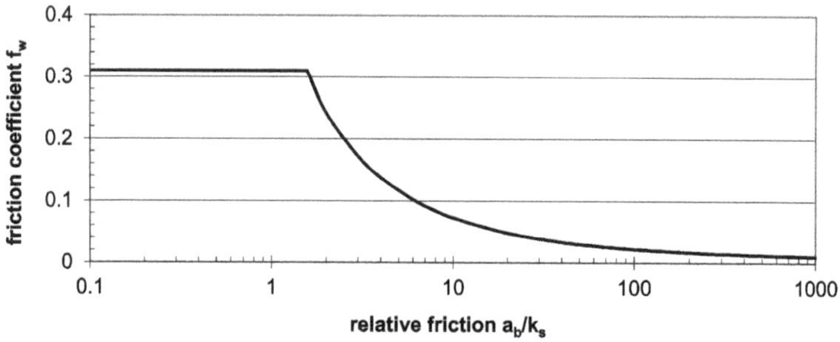

Fig. 3.17. Friction coefficient for waves f_w according to Swart (1977).

Table 3.4. Coefficients in Soulsby's wave friction coefficients for muddy beds.

Smooth: $f_w = B\mathrm{Re}_w^N$		Rough:
Laminar: $\mathrm{Re}_w \leq 5 \cdot 10^5$	Smooth turb: $\mathrm{Re}_w > 5 \cdot 10^5$	$f_w = 0.237(a_b/k_s)^{0.52}$
$B = 0.2$ and $N = -0.5$	$B = 0.0521$ and $N = -0.187$	

hydraulically rough and smooth conditions is not clear, and Soulsby suggests to compute the friction coefficient f_w for both smooth and rough conditions, taking the larger of the two, while in muddy systems, a hydraulically smooth assumption for assessing the threshold of motion is probably the better approach.

In case of combined waves and currents, the wave- and current-induced bed shear stresses need to be combined. Intuitively, adding the velocity vectors near the bed seems appropriate, leading to:

$$\vec{\tau}_{fw} = \vec{\tau}_f + \vec{\tau}_w \tag{3.46}$$

where $\vec{\tau}_f$ is the bed shear stress by (tidal) flow alone, $\vec{\tau}_w$ is the bed shear stress by waves alone and $\vec{\tau}_{fw}$ is the effective bed shear stress, combining the effects of waves and currents.

Soulsby and Clarke (2005) advocate an alternative approach. Their method accounts for the difference in boundary layer development under steady and oscillatory conditions (Fig. 3.13) yielding the vector sum of the individual bed shear stresses themselves. Comparison with data from the field and the laboratory shows that this approach gives better results, as argued in detail by Soulsby and Clarke (2005). Therefore, this method is advocated in this book. The wave-period mean bed shear stress τ_m under waves and currents for hydraulically rough and hydraulically smooth conditions reads (note that this is a scalar):

$$\tau_{m,\text{rough}} = \tau_f \left[1 + 1.2\left(\frac{\tau_w}{\tau_w + \tau_f}\right)^{3.2}\right] \tag{3.47a}$$

$$\tau_{m,\text{smooth}} = \tau_f \left[1 + 9\left(\frac{\tau_w}{\tau_w + \tau_f}\right)^{9}\right] \tag{3.47b}$$

The maximum effective bed shear stress $\tau_{e,\max}$ under waves and currents becomes:

$$\tau_{e,\max}^2 = (\tau_m + \tau_w \cos\alpha)^2 + (\tau_w \sin\alpha)^2 \tag{3.48}$$

in which α is the angle between waves and currents (near the bed, thus in the wave boundary layer). The rms-value of the effective bed shear stress $\tau_{e,\text{rms}}$ is given by

$$\tau_{e,\text{rms}}^2 = \tau_f^2 + \tau_w^2/2 \tag{3.49}$$

It is common to use the significant wave height H_s in computing bed shear stresses in the formulas for water-bed exchange processes. However, one may argue that a mean wave height would be more appropriate.

Wave-induced bed shear stresses are of course only relevant in case of shallow water; in deep water, orbital movements will not reach the seabed. As a rule of the thumb, waves affect the seabed if $h < L/2$, where h is water depth, and L is wave length. For deep water, $L = gT^2/2\pi = 1.56T^2$, where T is the wave period. Thus a good engineering rule is that waves affect the seabed if $h < 0.8T^2$, realizing that short waves (sea) have periods of typically 4–6 sec, whereas swell is characterized by periods of 10–15 sec.

In this book, we follow common practice, using H_s for assessing effective bed shear stresses under waves and currents, and use its mean value $\tau_{e,\text{rms}}$ in assessing surface erosion rates. However, for evaluating risks of liquefaction, the maximum bed shear stress $\tau_{e,\max}$ (and the equivalent maximum normal stress) is more relevant, as well as for assessing the onset of mass erosion (Sec. 6.4).

Determination of the bed shear stress from field or laboratory measurements is not straightforward either, especially, the unraveling of the fluctuating components of waves and turbulence leads to difficulties. We refer to Zhang *et al.* (2018) and Zhu *et al.* (2016) for further details.

3.3.3 *Effects of stratification on the bed shear stress*

Stratification-induced damping of turbulence reduces the vertical exchange of momentum, i.e. reduces $\nu_{T,\mathit{eff}}$, which also reduces the turbulent energy dissipation. As a result, the effective hydraulic drag of the turbulent water movement reduces, as was first suggested by Vanoni (1940), and later extensively elaborated upon by Einstein and Qian (1955). In the following decades many experimental and theoretical studies were carried out, and also field observations revealed significant drag reductions in sediment-laden flow (e.g. Beardsley *et al.*, 1995; Dong *et al.*, 1997; Gabioux *et al.*, 2005; Weilbeer, 2008; Le Hir *et al.*, 2000; Le Hir and Cayocca, 2002; Odd and Cooper, 1988; Wang *et al.*, 1998; Guan *et al.*, 1998).

Jiang (2012) analyzed tidal propagation in the fresh water part of the Yangtze River between Xuliujing and Jiangyin, a tidal gauge 80 km further up-estuary. His major observations are shown in Fig. 3.18, showing that during the dry season, tidal damping in up-estuary direction increases with increasing tidal range (i.e. larger tidal range, thus larger tidal velocities, hence larger drag), as is normally the case in estuaries. However, during wet season-spring tide, damping decreases with increasing tidal amplitude. This reverse trend can only be explained from the measured larger SPM-concentrations, mobilized by the larger velocities in the river: during wet season-neap about 100 mg/L, during wet-season spring about 800 mg/L. During the dry season, SPM-values do not exceed 50–100 mg/L. The observed reverse trend in tidal damping can be understood from an increase in effective Chézy coefficient from 60 $\mathrm{m}^{1/2}/\mathrm{s}$ at neap tide to 70 $\mathrm{m}^{1/2}/\mathrm{s}$. Thus, this relatively small reduction in effective bed shear stress (25%) has a profound effect on the on the tidal amplification, hence on tidal levels.[1]

In the following, we present a first-order assessment of the reduction of effective hydraulic drag by SPM, to obtain a feeling on the reductions which may be expected. Here we follow Barenblatt (1953), quantifying this drag reduction. Winterwerp *et al.* (2009) proposed modifying Barenblatt's model slightly, accounting for the free water surface in open channel flow:

$$\frac{u}{u_*} = \frac{1}{\kappa}\left[\ln\left(\frac{z}{z_0}\right) + \alpha_2\left(\frac{z}{\ell}\frac{z}{h_{ref}}\right)\right] \tag{3.50}$$

[1] Jiang studied extensively all other possible effects which may enhance or reduce the effective hydraulic drag in the river, including the formation of bed forms, and concluded that none of these other effects can explain the observations.

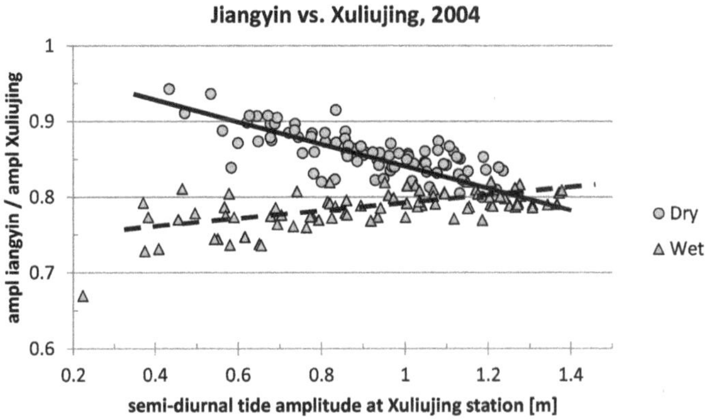

Fig. 3.18. Observed tidal amplification in the Yangtze River (China) between Xuliujing and Jiangyin (a station 80 km further up-estuary) for the dry (December–February) and wet season (July–September) conditions of 2004.

The first two terms of Eq. (3.50) reflect the classical logarithmic velocity profile (Eq. (3.20)). The second term of the right-hand side reflects the effects of stratification, with the Monin–Obukhov length scale $\ell \equiv \rho_b u_*^3/\kappa g \overline{\rho' w'}$, and α_2 and h_{ref} are parameters to be determined. Equ. (3.50) is integrated over the water depth, and α_2 and h_{ref} are determined through a large number of simulations with a 1DV model (see Winterwerp *et al.*, 2009), yielding:

$$\frac{\bar{u}}{u_*} \equiv \frac{C_{\text{eff}}}{\sqrt{g}} = \frac{C_0}{\sqrt{g}} + \frac{C_{SPM}}{\sqrt{g}} = \frac{C_0}{\sqrt{g}} + 4h\text{Ri}_*\beta \tag{3.51}$$

in which C_{eff} is the effective Chézy coefficient, C_0 is the Chézy coefficient for buoyant neutral conditions (no SPM), C_{SPM} is the effective hydraulic roughness by SPM-induced turbulence damping, and the bulk Richardson number Ri_* and Rouse number β (introduced in Sec. 3.4) are defined as

$$\text{Ri}_* \equiv \frac{(\rho_b - \rho_w)\, gh}{\rho_b u_*^2} \tag{3.52}$$

$$\beta \equiv \frac{\sigma_T W_s}{\kappa u_*} \tag{3.53}$$

Equation (3.51) is plotted in Fig. 3.19 together with the computational "data" points obtained from numerous simulations with a 1DV-model, and in which the Prandtl–Schmidt number σ_T is set to 2 upon calibration. Figure 3.19 shows that the effects of SPM induced stratification can be large in deeper water.

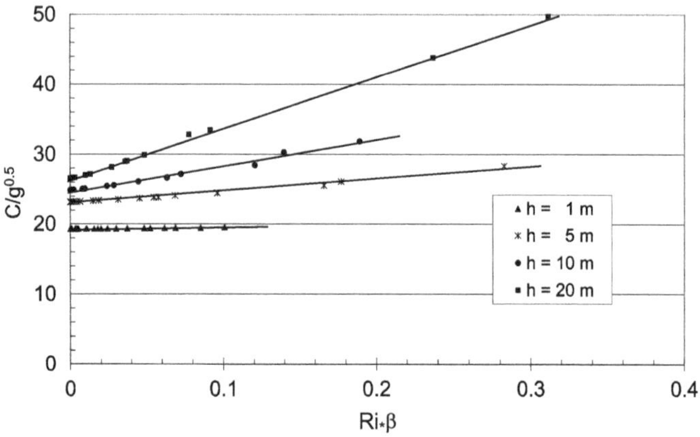

Fig. 3.19. Drag reduction by suspended sediment according to Eq. (3.50) with numerical data points from 1DV model.

3.3.4 *Some further thoughts on the bed shear stress*

Assessment of the bed shear stress, e.g. the appropriate friction coefficient is one of the more difficult aspects of hydraulic engineering in general, and of its modeling in particular. Below a number of issues which have to be considered:

(1) Natural open water systems are generally characterized by large spatial variability in bed composition, bed forms, depth, intertidal areas, vegetation, etc. (see also Fig. 4.11). Worse, these vary over shorter and longer time scales as well. It is virtually impossible to account for (all) this variability. More important is to be aware of the possible errors and judge the outcomes of the simulations in that respect.

(2) The distinction between skin friction and form drag from a conceptual point of view seems quite straightforward, until one dives into the details. How to treat ripples of a few cm? Which characteristic flow velocity do we use in determining the skin friction? For instance, consider a large and fairly smooth bed form. The flow accelerates at the windward side of the bed form, while decelerating at its lee side. As the bed shear stress does not scale linearly with velocity, asymmetry is induced, increasing when the bed forms themselves become asymmetric as well. If the bed forms are steep, flow separation will occur, affecting

water-bed exchange processes locally. Hence, what is the characteristic velocity in relation to the skin friction coefficient? For numerical modeling, it is essential to be aware of the differences between skin friction and form drag.

(3) (Vertical) Turbulent mixing is affected largely by bed forms and local flow separation. As turbulent mixing governs the (horizontal) transport of fine sediments, while its availability is governed by near bed processes (skin friction), one may have to use both the total drag and skin friction for assessing sediment transport.

(4) In muddy systems, bed forms are generally not formed, though there are some exceptions, in particularly when the bed is well-consolidated (see e.g. Dyer and Manning, 1999). Local irregularities can occur due to differences in bed strength. Close to the bed of muddy systems, thin layers of very soft mud are often present (fluff layer), which are likely to affect the effective drag, and possibly even the thickness and properties of the laminar sub-layer — possibly, the bed becomes hydraulically smooth (e.g. Chapter 7).

(5) Bed shear stresses under waves and currents have not yet been fully resolved. Moreover, waves generally induce ripples, which affect the bed shear stress further. The interaction between hydrodynamics (currents and waves) and bed forms is another issue which needs further exploration. One of the aspects that was not considered is the skewness and asymmetry of waves. Particularly, in (very) shallow water, waves are no longer sinusoidal.

(6) The effective hydraulic drag over very soft near-bed mud layers, i.e. fluid mud (Chapter 7) reduces significantly (C values up to 90–100 $\mathrm{m}^{1/2}/\mathrm{s}$). However, waves would be damped by viscous dissipation (Sec. 7.4). Rogers and Holland (2009) showed that, if not too large, such damping can be modeled through an increase in the wave friction coefficient. This implies the paradoxical effect of a large decrease in current-related drag and a large increase in wave-related drag.

These issues are potential reasons why numerical models cannot be considered as black boxes and require expert judgment for interpretation. In the second part of this book, we show that proper system understanding and a proper formulation of the engineering questions at hand are essential for making the proper assessments of roughness coefficients, and other modeling parameters.

3.4 Vertical mixing of particles

3.4.1 *Equilibrium profiles*

In the previous section, the hydrodynamics of the benthic boundary layer were discussed. Here, we continue with the distribution of sediment in the benthic boundary layer. The sediment concentration profile is a consequence of exchange between the water column and the bed (erosion and deposition) and of the mixing of sediment within the water column. This last aspect is considered here, the first one in Chapter 6.

In analogy to the Reynolds decomposition of Sec. 3.2.2, also the SPM-concentration can be decomposed into a turbulence mean part $\bar{c}$ and its turbulent fluctuation c'. Thus, similar to Eq. (3.13), the vertical SPM flux reads:

$$\overline{wc} = \overline{(\bar{w} + w')(\bar{c} + c')} = \bar{w}\bar{c} + \overline{\bar{w}c'} + \overline{w'\bar{c}} + \overline{w'c'} = \bar{w}\bar{c} + \overline{w'c'} \tag{3.54}$$

with sediment concentration c, vertical velocity w, fluctuations denoted by a $'$, and the averaging operator is indicated by an overbar. In analogy with the Boussinesq approach, an eddy diffusivity (turbulent diffusion) coefficient Γ_T is defined as

$$\overline{w'c'} = -\Gamma_{T,z}\frac{\partial \bar{c}}{\partial z} \tag{3.55}$$

Similar relations can be derived for the two horizontal coordinates. The relation between the eddy diffusivity and eddy viscosity is given by the turbulent Prandtl–Schmidt number σ_T:

$$\sigma_T \equiv \nu_T/\Gamma_T \tag{3.56}$$

In isotropic turbulence in homogeneous flow, σ_T attains a value of about 0.7 (see Gualtieri *et al.*, 2017). Note that under these conditions, the degree of freedom for a sediment particle is larger than for water molecules (or groups of molecules). For instance, the flow velocity components u' and v' can attain any value within the turbulence spectrum. But then the third velocity component w' is set by continuity. This restriction does not apply for the turbulent movement of particulate matter in a turbulent flow. For non-isotropic conditions, σ_T may deviate from 0.7. For instance, under stratified conditions, σ_T may even attain values larger than unity, as momentum may be transferred over the interface by internal waves, whereas turbulent mixing is considerably reduced, (see e.g. Gualtieri *et al.*, 2017). Van Maren *et al.* (2009b) used $\sigma_T = 2$ for modeling sediment transport in the hyper-turbid Yellow River. The effects of stratification can also be included through a damping term, which is a function of the (local) Richardson number.

Under the assumptions of uniform water density (no stratification by temperature, salinity nor sediment) and steady flow (with a logarithmic velocity profile), also the eddy diffusivity exhibits a parabolic shape. Further to Eq. (3.14), a typical value for the mean vertical eddy diffusivity yields $\Gamma_T \simeq 0.1hu_*$, implying that Γ_T scales with U.

The consequence of the parabolic eddy diffusivity profile on the vertical SPM-concentration distribution in open water was elaborated by Hunter Rouse (Rouse, 1939), and one refers to the Rouse profile for suspended sediment ever since. To derive at the Rouse profile, we start from the 1DV (one-dimensional in vertical direction) sediment balance equation:

$$\frac{\partial c}{\partial t} - \underbrace{\frac{\partial}{\partial z}(w_s c)}_{\text{advection}} - \underbrace{\frac{\partial}{\partial z}\left(\Gamma_{T,z}\frac{\partial c}{\partial z}\right)}_{\text{diffusion}} = 0 \tag{3.57}$$

with boundary conditions at the water surface

$$-w_s c - \Gamma_{T,z}\frac{\partial c}{\partial z} = 0 \tag{3.58}$$

and the sediment bed

$$-w_s c - \Gamma_{T,z}\frac{\partial c}{\partial z} = E - D \tag{3.59}$$

in which w_s is the settling velocity, E the erosion rate (Chapter 6) and D the deposition rate (Sec. 4.1).

Under steady conditions, the advection and diffusion terms in Eq. (3.57) balance. When also the water-bed exchange is zero, and upon substitution of the parabolic viscosity profile (Eq. (3.22)) one obtains:

$$w_s c = -\Gamma_{T,z}\frac{\mathrm{d}c}{\mathrm{d}z} = -\frac{\kappa u_*}{\sigma_T} z\left(1 - \frac{z}{h}\right)\frac{\mathrm{d}c}{\mathrm{d}z} \tag{3.60}$$

which can be rewritten as

$$\int_h \frac{\mathrm{d}c}{c} = -\frac{\sigma_T w_s}{\kappa u_*}\int_h \frac{h\mathrm{d}z}{z\,(h-z)} = -h\beta\int_h \frac{\mathrm{d}z}{z\,(h-z)} \tag{3.61}$$

in which $\beta(\equiv \sigma_T w_s/\kappa u_*)$ is the famous Rouse number. Integration, using separation of variables, gives

$$\ln c + \text{const} = \beta \ln\left(\frac{h-z}{z}\right) + \text{const}, \quad (0 < z \leq h) \tag{3.62}$$

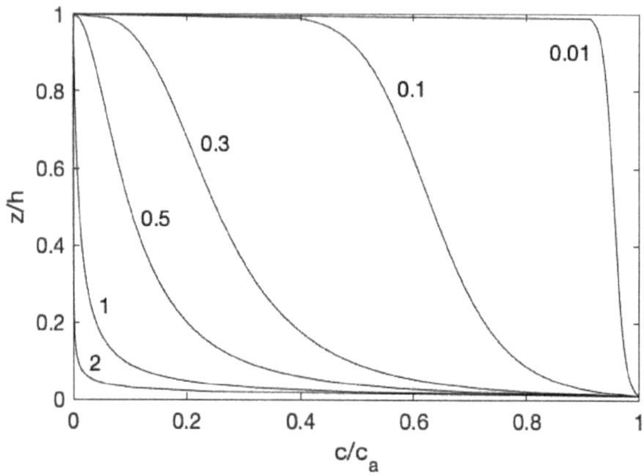

Fig. 3.20. Rouse profiles for different Rouse number values $\beta = 0.01; 0.1; 0.3; 0.5; 1; 2$. The lower the Rouse number, the more uniform the concentration distribution. A value of $a = 0.01h$ is used.

and after some rearrangement:

$$\frac{c}{c_a} = \left[\frac{a/h\,(1 - z/h)}{z/h\,(1 - a/h)}\right]^{\beta} = \left[\frac{a\,(h - z)}{z\,(h - a)}\right]^{\beta} \tag{3.63}$$

In this Rouse model, c_a is a reference SPM-concentration at height a above the bed. Mathematically, a and c_a are integration constants, whereas physically this means that a vertical SPM profile can only be constructed if the SPM-concentration is known at a particular point in that vertical. For sandy sediments, various methods exist for assessing a and c_a. Van Rijn (1993) for instance proposes to use the top of the bed load layer for a (saltation height) and the corresponding concentration for c_a.

In case of fine sediment, particles are well mixed throughout the water column. Mathematically, this implies $\beta \ll 1$, so that Eq. (3.63) can be integrated (analytically) again over the water depth. Thus, we obtain a relation between $c(z)$ and its depth-mean value $\bar{c}$:

$$c = \frac{\sin \pi\beta}{\pi\beta}\left(\frac{1 - z/h}{z/h}\right)^{\beta} \bar{c} \tag{3.64}$$

As an example, Rouse profiles are shown in Fig. 3.20 for a number of β-values. The profiles (especially for $\beta < 0.5$) show an almost linear SPM distribution over the major part of the water column, while close to the bed and near the water surface large gradients are found.

Table 3.5. Vertical SPM profiles for various viscosity profiles.

Viscosity profile	Eddy diffusivity	SPM profile
Constant	$\Gamma_{T,z} = \alpha_1 \kappa u_* h$	$\frac{c}{c_a} = \exp\left\{-\frac{w_s}{\alpha_1 \kappa u_*}\frac{z-a}{h}\right\}$
Linear	$\Gamma_{T,z} = \alpha_2 \kappa u_* z$	$\frac{c}{c_a} = \left(\frac{a}{z}\right)^{w_s/\alpha_2 \kappa u_*}$
Par/linear	$z/h < 0.5$: $\Gamma_{T,z} = \kappa u_* z\,(1 - z/h)$	$\frac{c}{c_a} = \left(\frac{h-z}{z}\frac{a}{h-a}\right)^{w_s/\kappa u_*}$
Par/linear	$z/h > 0.5$: $\Gamma_{T,z} = 0.25 \kappa u_* h$	$\frac{c}{c_a} = \left(\frac{a}{h-a}\right)^{w_s \beta \kappa u_*} \times \exp\left\{-\frac{4w_s}{\beta \kappa u_*}\left(\frac{z}{h} - 0.5\right)\right\}$

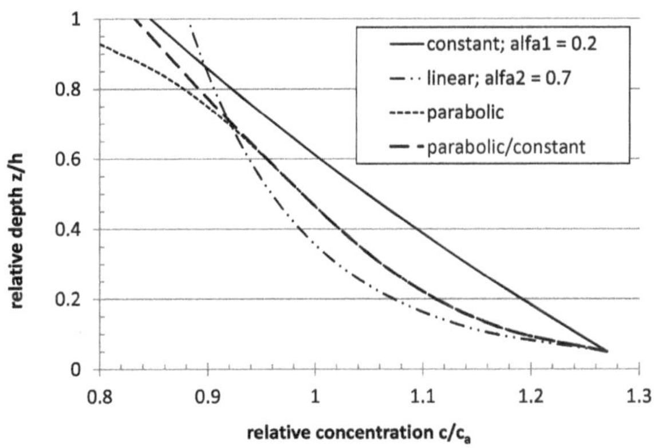

Fig. 3.21. Vertical SPM profiles as function of various eddy viscosity formulations.

3.4.2 *Deviations from the parabolic eddy diffusivity profile*

Measurements in open water/channel flow have shown that the eddy viscosity profile is not fully parabolic in the upper half of the water column, likely because of secondary circulations, (some) stratification, non-equilibrium conditions, or otherwise. Van Rijn (1993) presents some SPM-profiles as a function of a variety of eddy diffusivity profiles; see also Parker (2006). These are summarized in Table 3.5 and visualized in Fig. 3.21. The parameter β, introduced by Van Rijn reflects the effects of individual sediment particles on the eddy viscosity. Here, Van Rijn's β is set to unity.

3.4.3 *Deviations from the Rouse profile*

The derivation of the Rouse profile is based on various assumptions. These are however not always met in nature, such as:

(1) the flow is not steady;
(2) the water density is not uniform over the water column (stratification); and/or
(3) the amount of SPM changes over time as a result of erosion and deposition.

Deviations from these equilibrium assumptions induce asymmetries in SPM distributions, both in space (horizontal and vertical) and in time, yielding residual SPM transport as elaborated upon in Chapter 9. To understand how violation of these assumptions influences SPM profiles for non-stationary conditions, we first discuss the relevant time scales for establishing equilibrium profiles. These can be derived from the 1DV mass balance equation (Eq. (3.57)). The time scale for settling T_s follows from the $\partial c/\partial t$ and the advection term, whereas the time scale for vertical mixing T_m follows from comparing the diffusion term with the $\partial c/\partial t$ term:

$$T_s \sim \frac{h}{w_s} \tag{3.65}$$

$$T_m \sim \frac{h^2}{\Gamma_{T,z}} \sim \frac{h^2}{0.1hu_*} = \frac{10h}{u_*} \tag{3.66}$$

Which of these time scales dominate depends on the ratio between w_s/u_* (thus T_m/T_s), in analogy with the Rouse number β? If large, $T_s \ll T_m$ and the adaptation time scale is dominated by mixing, i.e. T_m. For small ratios, the settling time scale T_s is more relevant. In other words, for sediment with smaller settling velocities, T_s becomes dominant to attain an equilibrium profile, whereas for coarse sediment it is the mixing time T_m. These time scales determine the redistribution of a fixed amount of sediment over the water column to its equilibrium profile.

It is illustrative to elaborate on the magnitude of these time scales for two types of sediment, i.e. 200 μm sand and fines in a 5 m deep channel at 1 m/s flow velocity (hence $u_* = 0.05$ m/s for $C = 60\,\mathrm{m}^{1/2}/\mathrm{s}$), the results of which are given in Table 3.6. Hence, for fines considerable hysteresis in the vertical SPM-concentration profile is expected, whereas sand responds rapidly to changes in flow conditions.

Table 3.6. Typical time scales for sand and fines ($h = 5\,\mathrm{m}$, $U = 1\,\mathrm{m/s}$).

	w_s	T_s	T_m
Sand	2 cm/s	4 min	15 min
Fines	0.5 mm/s	2.8 hrs	15 min

The effects of hysteresis on SPM-concentration profiles for fines can be analyzed qualitatively as follows. Under tidal conditions, non-steady conditions occur under accelerating and decelerating flow, violating the first assumption above. In case of accelerating flow (u_* increases), the Rouse number decreases with time. This implies that the sediment should become more uniformly mixed over time. The instantaneous profile, however, lags behind (mixing time T_m) and is therefore less uniform than its equilibrium. The opposite occurs during decelerating flow: the concentration profile becomes less stratified than its equilibrium, lagging behind again, owing to the settling time T_s. Thus, the instantaneous profile always lags behind its equilibrium, but for fine sediment more so for decelerating than for accelerating flow, owing to $T_s \gg T_m$. Figure 3.22(a) sketches the profiles for accelerating and decelerating flow in case there is no exchange with the bed.

In case of sediment exchange with the bed, a second set of time scales becomes relevant as well, related to changes in total SPM in the water column. However, the response of the SPM distribution is different for capacity and below-capacity conditions, as elaborated below.

Deviations of the Rouse profile for sandy sediment (capacity conditions)

The transport of granular material, i.e. sand over alluvial beds, commonly occurs at capacity conditions (see also Chapter 1). Therefore, under tidal conditions with relative small accelerations/decelerations the sediment load in the water column varies only slowly around its capacity concentration C_{cap}. The depth-averaged mass balance at one point in the system reads:

$$\int^h \frac{\partial c}{\partial t}\mathrm{d}z = \alpha h \frac{\mathrm{d}c_a}{\mathrm{d}t} = -D + E = -w_s c_a + E \tag{3.67}$$

where D and E are the deposition (see Sec. 4.2) and erosion (see Chapter 6) rate, respectively, and the form factor $\alpha = \bar{c}/c_a$ parameterizes the vertical SPM-concentration distribution ($\alpha < 1$, see Galappatti and Vreugdenhil (1985) for further elaborations). Equation (3.67) yields a time scale for deposition T_D:

$$T_D \sim \frac{\alpha h}{w_s} \tag{3.68}$$

For sand, the vertical concentration profile is generally significantly stratified, hence $\alpha \ll 1$ and $T_D \ll T_s$.

Because of the slowly varying conditions, the sediment distribution exhibits near capacity conditions throughout, thus $w_s c_a \approx E$, as elaborated upon in many studies and textbooks on sandy systems (see e.g. Van Rijn, 1993). This implies also that the time scale for erosion T_E is of the same order as that of deposition:

$$T_E \sim \frac{\alpha h}{w_s} \tag{3.69}$$

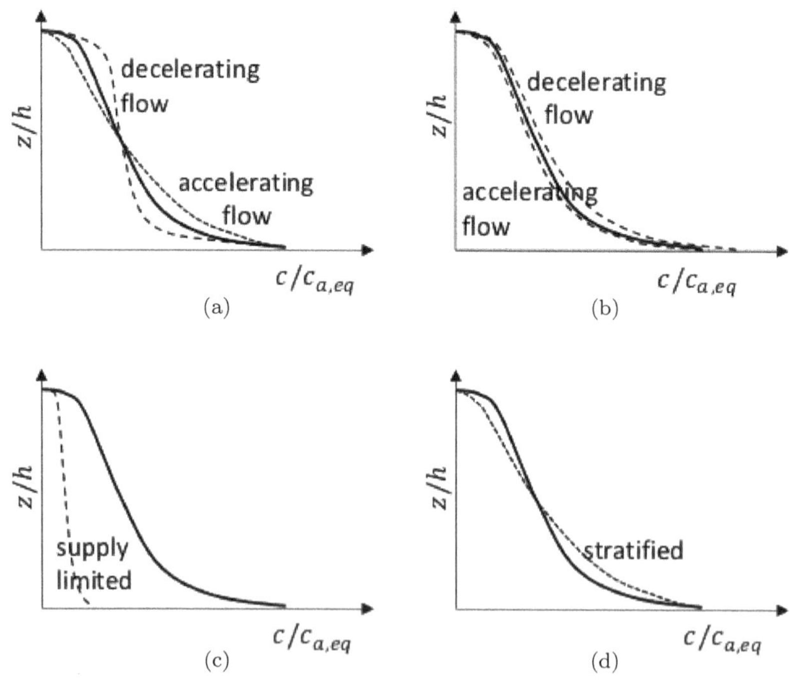

Fig. 3.22. Deviations from the equilibrium Rouse profile, indicated by the fat, solid lines. (a) Accelerating (dotted) and decelerating (dashed) flow without erosion/deposition. (b) Accelerating and decelerating flow with erosion/deposition for capacity conditions (sand). (c) Accelerating flow for supply-limited conditions (below-capacity). (d) The effects of (salinity-induced) stratification.

Thus it is concluded that for sandy systems, where the sediment is mainly transported in suspension, it is the vertical mixing time T_m which dominates the vertical concentration profiles. Settling, deposition and erosion do not alter the response of the SPM distribution much (Fig. 3.22(b)).

Deviations of the Rouse profile for fines under supply-limited conditions (below-capacity)

The capacity condition approach is not valid for fine sediments. We first consider situations for which the concentrations are much lower than capacity concentrations. Such conditions occur for non-alluvial beds, i.e. for situations where the erosion rate depends on the available erodible sediment, referred to as supply-limited conditions. Sandy seabeds with a limited amount of fines (like the North Sea), or tidal flats with a strong vertical gradient in bed strength are examples of supply-limited conditions.

The time scale for deposition T_D for fines is identical to that of coarser sediment, i.e. Eq. (3.68). However, vertical SPM profiles are much less stratified and α becomes close to unity. Therefore, for fines $T_D \approx T_s$.

At relatively low SPM-concentrations, sediment-induced stratification is too small to affect vertical mixing (see Sec. 7.1), as occurs in most natural systems with SPM-concentrations of a few 10s–100s mg/L (LCMS, Chapter 7). Deposition and erosion processes are then decoupled, and variations in SPM-concentrations are mainly governed by erosion, as T_s is large. The available sediment is relatively quickly resuspended (Secs. 6.5 and 9.2.2). The sediment concentration is hardly influenced by deposition ($E \gg D$).

Variations in SPM-concentration over a tidal cycle in such systems may measure a few 100 mg/L. For instance, the concentration would increase by 200 mg/L in 4 min for the conditions of Table 3.6, assuming $M = 10^{-3}\,\mathrm{kg/m^2/s}$ and $\tau_e = 0.7$ Pa (i.e. $E = 2\,\mathrm{g/m^2/s}$; see Chapter 7). This is significantly shorter than the vertical mixing time T_m, while the concentration remains well below C_{cap} (see Sec. 7.1). Even if erosion were to occur more gradually, the mass of sediment in the lower part of the water column would rise monotonically, as erosion (which scales with u_*^2) and vertical mixing (which scales with u_*) are both independent of c at these low-concentration conditions.

Elaborating on a balance between erosion and deposition rates is meaningless, as the erosion rate depends on the availability of sediment, which reduces in time. An equilibrium is obtained as soon as all mobilizable sediment is resuspended. An associated time scale for erosion therefore depends on the erosion rate and the available sediment. This system is therefore different from sandy systems at capacity conditions: the capacity concentration is reached before the available is resuspended.

The sediment concentration profile will approximately follow a Rouse profile (Fig. 3.22(c)).

Deviations of the Rouse profile for fines under alluvial conditions (near-capacity)

At high SPM-concentrations, we approach the HCMS conditions analyzed in Sec. 7.1, and hydrodynamics and sediment behavior become strongly coupled: (1) because of stratification, vertical mixing and bed shear stresses reduce, and (2) because of the high concentrations, w_s is reduced by hindered settling (Sec. 4.4). Under accelerating flow, stratification builds up as erosion scales with u_*^2 while mixing with u_*, hence erosion rates are larger than mixing rates. Also under decelerating flow, stratification builds up as the water column becomes saturated (Sec. 7.1). Thus, a strong hysteresis over the tidal cycle is expected. As a result, all parameters in the simple Eqs. (3.65), (3.66), (3.68) and (3.69) vary nonlinearly with flow conditions, become coupled, and cannot be assessed generically.

Other effects

In many cases, the assumption of uniform and constant water density is violated in estuaries. Stably stratified conditions (increase in density from surface to bed) result in reduction of vertical mixing. The eddy diffusivity $\Gamma_{T,z}$ reduces (Sec. 3.2.3). This results in larger concentration gradients, implying that the sediment is less homogeneous over the water column. Figure 3.22(d) sketches the effect of reduced mixing due to salinity-induced stratification on the sediment concentration profile.

Deviations from the parabolic eddy viscosity profile (Sec. 3.4.2) also affect the vertical SPM profile, deviating from the Rouse equilibrium profile. One example is a parabolic profile in the lower half of the water column and constant (equal to the half depth value) in the upper part, see Van Rijn (1993), Fig. 3.22(d). This results in the SPM profile of Fig. 3.22(d), which is more uniform near the surface (better mixed).

3.4.4 *Assessment of the settling velocity*

The Rouse profile can be used to determine the settling velocity from measured SPM-profiles, provided that the flow velocity (in fact u_*) is known. As an example, we provide results for measurements in the Passaic River in the USA, where u_*-values were obtained from numerical simulations. The results yield credible values, but exhibit a wide spread in settling velocity, which is to be attributed to a violation of the Rouse assumptions during (part of) the measurements, as explained above:

(1) The Passaic River is a tidal river, hence during part of the tide (slack water), the velocity profile is not logarithmic.
(2) Some of the data was collected in the salt intrusion region of the river, hence velocity profile and vertical mixing are affected by salinity-induced vertical stratification.
(3) The (local) SPM profile itself may not be in equilibrium because of longitudinal advection and water-bed exchange processes (settling-limited, Chapter 1).
(4) The settling velocity in the river actually exhibits a distribution, is not characterized by one value — we will elaborate on such settling velocity distributions in more detail in Sec. 4.1.

3.5 Turbidity and water quality

Sediment-laden water is often rather turbid, because the wave length of visible light (0.4–0.8 μm) is of the same order as the size of the finer sediment particles. These particles therefore scatter and reflect incoming light effectively. Turbidity is thus related to the finer fractions of SPM.

The wavelength of light relevant for biology almost overlaps with that of visible light (0.4–0.7 μm) — one refers to PAR (Photo-synthetically Active Radiation). Thus, fine SPM affects the light climate in aquatic ecosystems, often with unfavorable impact on the water quality, for instance through a reduction in primary production (the production of algae).

In this section, we present an order of magnitude assessment of the role of SPM on the underwater light climate. The reader is referred to Kirk (2010) for further details. The extinction of light is described with the Lambert–Beer model:

$$I_z(\lambda) = I_0(\lambda)\exp\{-K_d(\lambda)\, z\} \tag{3.70}$$

in which I_0 [W/m^2] is the reference irradiance (just below the water surface, corrected for reflections), I_z is the irradiance at z meter below the reference level, and K_d is the vertical light attenuation coefficient. As the latter is a function of the light's wave length, different frequencies in the light

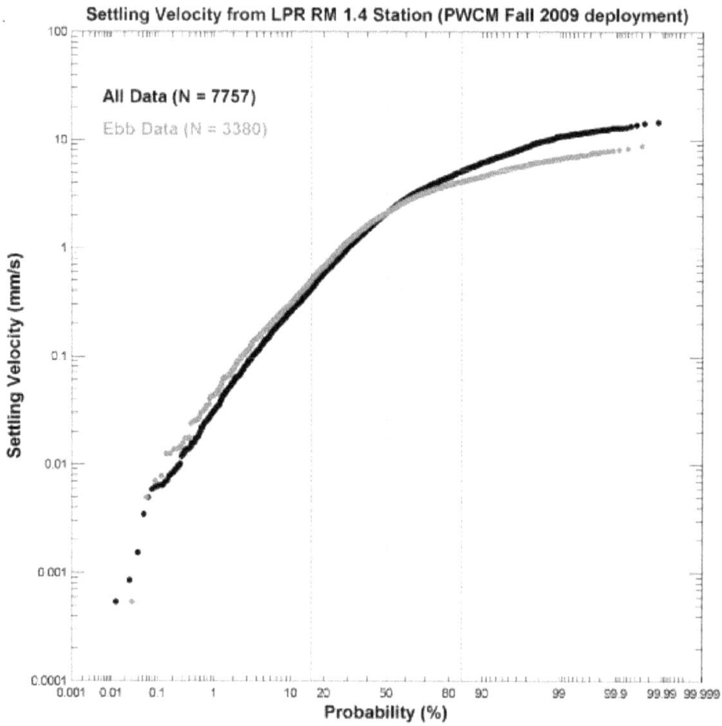

Fig. 3.23. Settling velocity distribution in the Passaic River (USA) obtained with Rouse method — courtesy R. Matthews.

spectrum are attenuated differently (Kirk, 2010). Here we ignore these subtleties, and discuss mean light attenuation only, averaged over the PAR-spectrum.

The attenuation coefficient K_d is affected by a number of substances in the water column, consisting of both dissolved and suspended matter. These various contributions are assumed not to affect each other. The overall attenuation coefficient K_d then becomes the linear sum of its components:

$$K_d = K_\infty + K_{cdom} + K_{phyt} + K_{det} + K_{im} \tag{3.71}$$

in which K_∞ is the background attenuation in an aquatic system (water inherent attenuation), K_{cdom} is the attenuation by colored dissolved organic matter (such as humic acids), K_{phyt} is the self-attenuation by phytoplankton, and K_{det} is the contribution by detritus (i.e. dead organic matter). Here we analyze the effects of inorganic fine SPM on the underwater light climate. This contribution is accounted for by K_{im} (inorganic matter), which

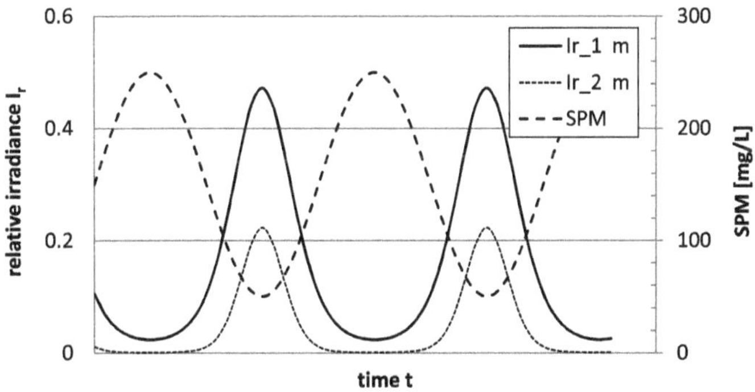

Fig. 3.24. Variation of relative irradiance $I_r = I_z/I_0$ at two depths with time for the plotted SPM variation, assuming $k_{im,m} \approx 0.015\,\text{m}^{-1}$.

is assumed to vary linearly with SPM-concentration: $K_{im} = k_{im} \times \text{SPM}$, with SPM in [mg/L].

Van Gils and Tatman (2003) calibrated k_{im} against observations in the North Sea, close to the Dutch coast, and found values of $k_{im,f} \approx 0.036\ \text{m}^{-1}$ for the finer SPM-fractions, values of $k_{im,c} \approx 0.0035\ \text{m}^{-1}$ for the "coarser" fractions, with a weighted mean of $k_{im} \approx 0.015\ \text{m}^{-1}$. The latter value is used in the assessment below.

Assume a turbid aquatic system where SPM-concentrations vary sinusoidally between 50 and 250 mg/L (typical values for parts of the North Sea). Then the relative irradiance $I_r = I_z/I_0$ varies also with time, as illustrated in Fig. 3.24 for two depths, i.e. $z = 1\,\text{m}$ and $z = 2\,\text{m}$. Here, other effects on the underwater light climate are ignored.

Figure 3.24 shows that only 50% of the light available near the water surface can penetrate to 1 m water depth at the lowest SPM-concentrations of 50 mg/L, while during periods with the higher SPM values, most light is attenuated. At 2 m below the water surface, only 20% of the available light is available for primary production at low SPM-values, while it becomes entirely dark during a long period of higher SPM values. The strong correlation between light climate and SPM values is a major reason for including these water quality effects in environmental impact assessments on infrastructural and managerial interventions in aquatic systems.

In the preceding sections, we discussed the vertical distribution of SPM in the water column (Rouse profile), showing that this profile may vary largely in space and time, depending on the hydrodynamic conditions and grain size (settling velocity). Hence, in theory, one should include these vertical effects in the assessment of Fig. 3.24 as well. However, we argued that in particular the finer fractions affect the underwater light climate. These fractions are fairly well mixed over the water column, with small vertical gradients. Given all other uncertainties in quantifying SPM values in open water systems, and in the light climate model Eq. (3.71) itself, such refinements would not pay off.

Further, we have deliberately omitted a time scale on the horizontal time-axis of Fig. 3.24 — this time scale is arbitrary on purpose, as it may represent a spring-neap cycle, seasonal variation, and/or other cyclic processes. However, owing to the fineness of the relevant SPM fraction, this time scale cannot represent intra-tidal effects, as fine SPM particles would not have time to settle, and their concentration in the upper part of the water column does not vary too much over a tidal cycle. As a result, the time scale in water quality models can be set to the time scale of the relevant biological processes, ignoring (semi-)diurnal variations in SPM.

Chapter 4

Settling and Deposition

4.1 The settling velocity of mud flocs

The settling velocity of sand is traditionally described by the Stokes settling formula. At equilibrium, the particle settling velocity attains a value such that the drag on a falling particle balances gravity, as sketched in Fig. 4.1.

The gravitational force F_g equals:

$$F_g = \alpha \frac{\pi}{6} D^3 g \left(\rho_s - \rho_w\right) \tag{4.1}$$

and the drag force F_d amounts to:

$$F_d = \beta \frac{\pi}{4} D^2 c_D \frac{1}{2} \rho_w w_s^2 \tag{4.2}$$

in which D is the particle diameter, w_s the settling velocity, c_D the drag coefficient and α and β are form parameters. In the Stokes regime, characterized by the particle Reynolds number $Re_p \left(\equiv Dw_s/\nu\right) < 1$, the flow is laminar and the form parameters are not too relevant, i.e. $\alpha \approx \beta \approx 1$. The Stokes regime generally holds for particles up to about $100\,\mu$m, and the drag coefficient is then $c_D = 24/Re_p$. At equilibrium ($F_d = F_g$), substituting these various values gives the famous Stokes formula for the settling velocity of a single particle in still water:

$$w_s = \frac{\left(\rho_s - \rho_w\right) g D^2}{18\mu} \tag{4.3}$$

Water temperature affects particle settling velocities through viscosity. This temperature dependency may be important in rivers, as the formation of bed forms is affected by the settling velocity (Vanoni, 1940). The viscosity of water as a function of temperature is approximated by:

$$\mu \approx 1.7473 - 0.0479T + 0.0005T^2 \; \left[10^{-3}\ \text{Pa} \cdot \text{s}\right] \tag{4.4}$$

with the temperature T in [°C]. This relation is plotted in Fig. 4.2, showing that (seasonal) variations in viscosity by a factor of two may occur.

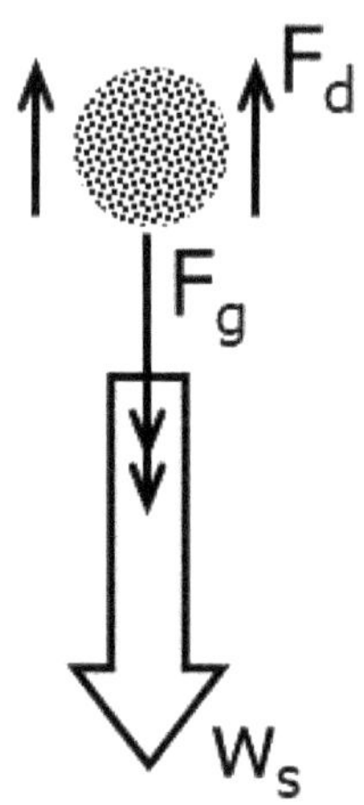

Fig. 4.1. Forces on a falling particle at equilibrium.

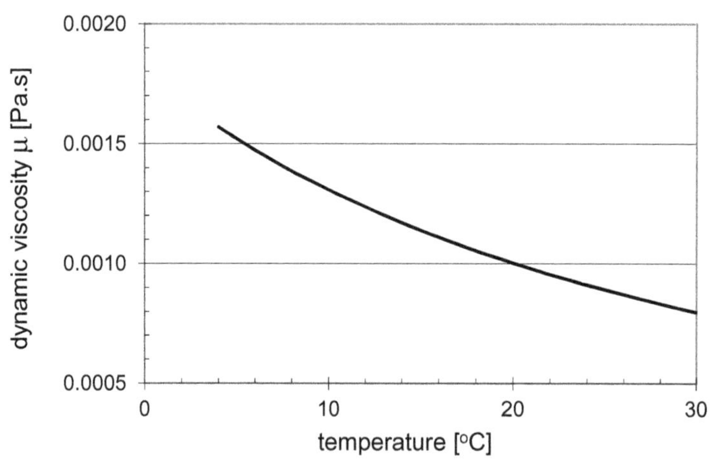

Fig. 4.2. Water viscosity as function of temperature.

Hence settling velocities in summer may be two times larger than in winter. For cohesive sediment, biological effects may also affect the floc size, thus settling velocity, see Sec. 4.3 and Chapter 8.

Beyond the Stokes-regime, the Stokes equation is no longer valid. The literature contains many formulas for such larger particles. Here we follow

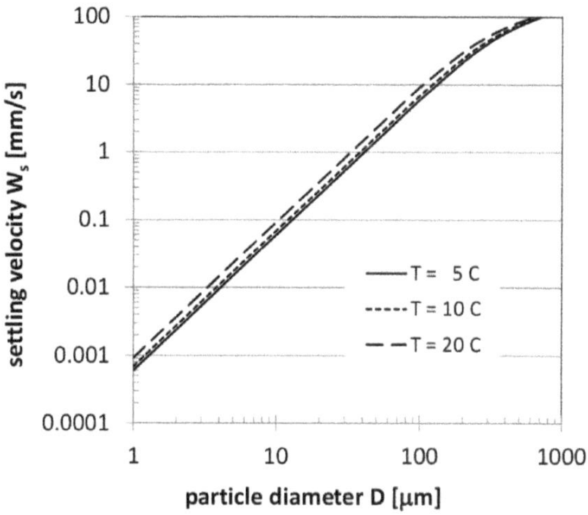

Fig. 4.3. Settling velocity for sand by Van Rijn (1993), Eq. (4.5).

Van Rijn (1993), which should be applicable for $100\,\mu\text{m} < D_{50} < 1\,\text{mm}$:

$$w_s = \frac{10\nu}{D_{50}}\left[\sqrt{1 + \frac{0.01\,(\rho_s - \rho_w)\,gD_{50}^3}{\rho_w \nu^2}} - 1\right] \tag{4.5}$$

with D_{50} is the median particle size. Van Rijn assumes a shape factor S_F, introduced by Corey (1949), defined as $S_F = c\big/\sqrt{ab}$, where a, b and c are the particle's longer, intermediate and shorter axes. Equation (4.5) is plotted in Fig. 4.3, together with the Stokes values for $D < 100\,\mu\text{m}$.

Self-similar flocs

Stokes' law is not directly valid for fine, cohesive sediment. Because of physico-chemical forces on the clay particles, chemistry of the ambient and pore water, and the presence of organic material (polysaccharides, etc.), fine sediment form flocs (see Sec. 2.1), which largely modifies the effective density of SPM. Figure 4.4 shows two photographs of mud flocs from the Merwede River (The Netherlands). The left panel shows a very open floc with many organic filaments, while the floc in the right panel is more dense.

Krone (1986) was likely the first to recognize a hierarchy in the build-up of these flocs. Here we follow Kranenburg (1994) quantifying this hierarchy with a self-similarity/fractal approach. The concept of self-similarity in two dimensions is sketched in the schematic of Fig. 4.5. We start from a single particle. This primary particle forms the flocs building stone with diameter

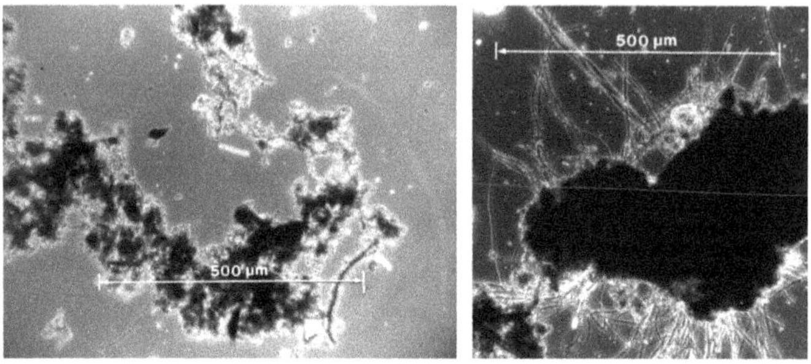

Fig. 4.4. Photograph of mud flocs from the Merwede River (The Netherlands).

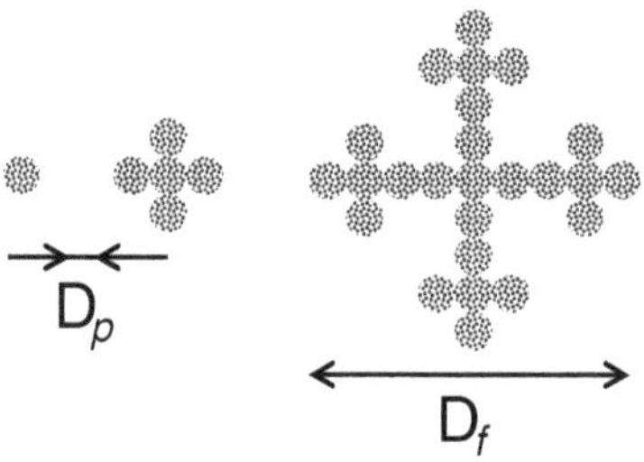

Fig. 4.5. Self-similarity in two dimensions.

D_p (this is NOT the size of the individual clay particles). Next, we form a regular cross of five primary particles, as in the middle panel of Fig. 4.5. This lowest class of a regular pattern can be interpreted as a flocculus (Krone, 1986). This pattern is repeated at all scales, as in the right panel of Fig. 4.5.

The diameter D_f of the latter floc consists of 9 primary particles, while its solids volume (i.e. the flocs mass) contains 25 primary particles. The relation between diameter and volume can be written as

$$\text{mass} = \text{length}^{n_f} \tag{4.6}$$

in which n_f is the so-called fractal dimension. The formal definition of the fractal dimension reads:

$$n_f = \lim_{L \uparrow \infty} \frac{\ln(N(L))}{\ln(L)} \tag{4.7}$$

where N is the number of particles with length scale L. Of course $1 \leq n_f \leq 3$, where $n_f = 1$ represents a string, and $n_f = 3$ a Euclidian (massive) body. Visceck (1992) presents a good introduction to self-similar structures.

In two dimensions, the floc in the right panel of Fig. 4.5 contains 25 primary particles, while in three dimensions, it would contain 49 particles. Note that a Euclidian "floc" with diameter $9D_p$ would contain $9^3 = 729$ primary particles. The fractal dimensions in 2D and 3D would yield:

$$\text{2D: } n_f = \frac{\ln 25}{\ln 9} = 1.46; \text{ and 3D: } n_f = \frac{\ln 49}{\ln 9} = 1.77 \tag{4.8}$$

In general, the number of primary particles N in a floc in 3D space is given by Meakin (1998):

$$N = \left(\frac{D_f}{D_p}\right)^{n_f} \tag{4.9}$$

Typical values of n_f for flocs in the water column amount to $1.8 \leq n_f \leq 2.2$ (Winterwerp and Van Kessel, 2003), while Maggi (2005) and Maggi *et al.* (2007) found that the fractal dimension reduces with floc size D_f relative to the primary particle: $n_f = 3(D_f/D_p)^{-0.1}$.

In the bed, fractal dimensions are much larger, i.e. 2.6–2.8 (Kranenburg, 1994; Merckelbach and Kranenburg, 2004a; see also Sec. 5.1). Qualitatively, it can be understood that the over-burden during consolidation squeezes the flocs, reducing the flocs voids (see also Sec. 2.1 and Fig. 5.4). However, this increase in fractal dimension is not yet fully understood.

The void ratio n (i.e. volume of voids in a given volume of soil) increases with decreasing fractal dimension. These voids are filled with water (we ignore gas) — the water content of flocs is therefore large. An overview of the (volumetric) water content (excess density) of flocs measured at various locations across the world was given by Van Leussen (1994). Figure 4.6 shows the water content for flocs as function of the flocs diameter D_f, assuming $D_p = 10\,\mu$m, using Maggi's relation. As flocs at equilibrium typically measure a few $100\,\mu$m, the water content of flocs in the water column measures 90% or more.

But also within the bed, flocs still contain 60–80% water. This implies that mud typically behaves as a fluid with "some" solids, requiring non-Newtonian stress–strain relations. Such relations are consistent with the Atterberg limits approach introduced in Sec. 2.2.

Floc settling velocity

In the following, floc volumes are related to the SPM's dry solids content c, or dry density ρ_{dry}, following Kranenburg (1994). The excess density $\Delta\rho_f$

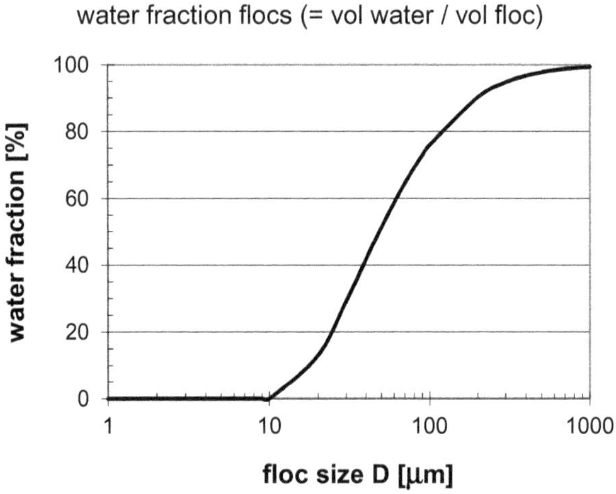

Fig. 4.6. Volumetric water content of flocs in relation to the floc diameter.

of a floc reads:

$$\Delta\rho_f = \rho_f - \rho_w = (\rho_s - \rho_w)\left[\frac{D_p}{D_f}\right]^{3-n_f} \tag{4.10}$$

where ρ_f is the density of a floc. The specific sediment density ρ_s should read as the specific density of the primary particles with diameter D_p, which is generally smaller than commonly used values of about 2600 kg/m^3, see below. From Eq. (4.10) the flocs volumetric concentration ϕ_f follows:

$$\phi_f = \left(\frac{\rho_s - \rho_w}{\rho_f - \rho_w}\right)\frac{c}{\rho_s} = \left(\frac{\rho_s - \rho_w}{\rho_f - \rho_w}\right)\phi_s = \frac{c}{\rho_s}\left[\frac{D_f}{D_p}\right]^{3-n_f} \tag{4.11}$$

Equation (4.11) shows that the volume of flocs can be orders of magnitude larger than the volume of the solid sediment (clay) particles themselves. With increasing concentrations, the flocs form a space-filling network, i.e. the volume concentration ϕ_f becomes unity. The mass concentration at space-filling is referred to as the gelling concentration c_{gel} (equivalent to the structural density in soil mechanics):

$$c_{gel} \equiv c_{\phi_f=1} = \rho_s\left[\frac{D_p}{D_f}\right]^{3-n_f} \tag{4.12}$$

which is a measure for the dry density of fluid mud (see Chapter 7). Typical values for c_{gel} are in the range of 100–200 g/L, i.e. $\phi_{s,gel} \approx$ 4–8%, which is

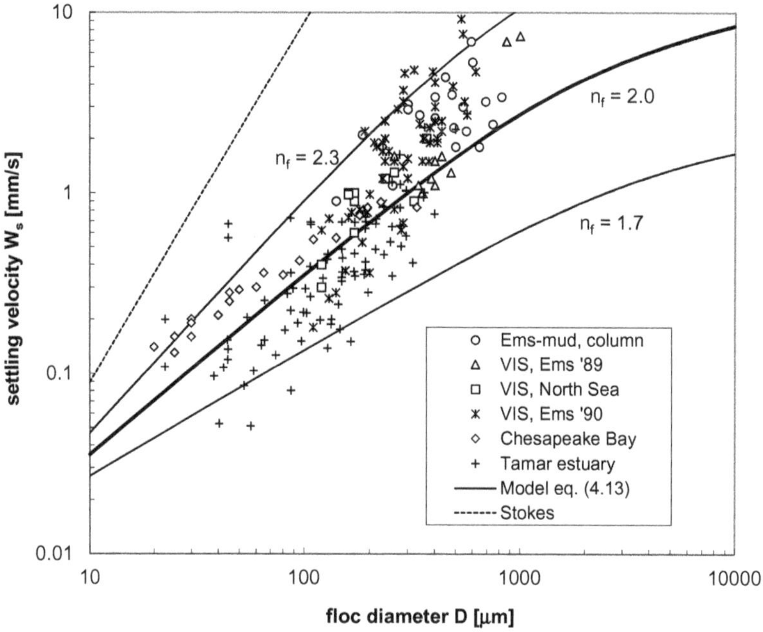

Fig. 4.7. Settling velocity Eq. (4.13) compared to various data and Stokes (Eq. (4.3)).

in the range of the transitional clay content ξ_0 for soils, predicted by the Atterberg limits (activity plot), see Fig. 2.18.

Substitution of Eq. (4.11) into Eq. (4.3), and using an empirical function for the drag coefficient in the Stokes regime $c_D = 24\left(1 + 0.15Re_p^{0.687}\right)/Re_p$ (e.g. Raudkivi, 1976; Graf, 1977), a settling velocity for a single fractal floc in still water is found:

$$w_s = \frac{1}{18}\frac{(\rho_s - \rho_w)\,g}{\mu}D_p^{3-n_f}\frac{D_p^{n_f-1}}{1 + 0.15\mathrm{Re}_f^{0.687}} \tag{4.13}$$

in which $\alpha = \beta = 1$, which is a good approximation in the Stokes regime ($Re_p < 1$). Figure 4.7 presents the settling velocity according to Eq. (4.13) with some laboratory and field data (after Winterwerp, 1999), showing that a mean value $n_f = 2$ captures the data reasonably well. Also the many studies on floc size in the laboratory and in-situ show typical values for flocs in the water column of $1.8 < n_f < 2.2$, as mentioned above.

As $D_f \gg D_p$, the settling velocity of flocs is orders of magnitude larger than that of the primary particles. Further, for $n_f = 2$, Eq. (4.13) shows that w_s scales with the floc diameter, and not with its diameter squared, as in Stokes' law.

As discussed above, D_p does not represent the size of the clay particles — primary particles consist of aggregates of many clay particles (Cuthbertson *et al.*, 2016; Khelifa and Hill, 2006). The specific density of primary particles should therefore be (considerably) lower than than the specific density of the solids themselves. However, experimental data on the specific density of such primary particles are currently not available.

Khelifa and Hill (2006) developed a model for the excess density of a floc composed of k poly-sized particles of diameter d_i, yielding:

$$\Delta\rho_f = (\rho_s - \rho_w)\frac{\sum\limits_{i=1}^{k} d_i^3}{\left(\sum_{i=1}^{k} d_i^{n_f}\right)^{3/n_f}} \tag{4.14}$$

They defined a parameter Θ as

$$\Theta = \sum_{i=1}^{k} d_i^3 \Bigg/ \left(\sum_{i=1}^{k} d_i^{n_f}\right)^{3/n_f} \tag{4.15}$$

which yields a slightly different formula for the settling velocity of fractal flocs with a poly-sized composition:

$$w_s = \frac{\alpha}{18\beta}\frac{(\rho_s - \rho_w)\,g}{\mu} D_p^{3-n_f} \frac{D_f^{n_f-1}}{1 + 0.15\mathrm{Re}_f^{0.687}}\Theta \tag{4.16}$$

which reduces to Eq. (4.13) for $\Theta = 1$, i.e. in case of mono-dispersed primary particles.

Quantification of the settling velocity

Four techniques exist to measure settling velocities *in situ*:

(1) *In-situ* camera: Various research instruments (digital camera) have been developed to monitor the *in-situ* settling of sediment flocs. Considerable post-processing is required to recognize and define the individual flocs, and to correct for the return (secondary) flow induced in the instruments settling chamber. The reader is referred to for instance Manning and Dyer (2002).

(2) LISST: The LISST-ST is a commercial instrument to measure the settling velocity of flocs in a settling column attached to the LISST, using laser diffraction techniques to determine particle size (Agrawal and Pottsmith, 2000) — it is widely in use.

Table 4.1. Typical values for median settling velocities.

From direct W_s measurements	$0.1 < w_s < 10$ mm/s
For siltation studies with constant W_s	$0.1 < w_s < 0.5$ mm/s
For siltation studies with variable (flocculation) W_s	$0.5 < w_s < 5$ mm/s
Low-flocculated SPM	$0.01 < w_s < 0.1$ mm/s
Unflocculated SPM and water quality	$0.001 < w_s < 0.1$ mm/s

(3) ADV: By assuming local equilibrium between vertical turbulent mixing and gravitational settling, the settling velocity distribution of flocs (particles) can be assessed from the mean and turbulent fluctuations of SPM-concentration (in fact the ADV-backscatter) and vertical turbulent velocity, as measured with an ADV (e.g. Fugate and Friedrichs, 2003).
(4) SPM-profiles: Assuming a Rouse profile, the settling velocity distribution of flocs can be determined from vertical SPM-profiles, as explained in Sec. 3.4.

Various techniques have been developed to determine settling velocities in the laboratory, as elaborated in Appendix A. Note that all techniques involve pre-treatment of the samples and/or disruption of the flocs during sampling, transport and storage.

For practical applications, it should be pointed out that a unique settling velocity of mud flocs does not exist. In fact, settling velocities can vary widely, as shown in Table 4.1. The reasons for this are:

(1) Sediment is characterized by a grain size (hence settling velocity) distribution.
(2) Application determines which size is important (Chapters 13–18):

    (a) water quality (the smaller particles), dredging volumes (the larger particles), etc.;
    (b) spatial and temporal scale of the problem at hand, for instance: intra-tidal SPM dynamics *vs* large-scale engineer problems;
    (c) type of modeling (3D vs. 2DH, resolution, see Sec. 12.1).

(3) The settling velocity may vary over time because of flocculation (see for instance Benson and French (2007) and Sec. 4.3) — this may also yield seasonal variations in settling velocity.
(4) Hindered settling plays a role at lager SPM concentrations (see Sec. 4.4).

4.2 The deposition rate of mud flocs

The definitions for deposition and sedimentation, used in this book, are summarized in Table 4.2. Moreover we use capital W_s when referring to a characteristic settling velocity, and small w_s when referring to a specific one.

The deposition rate D is traditionally defined as the (gross) flux of sediment onto the bed:

$$D = w_{s,b} c_b \tag{4.17}$$

in which $w_{s,b}$ is the near-bed settling velocity and c_b the near-bed SPM-concentration. However, in general, suspended sediment is characterized by a particle size distribution (psd), thus a settling velocity distribution. The deposition rate D therefore has to be integrated over the various sediment classes i, and Eq. (4.17) then becomes:

$$D = \int_{c_b} w_{s,b} \, \mathrm{d}c_b \approx \sum_{i=n} w_{s,b,i} c_{b,i} \tag{4.18}$$

This relation is not equal to $w_{s,50} c_b$, i.e. using the median settling velocity — errors up to 50% can be introduced, depending on the shape of the psd. However, such errors are often small in comparison to errors by other unknowns, and therefore generally not critical in engineering models.

Chapter 13 elaborates further on the modeling of the deposition rate as a function of model resolution and model setup, arguing that the use of two or three fractions usually suffices.

Critical shear stress for deposition

In the literature on cohesive sediment, the so-called Krone deposition formula is often used, named after Ray Krone (Ariathurai and Arulanandan,

Table 4.2. Definitions of settling, deposition, sedimentation and consolidation: D = deposition rate, S = sedimentation rate, E = erosion rate (e.g. Chapter 6).

Settling	Particles falling through the water column
Deposition	Particles falling/depositing on the sediment bed
Sedimentation	Net changes in bed level: $S = D - E$
Consolidation	Further compaction of a deposited bed

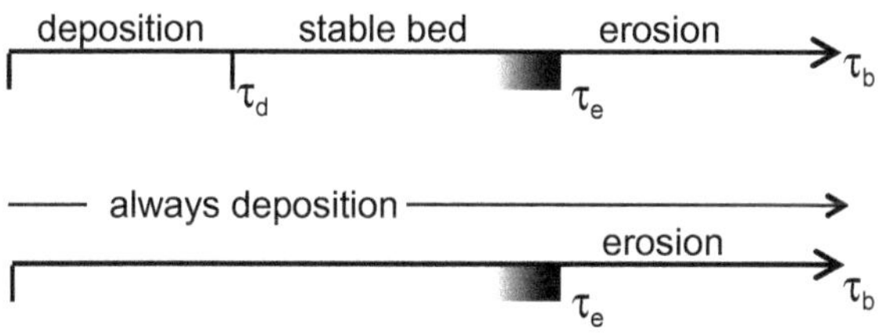

Fig. 4.8. Krone deposition formula (upper panel) yields stable bed conditions for $\tau_d < \tau_b < \tau_e$, whereas Eq. (4.17) (lower panel) represents continues deposition.

1978; Krone, 1962, 1993):

$$\begin{aligned} &D = w_s c\,(1 - \tau_b/\tau_d) \quad &&\text{for } \tau_b < \tau_d \\ &D = 0 &&\text{for } \tau_b > \tau_d \end{aligned} \tag{4.19}$$

where τ_d is known as the critical shear stress for deposition. Equation (4.19) implies stable bed conditions for $\tau_d < \tau_b < \tau_e$, where τ_e is the critical shear stress for erosion, see Sec. 6.1. This behavior is sketched in Fig. 4.8 (upper panel). Typical τ_d-values reported in literature are 0.05–0.1 Pa.

The physical explanation for introducing τ_d is floc breakup in the lower layers of the water column, where shear rates are large (Sec. 4.3), thus settling velocity $w_{s,b}$ low. This may indeed occur, but only when flocculation times are small compared to the tidal cycle (thus high SPM values, see Sec. 4.3), or compared to other relevant time scales, as explained in Sec. 4.3.

Application of Eq. (4.19) can also be useful in 2DH-engineering models in which the near-bed SPM-concentration is not resolved (e.g. Sec. 12.1).

Thus Eq. (4.19) may yield a useful parameterization of physical processes not resolved in a (numerical) model. In the applications of Chapters 14–18, the τ_d-concept is not used. For a further analysis on the background of Eq. (4.19) and its applicability, the reader is referred to Winterwerp (2007).

Reduced deposition

Van Kessel and Van Maren (2013) introduced the concept of reduced deposition:

$$D = p_d w_{s,b} c_b \tag{4.20}$$

in which p_d is the reduced deposition coefficient. Contrary to Eq. (4.19), reduced deposition occurs throughout the tidal cycle. This coefficient represents the parameterization of a number of physical processes not resolved in a numerical model:

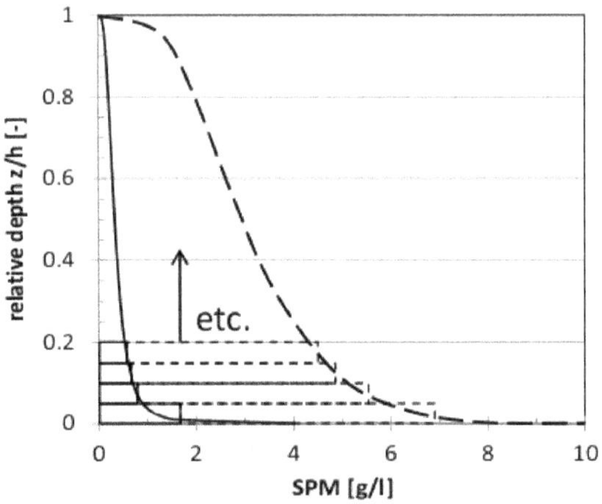

Fig. 4.9. Stepwise/discretized representation of actual SPM-profiles (solid lines for low, dashed lines for high SPM-concentrations) in a numerical model, leading to an underestimation of near-bed SPM values.

(1) Hindered settling close to the bed because of limited vertical resolution within a numerical model.
(2) Floc breakup close to the bed, as (partly) modeled by Eq. (4.19).
(3) Gradual strength development of deposited flocs, i.e. upon deposition some flocs cannot stick to the bed and are immediately resuspended.
(4) Horizontal/spatial resolution of a numerical model.
(5) Overall sediment balance within a numerical model.

Though the reduced deposition concept was born out of modeling needs (see Sec. 12.2 and Chapter 13), it has important physical and conceptual consequences as well. It is inevitable that physical processes very close to the bed cannot be resolved to their finest detail, as sketched in Fig. 4.9, showing the discretization of two arbitrary vertical SPM-profiles, typical for low- and high-concentrated environments, respectively. In a numerical model, SPM-values close to the bed are averaged over the lower computational grid cell, which has a number of implications:

(1) Even at low SPM-concentrations, near-bed SPM values may be high, reducing the settling velocity locally by hindered settling (e.g. Sec. 4.4) — in this case $p_d < 1$.
(2) As discussed above, shear stresses near the bed are large, breaking up flocs, thus reducing the settling velocity of the sediment. This process is relevant when flocculation times are small, otherwise flocs would not

have time to regrow higher in the water column — in this case $p_d < 1$. Of course, floc breakup can be explicitly modeled with a flocculation model (see Sec. 4.3).

(3) Over sandy beds, cohesive sediment forms a fluff layer upon deposition, and only a fraction of it may be entrained into the bed, e.g. Sec. 6.5. The fines in the fluff layer build up strength slowly, thus are resuspended easily — in this case $p_d < 1$.

(4) Over muddy beds, c_b can be considerably larger than the mean value within the lower computational grid — in this case $p_d > 1$.

(5) At high SPM-concentrations, hindered settling is important (e.g. Sec. 4.4), maintaining a pronounced vertical gradient in SPM, thus inducing interaction with the turbulent flow field (e.g. Chapter 7). These high concentration gradients cannot be maintained if the sediment escapes the computational domain by deposition. The concept of reduced deposition can help maintaining these high SPM gradients — in this case $p_d < 1$. Note that in a continuous modeling approach (e.g. Winterwerp, 2001; Le Hir *et al.*, 2000), deposition is not modeled, and all sediment is maintained within the computational domain. However, high resolution and a consolidation model are required in this approach (see Sec. 11.4).

Reduced deposition is therefore a function of shear, sediment type, and SPM-concentration. Especially, the relation with SPM is important, as this may induce a positive feedback mechanism leading to very high near-bed SPM-concentrations. The value for p_d should increase with SPM-concentration for three reasons (see Van Maren *et al.* (2020) for details):

(1) Higher SPM-concentrations are generally accompanied by stronger vertical SPM gradients, leading to suppression of turbulent mixing. This favors the formation of a high-concentration near-bed layer.

(2) At sufficiently large near-bed SPM-concentrations, sediment deposited on a bed no longer consolidates instantaneously (see also Sec. 7.2). Such deposits are easily erodible, at much smaller shear stresses (see also Sec. 7.3).

(3) Because of hindered settling, w_s decreases nonlinearly with c, reducing the deposition flux. On the other hand, the deposition flux $w_{s,b}c_b$ itself increases linearly with c_b. When discretized, near-bed SPM-concentrations are under-estimated (Fig. 4.9). Therefore, the actual deposition flux may be over- or under-estimated.

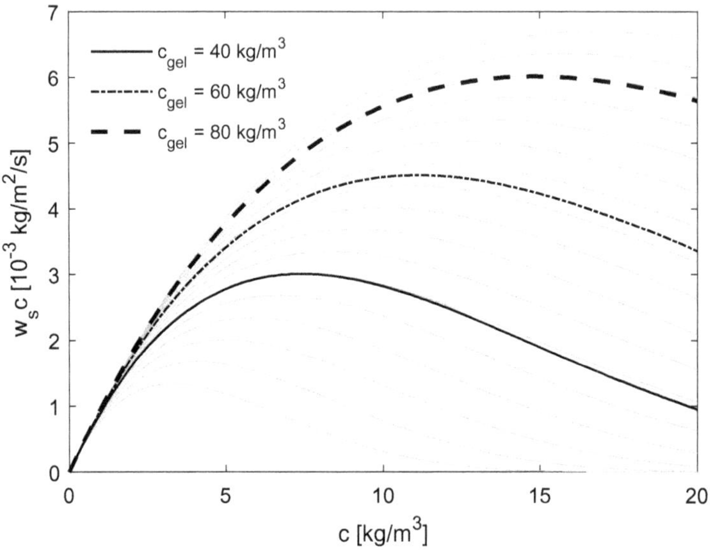

Fig. 4.10. Relation between deposition flux $w_s c$ and SPM-concentration c, for $c_{gel} = 40$, 60 and 80 kg/m^3 (black lines) at increments of 5 kg/m^3 (gray lines), accounting for hindered settling, using Eq. (4.34) with $n = 4.4$ (see Sec. 4.4). Modified from Van Maren *et al.* (2020).

The last argument can be elaborated further, using the hindered settling equation (4.34). Figure 4.10 shows the behavior of the deposition flux D as function of c. For the range in c_{gel} of 20–100 kg/m^3, D scales more or less linearly with c at SPM-concentrations of 3–16 kg/m^3. Hence at low concentrations, hindered settling does not contribute to p_d (and thus may be >1), but at high SPM-concentrations, limited vertical model resolution requires small values of p_d.

Numerical models typically have a spatial resolution of 10s to 100s and sometimes 1000s of meters, thus averaging/aggregating all bathymetrical and other features in and on the bed over a computational grid cell. As an example, Fig. 4.11 shows part of the North Sea bed in the Dutch coastal zone obtained from a 25 m resolution multi-beam survey (Damen *et al.*, 2018). On top of these bed forms, smaller ripples occur as well, which are not resolved, thus not visible. Deposition of fines likely takes place only in the troughs in between the bed forms. On their crests, local flow velocities (bed shear stresses) are likely too large for sedimentation. Of course high resolution models may resolve these problems (partly), though this is not always possible because of e.g. computational time.

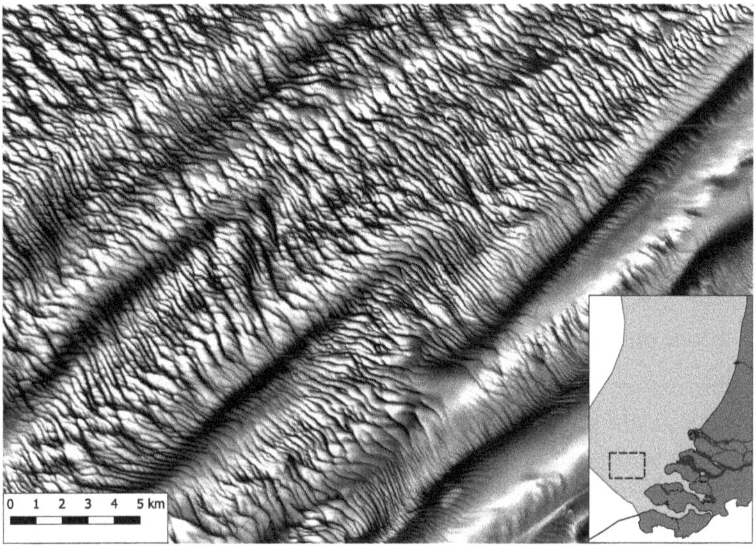

Fig. 4.11. Horizontal resolution issues in a numerical model to be resolved with the concept of reduced deposition with spatial variations in bed structure. North Sea bed structure from 25 m resolution multi-beam survey (courtesy E. Hendriks).

Application of the reduced deposition concept is then a feasible approach to account for these sub-grid effects.

Long-term modeling (years) of fine sediment in large spatial domains, such as estuaries and tidal inlets, requires consistency between the boundary conditions at the water-bed interface and the SPM-concentrations prescribed at the models open boundaries. Commonly, the erosion formula by Partheniades (Sec. 6.1) in conjunction with the deposition formula (4.19) — generally referred to as the Partheniades–Krone approach — for modeling water-bed exchange rates are used. In this approach, erosion rates are independent of SPM-concentrations in the water column, i.e. accumulation or depletion of fines stored in the bed is not accounted for, and SPM supply cannot become limited by depletion.

In the practice of morphostatic modeling, this implies that the amount of fines in the bed becomes either zero or infinite, depending on the parameter settings. In other words, the model has become over-determined. Thus, an additional condition is required for closing the model equations plus boundary conditions. A simple closure approach is the concept of reduced deposition, i.e. Eq. (4.20). A two-layer bed model, such as the buffer model presented in Sec. 12.2 (see also Van Kessel *et al.*, 2011a), is an even better option, as it keeps record of the fines stored within the bed.

We elaborate on the concept of reduced deposition and the buffer model further in Sec. 12.3, where we argue that an additional advantage of the reduced deposition concept consists of decoupling the effects of erosion and deposition rates during the calibration phase of a model study. In the remainder of this book we presume continuous deposition, using either Eq. (4.17) or (4.20), unless indicated otherwise.

4.3 Flocculation dynamics

The effective settling velocity of fine sediment is affected largely by flocculation: the settling velocity of mud flocs can be a few orders of magnitude larger than that of the primary particles, as argued before. Flocculation is the interplay between aggregation of smaller particles and breakup of larger particles. Flocculation is therefore a dynamic process, as a result of which floc sizes may change over time and in space. The reader is referred to e.g. Winterwerp and Van Kesteren (2004) and Mehta (2013).

Aggregation occurs when particles collide and stick together. How and why particles stick after colliding is treated in Sec. 2.1, discussing the role of the physico-chemical properties of the clays, organic matter (polysaccharides) and ambient/pore water. Collisions between particles in open water can be governed by three processes:

(1) **Brownian motion:** as clay particles are very small, thermal movements in open water may induce the particle to collide,
(2) **Differential settling:** particles with a large settling velocity can overtake slower settling particles: collisions between these particles may result in aggregation, and
(3) **Turbulent motion:** particles, carried by turbulent eddies, collide and form flocs.

Van Leussen (1994) quantified the relative contribution of these three mechanisms and concluded that in open water, only the last process, i.e. turbulent motion, is important. On the other hand, turbulent shear, induced by the turbulent motion may disrupt the flocs again, causing floc breakup.

This dynamic process of aggregation and breakup can be described with so-called flocculation models. At the end of this chapter we summarize a few models described in literature. First we present a more elementary dynamic model, developed by Winterwerp (1998, 2002), as this model captures the basic dynamic features of flocculation in a simple way. In its most basic, Lagrangian form (i.e. traveling with the mud flocs), this model reads:

$$\frac{\mathrm{d}D_f}{\mathrm{d}t} = k_A c G D_f^2 - k_B G^{3/2} D_f^2 (D_f - D_p) \tag{4.21}$$

where D_f and D_p are the size of the floc and the primary particle (building stone of the flocs), respectively, G is the shear rate at the Kolmogorov turbulent length scale λ_K (Eq. (3.9)), and k_A and k_B are aggregation and breakup coefficients (e.g. Winterwerp, 1998; see also Sec. 2.1). As discussed before, D_p should be interpreted as the smallest aggregate (flocculi) that can survive the stresses in a particular flow field, and may easily contain hundreds of clay particles (e.g. Fugate and Friedrichs, 2003; Verney *et al.*, 2011). The equilibrium floc size $D_{f,e}$ in Eq. (4.21) is

$$D_{f,e} = D_p + \frac{k_A c}{k_B \sqrt{G}} \tag{4.22}$$

In deriving Eq. (4.21), the following empirical observations have been used (e.g. Winterwerp, 1998):

$$D_{f,e} \propto \lambda_K, \quad W_{s,e} \propto c, \quad n_f = 2 \tag{4.23}$$

The first observation, i.e. $D_{f,e} \propto \lambda_K$, can be made plausible from the turbulent energy spectrum in Figure 3.3. From dimensional analysis (e.g. Levich and Tobias, 1963) one derives:

$$\tau_T \sim \mu \frac{\partial u'}{\partial z} \sim \frac{\sqrt{k_T}}{\ell} \sim k\sqrt{k^n} \sim k^{(1+n/2)} \tag{4.24}$$

where τ_T is the turbulent shear stress, k_T is turbulent kinetic energy, k the turbulence wave number, and ℓ is a turbulent length scale. In the inertial subrange (Fig. 3.3), $k_T \sim k^{-5/3}$, while in the dissipative subrange $k_T \sim k^{-3}$. Substitution into Eq. (4.24) yields $\tau_T \sim k^{1/6}$ and $\tau_T \sim k^{-1/2}$ in the inertial and dissipative subranges, respectively. In the inertial subrange, τ_T increases with k, while in the dissipation subrange, τ_T decreases with k. These two trends converge around the Kolmogorov scale λ_K, which explains $D_e \propto \lambda_K$.

The frequency of collisions scales with n^2, where n is the number of particles. However, at constant mass concentration c, the number of particles is inversely proportional to the floc diameter D_f, assuming $n_f = 2$. Thus it is plausible that the aggregation rate scales with c, which is in line with observations that $W_{s,e} \propto c$ (see also Fig. 4.17).

The settling velocity of mud flocs follows from a combination of the solution to the flocculation equation (4.21), or any other flocculation models, such as the population balance models below, and the settling velocity derived for fractal flocs, i.e. Eq. (4.13) — see also Winterwerp and Van Kesteren (2004).

Parameterization of flocculation

The effects of flocculation on the settling velocity is sometimes modeled as

$$W_s = W_0 c^n \tag{4.25}$$

where W_0 is a reference "settling velocity", c is the mass concentration, and $0 < n < 2$, depending on the kind of sediment (Van Leussen, 1994). Note that Eq. (4.25) does not account for the effects of turbulence on flocculation.

From a dimensional point of view, Eq. (4.25) is unattractive, while mathematically, this model does not yield a lower bound either. Therefore, the following alternative is proposed:

$$W_s = W_0 \max\{1, (1 + (c/c_r)^n)\} \tag{4.26}$$

where c_r is a reference concentration, to be determined empirically. Note that more complete algebraic relations for the settling velocity as function of the SPM-concentration are available today, such as Eq. (4.29), (4.30), or (4.31).

Flocculation time and equilibrium

Flocculation takes time, as accounted for in Eq. (4.21). The evolution towards equilibrium is shown in Fig. 4.12, starting from $D_p = 4\mu$m and $G = 7\,\text{s}^{-1}$ (test T71), $29\,\text{s}^{-1}$ (test T69) and $82\,\text{s}^{-1}$ (test T73), as predicted by Eq. (4.21), and where k_A and k_B were calibrated against laboratory data by Van Leussen (1994). The SPM-concentration in these column experiments amounted to about 1 g/L. The floc size evolution shown in Fig. 4.12 is characteristic for flocculation processes: after the start of a flocculation experiment, aggregation rates are very small for a long time. Only when flocs, significantly larger than the primary particles, are formed, the flocculation rate speeds up, almost exponentially. This behavior is explained by the fact that larger flocs collide more frequently than small particles.

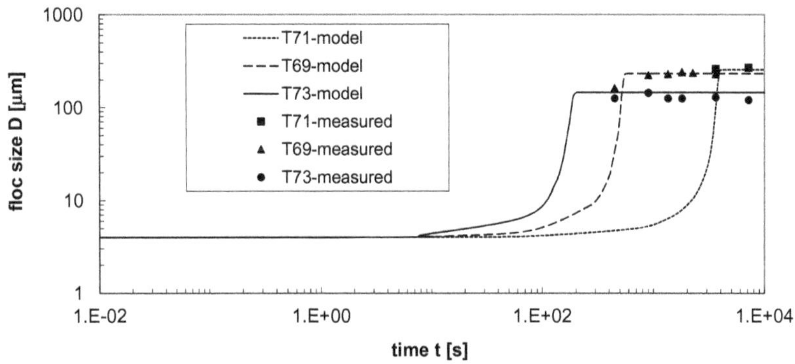

Fig. 4.12. Time development of floc size towards equilibrium for three flocculation experiments (data from Van Leussen, 1994 and solution to Eq. (4.21)).

From Eq. (4.21) also a characteristic time scale for flocculation T_f (aggregation) and floc breakup T_b is found (e.g. Winterwerp, 1998):

$$T_f \propto \frac{1}{k_A c G D_0} \tag{4.27}$$

$$T_b \propto \frac{k_B G^{3/2}}{2D_{f,e}^2} \tag{4.28}$$

where D_0 is the initial floc size at the start of a flocculation experiment. As generally $D_{f,e} \gg D_0$, $T_b \ll T_f$, which will appear to become important in Sec. 9.3.3. Equation (4.27) is plotted in Fig. 4.13, showing characteristic flocculation times (using $k_A = 14.6\,\mathrm{m^2/kg}$, e.g. Winterwerp, 1998) as a function of c for some typical G-values (Fig. 3.4). At low c and low G, the flocculation time amounts to hours to many days, hence flocs cannot attain equilibrium (within a tidal cycle). The floc size under such conditions is mainly determined by the highest stresses in the water column (hence near the bed at maximum tidal velocity).

On the other hand, when SPM-concentrations measure a few 100 mg/L, significant variations in floc size, hence settling velocity are to be expected over time, e.g. over a tidal cycle, and over the water depth.

While flocs grow, their settling velocity increases and may fall out of the observational domain. This behavior is qualitatively sketched in Fig. 4.14. Starting at large G, the equilibrium floc size $D_{f,e}$ increases with decreasing G, in line with Eq. (4.24). However, below a certain shear rate, $D_{f,e}$ reduces with G. This reduction is the result of four interrelated processes:

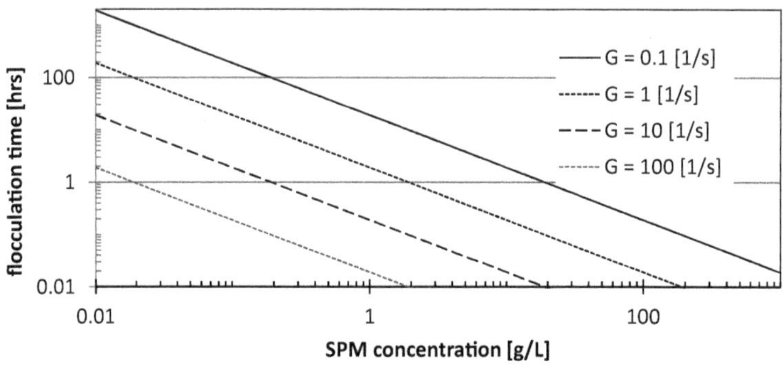

Fig. 4.13. Time scales for flocculation towards equilibrium floc size.

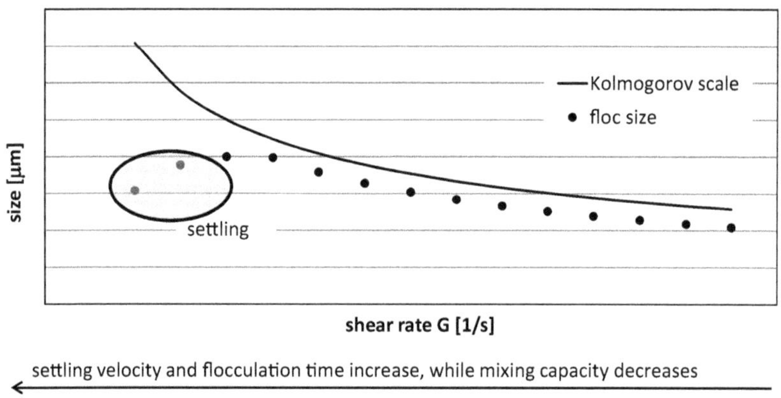

Fig. 4.14. Schematic of equilibrium floc size as a function of shear rate, showing decreasing size when the flocs become to large to be kept in suspension.

(1) with decreasing G, floc size increases, as predicted by Eq. (4.22),
(2) with increasing floc size, the flocs settling velocity increases, as predicted by Eq. (4.13), i.e. $w_s \propto D_f$,
(3) with decreasing G turbulent mixing decreases ($\Gamma_z \propto G$) — thus flocs fall out of the suspension as D_f increases and Γ_z decreases, and
(4) as flocs fall out of the suspension, SPM-concentrations decrease, further reducing $D_{f,e}$, according to Eq. (4.22), i.e. $w_s \propto c$.

Winterwerp (1998) derived an explicit formula describing this reduction in floc size with shear rate in response to settling.

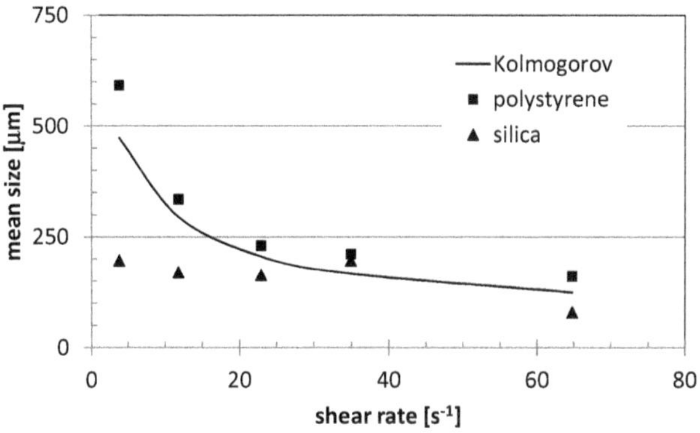

Fig. 4.15. Floc size as a function of turbulent shear rate for buoyant and non-buoyant particles, and comparison with the Kolmogorov turbulent length scale; Mietta *et al.* (2009).

An elegant experiment was carried out by Mietta *et al.* (2009) with flocculating polystyrene and silica particles in a mixing jar at a NaCl salinity of 55 ppt. The buoyant silica particles have a zeta potential of $-4\,\mathrm{mV}$ and specific density of $2000\,\mathrm{kg/m^3}$, while these parameters for the non-buoyant polystyrene particles are $-1.3\,\mathrm{mV}$ and $1055\,\mathrm{kg/m^3}$. These low zeta potentials suggest easy flocculation (Sec. 2.1). Figure 4.15 shows that the non-buoyant polystyrene flocs continue to grow with decreasing G, proportional to the Kolmogorov scale, while the buoyant silica particles fall out of the suspension to the bottom of the jar — the silica floc sizes cannot follow the Kolmogorov scale when they become too large.

Quasi-equilibrium

Also algebraic, quasi-equilibrium models have been developed, which describe equilibrium, or near-equilibrium floc sizes in open water. Soulsby *et al.* (2013) derived a semi-empirical formula for the settling velocity $w_{s,\mu}$ and $w_{s,M}$ of micro- en macro-flocs, respectively (e.g. Dyer and Manning, 1999), based on floc data from a large number of rivers. The transition between micro- and macro-flocs is set at 160 μm, while the micro-flocs have a substantially larger density than the macro-flocs. Likely this is because micro-flocs are more or less in equilibrium with bed sediment (and thus subject to consolidation processes), i.e. these micro-flocs frequently settle on the bed and are eroded from the bed, whereas macro-flocs stay much longer in the water column. The settling velocity for micro- and macro-flocs reads:

$$w_{s,\mu} = 0.537\left(\frac{u_*^3\zeta d_1^4}{\nu^3 z}\right)^{0.39} g\sqrt{\frac{\nu z}{u_*^3\zeta}}\exp\left\{-\left(\frac{u_{*s\mu}}{u_*\sqrt{\zeta}}\right)^{0.66}\right\} \tag{4.29}$$

$$w_{s,M} = 0.095\left(\frac{u_*^3\zeta \bar{d}_\mu^4}{\nu^3 z}\right)^{0.166} g\phi_s^{0.22}\sqrt{\frac{\nu z}{u_*^3\zeta}}\exp\left\{-\left(\frac{u_{*sM}}{u_*\sqrt{\zeta}}\right)^{0.463}\right\} \tag{4.30}$$

in which z is distance from the bed, $\zeta = z/h$, with h is water depth, ϕ_s is volumetric SPM-concentration of the solids, u_* and ν are shear velocity and viscosity, and the four calibration parameters were found to be $d_1 = 10^{-5}$ m, $\bar{d}_\mu = 10^{-4}$ m, $u_{*s\mu} = 0.025$ m/s, and $u_{*sM} = 0.067$ m/s.

Winterwerp *et al.* (2006a) derived a heuristic formula for the settling velocity of representative flocs (D_{50}). As the model coefficients were tuned with data from macro-flocs, this formula would represent settling velocities of macro-flocs only:

$$w_{s,M} = \left[0.007\frac{c^{1/2q}}{\tau^{3/8}} - \left(0.007\frac{c^{1/2q}}{\tau^{3/8}} - \alpha'' D_0\right)\exp\left\{-\frac{5\tau^{9/8}h}{n_f}\right\}\right]^{n_f-1} \tag{4.31}$$

Here the parameter α'' is defined:

$$\alpha'' = \frac{\alpha}{18\beta}\frac{(\rho_s - \rho_w)\,g}{\mu}\frac{D_p^{3-n_f}}{1 + 0.15Re_f^{0.687}} \approx \frac{(\rho_s - \rho_w)\,gD_p^{3-n_f}}{18\mu} \tag{4.32}$$

in which the parameter q is defined as $q = (n_f - 1)/2m$, and τ is the local (depth and time dependent) shear stress in the water column. From calibration, $m \approx 0.2 - 0.4$ and $n_f \approx 2.2$. The inhomogeneity of dimensions in Eq. (4.31) stems from substitution of the empirical coefficients into the equation.

These quasi-equilibrium models have a fairly stiff behavior in sediment transport models, and should therefore be used with care. For instance, sedimentation rates in harbor basins will be substantially miscalculated, as these models predict too large settling velocities in the basin, as the flocs are computed to attain local equilibrium values immediately, whereas in reality the floc size is largely determined by the ambient water conditions.

Population balance equations

Equation (4.21) describes the evolution of a characteristic floc size, containing no information on the size distribution, however. Therefore more sophisticated flocculation models have been developed, varying from highly advanced to very empirical ones. The more sophisticated models are based on the Population Balance Equation (PBE), e.g. Maggi (2005); Verney *et al.* (2011). A PBE describes the variation of the number of flocs n_k in floc class k as a function of production (gain) of flocs in that class by aggregation from smaller flocs (G_A) and breakup of larger flocs (G_B), and loss of flocs from that class by aggregation into larger flocs (L_A) and breakup into smaller flocs (L_B). Here we present Maggi's model:

$$\frac{\mathrm{d}n_k}{\mathrm{d}t} = \underbrace{\sum_{i=1}^{N-1} \sum_{\substack{j=i+1 \\ j+i=k}}^{N} \alpha_{i,j}\Lambda_{i,j}n_i n_j}_{G_A} - \underbrace{n_k \sum_{j=1}^{N} \alpha_{k,j}\Lambda_{k,j}n_j}_{L_A} + \underbrace{\sum_{j=1}^{N} \gamma_{j,k}B_j n_j}_{G_B} - \underbrace{B_k n_k}_{L_B} \tag{4.33}$$

where N is number of classes with n flocs per class, α is a collision frequency, Λ is an aggregation kernel and B is a breakup kernel. PBE-models have to be calibrated against data assessing α, Λ and B.

In the case of many classes, PBE-models are computationally expensive, in particular when coupled to three-dimensional sediment transport models. Therefore, more simple models have been developed, such as the two-fraction model by Lee *et al.* (2011), which are more suitable for engineering studies.

All flocculation models presented here assume free falling or free moving particles, hence are applicable for fairly low SPM-concentrations. At larger concentrations, collision mechanisms change, and the models presented here are no longer valid. Yet, no flocculation models for such higher concentrations have been developed yet, as far as known.

Another lacuna in our knowledge concerns aggregation and breakup processes under the influence of waves. One may argue that the irrotational motion of the waves outside the wave boundary layer would not induce shear rates that can affect flocculation. On the other hand, shear rates within the wave boundary layer are very high, likely breaking up flocs efficiently. Also under breaking waves, shear stresses are high, which must affect floc sizes. However, flocculation studies under wave action have never been carried out, as far as known.

Flocculation dynamics are also affected by biological processes. As explained in Secs. 2.1 and 8.2, various organisms in aquatic systems produce organic polymers, gluing particles together, thus affecting the aggregation coefficient k_A (see Eq. 4.21). For larger sticking efficiencies (large k_A), the flocculation of cohesive sediment falls within the class of Diffusion

Limited Cluster-Cluster Aggregation (DLCCA, e.g. Vicsek (1992); Meakin, 1986). Flocs are then generally very open and fractal dimensions are low ($1.7 < n_f < 1.9$). Moreover, some organisms form flocs while digesting organic matter (e.g. fecal pellets by mussels).

4.4 Hindered settling

As explained in Sec. 4.1, mud flocs occupy large volumes at relatively small mass concentrations, i.e. $\phi_f \gg \phi_s$, see Eq. (4.11). Therefore, at already moderate SPM-concentrations (but large ϕ_f), flocs hinder each other while settling, as sketched in Fig. 4.16.

Richardson and Zaki (1954) were the first evaluating this hindered settling effect. On the basis of dimensional analysis and a series of fluidization experiments, they arrived at their famous hindered settling formulation:

$$w_s = w_{s,0}(1 - \phi_s)^n \tag{4.34}$$

where ϕ_s is the sediment solid fraction. Later, Baldock *et al.* (2004) specified the coefficient n in a simple formula:

$$n = 4.4(D_{ref}/D_{50})^{0.2} \tag{4.35}$$

in which the reference diameter D_{ref} amounts to 200 μm. Equation (4.34) has been slightly modified for assessing hindered settling effects of flocs of cohesive sediment in open water:

$$w_s = w_{s,0}(1 - \phi_f)^n \tag{4.36}$$

in which w_s and $w_{s,0}$ are the effective settling velocity of flocs and the settling velocity of a single floc in still water, respectively, and the volumetric

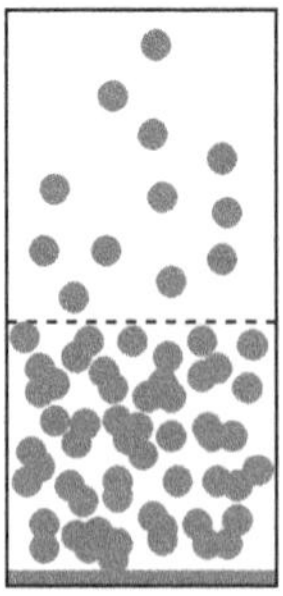

Fig. 4.16. Hindered settling and interface (lutocline) formation by self-stratification (see also Sec. 7.1 on lutoclines and SPM induced stratification).

floc concentration ϕ_f is defined as $\phi_f = c/c_{gel}$, while the exponent n attains values of $n = 4$–5.

Later Winterwerp (1999); Winterwerp and Van Kesteren (2004) and Dankers and Winterwerp (2007) elaborated on the work by Richardson and Zaki (1954) and proposed:

$$w_s = w_{s,0} \frac{(1-\phi_f)^2 (1-\phi_s)}{1+2.5\phi_f} \tag{4.37}$$

assuming that each flocs settles in the remainder of the suspension with augmented bulk density (the $(1-\phi_s)$-term) and viscosity (the $(1+2.5\phi_f)$-term), affected by the return flow induced by the other settling flocs (the $(1-\phi_f)$-term). The exponent "2" originates from laboratory experiments, accounting for non-linear effects.

Equation (4.37) is fit to a series of settling velocity data in Fig. 4.17, tuning $w_{s,0}$ and c_{gel}, showing that hindered settling effects become prominent at SPM-concentrations of a few g/L only.

In Chapter 13, we explain that hindered settling is often an important ingredient in numerical modeling of fine sediment transport, in particular in engineering models predicting siltation rates in harbor basins and fairways.

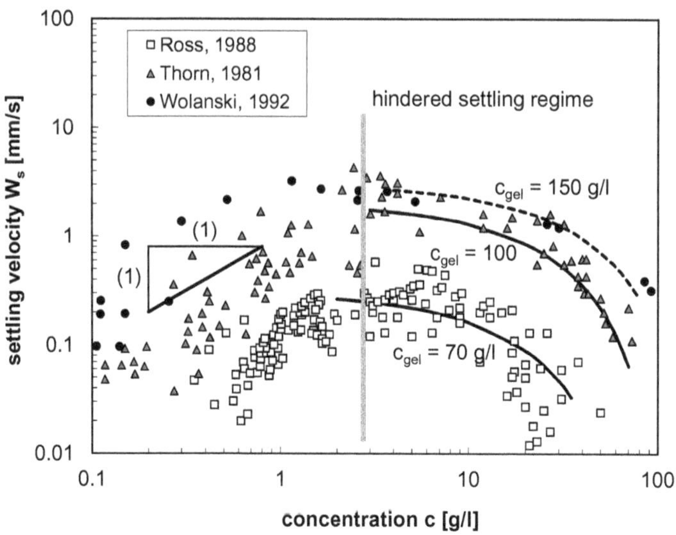

Fig. 4.17. Effective settling velocity as a function of SPM-concentration — comparison of Eq. (4.37) with data. The left part of the data supports Eq. (4.23, i.e. $w_s \propto c$).

In soil mechanics, the gelling point/concentration is known as the structural density, reflected by a kink in the settling curve, the so-called point of contraction. Beyond this point, consolidation starts. In Sec. 5.1, we argue that the smooth transition from hindered settling to consolidation is not yet understood from a physical point of view.

Some authors (e.g. Camenen and Van Bang, 2011) tried to resolve this transition by introducing a reference concentration C^* in Eq. (4.36) instead of c_{gel}. As $C^* > c_{gel}$, this hindered settling formula captures a part of the consolidation phase. We do not advocate this approach for a few reasons:

(1) the problem of finding a transitional formula coupling hindered settling and consolidation is not resolved, but postponed to another phase of the settling/consolidation process,
(2) settling and consolidation are fundamentally different processes, and need a different physical/mathematical approach, and
(3) the very practical fractal formulation of the consolidation phase (Sec. 5.1) becomes obsolete — this is our major objection against this approach.

Kynch and the settling curve

Substitution of Eq. (4.37) into the vertical mass balance equation (3.57) and omitting vertical mixing (the diffusion term in Eq. (3.57)) yields Kynch (1952) equation, which can be rewritten as

$$\frac{\partial \phi_f}{\partial t} - \frac{\partial}{\partial z}\left(w_{s,0}\mathrm{f}\left(\phi_f\right)\phi_f\right) = \frac{\partial \phi_f}{\partial t} - w_{s,0}\mathrm{F}\left(\phi_f\right)\frac{\partial \phi_f}{\partial z} = 0 \tag{4.38}$$

where $\mathrm{f}(\phi_f)$ represents a hindered settling function (Eq. (4.37)), and $\mathrm{F}\left(\phi_f\right) = \mathrm{d}\left(\phi_f \mathrm{f}\left(\phi_f\right)\right)/\mathrm{d}\phi_f$ (Kranenburg, 1992). The solution of the hyperbolic Kynch equation can be obtained by integration along the characteristics $\mathrm{d}z/\mathrm{d}t = w_{s,0}\mathrm{F}$. When the characteristic lines converge, concentrations gradients increase, and ultimately a stable interface (i.e. a shock wave) develops. As in general $\mathrm{d}\phi_f/\mathrm{d}z < 0$ (z is positive upward), characteristics converge when $\mathrm{dF}/\mathrm{d}\phi_f < 0$.

The hindered settling function F for Eq. (4.37) becomes:

$$\begin{aligned}\mathrm{F} = \frac{\mathrm{d}\left(\mathrm{f}\phi_f\right)}{\mathrm{d}\phi_f} &= \frac{\left(1-\phi_f\right)^2\left(1-\phi_s\right)}{1+2.5\phi_f} - \frac{2\phi_f\left(1-\phi_f\right)\left(1-\phi_s\right)}{1+2.5\phi_f}\\ &\quad - \frac{\left(1-\phi_f\right)^2}{\left(1+2.5\phi_f\right)}\frac{c_{gel}}{\rho_s} - 2.5\phi_f\frac{\left(1-\phi_f\right)^2\left(1-\phi_s\right)}{\left(1+2.5\phi_f\right)^2}\end{aligned} \tag{4.39}$$

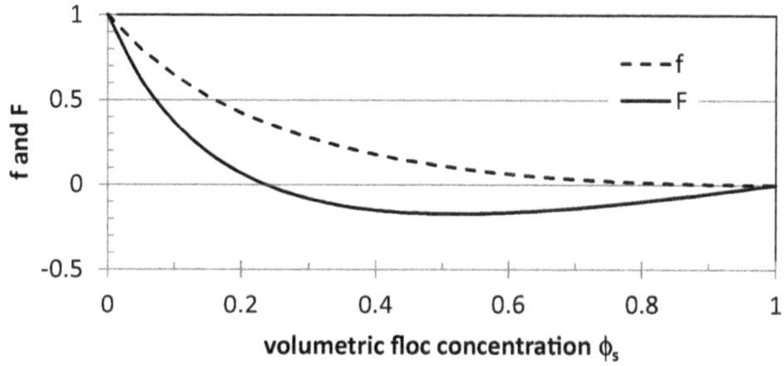

Fig. 4.18. Hindered settling functions f and F for Eq. (4.37), assuming $c_{gel} = 100\,\mathrm{g/L}$.

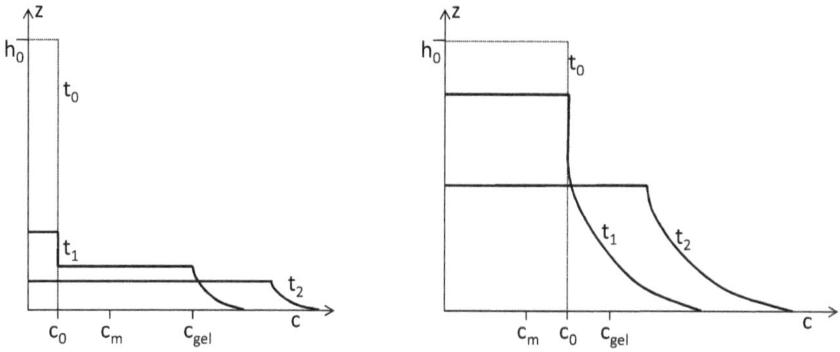

Fig. 4.19. Sketch of hindered settling sediment-water mixture and bed formation for $c_0 < c_m$ (left panel) and $c_0 > c_m$ (right panel), and first phase of consolidation ($t = t_2$).

which is plotted in Fig. 4.18. F shows a minimum at $\phi_{f,m}$, which, for Eq. (4.37) amounts to $\phi_{f,m} = 0.5$ (i.e. $c_{min} = 50\,\mathrm{g/L}$ for $c_{gel} = 100\,\mathrm{g/L}$).

For a settling sediment-water mixture (such as in a settling column), the analysis above predicts that at initial volumetric concentrations $\phi_{f,0} < 0.5$ two interfaces develop, i.e. an interface at the top of the settling mixture (which always occurs), and one at the "bed"-mixture interface. When $\phi_{f,0} > 0.5$, only the upper interface develops. This behavior is sketched in Fig. 4.19, after Kranenburg (1992) and Dankers (2006), but using mass concentration instead of volumetric floc concentration. At $t = t_0$, the sediment-water mixture is homogeneously mixed over the water depth h_0. The time t_1 represents an arbitrary moment within the hindered settling phase. The left panel of Fig. 4.19 shows two interfaces. The concentration above the lower interface remains at the initial concentration c_0, while just below that interface, c cannot exceed the gelling concentration c_{gel}. At the point of contraction (see Appendix A, Fig. 5.2), the two interfaces meet, and the concentration at (in fact just below) that single interface amounts to c_{gel}. Lower in the bed, consolidation has commenced.

In the right panel of Fig. 4.19, c in the upper part of the water column also remains at its initial value, but no second interface is formed. Figure 4.19 also shows the concentration profile for the first phase of consolidation (i.e. $t = t_2$). Note that this profile is still concave-up, while the concentration just below the interface now can exceed c_{gel} — see also Sec. 5.1.

Chapter 5

Consolidation and Strength

Much literature is dedicated to the subject of consolidation; see e.g. Terzaghi (1943); Gibson *et al.* (1967); Been and Sills (1981). In this chapter, we elaborate on the consolidation of (soft) cohesive sediment, i.e. soft mud. In natural systems, consolidation of soft deposits exhibits large deformations, at relatively low stresses. We treat fully saturated conditions only, and assume deformations in the vertical direction only.

Section 5.1 discusses consolidation from dilute suspensions, referred to as virgin consolidation — consolidation would follow the Virgin Consolidation (Compression) Line (VCL) of Fig. 2.24. However, during consolidation (small) lateral stresses develop, and the stress path follows the so-called K_0-line, and we refer to normal consolidation (see end of Sec. 5.1, Eq. (5.35); Schofield and Wroth, 1968). At larger initial concentrations, the consolidation process starts to deviate (partly) from these normal consolidation conditions, as explained in Sec. 5.3. We refer to the consolidation of slurries, where a slurry has an initial dry density beyond the gelling point. To understand the difference between normal consolidation and the consolidation of slurries, strength and stress development in consolidated soils are elaborated upon in Sec. 5.2.

5.1 Normal consolidation

In this section, the consolidation of fairly diluted suspensions with $c_0 < c_{gel}$ is discussed. The consolidation equations are given in Eulerian coordinates, obtaining an advection–diffusion equation consistent with the (vertical) mass balance equation for SPM in the water column, enabling the continuous modeling approach described in Sec. 11.4. The reader is further referred to e.g. Toorman (1996) and Grasso *et al.* (2015).

We analyze the settling and consolidation of a dilute suspension in a settling column, as this setup more or less represents processes in natural

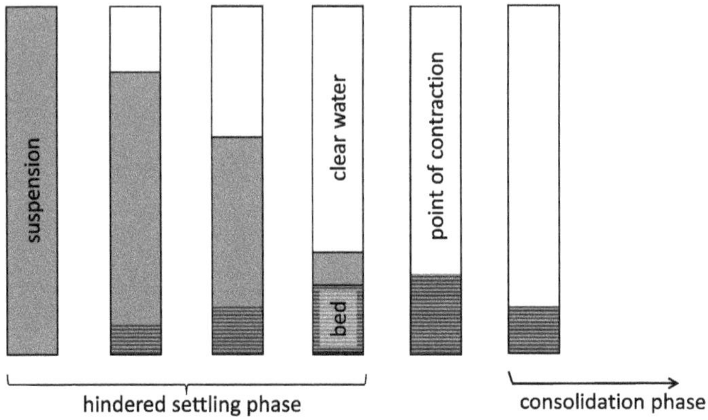

Fig. 5.1. Sketch of settling columns with time variation in interface and bed formation.

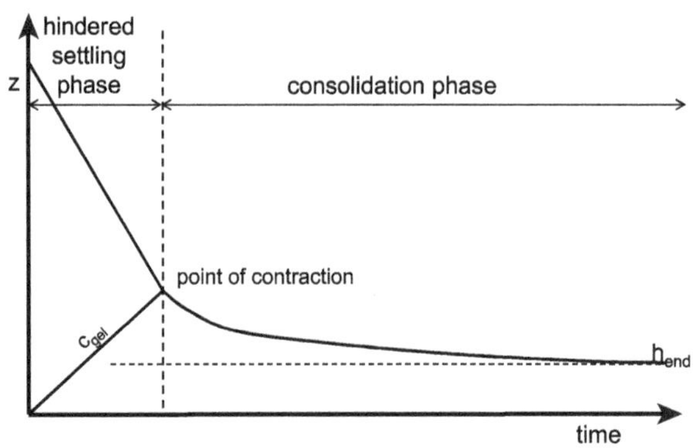

Fig. 5.2. Sketch of interface evolution (settling curve) for the settling experiments in Fig. 5.1.

systems. Such a settling experiment is sketched in Fig. 5.1, while the development of interfaces with time is sketched in Fig. 5.2. The upper interface lowers with time, consisting of a hindered settling (see Sec. 4.4) and a consolidation phase. The transition between these phases occurs at the so-called point of contraction. Further to the discussion in Sec. 4.4, bed formation in the hindered settling phase (i.e. left of the point of contraction) may or may not induce a second interface at the bed-suspension transition. Figure 5.2 sketches the formation of such a second interface. The dry density (or solids mass concentration of the soil) along that interface (i.e. the characteristic representing bed formation) equals c_{gel}.

To analyze the stresses in a soil, and its state of consolidation, Terzaghi (1943) introduced the effective stress concept:

$$\sigma_{zz} = p_f + \sigma_{zz}^{sk} = p_h + p_e + \sigma_{zz}^{sk} \tag{5.1}$$

in which σ_{zz} is the total normal stress in (vertical) z-direction, p_f is the total water pressure, p_h is the hydrostatic pressure, p_e is a possible excess pore water pressure and σ_{zz}^{sk} is so-called effective stress in the z-direction (in soil mechanics generally denoted as σ'_v). Note that the effective stress does not represent grain–grain stresses, as only normal stresses are accounted for. However, it is indirectly a measure of the strength of the soil (Sec. 2.2) — the effective stress is merely a convenient mathematical representation of the stresses in the bed. The water pressure p_f consists of a hydrostatic part p_h and its deviation p_e, the so-called excess pore water pressure.

Figure 5.3 shows the vertical stress distribution of a normally consolidated suspension (see below) at equilibrium, i.e. $p_e = 0$. The left panel shows the stresses in case the deposit is fully submerged. The water/pore pressure equals atmospheric pressure p_a at the water surface Z_w, hence $p_{f,Z_w} = p_{h,Z_w} = p_a \approx 0$. The total stress starts to deviate from the water pressure at/below the water–soil interface Z_s, following the hydrostatic stress distribution $\sigma_{zz}(z) = \int_{Z_s}^{z} \rho_b g z \mathrm{d}z'$. The left panel of Fig. 5.3 also depicts the effective stress, i.e. the difference between the total stress and the water pressure.

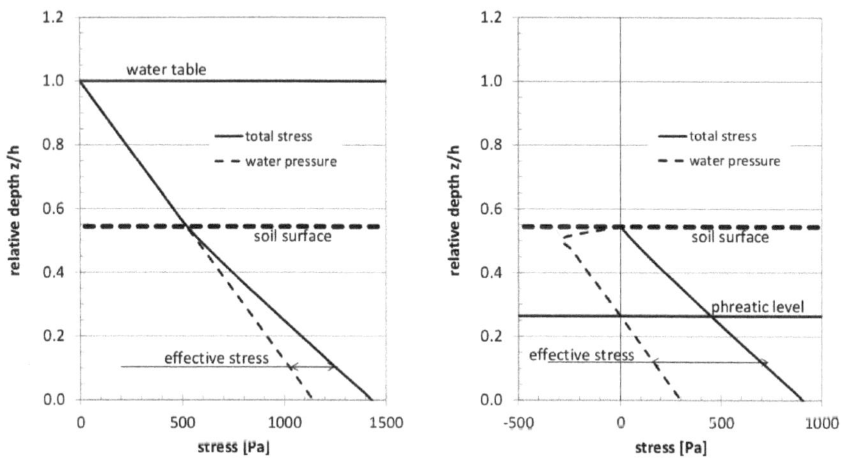

Fig. 5.3. Terzaghi's concept showing vertical stress distribution in an equilibrium deposit, left panel: fully submerged; right panel: partly emerged, but fully saturated.

The right panel of Fig. 5.3 reflects the situation for partly emerged, but fully saturated conditions. Emergence refers to the ambient water level — we refer to the water table. Within the soil, one generally refers to the phreatic level Z_f. At this level, the water pressure should equal the atmospheric pressure. We refer to the "phreatic level", though in literature this level often refers to the transition between the saturated and unsaturated (vadose) zone. Hence, the water pressure between Z_s and Z_f is negative, inducing capillary suction, or pseudo-cohesion, as discussed in Sec. 2.2. Following Terzhagi's concept, the effective stress σ_{zz}^{sk} increases by $g\rho_w(Z_w - Z_f)$, relative to the case of a fully submerged deposit, assuming fully saturated conditions. As the effective stress should be zero at the soil surface, the suction pressure reduces to zero at the soil surface as well — in the following we ignore this. The importance of this under-pressure with respect to consolidation is discussed below.

In the hindered settling phase, $\sigma_{zz}^{sk} = 0$. Hence, if one would define an excess pore water pressure in the hindered settling phase, this would amount to $p_e = gz(\rho_b(z) - \rho_w)$, where ρ_b is the mixture bulk density (varying over depth) and $gz\rho_w$ is the hydrostatic water pressure for sediment-free water. In a fully consolidated bed (i.e. at equilibrium), $p_e = 0$, and $p_f = p_h = \rho_w gz$ (hydrostatic pressure distribution).

Gibson's equation in Eulerian coordinates

In the following, we present Gibson's equation (Gibson *et al.*, 1967) for one-dimensional vertical (1DV) consolidation with large deformations in a settling/consolidation column (i.e. Fig. 5.1). This approach is considered to be state-of-the-art. However, instead of using Gibson's material coordinates (a Lagrangian coordinate system), we follow Merckelbach and Kranenburg (2004b) (see also Winterwerp and Van Kesteren, 2004), and use Eulerian coordinates. The advantage of using material coordinates is that the sharp concentration gradient at the sediment–water interface is implicitly resolved if the total mass of sediment is constant. In open water, water-bed exchange processes occur continuously. Transformation of these exchange processes into material coordinates becomes cumbersome, thus losing the advantage of using material coordinates.

In Eulerian coordinates, the consolidation equation attains the form of an advection–diffusion equation, and can therefore easily be implemented into existing sediment transport models. It is illustrative to derive this equation; see also Merckelbach and Kranenburg (2004b) or Winterwerp and Van Kesteren (2004). Winterwerp and Van Kesteren

(2004) also present a model for mud–sand mixtures. However, here we focus on the consolidation of mud only. We start from the sediment balance of a settling/consolidating water–sediment mixture in a one-phase description (which is again the simple wave equation as elaborated by Kynch, Sec. 4.4):

$$\frac{\partial \phi_s}{\partial t} + \frac{\partial}{\partial z}\left(w_s \phi_s\right) = 0 \tag{5.2}$$

in which w_s is the settling velocity of the particles and ϕ_s ($\equiv c/\rho_s$) is their volumetric concentration. Because of continuity, settling particles induce an opposite pore water flow with velocity w_f:

$$\phi_s w_s + \left(1 - \phi_s\right) w_f = 0 \quad \text{or} \quad \left(w_f - w_s\right)\left(1 - \phi_s\right) = -w_s \tag{5.3}$$

This pore water flow velocity can be modeled with Darcy's law:

$$\left(w_f - w_s\right)\left(1 - \phi_s\right) = -\frac{k}{\rho_w g}\frac{\partial p_e}{\partial z} \quad \text{or} \quad w_s = \frac{k}{\rho_w g}\frac{\partial p_e}{\partial z} \tag{5.4}$$

in which k [m/s] is the so-called hydraulic conductivity. The excess pore water pressure p_e yields the difference between the actual total water pressure p_f and the hydrostatic pressure:

$$\frac{\partial p_e}{\partial z} = \frac{\partial p_f}{\partial z} - g\rho_w \tag{5.5}$$

Combining with the effective stress concept (Eq. (5.1)):

$$\frac{\partial p_f}{\partial z} = \frac{\partial \sigma_{zz}}{\partial z} - \frac{\partial \sigma_{zz}^{sk}}{\partial z} \tag{5.6}$$

we obtain:

$$\begin{aligned} w_s &= \frac{k}{\rho_w g}\frac{\partial p_f}{\partial z} + k = \frac{k}{\rho_w g}\frac{\partial \sigma_{zz}}{\partial z} - \frac{k}{\rho_w g}\frac{\partial \sigma_{zz}^{sk}}{\partial z} + k \\ &= -\frac{\rho_s - \rho_w}{\rho_w} k\phi_s - \frac{k}{\rho_w g}\frac{\partial \sigma_{zz}^{sk}}{\partial z} \end{aligned} \tag{5.7}$$

Substitution into Eq. (5.2) gives the Gibson equation in Eulerian coordinates:

$$\frac{\partial \phi_s}{\partial t} - \frac{\partial}{\partial z}\left(\frac{\rho_s - \rho_w}{\rho_w} k\phi_s^2\right) - \frac{\partial}{\partial z}\left(\frac{k\phi_s}{\rho_w g}\frac{\partial \sigma_{zz}^{sk}}{\partial z}\right) = 0 \tag{5.8}$$

which can be rewritten as

$$\frac{\partial \phi_s}{\partial t} - \frac{\partial}{\partial z}\left(\frac{\rho_s - \rho_w}{\rho_w} k\phi_s^2\right) - \Gamma_c \frac{\partial^2 \phi_s}{\partial z^2} = 0 \tag{5.9}$$

We use a fractal model, assuming self-similarity, for the hydraulic conductivity k and effective stress σ_{zz}^{sk}, further to Kranenburg (1994) and Merckelbach and Kranenburg (2004b), yielding power-law relations for the material parameters:

$$k = K_k(\phi_s)^{-\frac{2}{3-n_f}} \tag{5.10}$$

$$\sigma_{zz}^{sk} = K_p(\phi_s)^{\frac{2}{3-n_f}} \tag{5.11}$$

in which n_f is the fractal dimension of the consolidating soil. The material parameters K_k and K_p have to be determined from experiments, as described in Appendix A.8; see also Sec. 5.2. Typical values for these material parameters are $K_p \approx 10^8$–10^{10} Pa and $K_k \approx 10^{-12}$–10^{-14} m/s, and for the fractal dimension in the soil $2.6 < n_f < 2.8$. The consolidation coefficient Γ_c [m^2/s] in Eq. (5.9) appears as a constant "diffusion coefficient", using the fractal relations (5.10) and (5.11), and reads:

$$\Gamma_c = \frac{2}{3 - n_f} \frac{K_k K_p}{g\rho_w} \tag{5.12}$$

Typical values for Γ_c are 10^{-10}–10^{-8} m^2/s for natural sediment.

The consolidation coefficient

The Gibson equation (Gibson *et al.*, 1967) has been derived for consolidation under large deformation rates and can therefore not directly be compared with the Terzaghi consolidation model. Terzaghi introduced a consolidation coefficient c_v, while Gibson derived at a consolidation coefficient c_F. The relation between these coefficients and Γ_c can be established in case of large stresses, assuming:

$$\frac{\partial \sigma_{zz}}{\partial t} \ll \frac{\partial \sigma_{zz}^{sk}}{\partial t} \wedge \frac{\partial p_e}{\partial t} \quad \text{and} \quad \frac{\partial \sigma_{zz}}{\partial z} \ll \frac{\partial \sigma_{zz}^{sk}}{\partial z} \wedge \frac{\partial p_e}{\partial z} \tag{5.13}$$

In that case, Eq. (5.8) reduces to a simple diffusion equation:

$$\frac{\partial \phi_s}{\partial t} - \frac{\partial}{\partial z}\left(\frac{k\phi_s}{\rho_w g}\frac{\partial \sigma_{zz}^{sk}}{\partial z}\right) = \frac{\partial \phi_s}{\partial t} - \frac{\partial}{\partial z}\left(\frac{k\phi_s}{\rho_w g}\frac{\mathrm{d}\sigma_{zz}^{sk}}{\mathrm{d}\phi_s}\frac{\partial \phi_s}{\partial z}\right) = 0 \tag{5.14}$$

This equation has the same form as Terzaghi's famous consolidation equation:

$$\frac{\partial p_e}{\partial t} - c_v \frac{\partial^2 p_e}{\partial z^2} = 0 \tag{5.15}$$

With Eq. (5.13) this can be rewritten as

$$\frac{\partial \sigma_{zz}^{sk}}{\partial t} - c_v \frac{\partial^2 \sigma_{zz}^{sk}}{\partial z^2} = 0 \tag{5.16}$$

Terzaghi's consolidation coefficient c_v is defined as

$$c_v = \frac{k}{\rho_w g m_v} \tag{5.17}$$

with the coefficient of volumetric compressibility m_v:

$$\mathrm{d}\varepsilon = -m_v \mathrm{d}\sigma_{zz}^{sk} \tag{5.18}$$

while the volumetric strain ε reads

$$\mathrm{d}\varepsilon = \frac{\mathrm{d}\phi_s}{\phi_s} \tag{5.19}$$

Thus we find:

$$c_v = \frac{k\phi_s}{\rho_w g} \frac{\mathrm{d}\sigma_{zz}^{sk}}{\mathrm{d}\phi_s} \tag{5.20}$$

From which follows that $\Gamma_c = c_v$. Gibson *et al.* (1967) show that their full model, given in material coordinates, also reduces to a similar diffusion equation in Eulerian coordinates:

$$\frac{\partial e}{\partial t} - c_F \frac{\partial^2 e}{\partial z^2} = 0 \tag{5.21}$$

in which e is the void ratio ($\phi_s = 1/(1+e)$ and the consolidation coefficient c_F is assumed constant (i.e. linearized). Gibson *et al.* (1967) show that:

$$c_v = \frac{\phi_{s,0}}{\phi_s} c_F \tag{5.22}$$

in which $\phi_{s,0}$ is the volumetric concentration of the solids at the start of (a) consolidation (experiment). In case of small deformations, $c_v \approx c_F$.

Barciela-Rial *et al.* (2020) indeed found values of Γ_c close to c_v from consolidation experiments with fresh water Lake Markermeer mud.

Effect sand

The fractal material relations Eq. (5.10) and (5.11) assume a floc skeleton, thus mud only. Small amounts of sand, though, would affect these relations. In case sand is passive, i.e. does not contribute to the soil skeleton (Merckelbach and Kranenburg, 2004a; Winterwerp and Van Kesteren, 2004, Sec. 2.2), the volumetric sediment (mud) concentration ϕ_s should be corrected for that sand content:

$$\phi_{s,\mathit{eff}} \Rightarrow \frac{\phi_s^m}{1-\phi_s^s} \tag{5.23}$$

in which $\phi_{s,\mathit{eff}}$ represents the effective mud content, and the superscripts $\cdot^m$ and $\cdot^s$ represent the mud and sand fraction, respectively.

The fractal dimension n_f in consolidating soil is much larger than for flocs in the water column, typically measuring $2.6 < n_f < 2.85$, which is related to the collapse of the floc structure, see also Secs. 5.2 and 2.1. Note that mathematically, this effect is very large, owing to the power $2/(3-n_f)$. Partheniades (1986) provided a qualitative explanation for the compaction of flocs during bed formation/consolidation, sketched in Fig. 5.4.

Equilibrium conditions

The consolidation time is determined mainly by the third (diffusion) term of Eq. (5.9). The general solution to the diffusion equation (5.9), omitting the second (advection) term reads:

$$\frac{\phi_s(z)-\phi_{s,eq}}{\phi_{s,0}-\phi_{s,eq}} = \text{erf}\left\{\frac{z}{\sqrt{4\Gamma_c t}}\right\} \tag{5.24}$$

where subscripts $\cdot_0$ and $\cdot_{eq}$ refer to initial and equilibrium conditions, respectively. Equation (5.24) shows that the time scale for consolidation

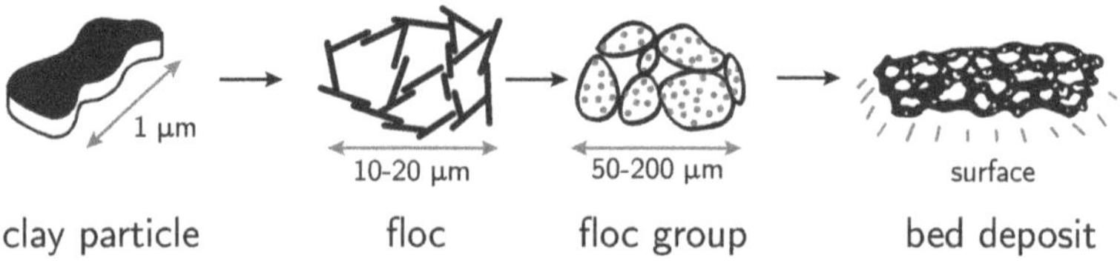

Fig. 5.4. Aggregation in the water column and collapse of floc structure during bed formation upon deposition — after Partheniades (1986, 2009).

T_c (i.e. the e-folding time) becomes:

$$T_c \sim \frac{\delta^2}{\Gamma_c} \tag{5.25}$$

where δ is the initial thickness of the consolidating sample.

Merckelbach and Kranenburg (2004b) showed that the final thickness h_{eq} of a consolidated bed is given by (see also Appendix A.8):

$$h_{eq} = \frac{nK_p}{(n-1)(\rho_s - \rho_w)g}\left[\frac{(\rho_s - \rho_w)g}{K_p}\zeta_m\right]^{\frac{n-1}{n}} \tag{5.26}$$

with $n = 2/(3 - n_f)$ and the Gibson height ζ_m represents the total amount of muddy sediment:

$$\zeta_m = \int_h \phi_s \mathrm{d}z = h_0\overline{\phi_{s,0}} \tag{5.27}$$

Equation (5.26) can be used for assessing the mean concentration at equilibrium as a function of initial conditions, defining:

$$\overline{\phi_{s,eq}} = \frac{h_0}{h_{eq}}\overline{\phi_{s,0}} \tag{5.28}$$

where an overbar indicates averaging over the thickness of the consolidating/consolidated layer. Substitution of Eq. (5.27) into Eq. (5.28) gives:

$$\overline{\phi_{s,eq}} \propto \left(h_0\overline{\phi_{s,0}}\right)^{\frac{3-n_f}{2}} \tag{5.29}$$

As typical values for the fractal dimension yield $2.7 < n_f < 2.8$, we find:

$$\overline{\phi_{s,eq}} \propto \left(h_0\overline{\phi_{s,0}}\right)^{0.1\text{–}0.15} \tag{5.30}$$

This implies that the final (bulk) density of a consolidating bed is not very sensitive to the initial conditions (see also experiments by Van Rijn and Barth (2018)). However, this does not hold for the final strength of the soil. This can be understood from a force balance. At equilibrium, $p_e = 0$, and the water pressure is hydrostatic. For a normal consolidated bed, the strength of the soil at an arbitrary level z_1, see Fig. 5.5, is equal to the stress by the mass of solids above z_1. As an example, let us assume a 0.2 m thick layer above this reference level, at a dry density of 400 kg/m^3, which would yield a stress, hence strength of about 800 Pa at that level.

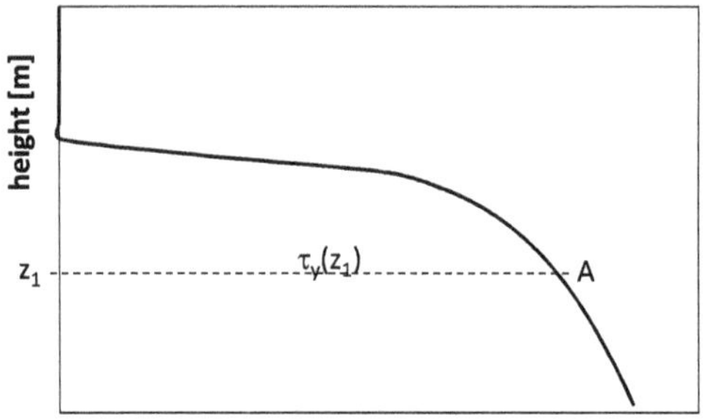

Fig. 5.5. Dry density profile for a normally consolidated bed at equilibrium.

This example further suggests that also the yield strength of the soil can be described with a fractal model (Kranenburg, 1994):

$$\tau_y = K_y (\phi_s)^{\frac{2}{3-n_f}} \tag{5.31}$$

where K_y is to be determined experimentally (vane tests, see Appendix A). Again, small amounts of sand in the soil would require modification of Eq. (5.31) using Eq. (5.23).

The Gibson equation implies that the strength at level z_1 in Fig. 5.5 would remain constant after equilibrium is reached, provided no additional load (surcharge) is brought onto the soil. However, from practice we know that after equilibrium, slow deformations can still occur, known as secondary consolidation, caused by creep. Creep can be understood from a microscopic perspective. Because of Brownian motion, or otherwise, bed particles may be rearranged, attaining a more stable arrangement. As a result, further compaction and increase in strength occur (sometimes up to 25%, see e.g. Das, 2008). Thus creep can be regarded as plastic deformation of a soil at macro-scale, induced by micro-scale processes.

In soil mechanics literature, a formal definition of the degree of creep is used, based on the degree of settling at 90% primary consolidation — this is however beyond the scope of this book on soft sediment.

Also biochemical effects may occur in a soil, for instance in response to changes in redox (Razakamanantsoa and Djeran-Maigre, 2016), mineral precipitation and cementing by reduced iron Fe^{3+} (Bronick and Lal, 2005; Rakotonimaro *et al.*, 2017), and interaction with organic materials (bio-binding, Adachi (2019)). Organic material is produced by bacteria and phytoplankton (algae), while bacteria also play a major role in oxidizing or reducing organic material. The soil may become stronger or weaker, depending on whether interparticle bonds become stronger or weaker.

Equation (5.31) can also be used to assess the strength of a soft, consolidating muddy soil, for instance fluid mud (see Chapter 7). For such soft mud, viscosity is also important. Kranenburg (1994) proposed a formulation for the viscosity of soft muds, again assuming self-similarity:

$$\mu = \mu_w \phi_f^{2(1+a_\mu)/3} \tag{5.32}$$

He showed that $a_\mu < n_f/(3-n_f)$, while typical values for a_μ are in the range of $2 < a_\mu < 4$ (Kranenburg found $a_\mu = 3.25$ for mud from the Scheldt River). This implies that the viscosity of mud increases slower with sediment concentration (dry density) than its strength.

Hindered settling and consolidation

The consolidation equation (5.9) can be combined with the hindered settling equation (4.37) (and also with turbulent mixing, as discussed in Chapter 11) as follows:

$$\frac{\partial \phi_s}{\partial t} - \frac{\partial}{\partial z}\left(\Xi_s \phi_s\right) - \Gamma_c \frac{\partial^2 \phi_s}{\partial z^2} = 0 \tag{5.33}$$

in which the heuristic function Ξ_s accounts for hindered settling and hydraulic conductivity, and its transition:

$$\begin{aligned} \Xi_s &= f_{hs} + \frac{f_c}{1+\eta f_c}, \quad \text{with} \\ f_{hs} &= w_{s,0}\frac{\left(1-\phi_f\right)^2\left(1-\phi_s\right)}{1+2.5\phi_f}, \quad \text{and} \\ f_c &= \frac{\rho_s-\rho_w}{\rho_w} k \phi_s \end{aligned} \tag{5.34}$$

The empirical settling parameter η (typical values are 10^4–10^6) is introduced to obtain a smooth transition between hindered settling and consolidation — to the knowledge of the authors, no physics-based function has yet been derived for this transition. Figure 5.6 presents an example of the settling function Ξ_s for $\eta = 200{,}000$, showing a fairly smooth transition between (hindered) settling and consolidation. The small irregularities in the Ξ_s-curve around $c = 100$ g/L only slightly affect the settling/consolidation behavior, owing to the abrupt transition between the settling and consolidation formulations in Eq. (5.34). It is stressed that no unique value for η exists. Thus η has to be found by trial and error until

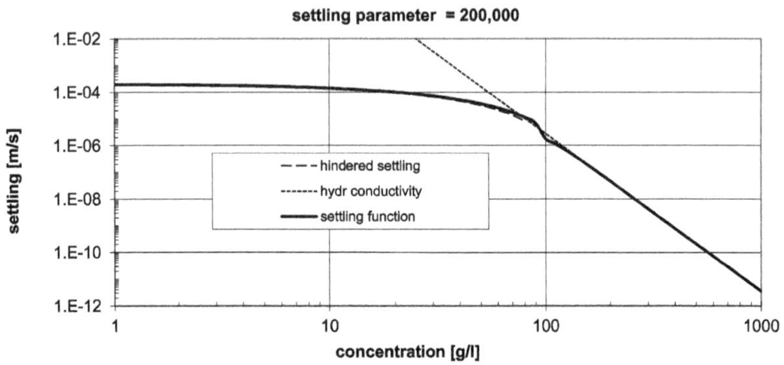

Fig. 5.6. Example of settling function Ξ_s, with $w_{s,0} = 0.2\,\text{mm/s}$, $c_{gel} = 120\,\text{g/L}$, $n_f = 2.71$ and $K_k = 1.0\ 10^{-14}\,\text{m/s}$.

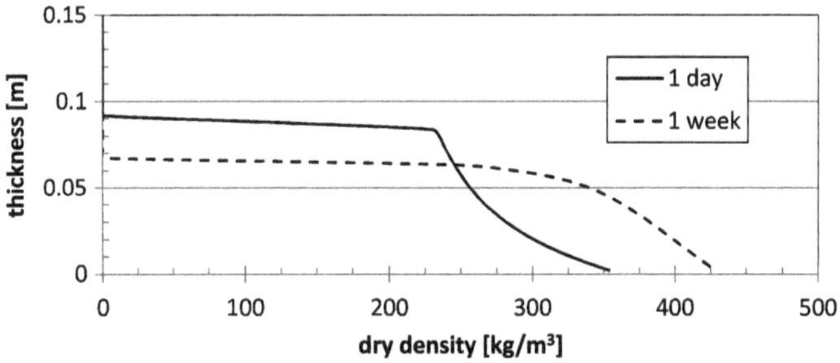

Fig. 5.7. Computed vertical concentration profile after one day and one week (almost equilibrium), showing development from a concave-up profile to a convex-up dry-density profile.

a smooth relation for the transition between settling and consolidation is obtained.

Figure 5.7 presents the results of a 1DV consolidation simulation, showing vertical dry density profiles after one day, i.e. during the initial phase of consolidation (see also Fig. 4.19), and the profile after one week, i.e. almost at equilibrium. The first profile has a concave-up shape. This is the result of the self-weight consolidation process: the near-bed sediment is compacted first, and all pore water squeezed out has to pass through the sediment above. Consolidation of the upper part of the bed is therefore hindered. Mathematically, this concave-up profile follows from the solution of the hyperbolic Kynch equation (4.38). The second profile has a convex-up profile (parabolic consolidation equation), which follows from the equilibrium solution of Eq. (5.9). Note that Gibson's equation predicts that this equilibrium solution is independent of the initial conditions — this has important implications for the consolidation of slurries, as discussed in the next sections.

Normal consolidation and the CSM

A bed that is formed by settling from a dilute suspension ($c_0 < c_{gel}$) and by self-weight consolidation, as described above, is referred to as a "virgin bed" and the state of consolidation is referred to as "normal consolidation". At equilibrium, the strength of a soil at any location within the bed equals the load carried by the soil, as sketched in Fig. 5.5.

Further to Secs. 2.3 and 5.1, this equilibrium state is represented by the K_0-line in the critical state model, see Fig. 5.8 (Schofield and Wroth, 1968). The subscript $\cdot_0$ refers to zero horizontal strains, and K_0 is also known as the coefficient of earth pressure at rest. However, the K_0-line does not collapse onto the virgin compression line, because of internal friction in the consolidating soil ($q \neq 0$). This is further quantified in the empirical

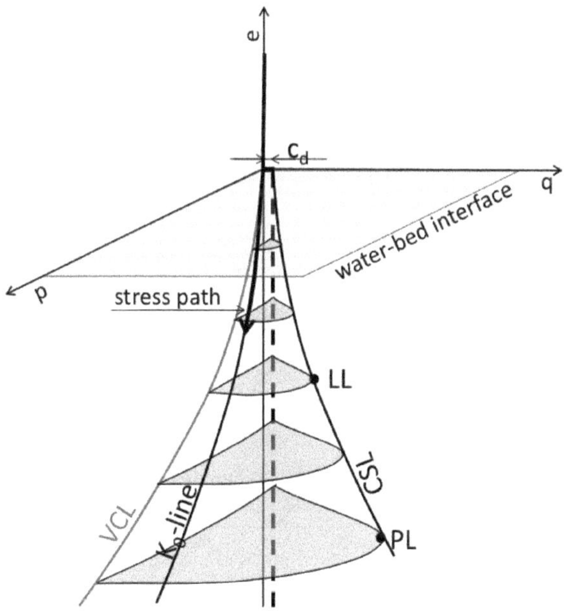

Fig. 5.8. Consolidation depicted as a stress path by the solid arrow in the CSM (see also Sec. 2.3) with the K_0-line indicating natural consolidations. Beyond the point of contraction, effective stress develop, i.e. $q > 0$.

relation Eq. (5.35), from Schofield and Wroth (1968):

$$K_0 \equiv \frac{\sigma_{xx}^{sk} \wedge \sigma_{yy}^{sk}}{\sigma_{zz}^{sk}} \approx 1 - \sin\phi_d \tag{5.35}$$

where ϕ_d is the internal friction angle under drained conditions. For $\phi_d =$ 10–30°, we find horizontal stresses of about 20–50% of the vertical stress σ_z. Equation (5.35) suggests that $K_p/K_y = K_0$, i.e. Eqs. (5.11) and (5.31). Indeed, Winterwerp and Van Kesteren (2004) found values of K_y about 30–50% smaller than K_p.

The K_0-line is found at the wet side of the critical state line: during consolidation (i.e. plastic deformation), water is squeezed out of the soil.

Figure 5.9 shows a two-dimensional representation of the CSM. The K_0-line is parallel (see also Sec. 5.2) to and in between the VCL- and CSL-lines. The slope of all these lines represents the compression coefficient C_c. As discussed in Sec. 2.3, swelling (i.e. the uptake of water by elastic rebound) would occur at a smaller rate, following the swelling line (SL), with a slope C_s, the swelling coefficient.

The stress path for self-weight consolidation, starting from a dilute sediment–water mixture ($c_0 < c_{gel}$) is also sketched in Fig. 5.9. Effective stresses in the dilute suspension are zero, and the stress path only becomes

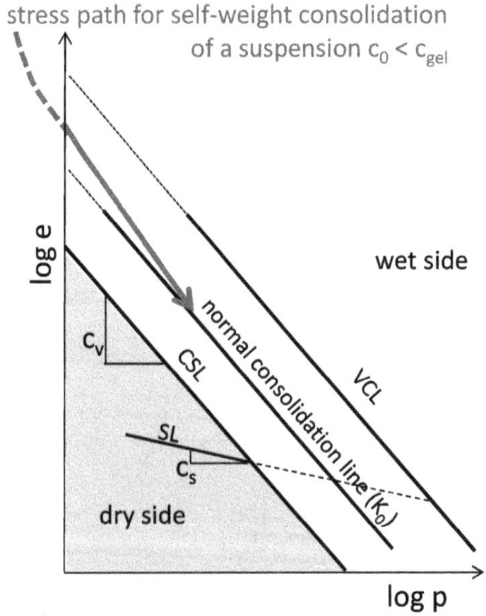

Fig. 5.9. Two-dimensional representation of stress path for normal consolidation.

meaningful when some effective stresses have developed, i.e. beyond the point of contraction. The p–e relation gradually approaches the K_0-line. Deeper in the soil, larger p at smaller e develop.

The K_0-line thus represents all stress–volume conditions in a normal consolidated bed. This line is unique for different soils, thus sediment composition (clay minerals and content, organic matter, etc.). However, the idealized CSM predicts that a K_0-line always runs parallel to the CSL for a specific sediment, as discussed in the next section.

5.2 Over-consolidation and stress history

In nature, normal consolidation states are rare. These may be found for instance when large amounts of sediments enter a navigation channel, forming fluid mud (Chapter 7). In general, mud deposits are in a so-called over-consolidated state. This state is best understood by analyzing the response of a mud sample in an oedometer (incremental loading test), as sketched in Fig. 5.10. In an oedometer, a homogeneously mixed mud sample is placed in between two filter stones, and compressed under drained conditions, while measuring stresses and compaction (settling). After equilibrium is attained, the compression load can be increased (loading step) or removed/reduced (unloading step).

State A in Fig. 5.10 represents the initial, undisturbed and drained configuration. In state B, a load σ is supplied, compacting the sample, while generating pore water over-pressures. This compaction consists of an elastic and a plastic, irreversible component (Schofield and Wroth, 1968). At equilibrium, when excess pore water pressures are fully dissipated, the

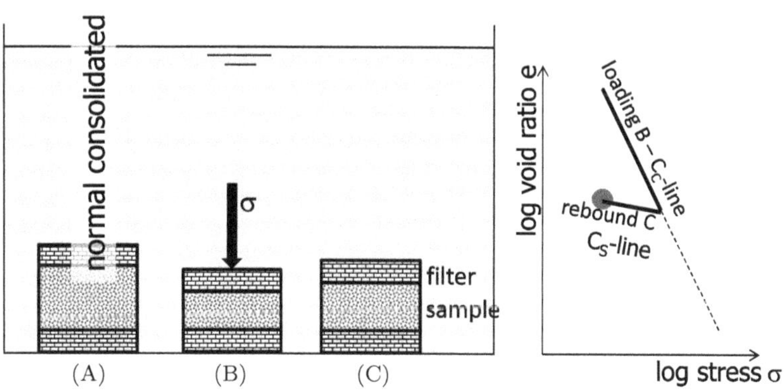

Fig. 5.10. Schematic of loading and unloading with an oedometer.

stress condition is found on the loading (compression) line indicated in the stress–volume diagram in the right panel. The slope of this line in a log-linear diagram is known as the compression index C_c, and we continue to use this terminology in our modified scheme with power-law relations. As $p_e = 0$ at equilibrium, the effective stress in the sample thus becomes equal to the applied load minus the hydrostatic pressure:

$$\sigma_{zz}^{sk} = \sigma - p_h \tag{5.36}$$

Because water is assumed to be incompressible, the rate at which equilibrium is attained depends primarily on the permeability of the soil sample.

In state C, the load σ is removed. This induces a water under-pressure in the sample, initially equal to the previously applied load:

$$p_e + p_h = -\sigma_{zz}^{sk} = -\sigma + p_h \tag{5.37}$$

Thus a pseudo-cohesive state is generated (Sec. 2.2). Dissipation of this under-pressure involves inflow of water, and subsequent expansion (elastic rebound) of the sample — in soil mechanics, one refers to swelling. As compression involves plastic deformations as well, the swelling rate is generally (much) smaller than the compression rate. This swelling is reflected by the rebound line C in the right panel of Fig. 5.10, which has a (much) smaller slope than the compression line, as argued above. In a log-linear diagram, this slope is the swelling index C_s. The ratio C_s/C_c generally lies between 1/7 and 1/3 (Lambe and Whitman, 1979). The dissipation of pore water excess pressures can be described with Terzhagi's model (Eq. (5.15)). Note that the Gibson model (Eq. (5.9)) is not suitable to model expansion of the soil, as it does not contain an elastic component. The equilibrium stress (solid point in Fig. 5.10) upon swelling is not known in advance, and has to be established experimentally.

Note that if loads become very large, the permeability may decrease so much, that water under-pressures upon unloading is maintained for a very long time (the tar sands in Canada are an extreme example, thus centuries).

The state of over-consolidation is defined in soil-mechanics by the over-consolidation ratio (OCR):

$$\mathrm{OCR} = \frac{p_{K_0}}{p} \tag{5.38}$$

where p_{K_0} is the maximal achieved isotropic stress for normal consolidated soil (stress state along the K_0-line), and p the actual isotropic stress. For over-consolidated soils, OCR > 1, which thus reflects the stress history of the soil.

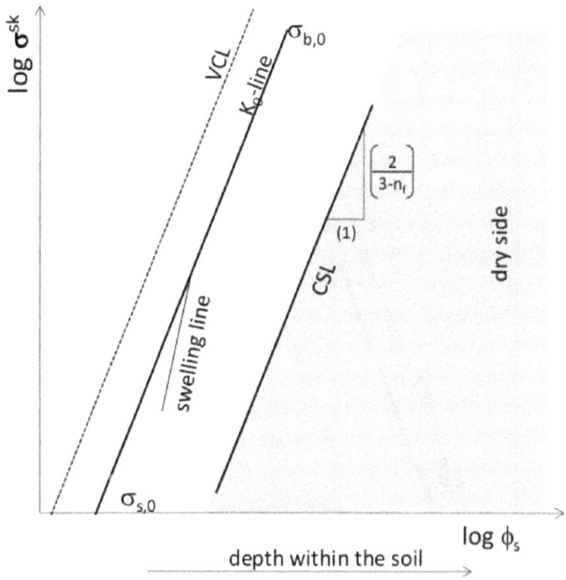

Fig. 5.11. Stress state of Fig. 5.10 in a σ_{zz}^{sk} vs. ϕ_s diagram.

Over-consolidation and self-similarity

As the mechanical behavior of soft soil can be described with fractal (power law) relations, the stress–volume diagram of Fig. 5.10 is redrawn in Fig. 5.11 as a $\sigma^{sk} - \phi_s$ double logarithmic diagram. The virgin consolidation line (VCL), the normal consolidation line (K_0-line) and the critical state line (CSL) appear as straight, parallel lines (see also Sec. 2.3). These lines are fairly steep, as their slope is controlled by the exponent of Eq. (5.11), with $n_f \approx 2.7$–2.8. Also the swelling line is drawn, with an even steeper slope.

The vertical effective stress distribution of the normal consolidation density profile of Fig. 5.5 is reflected by the K_0-line in Fig. 5.11. The horizontal ϕ_s axis therefore reflects also depth within the soil, and $\sigma_{s,0}$ and $\sigma_{b,0}$ represent the effective stresses at the bed surface and its base. At the soil surface, $\sigma_{s,0} = 0$, which, of course, cannot be represented properly in this double logarithmic diagram.

Next, we analyze the case that all sediment above z_1 in Fig. 5.5 is removed (indicated by A), while the sample remains under water at constant water level. As a result, the total load σ_{zz} reduces, while the effective stress σ_{zz}^{sk} initially remains constant. Equation (5.1) then predicts the development of a water under-pressure p_e, as also the hydrostatic water pressure

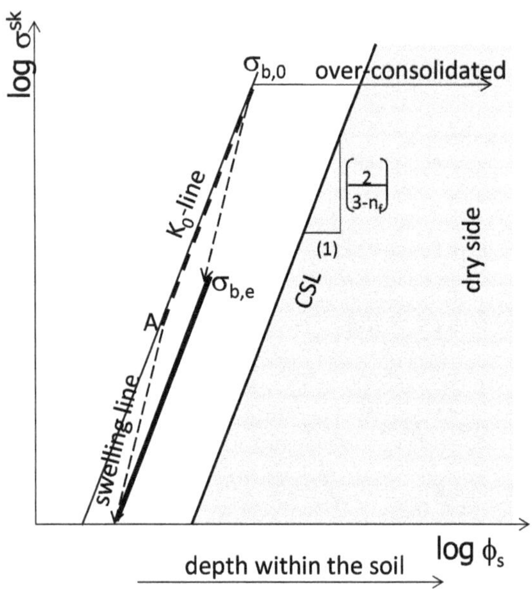

Fig. 5.12. Schematic stress path towards an over-consolidated state by load removal.

p_h remains constant. Therefore, the sample will swell along the swelling lines in the stress diagram of Fig. 5.12. Because all swelling lines run parallel, the new stress state at equilibrium runs parallel to the original K_0-line, but to the right. As the total load σ_{zz} reduced, the effective stresses at equilibrium (dissipation of p_e) are smaller than initially ($\sigma^{sk}_{zz,e} < \sigma^{sk}_{zz,0}$), while at the soil surface, these remain zero.

Also if the sample would be dried, under-pressures (suction) are induced in the soil. Assuming that full-saturated conditions are maintained, the development of under-pressure induces an increase in effective stress (Eq. (5.1); pseudo-cohesion), while the solids concentration ϕ_s increases. The underlying stress path in Fig. 5.13 is different from the one induced by load removal, as the swelling lines are not followed. If the under-pressures are not compensated by the inflow of water, effective stresses remain higher than the initial ones. However, if to-flow of water would be permitted, the soil would swell again along the swelling lines — the thick line of Fig. 5.13 would lower and shift to the right, with $\sigma^{sk}_{zz,0} = 0$.

Hence drying and load removal both result in over-consolidated conditions, with larger ϕ_s at the same effective stress. This implies a shift along

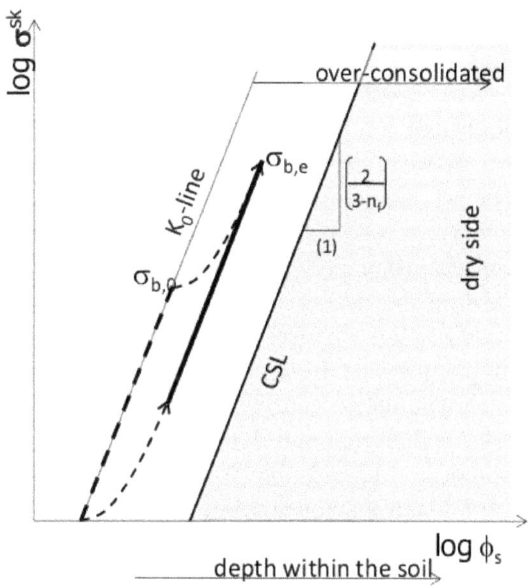

Fig. 5.13. Schematic stress path towards an over-consolidated state by drying.

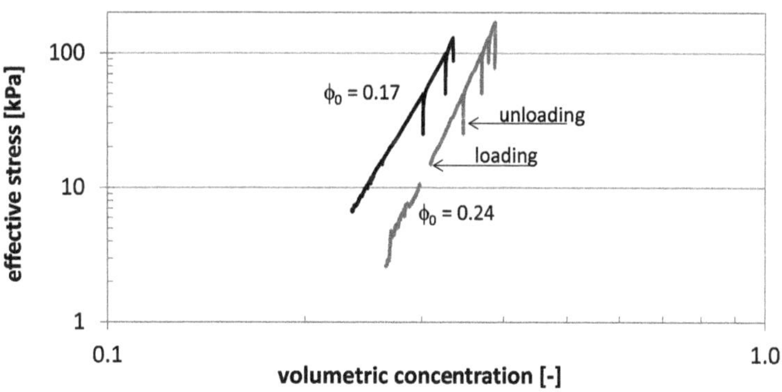

Fig. 5.14. Stress–volume relation for loading–unloading cycles in a CSR-instruments for two different initial water contents (courtesy Barciela Rial, 2019).

the e-line to smaller e (larger ϕ_s) in the CSM in Fig. 2.24, thus larger yield planes and an increase in strength.

This is further illustrated in Fig. 5.14, showing the results of two experiments on a mud sample from the fresh water Lake Markermeer, carried out with a so-called constant strain rate (CSR) instrument Barciela Rial (2019).

Contrary to the oedometer, the (homogeneously mixed) sample is squeezed by a piston at constant rate (here 0.0012 mm/min), while total stress and pore water pressures are measured. The CSR-test is carried out under drained, non-equilibrium conditions, and the effective stress is determined from the difference between the total stress and the excess pore water pressure, the latter being non-zero. The first experiment was carried out at *in-situ* water content, on a homogeneously mixed sample, yielding an initial volumetric concentration $\phi_0 = 0.17$. Consistent with the assumption of self-similarity, the data shows power law behavior. We have no information whether this sample was initially over-consolidated or not, but likely it was.

Next, this soil sample was dried at room temperature, then remolded (yielding an initial volumetric concentration $\phi_0 = 0.24$), and a second CSR-test was performed. Again the data show power law behavior, running parallel to the $\phi_0 = 0.17$ data, but to the right. The small deviation at the start of the experiment may be attributed to experimental inaccuracies. Thus drying results in an over-consolidated state, as mentioned above.

During the experiments on both samples, the piston was released a number of times, as depicted in Fig. 5.14. The effective stress reduced considerably, at almost constant ϕ_s though. As the experiments were stopped before p_e became negative, p_e remained positive throughout the tests. The (almost) vertical unloading response, therefore, reflects elastic effects only. These are very small though (see Barciela Rial (2019) for details) for this soft sediment. Swelling due to dissipation of pore water under-pressures therefore did not occur.

The above has important consequences for assessing material parameters. As an example, data by Barciela Rial (2019) are presented in Fig. 5.15,

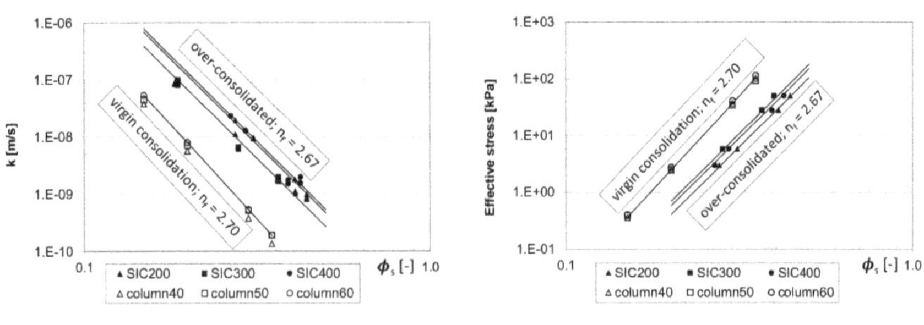

Fig. 5.15. Hydraulic conductivity (left panel) and effective stress (right panel) measured with settling column and SIC tests, indicating initial concentration for each tests (courtesy Barciela Rial, 2019) — see also the end of this chapter.

showing the results of the hydraulic conductivity k and effective stress $\sigma^s k_z z$ measured in a settling column (see Appendix A.8) and SIC-instrument, again on samples from Lake Markermeer. Though these methods may induce some differences in the k- and $\sigma^{sk}_{zz,0}$-values, the major differences in the measured parameters (order of magnitude) are to be attributed to the differences in initial conditions, i.e. the stress history. The settling column tests reflect normal consolidation conditions with initial concentrations below the gelling point (i.e. $C_0 = 40$, 50 and 60 g/L), whereas the SIC-tests were carried out on over-consolidated samples, with initial dry densities of 200–400 g/L, obtained by gently drying Lake Markermeer mud samples.

These results can be understood from a virtual experiment in which the sediment mass above level "A" in Fig. 5.5 is removed, while the water level is kept constant, as before. As the total stress σ_{zz} is reduced by the mass removed, while the total hydrodynamic stress p_h does not change, the effective stress σ^{sk}_{zz} reduces by the load removed, further to the effective stress equation (5.1). Hence, the effective stress line shifts to the right, or down, as in Fig. 5.15.

Upon removal of the load, the remaining sediment will undergo some elastic rebound (swelling). As a result, the structure of the soil opens a bit, which is reflected by the slightly smaller fractal dimension measured on the over-consolidated sample, i.e. $n_f = 2.67$ vs. 2.70 for the normally consolidated sample, Fig. 5.15. This would imply an increase in hydraulic conductivity, as measured. The effect is quite large, owing to the nonlinearity of the material parameters (Eq. (5.10)).

The role of sand and organic material

Figure 5.16 presents two series of experiments on other samples from Lake Markermeer — these experiments were carried out with an oedometer (Barciela Rial, 2019). In the first series, various amounts of sand were added to a mother sample (0, 10, 20, 40 and 70%), while the initial gravimetric water content with respect to the $< 64\,\mu$m fraction was kept constant at around $W_0 \approx 1.4$.

When plotted against the effective volume concentration $\phi_{s,eff}$ (see Eq. (5.23)), the data collapse at a slope $n_f = 2.8$, except for the 70%-sand data.

Next, the organic matter of the 40%-sand sample was removed by heating (105°C in oven, referred to as "Treated"), after which the sample was rewetted to $W_0 = 0.9$ and 0.7. The results of the oedometer tests on these two samples are also presented in Fig. 5.16, again showing power-law

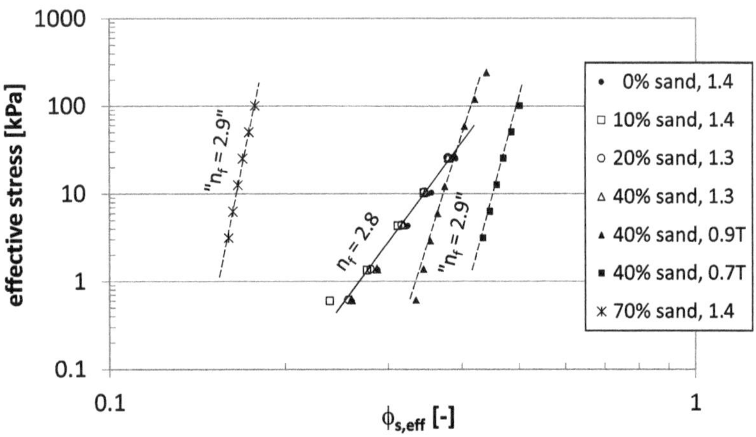

Fig. 5.16. Stress–volume relations measured in oedometer for fresh water sediment at various sand contents, initial water content (1.4, 0.9, etc.) and removal of organic matter (T) (courtesy Barciela Rial, 2019).

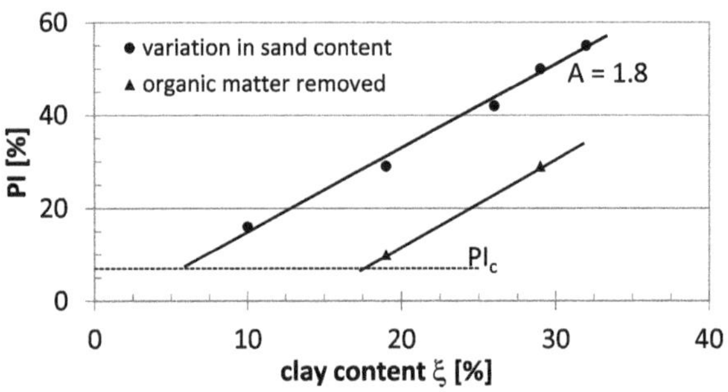

Fig. 5.17. Activity of fresh water sediment (courtesy Barciela Rial, 2019).

behavior, at much steeper slopes though, with $n_f \approx 2.9$. It is remarkable that these lines run parallel to the 70%-sand results.

The results of the 70%-sand and 40%-sand&treated samples can be understood from the activity plot in Fig. 5.17. The upper activity line reflects the samples with variations in sand content, showing a critical clay content $\xi_0 = 6\%$ (Sec. 2.2), and an activity $A = 1.8$, implying the presence of montmorillonite.

The activity line for the treated samples (removal of organic matter, drying and rewetting) reveals the same activity, but runs further to the right with $\xi_0 = 17\%$.

Further, Barciela Rial (2019) found a silt–clay ratio of 2 for this sediment, which implies cohesive behavior for sand contents below 80% for the untreated and 50% for the treated samples. In other words, the 70%-sand and 40%-sand and treated samples are close to the transition between cohesive and non-cohesive behavior. This also explains why n_f is so high ($n_f = 3$ for granular material), though the power-law behavior for these sandy samples itself is remarkable.

These results suggest that the activity is still governed by the clay minerals in the soil, but that the transition between cohesive/non-cohesive behavior depends strongly on the organic matter in the soil.

Intertidal mudflats

Intertidal mudflats host important ecosystems (Chapter 8) and play a critical role in estuarine and coastal fine sediment dynamics by influencing residual transport (e.g. Sec. 9.2.3), and by temporarily or permanently sequestering fines. A fair amount of field research related to fine sediment transport takes place on mudflats, such as the erodability of *in-situ* deposits.

The dynamics of groundwater flow in intertidal flats has been studied extensively through field and laboratory experiments, and numerical modeling, in particular for sandy substrates, e.g. beaches (Brakenhoff *et al.*, 2019; Mango *et al.*, 2004; Pollock and Hummon, 1971; Riedel *et al.*, 2010). However, the relevant soil mechanics processes have not been addressed. Although the considerations in the previous sections may shed some light on these processes, quantitative information is scarce and difficult to acquire (see also Sec. 6.3.1). The soil mechanical analysis below is therefore hypothetical pending validation.

We start this analysis with a hypothetical intertidal mudflat in soil mechanical equilibrium, i.e. at final density and strength profiles. The tide-induced pressure head fluctuations induce a down-slope residual pore water flow, as sketched in Fig. 5.18. The phreatic level within the mudflat, for

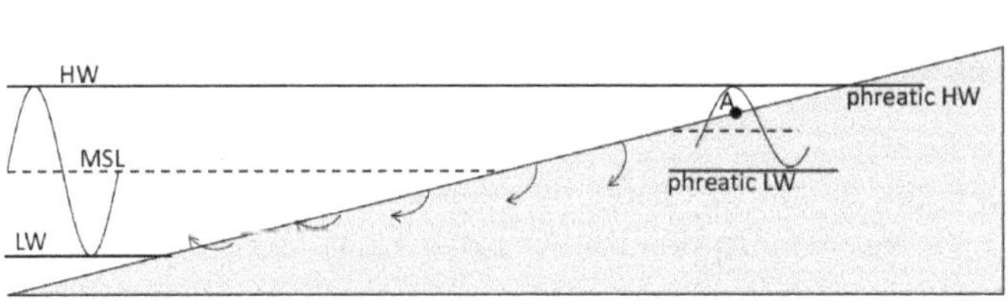

Fig. 5.18. Schematic of tide-induced variations in phreatic level within intertidal mudflat with residual pore water flow.

instance at location A, is expected to follow the tidal water level variations, but at smaller amplitude, and delayed in time because of the low permeability of the flats substrate. Moreover, the suction head rises up-flat owing to the asymmetry in ingress and egress of pore water Riedel *et al.* (2010).

Next, it is assumed that a few cm of fresh mud is deposited on this mudflat during high tide. During falling tide, this deposit emerges, while the phreatic level lowers below the mudflat surface, inducing capillary suction (under-pressures), as depicted in Fig. 5.3 (right panel). These under-pressures induce additional compaction of the soil beyond its virgin consolidation state: the soil becomes over-consolidated. This compaction/ consolidation can be quantified with Gibson's equation. At equilibrium, Eq. (5.2) reduces to

$$\phi_s\left(\frac{\rho_s-\rho_w}{\rho_w}\right)=-\frac{1}{g\rho_w}\frac{\partial\sigma_{zz}^{sk}}{\partial z} \tag{5.39}$$

Substituting from the fractal model (5.11),[1] Eq. (5.39) can be integrated from z to Z_s:

$$\phi_s(z)=\left[\frac{n-1}{n}\frac{(\rho_s-\rho_w)\,g}{K_p}(Z_s-z)+\left(\frac{\rho_w g\Delta h}{K_p}\right)^{\frac{n-1}{n}}\right]^{\frac{1}{n-1}} \tag{5.40}$$

using the boundary condition

$$\sigma_{zz}^{sk}\Big|_{z=Z_s}=\rho_w g\,(Z_s-Z_f)=\rho_w g\Delta h \tag{5.41}$$

where $n=2/(3-n_f)$ and Δh is the suction head between the soil surface Z_s and the phreatic level Z_f (Fig. 5.3). The final equilibrium thickness h_{end} of the deposit is obtained from the integration of Eq. (5.40) over the thickness of the deposit, which yields:

$$h_{end}=\frac{n}{n-1}\frac{K_p}{(\rho_s-\rho_w)\,g}\left[\left(\frac{\rho_w g\Delta h}{K_p}-\frac{(\rho_s-\rho_w)\,g}{K_p}\zeta_m\right)^{\frac{n-1}{n}}\right.$$
$$\left.-\left(\frac{\rho_w g\Delta h}{K_p}\right)^{\frac{n-1}{n}}\right] \tag{5.42}$$

This converges to the solution by Merckelbach and Kranenburg (2004a) for $\Delta h=0$, see also Appendix A, Eq. (A.14). As an example, the final

[1]In the following, it is presumed that the soil structure, i.e. n_f, does not change during further compaction by suction nor evaporation.

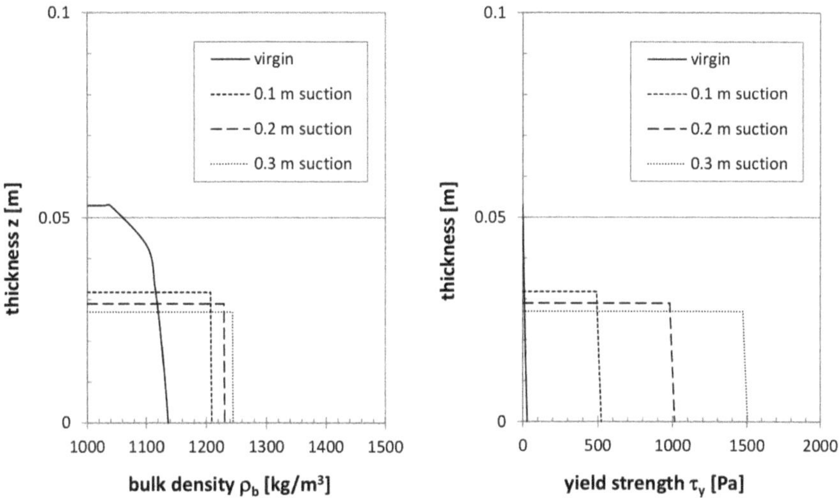

Fig. 5.19. Equilibrium density and strength profile as function of suction head Δh.

compaction state of a thin mud deposit is elaborated, assuming that capillary suction increases up-flat. The results are presented in Fig. 5.19, using the following parameters: $n_f = 2.7$, $K_p = 10^9$ Pa, $K_k = 10^{-13}$ m/s, thus $\Gamma_c = 5 \cdot 10^{-8}$ m^2/s, and further assuming $K_y = 0.5K_p$ and a Gibson height $\zeta_m = 2$ mm (implying a few cm thick deposit).

Apparently, a small under-pressure of a few 0.1 m has a significant effect on the density and strength profile, whereas the vertical density gradients almost disappear. The latter can be understood from the first term in the right-hand side of Eq. (5.42), which is much larger than the second one (with the ζ_m-term). The permeability of the soil decreases by an order of magnitude from $k = 1.3 \cdot 10^{-7}$ m/s for normally consolidated conditions to $k = 3.0 \cdot 10^{-8}$ m/s for $\Delta h = 0.3$ m. The consolidation time for this layer amounts to about $\delta^2/\Gamma_c = 0.05^2/5 \cdot 10^{-8} \approx 14$ hrs, hence about two low water cycles of a semi-diurnal tide. Note that the suction pressure depend on the tidal range, and can therefore be much larger than in this example.

As the entire mudflat, built up over time, thus has experienced larger or smaller capillary suction, one may conclude that the entire mudflat above LLWS is in an over-consolidated state. As the capillary rise in fine-grained mud flats can be several meter, the mudflat can remain fully saturated.

However, evaporation may induce non-saturated conditions, as encountered frequently in the field, in particular higher on the intertidal.

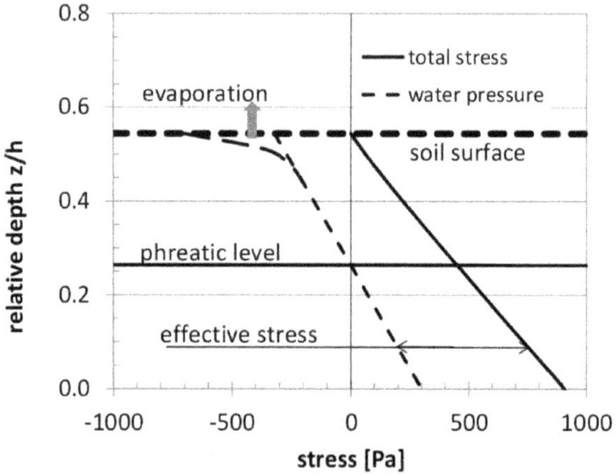

Fig. 5.20. Schematic of pore water and effective stress induced by phreatic level and evaporation.

Evaporation induces additional suction (Fig. 5.20). The potential evaporation rate $V_{e,p}$ [m/s] can be established to first order with the Harbeck and Meyers (1970) equation:

$$V_{e,p} \approx (25 + 19V_2)\,\frac{\zeta_{v,s} - \zeta_v}{\rho_w} \tag{5.43}$$

where ζ_v is the local humidity (mass vapor/mass dry air), $\zeta_{v,s}$ its value for saturated conditions, and the first term of the left-hand side is an empirical term accounting for ventilation by wind, with V_2 [m/s] the wind velocity (at 2 m height, with a correction for hours → seconds). The humidity for saturated conditions is a function of temperature, and varies between 0.005 and 0.020 $\mathrm{kg_v/kg_a}$ in the 5–25°C range. Thus, for wind speeds between 5 and 20 m/s (Bft 3–8), $V_{e,p} \approx 10^{-7}$ m/s, assuming the higher wind speeds at lower temperature, and $\zeta_v = 0.8\zeta_{v,s}$, characteristic values over wetlands in temperate climate zones. This would yield about 1 mm water loss after three hours of evaporation, inducing further compaction of the deposit.

The evaporation-induced suction pressure varies over the depth within the mudflat, affected by evaporation. Hence the 2.7 cm thick denser over-consolidated deposit of Fig. 5.19 would have to thin by 1 mm to supply the evaporated water. Its density would then increase from 1244 to 1254 kg/m^3, i.e. ϕ_s would increase from 0.148 to 0.154. Using Eqs. (5.11) and (5.10), this would imply an increase in effective stress by almost 900 Pa, and a reduction in hydraulic conductivity from $3.3 \cdot 10^{-8}$ to $2.6 \cdot 10^{-8}$ m/s. Thus evaporation has two consequences:

(1) The near-surface additional compaction forms an over-consolidated "crust" atop the mudflat, with an additional strength of several 100 Pa during low water. By comparison, flow- and wave-induced bed shear stresses amount to a few Pa only.
(2) This crust also reduces the near-surface permeability, reducing the actual evaporation and consolidation rates, and increasing the air-entry value. The mudflat remains saturated and wet for a long time.

Higher on the intertidal, above MHW, mud deposits tend to dry out, particularly during summer. This is characterized by a hard crust crisscrossed by prominent cracks forming a panther-skin pattern. From a fully saturated condition, the deposit dewaters and desiccates, while air fills (part of) the pores. This three-phased drying of soil is traditionally modeled with the so-called soil-water characteristic curve (SWCC, see Van Genuchten, 1980) and the shrinkage curve (SC, see Fredlund *et al.*, 2002; Vardon *et al.*, 2015) in conjunction with appropriate boundary conditions describing evaporation rates. A limitation is that these models are highly empirical, and have never been applied to thin, loose intertidal deposits.

The effects of biota can be critically important; in summer, large matsses of microphytobenthos develop, stabilizing the deposits by secreted EPS. The role of microphytobenthos on erodability has been investigated in Sec. 8.1. However, they also affect pore water dynamics and evaporation rates, thus affecting soil processes and consolidation rates. The authors are not familiar with studies on these processes.

Higher on the intertidal, vegetation emerges and affects the stability of deposits in various ways (Barciela Rial, 2019):

(1) roots withdraw water from the soil, enhancing dewatering and compaction;
(2) roots provide armoring to the soil, enhancing soil strength, thereby reducing/preventing compaction;
(3) degenerated roots provide drainage channels within the soil, enhancing dewatering;
(4) vegetation biomass induces an additional load (surcharge) on the soil, enhancing compaction.

It is not known which process dominates and under what conditions. In general, vegetated mudflats in temperate zones tend to be stable, and may erode only through wave-induced mass erosion (cliff erosion), as discussed in Sec. 6.4.

Summary on stress history

The oedometer and CSR- and SIC-instrument, and the underlying theory have been developed for soils much denser and at much higher stress levels than generally occurring in open water. The use of these instruments for soft soils should therefore be analyzed with care. Yet, important conclusions can be drawn:

(1) Natural soils, particularly on the intertidal are generally in an over-consolidated state.
(2) Such an over-consolidated state can be attained by removal of over-burden, by drying and/or by evaporation.
(3) Removal of over-burden induces water under-pressures within the soil, while drying does not.
(4) Drying (at fully saturated conditions) induces plastic deformations, but the soil floc structure is not largely altered.
(5) Though we have little data only, the activity of a soil appears to be unaffected by organic matter. However, the critical clay content, reflecting the cohesive/non-cohesive transition appears to be strongly affected by organic matter. This might imply seasonal variations, at least in temperate climates.
(6) The cohesive/non-cohesive transition is also reflective for the degree of plasticity of a soil, because of the water-binding capacity of organic matter. For determining the Atterberg limits, drying of natural soils may therefore not be done by heating, as in that case organic matter will be affected, as indicated in Appendix A.
(7) Stresses within a normal consolidated soil follow the K_0-line. Undisturbed samples from the field and/or mixing (see also Sec. 5.3) prior to testing generally exhibit over-consolidated characteristics. Though the self-similar structure would remain unaffected (constant n_f), the stress levels themselves are likely seriously affected, modifying the K_p- and K_y-parameters. This implies that it may not be straightforward to use results from e.g. oedometer tests to predict the consolidation behavior of soft soils.
(8) Over-consolidation and the generation of pore water under-pressures (pseudo-cohesion) may strongly affect the erodability of a soil, as discussed in Sec. 6.3.

(9) Soil mechanical theory on over-consolidation may be applicable to determine the degree of over-consolidation of soft soils as well (OCR, Eq. (5.38)), which may then be used to obtain the relevant material parameters at low (normal-consolidation) stresses. However, this theory has to be elaborated upon, and tested for soft this mud layers, which has not yet been done, to the knowledge of the authors.

Understanding the features of over-consolidation is in particular important when dealing with soft soils, as discussed in the next section.

5.3 Consolidation of slurries

Let us mix a normal consolidated layer of soft sediment homogeneously over its thickness, as in Fig. 5.21, thus at constant water content. The upper part then becomes over-consolidated, as it is denser/stronger than necessary to carry the sediment mass above. The resulting mixture is referred to as a slurry. In the following, all mixtures with a mean initial layer-averaged dry density beyond c_{gel} are referred to as slurries.

In the upper part of the slurry, the density is now above the equilibrium density, hence the strength of the soil exceeds the mass to be carried: this part of the soil becomes over-consolidated. The lower part of the soil, however, returns to a consolidating state, as its density is below equilibrium.

This is further illustrated in Fig. 5.22, showing the total, hydrostatic and effective stress distribution in the normal consolidated soil of Fig. 5.21. After vertical mixing, maintaining the overall water content, the effective stress becomes homogeneous over the vertical. Thus pore water under- and overpressures are induced in the upper and lower part of the soil, respectively, owing to:

$$p_e = \sigma_{zz} - \sigma_{zz}^{sk} - p_h \tag{5.44}$$

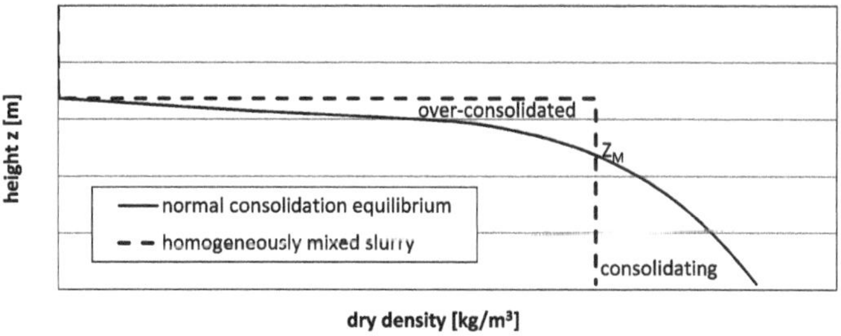

Fig. 5.21. Preparation of slurry from consolidated suspension at equilibrium.

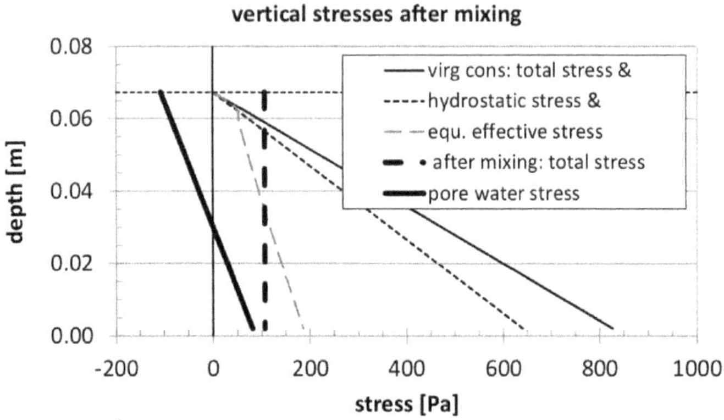

Fig. 5.22. Vertical stress distribution in normal consolidated soil and a slurry directly after mixing.

This stress state directly after mixing would be reflected by a dot in the CSM-diagram, as indicated in Fig. 5.23. Upon further consolidation, the lower part of the soil would follow the stress path along the K_0-line to larger stresses, whereas the upper part would follow the swelling line. Thus, the overall stress state in no longer parallel to the K_0-line, thereby changing the fractal dimension as well.

Also the hydraulic conductivity in the upper over-consolidated part of the slurry will be lower than for normal consolidated conditions. This implies that this upper part acts as a crust with respect to the consolidating lower layer, and settling curves will become more or less linear.

A remarkable example of the role of crust formation was found by Te Slaa (2020), who carried out settling and deposition experiments with graded silts ($D_{10} = 13\,\mu$m; $D_{50} = 36\,\mu$m; $D_{90} = 75\,\mu$m) in a settling column. Te Slaa measured vertical profiles of excess pore water pressure over a period exceeding 200 hrs, as shown in Fig. 5.24. The measured vertical p_e-profiles are the result of a layer of fine silt on top of the rest of the silt deposit. This crust is formed by differential settling of the silt mixture. Pore water pressure gradients equalize over depth below this crust (irregular patterns are due to measuring inaccuracies).

In practice, slurries are made by e.g. mechanically diluting and pumping of soils, as common in the dredging industry. Such mechanical treatment will break up the floc structure, and the solid point of Fig. 5.23 will move

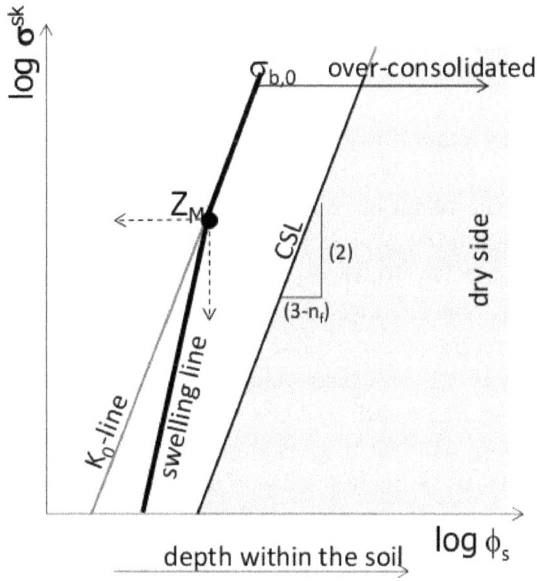

Fig. 5.23. Initial vertical stress distribution in a slurry (solid point) and some time after mixing, thus in the consolidation phase. Floc breakup by mechanical mixing and dilution by adding water are indicated by dotted arrows.

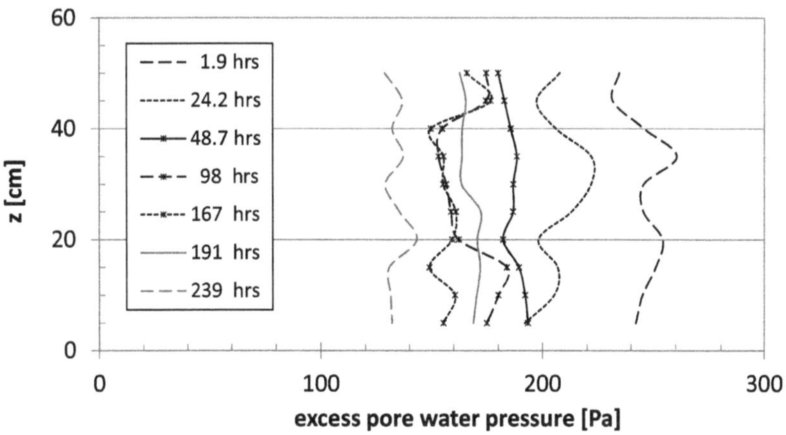

Fig. 5.24. Vertical excess pore pressure profiles in a compacting graded silt mixture. The crust is very thin (mm's) and not visible in the data (courtesy Te Slaa (2020).

down, at constant ϕ_s. Note that the K_0-line for this mechanically stirred slurry will also change, likely becoming steeper as well. Dilution would move the initial point to the left.

The 1DV-slurry consolidation model

In the following, we elaborate further on the consolidation of slurries using the 1DV-consolidation model, which has to be modified for the upper, over-consolidated part (for the original 1DV consolidation model, see Winterwerp and Van Kesteren, 2004). Dissipation of pore water under-pressures can be modeled with Terzaghi's equation, which is obtained by omitting the advection term (second term) of Eq. (5.9), bearing in mind that the porosity $n = 1 - \phi_s$. Instead of the consolidation coefficient Γ_c, the swelling coefficient Γ_s is used.

However, water still has to pass this over-consolidated crust, and the advection term has to be maintained — if not the pore water under-pressure would only be homogenized within the slurry. The 1DV-consolidation model for slurries is therefore identical to the Gibson model, Eq. (5.9), replacing Γ_c by Γ_s in the over-consolidated zone:

(1) For normal consolidation: $\Gamma_c = \Gamma_{c,n}$, where $\Gamma_{c,n}$ is the consolidation coefficient for normal consolidating mud, as in Eq. (5.12).
(2) When over-consolidated: $\Gamma_c = \alpha_{sw}\Gamma_{c,n}$, where $\alpha_{sw}(\ll 1)$ is a coefficient representing the crust-like behavior where the slurry is over-consolidated, accounting for swelling. We refer to the 1DV-slurry consolidation model.

Prognostic 1DV simulations of consolidating slurries

Next, we discuss the consolidation of a slurry, defined as a sediment-water mixture at an initial concentration greater than the gelling concentration. We treat fully saturated conditions only, i.e. consolidation takes place under water only. We use a 1DV consolidation model, with the modifications above, solving the consolidation model Eq. (5.9) for illustrating the effect of initial concentration on consolidation. We start from virgin conditions, i.e. a dilute suspension with an initial vertically homogeneous concentration $C_0 = 60$ g/L ($< c_{gel}$); the relevant material parameters are summarized in Table 5.1. After two hours, the soil–water interface is found at 15.2 cm. The sediment profile is then homogeneously mixed over this 15.2 cm, obtaining a slurry with an initial concentration of $C_0 = 161$ g/L, see Fig. 5.25. Hence, the upper part of the slurry is in a relatively over-consolidated state, as shown in Fig. 5.25.

Table 5.1. Parameter settings for consolidation simulations.

	From virgin cons. exp.	From SIC300-test
C_0	60 g/L	300 g/L
h_0	0.41 m	
$W_{s,0}$	3.55 mm/s	
c_{gel}	80 kg/m^3	
n_f	2.70	2.67
K_k	$1.74 \cdot 10^{-13}$ m/s	$4.0 \cdot 10^{-12}$ m/s
K_p	$1.26 \cdot 10^7$ Pa	$1.2 \cdot 10^7$ Pa
$\Gamma_{c,n}$	$1.48 \cdot 10^{-9}$ m^2/s	$4.11 \cdot 10^{-8}$ m^2/s
α_{sw}	0.15	0.15
$C_{0,slurry,1}$	161 kg/m^3	
$h_{0,slurry,1}$	0.15 m	
$C_{0,slurry,2}$	373 kg/m^3	
$h_{0,slurry,2}$	0.063 m	

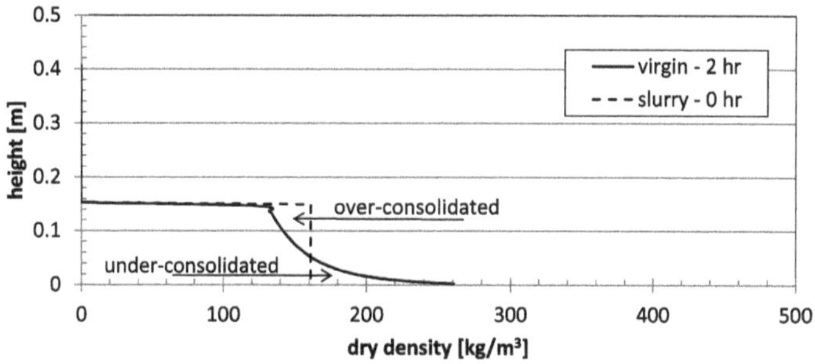

Fig. 5.25. Preparation of slurry from consolidating suspension.

Figure 5.26 shows a few computed vertical dry density profiles during the consolidation of this dilute suspension (virgin consolidation) and of the slurry. The slurry simulation starts two hours after the virgin consolidation experiment. Three hours later (i.e. $t_{virgin} = 5$ hrs and $t_{slurry} = 3$ hrs), the suspension has consolidated further. This is further illustrated by the computed settling curve, e.g. Fig. 5.27, showing some differences in the rate of consolidation for the two cases. These differences become smaller over time, being negligible after about 20 hrs. The dry density profiles at equilibrium (after about two weeks, Fig. 5.26) for the two cases are identical, as argued above (equilibrium profile independent on initial conditions).

The difference in consolidating rate for the two cases can be understood from Fig. 5.25: the upper part of the slurry is (slightly) over-consolidated, hence its permeability is smaller than under virgin consolidation. Pore

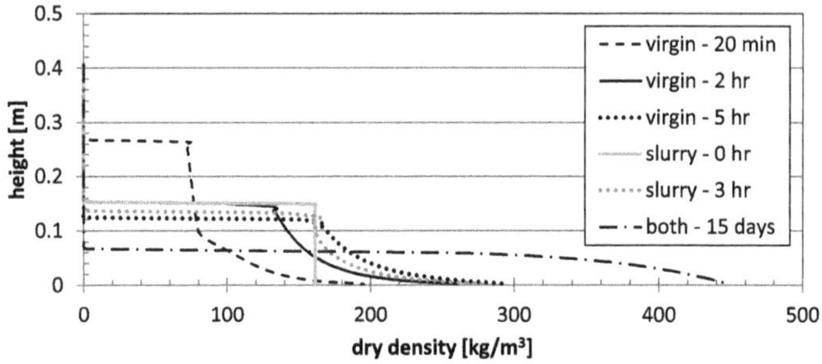

Fig. 5.26. Computed vertical density profiles for a consolidating dilute suspension ($C_0 < c_{gel}$) and for a slurry ($C_0 > c_{gel}$).

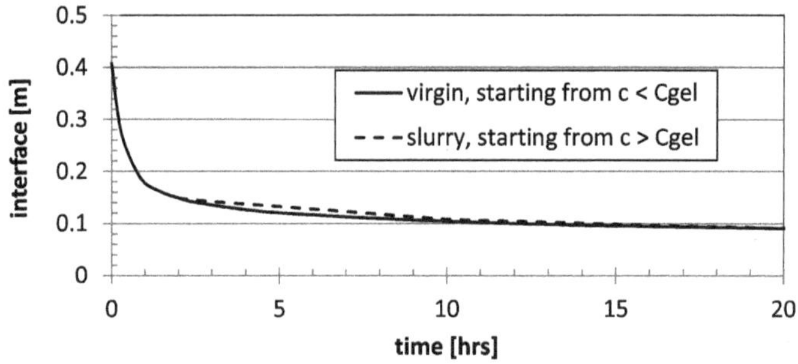

Fig. 5.27. Computed settling curves for a consolidating dilute suspension ($C_0 < c_{gel}$) and for the slurry of Fig. 5.26.

water, expelled from the lower part of the bed, escapes from the consolidating slurry at a lower rate than from the virgin consolidating bed. Only when the stress balance within the consolidating slurry is restored, further consolidation of the upper layers of the slurry is possible.

The effect of reduced permeability by over-consolidation and swelling is further shown in Fig. 5.28. This time, the slurry is obtained by mixing the equilibrium dry density profile resulting from the normal consolidation experiment over its equilibrium depth (e.g. Table 5.1). The initial dry density of the slurry then amounts to 373 kg/m^3 (i.e. $\rho_b = 1250\,\text{kg/m}^3$ at a thickness of 6.3 cm). As the upper centimeters of the slurry cannot swell towards the normal equilibrium profile, the final thickness of the slurry is smaller, at a larger dry density though. These equilibrium conditions are attained after about 25 days, considerably longer than for the virgin consolidation.

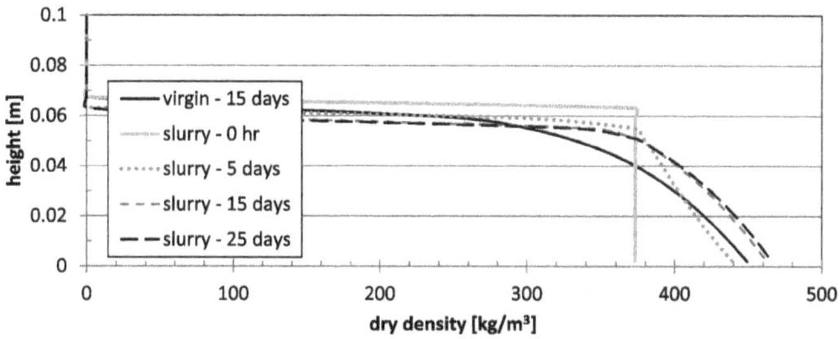

Fig. 5.28. Computed dry density profiles of a consolidating slurry, prepared by mixing the equilibrium dry density profile of the virgin bed over its thickness.

Experimental validation

Van den Bosch (2016) carried out consolidation experiments with slurries from fresh water mud from Lake Markermeer (see also Barciela Rial, 2019; Barciela-Rial *et al.*, 2020). Samples were collected from the lake, with a mud concentration of about 400 g/L. These were diluted with lake water to obtain slurries with C_0 = 300, 200, 100, 60, 50 and 40 g/L. The consolidation experiments were carried out in transparent columns of 8 cm diameter and initial suspension height of 40 cm. Settling tests on the three lower concentration suspensions were carried out to determine c_{gel} and the other material parameters given in Table 5.1 — these results are not further discussed, as they are qualitatively similar to those of Sec. 5.1 (see Van den Bosch, 2016 and Barciela Rial, 2019 for details).

The settling curves for the experiments with C_0 = 400, 300, 200 and 100 g/L are shown in Fig. 5.29. These settling curves exhibit a linear part, which is NOT to be interpreted as hindered settling. The linear part is explained from the crust formation (over-consolidation) discussed above. The (initially constant) hydraulic conductivity of this crust governs the settling rate of the slurry.

At the end of each consolidation experiment, the dry density profile was measured with an Ultrasonic High Concentration Meter (UHCM, Deltares-Berkhout, 1994). The profile after 67 days for the C_0 = 300 g/L test is shown in Fig. 5.30. Irregularities in the profile are attributed to inaccuracies of the UHCM.

Next, the consolidated mud was removed from the columns and gently homogenized. Hydraulic conductivity and effective stress where measured with a SIC-device (Seepage Induced Consolidation; see also Liu and Znidarčić (1991). The material parameters from these measurements are also given in Table 5.1.

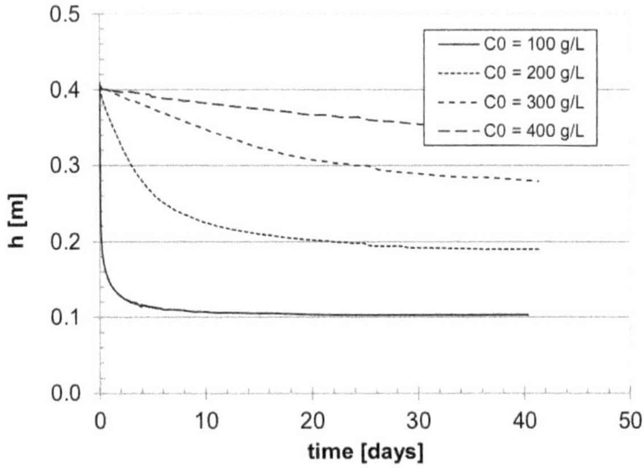

Fig. 5.29. Measured settling curves for consolidating slurries with a variety of initial mud concentrations.

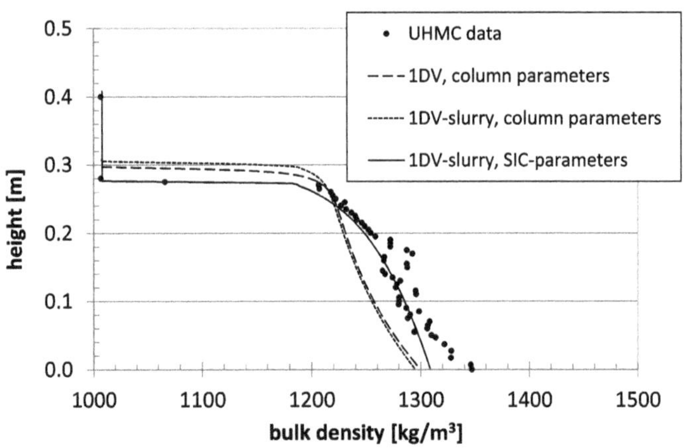

Fig. 5.30. Computed dry density profiles of a consolidating slurry for $C_0 =$ 300 g/L and comparison with experiment — after 67 days.

Next three series of 1DV consolidation simulations were carried out, using the parameters of Table 5.1. The results of these simulations are compared with the $C_0 = 300\,\mathrm{g/L}$ consolidation experiment:

(1) simulation of the consolidation experiment with the traditional 1DV model and material parameters obtained from the virgin consolidation experiment in the settling column;

(2) simulation of the consolidation experiment with the 1DV-slurry model and material parameters obtained from the virgin consolidation experiment;
(3) simulation of the consolidation experiment with the 1DV-slurry model and material parameters obtained from the SIC-tests.

The model results are presented in Fig. 5.30 (see also Barciela Rial (2019); Van den Bosch (2016). The data are well-predicted with the 1DV-slurry model, using material parameters measured for a homogenized sample from the present consolidation experiment, using the SIC-instrument.

Using the parameters from the virgin consolidation experiment in the settling column shows a difference between model predictions and data. The difference between the simulations with the traditional 1DV model and the 1DV-slurry model, using the column-obtained parameters, is not large. Apart from the deviations from the data, the computational results, using the column parameters, also show that no equilibrium is attained after 67 days, which follows from the following conditions:

(1) The density distributions still exhibit a concave-up profile, typical for the Kynch-phase of settling process (see Sec. 5.1).
(2) The bed thickness computed with the 1DV-slurry model is higher than that with the traditional 1DV consolidation model.

The (relatively) large differences between the simulations with the column and SIC-parameters are to be attributed to the degree of over-consolidation of the 300 g/L slurry. As a result, the material parameters derived from the virgin consolidation experiments are not representative for this over-consolidated slurry; see also Barciela Rial (2019) for further details, and discussions in the previous section.

These results have been obtained for laboratory conditions. In practice, slurries of much larger thickness are formed, often breaking the flocs, and diluting the mixture. Though the effects discussed above do not scale linearly with initial thickness, the effect of crust formation and non-representative material parameters may be large, both with respect to the final conditions (thickness, density and strength) and to the consolidation/equilibrium time in particular. As no universal recipes can be given, care should be taken when working with slurries.

Chapter 6

Erosion of Cohesive Sediment: Pick-Up Functions at Below-Capacity Conditions

This chapter describes the erosion of sedimentary beds. In case of non-equilibrium/non-capacity conditions, erosion can only be modeled with pick-up (or entrainment) functions, as explained in Chapter 1. For capacity conditions, sediment transport formulas may be used as well. Capacity conditions imply that the amount of sediment carried by the turbulent flow equals the transport capacity of that flow. Though capacity conditions may exist in the case of fine, cohesive sediment (e.g. Winterwerp, 2002; Winterwerp and Van Kesteren, 2004), practical (engineering) problems with respect to fine sediment generally relate to below-capacity conditions. This chapter focuses on below-capacity conditions, while capacity conditions are touched upon in Chapter 7.

Natural beds generally consist of sediment mixtures. As in Fig. 1.5, we distinguish between sand–mud mixtures with a granular (sand) skeleton and mud–sand mixtures, which are cohesive. The transition between these two types of beds occurs at a critical clay ($<2\,\mu$m) content of 5–10%, see Sec. 2.2, depending on the clay mineral at hand. Because of sorting by hydrodynamic gradients, beds are generally either sandy with a few percent of fines, or muddy with little sand, i.e. well away from the cohesive/non-cohesive transition. This is sustained by data on sediment composition in the Western Scheldt and the Wadden Sea (both The Netherlands) and in the mouth of the Yangtze River (China). In this chapter, we therefore treat sedimentary beds as either fully cohesive or fully non-cohesive, the latter with a granular (sandy) skeleton/substrate. Mixtures with coarser sediment, such as gravel, are not addressed.

As a rule, sediment composition tends to vary significantly in space and time. The low fringes and intertidal areas of estuaries and bays are often very muddy, whereas their tidal channels are mainly sandy. The transition between these sediment occurrences is generally very abrupt. Furthermore, considerable variations over time may occur, in particular in temperate climates with large seasonal variations in wind and wave action, whereas biological effects are also characterized by significant seasonal variations (Chapter 8). Understanding such spatial variations is at the heart of good modeling practice, as discussed in Chapter 13.

The sedimentary beds in this chapter are fairly recently formed, and are in a dynamic state — we do not address heavily over-consolidated clay beds, formed by e.g. geological processes. Sandy, granular beds often exhibit bed forms, such as ripples, dunes, etc. (Raudkivi and Witte, 1990; Van Rijn, 1993; Vanoni, 1940). These bed forms not only affect the overall hydrodynamics (effective hydraulic roughness, see Sec. 3.3), but likely play an important role also in mixing fines into an otherwise sandy bed (Sec. 6.6).

In this section, we start with the more traditional formulae for erosion, found in literature. Next, we present a conceptual description of erosion from a soil mechanical perspective with the aim to relate erosion rates to measurable soil-mechanical properties, distinguishing between cohesive and non-cohesive beds. How to account for the spatial variability of these erosion rates (thus bed properties) in numerical sediment transport models is treated in Chapters 12 and 13.

In tidal waters, fine cohesive sediment settles around slack water, temporarily forming a thin layer of fluffy material on the bed. This so-called "fluff layer" plays a crucial role in the net transport of fines in estuaries and tidal bays, as discussed in Chapter 9. A conceptual description of this fluff layer is presented in Sec. 6.5.

The state and properties of the bed are also largely affected by biological activity. We give a brief summary of biological effects in Chapter 8, summarizing their parameterization in Sec. 6.7, recognizing that a separate book may be dedicated to this subject.

The last section of this chapter presents an example of the role of navigation on the stability of sedimentary beds. Though the literature contains some (empirical) formulas describing scour by shipping and its propeller, no general formulas exist (to the knowledge of the authors) on implementing ship-induced erosion into large-scale fine sediment transport models.

6.1 Traditional erosion formulas

In the 1960s, Partheniades carried out erosion experiments on marine cohesive sediment in a systematic way; this work was summarized in

Partheniades (2009). Partheniades started his work with mud from San Francisco Bay in a straight, recirculating flume with Bay water. In his later work, Partheniades mainly studied artificial clays (in particular industrial kaolinite). For these experiments, he used the so-called rotating annular flume, which is a circular flume with a rotating lid driving the flow. The flume itself can rotate in opposite direction to minimize secondary currents, e.g. Partheniades (2009).

Partheniades' first series of erosion experiments were carried out on a so-called placed, remolded bed at *in-situ* density. The second series was done on so-called deposited beds, which were formed through deposition and subsequent consolidation from a cohesive sediment suspension. Deposited beds are characterized by vertical gradients in sediment properties, whereas placed beds are more homogeneous in density. Thus, both beds must have been in a (partly) over-consolidated state (Sec. 5.2). None of these beds contained sand.

Both experiments showed a small, non-zero erosion rate at the smallest bed shear stresses applied. Partheniades assumed a Gaussian bed shear stress distribution, obtaining the following formula for the erosion rate E [kg/m^2/s]:

$$E = \frac{AD_{50}\rho_s}{t^{(\tau_b)}} \left[1 - \frac{1}{\sqrt{2\pi}} \int_{-\frac{c}{k\eta_b\bar{\tau}_b}-\frac{1}{\eta_b}}^{\frac{c}{k\eta_b\bar{\tau}_b}+\frac{1}{\eta_b}} \exp\left\{ -\frac{\omega^2}{2} \right\} \mathrm{d}\omega \right] \tag{6.1}$$

in which A and k are empirical coefficients, D_{50} is the median diameter of the flocs in/on the bed, $t^{(\tau_b)}$ is the period that the time-varying bed shear stress exceeds the cohesive forces within the bed, c is the cohesion due to interparticle forces, $\bar{\tau}_b$ is the mean bed shear stress with $\eta_b\bar{\tau}_b$ its variance, and ω is a dummy variable. Similar results were found by Croad (1981), though he derived different coefficients.

Homogeneous beds

Equation (6.1) was parameterized by Kandiah and Arulanandan (1974) and Ariathurai and Arulanandan (1978), and is generally referred to as Partheniades' erosion formula in acknowledgment of his pioneering work:

$$E = M \left(\frac{\bar{\tau}_b - \tau_e}{\tau_e} \right) \quad \text{for } \bar{\tau}_b > \tau_e \tag{6.2}$$

in which M [kg/m^2/s] is an erosion flux parameter, $\bar{\tau}_b$ is the (turbulence) mean bed shear stress, and τ_e the critical (threshold) shear stress for erosion. For $\bar{\tau}_b < \tau_e$, $E = 0$. The difference $\tau_b - \tau_e$ is generally referred to as the excess bed shear stress.

Note that putting τ_e in the denominator of Eq. (6.2) is attractive from a dimensional point of view. However, when analyzing experimental results, small errors in τ_e may imply large variations in M.

Partheniades' erosion formula (6.1) has been generalized (Kandiah and Arulanandan, 1974; Mehta, 1991; Aalderink *et al.*, 1985; Harrison and Owen, 1971; Lick, 1982; Sheng, 1986), and is being used (in mathematical models) throughout the world, no doubt because of its simplicity:

$$E = M\left(\frac{\bar{\tau}_b - \tau_e\left(z,t\right)}{\tau_e\left(z,t\right)}\right)^n \quad \text{for } \bar{\tau}_b > \tau_e \tag{6.3}$$

where t is time and z is the vertical coordinate within the bed. The exponent n is generally set to unity, though other values have been used (Harrison and Owen, 1971; Kusuda *et al.*, 1984). The critical bed shear stress for erosion τ_e is generally assumed to be a constant material parameter, but may vary over depth. Values given in literature are: 0.1 Pa $<$ $\tau_e < 5$ Pa. In addition, the erosion flux parameter M may vary with time and depth, but is commonly kept constant as well; typical values read: 0.01 g/m^2/s $< M <$ 0.5 g/m^2/s. Whitehouse *et al.* (2000) present a summary of parameter values for mud from English estuaries. However, τ_e and M can be (much) smaller or larger, in particular for consolidated clay deposits and/or under the influence of biota. For instance, Le Hir *et al.* (2007) found high values of τ_e up to 4 Pa depending on the chlorophyll-*a* content — see also Sec. 6.7.

Equation (6.2) is commonly applied to well-consolidated, homogeneous beds, and τ_e and M are assumed constant throughout the bed. This type of erosion is sometimes referred to as unlimited (type II) erosion in the literature, and is studied in the laboratory on so-called placed beds. These are made from a homogenized mud slurry; the slurry is brought into the experimental facility, and its surface is leveled manually; sediment properties are more or less constant with depth. Likely, (part of) such beds are in an over-consolidated state (see Sec. 5.3), affecting measured erosion rates, while further consolidation does not occur. Likely, the variations in the degree of over-consolidation of the muds used in the experiments is one reason for the large variability in parameters reported in literature.

Stratified beds

Limited erosion (type I) is characterized by erosion rates decreasing in time at constant forcing — the bed is stratified. At the depth where the bed shear stress reaches the critical bed shear stress, erosion stops. This type of erosion is therefore also called depth-limited or supply-limited erosion, and can be studied in the laboratory with beds obtained through sedimentation and/or consolidation from a relatively dilute suspension (so-called deposited beds) — here we refer to erosion-limited conditions (e.g. Chapter 1). If formed from a dilute suspension, at equilibrium these deposited beds must be at, or close to a normal consolidated state (e.g. Sec. 5.1). If made from a slurry, the upper part of the bed is likely in an over-consolidated state (see Sec. 5.3).

Mehta and Partheniades (1982) carried out erosion experiments on such stratified beds, and proposed the following erosion formulation:

$$E = E_f \exp\left\{\alpha\left(\frac{\tau_b - \tau_e(z)}{\tau_e(z)}\right)^{\beta}\right\} \tag{6.4}$$

where E_f [kg/m^2/s] is referred to as the floc erosion rate, α and β are material-dependent parameters, and $\tau_e(z)$ is a depth-varying critical shear stress for erosion. This formula was applied by e.g. Parchure and Mehta (1985) and Amos *et al.* (1992a). Typical values for the various parameters are: $0.003\,\text{g/m}^2/\text{s} < E_f < 5\,\text{g/m}^2/\text{s}$; $0.5 < \alpha < 1$; $5.0 < \beta < 15.0$ and $0.01\ Pa < \tau_e < 0.1\ Pa$.

Sanford and Maa (2001) showed that Eqs. (6.4) and (6.2) would yield identical erosion rates under the assumptions that in Eq. (6.2) $\tau_e = \tau(z)$, $d\tau_e/dz =$ constant, and $M = M(z) = \eta'\rho_b$, where bulk density ρ_b and η' are constant. They argue further that the time scale of the eroding forces in relation to that of erosion determines whether erosion rates are unlimited (type II) or depth-limited (type I). This observation is qualitatively similar to the analysis in Sec. 6.3 on the role of the permeability of the soil, as discussed in Sec. 6.3.1.

Measuring erodability

Many efforts have been carried out to measure erosion rates and/or erodability *in-situ*. Winterwerp and Van Kesteren (2004) give a detailed overview. Most techniques can only be used on the dry intertidal, such as the CSM (Cohesive Strength Meter, e.g. Paterson, 1989; Tolhurst *et al.*, 1999 and the ISIS (Instrument for Measuring Erosion Shear Stress *In-situ*, Williamson and Ockenden, 1996), which both measure the onset of erosion, and the ISEF (*In-situ* Erosion Flume, Houwing and Van Rijn (1998)), which measures onset and rates of erosion. As discussed in Chapter 5, these measurements provide data on over-consolidated muddy beds.

Measuring onset and rates of erosion under water has been done with the Sea Carousel (Maa *et al.*, 1993) and the Benthic Annular Flume (Widdows *et al.*, 1998). Both instruments isolate a part of the bed, and drive an erosive current at increasing velocity, while measuring SPM-concentrations of the sediment eroded from the bed. As far as known to the authors, these instruments are no longer in use, no doubt because of the complicated logistics involved.

The American Corps of Engineers continues using the Sedflume, in which undisturbed sediment cores are placed — the Sedflume itself is operated onshore. The flow in that flume is increased in steps, and the core is pushed upward into the flow (e.g. McNeil *et al.* (1996). Ifremer applies a similar instrument, the ERODIMETRE Le Hir *et al.* (2006).

Erosion experiments are also performed on undisturbed cores with a U-GEMS, which is a modification of the Gust microcosm (Gust and Müller, 1997). This microcosm consists of a closed 0.1 m diameter cylinder, with a spinning disk to drive a shear flow and a central suction to generate a nearly uniform shear stress across the core. The U-GEMS is suitable for measuring the erodability of fine sediment up to about 1 Pa. The microcosm can be calibrated with sand of various diameter using the Shield or Van Rijn diagram for the onset of mobility (Van Rijn, 1984).

Erosion parameters have also been measured in the laboratory on so-called deposited and placed beds (e.g. Sec. 6.1), using (rotating) annular or straight flumes. Such experiments can be very useful for enhancing our scientific understanding of erosion processes. However, they do not give useful information on the actual erosion rates in nature, owing to the large disturbance of the sediment structure when sampled and brought into the experimental facility.

The majority of these techniques are cumbersome, do not provide in-situ information, or only on the (over-consolidated) intertidal. We believe that measuring erodability on cores is doable from a cost-benefit point of view, while providing reasonable information on soil behavior, though at isolated points only.

6.2 Modes of erosion

Traditionally, four modes of erosion of cohesive sedimentary beds have been recognized:

(1) Floc erosion, implying small erosion rates at shear stresses below τ_e — see also the experiments by Partheniades.

(2) Surface erosion, which is generally referred to as the erosion processes observed in laboratory experiments, and described with Eqs. (6.2) and (6.4).
(3) Mass erosion, when large lumps of sediment are torn from the bed — cliff erosion on intertidal areas is a form of mass erosion. Mass erosion is also known as bulk erosion.
(4) Turbulent entrainment, which can be regarded as erosion of fluid mud layers through one-way turbulent mixing with the ambient water; this is treated in Sec. 7.5.

The first three erosion mechanisms may occur on normal-consolidated and over-consolidated beds. Erosion mechanism (4) is relevant for "under-consolidated beds", e.g. layers of fluid mud.

In this chapter, we argue that surface erosion can be considered as drained failure of the bed, whereas mass erosion is caused by undrained failure, see Fig. 6.1. This is therefore our definition, as sketched in the schematic of Fig. 6.1:

(1) Surface erosion is caused by drained failure of the soil,[1] and
(2) Mass erosion is caused by undrained failure of the soil.

Drained failure implies that water has to enter a soil: the soil has to swell at macro-scale (this is not swelling of the clay particles themselves, as discussed in Sec. 5.2). The swelling rate (elastic rebound) therefore determines the rate of surface erosion.

Whether or not mass erosion can occur depends on the undrained shear strength of the soil in relation to the hydraulic stresses (the Péclet number, see Sec. 2.2). The rate of mass erosion is governed by inhomogeneities

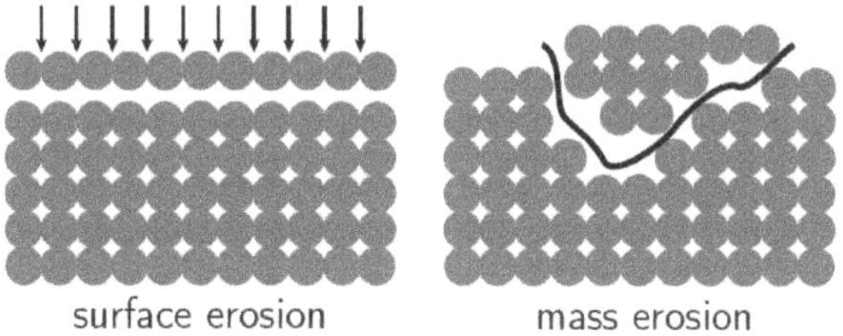

Fig. 6.1. Surface erosion is a drained process (left panel) and mass erosion is an undrained process (right panel).

[1]Note that partly drained conditions can occur also (generally occur) — see below.

and faults in the soil, and is therefore stochastic by nature, hence not predictable.

Floc erosion is also a drained process and results from the stochastic spatial distribution of bed strength and the stochastic nature of the turbulent stresses on the bed. As such, floc erosion is a special form of surface erosion.

In terms of the diagram of Fig. 1.4, floc, surface and mass erosion are relevant for erosion-limited beds.

All erosion formulas are based on gradients in strength over the vertical. However, gradients in the horizontal plane are generally much larger. We discuss the implication of this important observation with respect to modeling in Sec. 12.2. However, from a conceptual point of view, erosion-limited conditions reflecting horizontal inhomogeneities may be parameterized by vertical gradients in an erosion model.

Further to traditional erosion literature, we introduce the "fluff layer" concept. This fluff layer is a thin unconsolidated layer of fine cohesive sediment on top of an otherwise fairly solid bed, and is mainly formed around slack water. Erosion of this layer is therefore governed by turbulent mixing, rather than by soil mechanical processes. In that sense, erosion of a fluff layer is similar to the turbulent entrainment processes discussed in Sec. 7.5.1. As discussed in Sec. 9.2.2, net fine sediment transport by so-called scour lag is governed by the timing at which sediment from the fluff layer is picked up, rather than by the erosion rate itself. The critical shear stress for erosion of the fluff layer is therefore a more critical parameter than its erosion rate itself.

In the above, it is implicitly presumed that the sedimentary bed behaves as a cohesive soil (e.g. Sec. 2.2). When the bed is granular, the entrainment of fines from the soil granular skeleton is governed by the erosion rate of the coarser particles. The bed is thus armored with respect to the erosion of fines, and supply-limited conditions prevail (Fig. 1.4). However, the fines in the bed may affect the erosion rate of the granular material itself, as the soil permeability decreases when the amount of fines in the soil increases (up to a certain value), e.g. Jacobs (2011); Van Ledden (2003). Van Ledden (2003) therefore modified the critical shear stress for erosion of the granular fraction (Sec. 6.6), elaborated further upon by Jacobs (2011). This approach is currently used in numerical models, see Sec. 6.6.

6.3 Surface and floc erosion of a cohesive bed

In this section, we discuss the processes relevant for surface erosion, which by (our) definition is a drained process. If the bed would be in an over-consolidated state, the upper part of the bed swells till the critical state is locally reached (Secs. 2.2 and 5.2). Deformation (failure) of the bed then requires minimum work by the turbulent flow, as only the cohesive bonds between the particles have to be broken. The under-pressure in over-consolidated soils induces water inflow into the soil skeleton. The dissipation rate of these under-pressures is governed by the consolidation coefficient of the soil (Schofield and Wroth, 1968; Terzaghi, 1943). Further to the discussion in Sec. 2.2; Fig. 2.23 and Sec. 5.2, limited drainage may occur if erosion rates are too large, inducing (small) pore water under-pressures, reflected as pseudo-cohesion.

When the bed is in a consolidating state, with densities below those of a normally consolidated bed (i.e. in between the K_0- and the VCL-line introduced in Sec. 5.1, Fig. 5.8), failure of the bed (erosion) does not require the soil to swell towards its critical state. Erosion of consolidating soils can then be regarded as a turbulent entrainment process, as these soils behave as a (non-Newtonian) fluid (therefore the name fluid mud) — see Chapter 7. Erosion of consolidating soils therefore differs fundamentally from normally or over-consolidated soils.

In this chapter, we discuss the erosion of normally or lightly over-consolidated soils only (i.e. OCR ≥ 1; but $p > p_{\text{CSL}}$). Part of the contents of this chapter has been published earlier by Winterwerp *et al.* (2012) and described in Winterwerp and Van Kesteren (2004).

6.3.1 *Critical shear stress for erosion*

All experimental and field data reveal that substantial erosion of a cohesive bed only occurs when the hydrodynamic forcing exceeds a threshold, referred to as the critical shear stress for erosion τ_e, as introduced in Sec. 6.1. For surface erosion, the relevant eroding hydrodynamic force is the (turbulence) mean bed shear stress $\bar{\tau}_b$ (we discuss the "mean" in Sec. 6.3.3), though normal stresses play an important role as well, e.g. Sec. 6.3.2.

Under drained conditions, the strength of a soil at its very surface is likely slightly larger than the drained shear strength of the soil c_d, as the particles themselves induce some normal stresses. This was discussed in

Sec. 2.2, and sketched in Fig. 2.14. c_d is also known as the true cohesion c of the soil.

However, if erosion rates are large, (partly) undrained conditions may prevail (see Fig. 2.23), and the strength of the soil is affected by pseudo-cohesion c_{ps}. Therefore, the critical shear stress for erosion consists of two parts, true cohesion and pseudo-cohesion:

$$\tau_e = c_d + c_{ps} \tag{6.5}$$

where c_{ps} reflects the apparent strength/cohesion of a soil. The drained shear strength c_d, or true cohesion of a soil is a function of the material properties of the soil and its pore water (i.e. salinity, pH, ...), the sediment mineralogy (clays) and the effects of organic material, such as organic polymers (EPS, etc.), including its redox state. However, c_d would not vary over depth or with density, as it is a material parameter.

As discussed in Sec. 2.2 and Appendix A.6, it is not difficult to measure the soil's vane strength, which provides the undrained strength of a soil. In the following, we discuss how strength measurements deeper in a soil can be extrapolated to strength values at the soil surface, thus assessing the drained strength of a soil c_d from c_u. The role of pseudo-cohesion is discussed further below.

Vane strength

Figure 6.2 presents the measured peak and undrained shear strength, τ_p and c_u, of mud from Lake Ketel (fresh water) and IJmuiden Harbor (saline water), both in The Netherlands, as assessed from vane strength measurements. The τ_p values are a bit higher than the c_u values, as should be the case with vane tests. Assuming a self-similar (fractal) structure of the soil, we can use Eq. (5.31) to analyze the data of Fig. 6.2. Thus we find $\tau_y = 3.9 \cdot 10^4 (\phi_s)^{6.33}$, i.e. $K_y = 3.9 \cdot 10^4$ Pa and $n_f = 2.7$ — the latter value agrees well with fractal dimensions assessed from consolidation experiments on the same sediment.

Determining c_d would imply extrapolation of Fig. 6.2 to the very small ϕ_s-values at the soil surface, which is of the order of the gelling concentration c_{gel}, i.e. $\phi_s = 0.03$–0.06. This would yield $c_d = 0.01$–0.7 Pa for these particular muds, which is in the range of values reported in literature.

Vane strength and Atterberg limits

For our analysis, it is convenient to plot the data of Fig. 6.2 as a function of void ratio $e (= (1 - \phi_s)/\phi_s)$, so that the critical state model (CSM) can be used to assess c_d — see Sec. 2.3, e.g. Schofield and Wroth

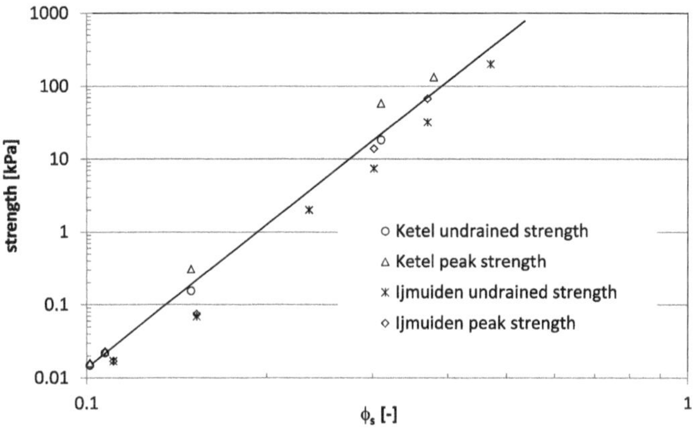

Fig. 6.2. Vane strength measurements on muds from Lake Ketel and IJmuiden Harbor (both in The Netherlands) showing peak and undrained shear strengths, as a function of volumetric solids concentration ϕ_s, indicating a power-law relation.

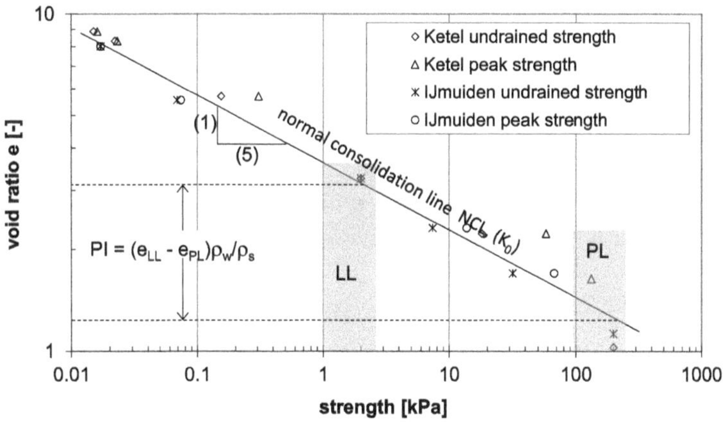

Fig. 6.3. Vane strength measurements on muds from Lake Ketel and IJmuiden Harbor (both in The Netherlands) showing peak and undrained shear strengths as a function of void ratio e.

(1968); Winterwerp and Van Kesteren (2004); Winterwerp *et al.* (2012). For instance, Lambe and Whitman (1979); Whitlow (1990) suggest that in theory c_d at the bed surface can be obtained from an extrapolation of strengths measured as function of the void ratio e.

Figure 6.3 presents the data of Fig. 6.2 in such an e–p diagram (e.g. Sec. 2.3). As the samples were taken from natural deposits under water, they are likely close to the normal-consolidated state. The data in Fig. 6.3 yield a power law relation, as in Fig. 6.2, justifying our modification of the CSM, as discussed in Sec. 2.3. The 1:5 slope of the power-law fit in Fig. 6.3 represents the compression coefficient of the soil, yielding $C_c = 0.2$ for these soils.

Soil mechanical literature (e.g. Lambe and Whitman, 1979, see also Winterwerp and Van Kesteren, 2004) suggests that the undrained shear strength at the plastic limit amounts $c_{u,\mathrm{PL}} = 100$–$200\,\mathrm{kPa}$, whereas at the liquid limit, $c_{u,\mathrm{LL}} = 1$–$2\,\mathrm{kPa}$. The slope of the experimental fit in Fig. 6.3 then relates void ratio to PI ($\equiv$PL − LL):

$$\mathrm{PI} = \frac{\rho_w}{\rho_s}\left(e_{\mathrm{LL}} - e_{\mathrm{PL}}\right) \tag{6.6}$$

The data of Fig. 6.3 have been obtained from various mud samples at a variety of dry density (water content). Therefore, Fig. 6.3 can be interpreted also as the vertical strength distribution within a normal-consolidated bed, with the higher strengths lower in the bed, of course.

In the e–p diagram (see Fig. 6.4), swelling (elastic rebound) follows the so-called swelling line, the slope of which is defined by the swelling coefficient C_s, which is generally a factor 1/7–1/3 smaller than C_c (Lambe and Witman, 1979) — see also Sec. 5.3.

Extrapolation to the bed surface

In the following, we start from an arbitrary point in the bed (but near its surface) with void ratio e_1, see Fig. 6.4. Next, we elaborate on the

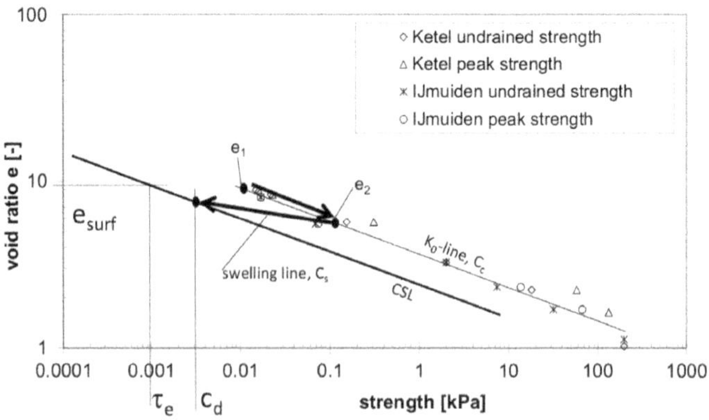

Fig. 6.4. Determination of undrained strength at a virtual bed surface relative to an arbitrary depth within the bed.

stresses when a small layer (mass) of sediment is removed by e.g. erosion, thus the bed becomes over-consolidated. The void ratio now becomes e_2, while the strength at e_2 becomes $p_2 = p_1(e_1/e_2)^{-C_c}$. Such a state of over-consolidation can also be induced by external stresses, for instance lift forces by turbulent flow induce under-pressures within a bed (e.g. Jacobs (2011) and Sec. 6.3.2). As the soil is now over-consolidated, it starts to swell, following the swelling line, until the critical state line (CSL) is reached, as sketched in Fig. 6.4. The strength at the bed then becomes $p_{\mathrm{CS}} = p_2(e_2/e_{\mathrm{CS}})^{C_s}$. The crossing of the swelling line and the critical state line represents the void ratio at the critical state, thus the accompanying drained strength c_d — for the example in Fig. 6.4, $c_d \approx 3$ Pa.

As along the NCL, e_1–e_2 is proportional to PI (Fig. 6.3), we obtain a relation for c_d as a function of the plasticity index PI:

$$c_d = \gamma_{er} \mathrm{PI}^{\beta} \tag{6.7}$$

where γ_{er} is a proportionality parameter, and $\beta = C_s/C_c$. Note that we have explicitly assumed that the bed is at a normal consolidated state, e.g. Sec. 5.2 and Fig. 5.14.

The effect of pseudo-cohesion

In 1982, Mehta and Partheniades (1982) carried out erosion experiments in an annular flume, and found an increase of τ_e with depth in the soil. These findings were confirmed by Amos *et al.* (1992a,b) through erosion experiments with a sea carousel in the Bay of Fundy. These and other findings were summarized and formalized by Sanford and Maa (2001) to account for such vertical gradients in τ_e, as discussed above.

An illustrative field experiment was carried out by Zhu *et al.* (2019) on an almost permanently submerged mudflat, measuring erosion rates, hydrodynamic conditions and sediment concentrations. The results for a single tide during a storm are presented in Figs. 6.5(a)–(c), showing that the bed was eroded by about 2 cm in a period of approximately 3 hrs. Comparing Fig. 6.5(b) with 6.5(a) gives a measure for τ_e at which erosion started. From bed shear stress computations and measured bed level changes during five subsequent days (results not presented, see Zhu *et al.*, 2019) the vertical profile of critical shear stress in Fig. 6.5(d) is constructed.

These measurements allow for an order of magnitude assessment of the Péclet number Pe_w, Eq. (2.7). V is about 2 cm divided by 3 hrs, ℓ is about 2 cm and c_v is typically about 10^{-8} m^2/s. This would give $\mathrm{Pe}_w \approx 4$, implying, further to Fig. 2.23, that some pore water under-pressures are expected to build up during erosion. In other words, the erosion observed in Fig. 6.5

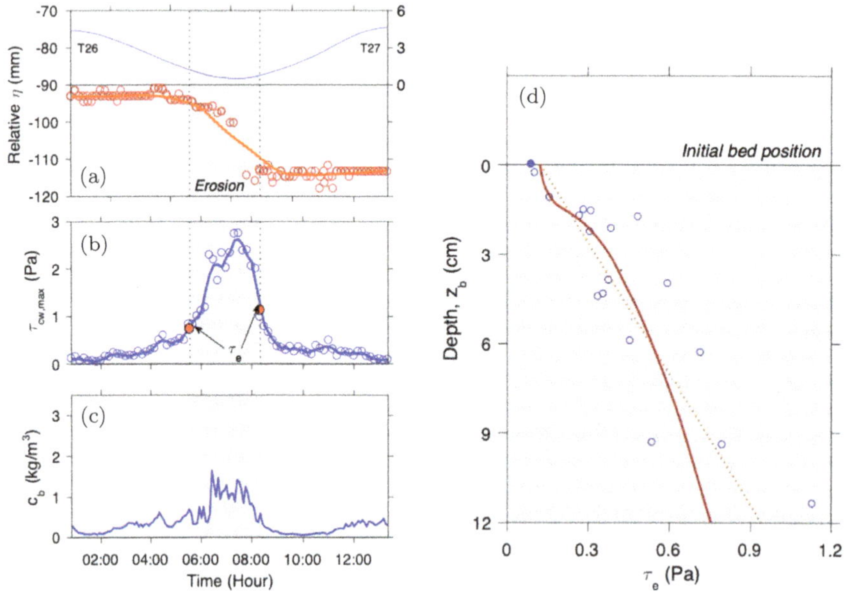

Fig. 6.5. (a) Bed level change during a single tide; (b) bed shear stress by currents and waves; (c) sediment concentration; (d) critical bed shear stress for erosion as determined from panel (b) (redrawn from Zhu *et al.*, 2019).

is likely affected by pseudo-cohesion. As an example, erosion of 1 mm under fully undrained conditions from a bed with a bulk density of 1300 kg/m^3 (i.e. $\rho_{dry} \approx 500\,\mathrm{kg/m^3}$) would induce a pore water under-pressure of about 5 Pa. This explains why τ_e increases with depth (and soil dry density), whereas $c_d \approx 0.2$ Pa is a material parameter, which does not vary with depth and c_{pc} increases to about 0.5 Pa, considerably smaller than the estimated 5 Pa.

Calibration of critical shear stress for erosion

From the above we conclude that τ_e in Eq. (6.5) consists of a part which reflects the true cohesion of a soil, thus is a function of PI, which is a material parameter, and a part which is a function of the erosion process itself, via the Péclet number Pe_w:

$$\tau_e = c_d(\mathrm{PI}) + c_{ps}(Pe_\mathrm{w}) \tag{6.8}$$

Unfortunately, the effects of pseudo-cohesion and their temporal and spatial variations are very difficult to measure in the real world. In the case of the data of Fig. 6.5, these effects are relatively small though, and only measurable with dedicated *in-situ* experiments. For practical applications, we therefore recommend a more simple model for the critical shear stress for erosion:

$$\tau_e = c_d = \gamma_{er}\mathrm{PI}^{\beta} \tag{6.9}$$

accepting that variations of τ_e with depth cannot be addressed. Note that this pragmatic approach is consistent with our advocated modeling approach to deal with the large, yet unknown, variations over the horizontal space in sediment properties and bed structure.

Jacobs (2011) carried out erosion experiments in the ERODIMETRE, an erosion device developed at Ifremer (Le Hir *et al.*, 2006). The measured values for the onset of erosion are presented in Fig. 6.6, together with the data by Smerdon and Beasley (see below), experiments carried out at Deltares (Lake Ketelmeer, Kembs Reservoir and IJmuiden Harbor, see Fig. 6.3 and Winterwerp and Van Kesteren (2004), and data by Torfs (1995, with kaolinite, montmorillonite and natural mud). For the latter, some assumptions were made for the activity of the minerals applied in the experiments; for details the reader is referred to Jacobs (2011). To account for low-plasticity sediment, Jacobs (2011) proposed a synthetic plasticity index, using the mud activity A: $\mathrm{PI}^* = A\left(\xi^{cl} - \xi_0\right)$, e.g. Fig. 2.18, referred to as the generalized plasticity index (see also Sec. 2.2).

Figure 6.6 shows that all data follow more or less the power law proposed by Eq. (6.9). We ignore data at very low PI (PI $< 7\%$, see also

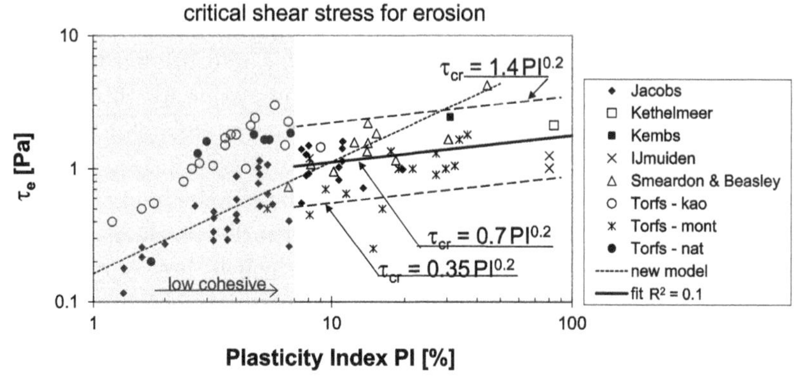

Fig. 6.6. Critical shear stress for erosion τ_e as a function of the Plasticity Index PI; experimental data and relation (6.9) and (6.10). For PI $< 7\%$, samples are low to non-cohesive. Power-law fit only through data with PI $> 7\%$, and effects of pseudo-cohesion are ignored.

Fig. 2.17), as the soil is then non- to low-cohesive. Thus a reasonable fit is obtained (note that PI is in %):

$$\tau_{cr} = \gamma_{er}\,\mathrm{PI}^{\beta} \text{ with } \beta = 0.2 \text{ and } \gamma_{er} = 0.7 \text{ Pa } (0.35 < \gamma_{er} < 1.4 \text{ Pa}) \quad (6.10)$$

It is encouraging that the exponent β in Eq. (6.10) matches the slope of 0.2 of Fig. 6.3, which, moreover lies in between 1/7 and 1/3, consistent with soil mechanical theory. However, the data in Fig. 6.6 show considerable scatter ($R^2 \approx 0.1$), which may (partly) be due to difficulties in measuring τ_e and ξ^{cl} accurately (see also Jacobs, 2011), and/or may be due to the preparation of sedimentary beds in the laboratory, leading to over-consolidation, as discussed in Sec. 5.1 (Fig. 5.9) and/or may be attributed to the effects of pseudo-cohesion.

Further, the data by Smerdon *et al.* (1959), plotted in Fig. 6.6, show also a power relation with the PI, though with a considerable steeper slope ($\beta = 0.84$). However, Smerdon and Beasley carried out erosion experiments on (heavily) over-consolidated soils ($c_u = 0.1$–10 kPa, implying $p < p_{CS}$), while no information is available on sample preparation. Hence, these data may have been affected largely by pseudo-cohesion effects.

In traditional literature, the critical shear stress for erosion is often given as a power law of the bulk or dry density of a soil, i.e. $\tau_e \propto \rho_b^{\beta_1}$ or $\tau_e \propto \rho_{dry}^{\beta_2}$, with $\beta_{1,2} = 2.3$–2.5 (see for instances Harrison and Owen, 1971; Krone, 1986; Migniot, 1968; Whitehouse *et al.*, 2000). These density-related formulas are conceptually identical to the τ_e-formulation derived in this section, as explained by the relation between PI and e, Eq. (6.6). For fractal dimensions of the bed of $n_f = 2.6$–2.8, we expect $\beta_{1,2}$-values of 5–10. As discussed above, these values are consistent with Eq. (6.10), but higher than the $\beta_{1,2}$-values from literature. We have no explanation for this difference in sensitivity of τ_e to the dry density, though may speculate on the role of pseudo-cohesion.

Finally it is to be understood that the Atterberg limits implicitly include the effects of sediment composition (sand–mud ratio, which also affects PI and ρ_{dry}) and of organic material (e.g. EPS), provided these polymers were not lost by heating the soil samples. In other words, the effects of sediment composition and organic material are implicitly accounted for in Eq. (6.10), thus do not explain the discrepancies in $\beta_{1,2}$ discussed above.

Erosion threshold on intertidal areas

If one walks over a muddy intertidal flat from the marsh, over the mudflat to beyond the low water line, the apparent strength of the sediment decreases markedly. In the marshes, the bed is stiff and sediment stability is affected largely by vegetation, as described in Sec. 6.7: the threshold for erosion is much larger than predicted by Fig. 6.6. Below low water, the mud feels very soft, almost fluffy, and the critical shear stress for erosion is given by Eq. (6.10), which may have to be corrected though for the effects of pseudo-cohesion.

On the intertidal, the muddy bed is still very soft, though stiffer than below low water. Measurements showed that the critical shear stress for erosion can be much higher than follows from Fig. 6.6. This is likely caused by two processes — it is not known whether these compete or reinforce each other:

(1) Biological processes, such as cementing by EPS, as discussed in Sec. 8.1.
(2) Soil mechanical processes, such as over-consolidation and crust formation, as discussed in Sec. 5.2.

The state of over-consolidation (OCR, Sec. 5.2) is not predictable, as it depends on the history of the mud flat, such as the cumulative hours of drying, sunshine, winds, possibly rainfall, and the role of biota (Chapter 8). Desaturation, cementing and crack formation will affect the soil mechanical properties further. In numerical modeling studies, the critical shear stress for the erosion of intertidal mud flats has therefore to be determined by best-engineering practices, or trial and error.

6.3.2 *Surface erosion rate*

We have defined surface erosion as a drained process of a normal- or (slightly) over-consolidated soil. Pore water under-pressures can be induced in this soil by turbulent pressure fluctuations (lift forces) and/or by removal of material from the bed (see also Jacobs, 2011). It is illustrative to assess the order of magnitude of these under-pressures. Turbulence-induced under-pressures follow on ejection events by turbulent motions (Nezu and Nakagawa, 1993). Snarski and Lueptow (286) and Buschmann and Gad-El-Hak (2010) report values of turbulence-induced peak stresses normal to the wall of about 10 times the mean bed shear stress, i.e. of the order of 10 Pa at the onset of erosion ($\tau_e = 1$ Pa, e.g. Fig. 6.6), equivalent to about 1 mm water pressure. Thus turbulence can generate pore pressure fluctuations within a bed with an amplitude of about 10 Pa.

The erosion flux of Fig. 6.5 amounts to about 2 cm/3 hrs $\approx 2 \cdot 10^{-6}$ m/s, or $E \approx 10^{-3}$ kg/m^2/s at $\rho_b = 1300$ kg/m^3 ($\rho_{dry} = 500$ kg/m^3). This would

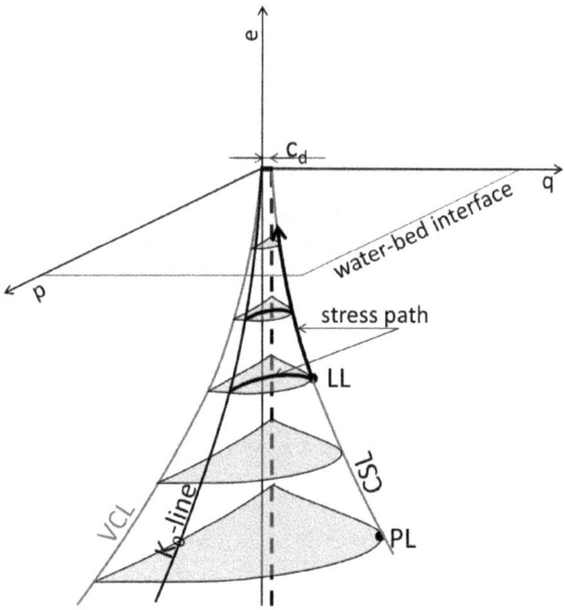

Fig. 6.7. Conceptual stress paths (indicated by arrows) for surface erosion in the CSM at various depth within the bed — at large e, c_u decreases to c_d.

induce an under-pressure at a rate of 0.001 mm water pressure per second, or 0.01 Pa/s, thus 10 Pa after 15 min of erosion (assuming no dissipation of these under-pressures).

These numbers show that the under-pressures are quite small, basically impossible to measure. Yet, these small stresses govern the erosion of soils.

While under-pressure builds up, stresses move from the K_0-line to the CSL-line, at constant e, in case of normal consolidated soils. These stress paths are sketched in the critical state diagram of Fig. 6.7 in two failure planes, one close to/at the bed surface, the second a bit deeper within the soil.

The dissipation of pore water gradients can be described with Terzaghi's consolidation model Eq. (5.15), which includes the consolidation coefficient c_v; the relation between c_v and Γ_c is given by Eq. (5.12). This dissipation is accompanied by swelling (see Sec. 6.3.1), increasing the void ratio e, attaining the undrained shear strength c_d at the beds surface (as c_u then reduces to c_d). Note that work against gravity for this expansion of the soil is small at the bed surface, as the excess density of the flocs near the bed is small (Sec. 4.3). Deeper in the bed, conditions are still undrained, and e remains constant. The stress path during swelling close to the bed surface is also sketched in Fig. 6.7 by arrows. These paths end at the CSL, as deformations at the critical state require minimal energy.

A typical time scale for pore water pressure dissipation T_s follows from Terzaghi's consolidation equation (5.15):

$$T_s \propto \frac{z^2}{\pi c_v} \tag{6.11}$$

Differentiating Eq. (6.11) to z gives the dissipation (swelling) rate V_s:

$$V_s = \frac{\mathrm{d}z}{\mathrm{d}t} = \frac{\mathrm{d}\delta_s}{\mathrm{d}t} = \frac{\pi c_v}{2\delta_s} \approx \frac{c_v}{\delta_s} \tag{6.12}$$

where δ_s reflects the "swelling front". This swelling is sketched in Fig. 6.8. The soil strength at the bed surface amounts to c_d and deeper in the bed, beyond the swelling front to c_u. Sediment can be eroded over a layer where $\tau_b > c_d$ $(=\tau_e)$, ignoring the effects of pseudo-cohesion. This layer is very thin (note that the schematic of Fig. 6.8 is not to scale), and stress gradients within the bed are therefore high during erosion.

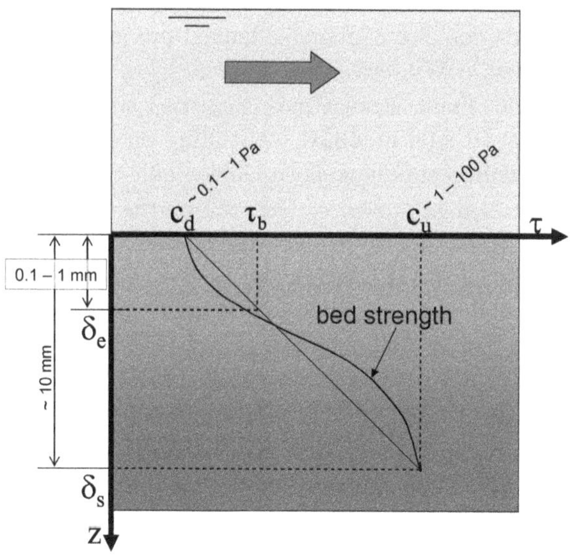

Fig. 6.8. Schematic of surface erosion as a swelling process of normally-consolidated soil — figure not to scale.

Assuming a linear strength distribution within the bed, Fig. 6.8 yields:

$$\delta_e \approx \frac{\tau_b - c_d}{c_u - c_d}\delta_s \approx \frac{\tau_b - c_d}{c_u}\delta_s = \frac{\tau_b - c_d}{c_u}\frac{c_v}{V_e} \approx \frac{\tau_b - \tau_e}{c_u}\frac{c_v}{V_e} \tag{6.13}$$

in which we assume $\mathrm{d}\delta_e/\mathrm{d}t = \mathrm{d}\delta_s/\mathrm{d}t$. Further, we assume that the sediment bed is composed of flocs, and that during surface erosion, layers of flocs are removed one by one. In other words, we assume that the erosion depth δ_e scales with the median size of the primary particles $D_{p,50}$ in the bed, corrected for interstitial water:

$$\delta_e \approx \alpha_\delta D_{p,50}/\phi_s \tag{6.14}$$

where α_δ is a calibration parameter, representing the number (layer thickness) of primary particles picked up during erosion. Then the surface erosion flux E [kg/m^2/s] follows from the erosion velocity V_E ($= V_s$), multiplied with the dry bed density ρ_{dry}:

$$E = V_E\rho_{dry} = \frac{c_v \rho_{dry}\phi_s}{\alpha_\delta D_{p,50}}\frac{\tau_b - \tau_e}{c_u} = M_E\left(\tau_b - \tau_{er}\right) \tag{6.15a}$$

$$M_E = \frac{c_v \rho_{dry}\phi_s}{\alpha_\delta D_{p,50} c_u} \quad \text{for } \tau_b > \tau_e \tag{6.15b}$$

in which the erosion parameter M_E has the dimension [s/m]. Note that Jacobs (2011) suggests to relate the erosion depth δ_e to the floc size of the bed. However, this floc size cannot be defined properly, and we prefer the approach presented in Winterwerp *et al.* (2012).

The undrained shear strength c_u increases with depth. Therefore, c_u-values may have to be modified, depending on the sample density at which the c_u measurements have been carried out. Such correction can be done with the fractal approach of Fig. 6.2 or the power-law relation of Fig. 6.3, relating c_u to the soil dry density ρ_{dry} of Eq. (6.15).

The erosion model (6.15) is compared with data in Fig. 6.9 for $\alpha_\delta = 10$, showing a reasonable agreement between model and data, with the larger deviations for the softer sediment though. It is noted that if drainage is not complete ($Pe_\mathrm{w} > 0$), erosion rates become smaller. However, as long as the induced pseudo-cohesion remains small, the above analysis holds, i.e. surface erosion can be regarded as a (almost fully) drained process. The reader is referred to Winterwerp *et al.* (2012) for further details on the experimental procedures and values.

The erosion model (6.15) shows the same linear relation with τ_b as the traditional Partheniades model Eq. (6.2). If a nonlinear strength distribution within the soil (Fig. 6.8) would be assumed, a nonlinear erosion formula

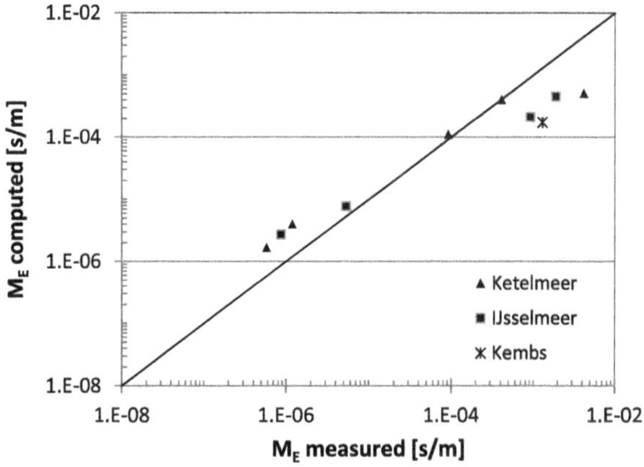

Fig. 6.9. Comparison of erosion model (6.15) with various data obtained in an erosion flume in the laboratory for $\alpha_\delta = 10$ (data from internal reports Delft Hydraulics, e.g. Winterwerp and Van Kesteren, 2004).

would be obtained, as Eq. (6.3). Further it is noted that the bed shear stress τ_b in Eq. (6.15) has not yet been specified, which is the subject of the next section.

6.3.3 *Floc erosion*

Floc erosion is a drained process, resulting from the stochastic distribution of bed strength (true cohesion) and turbulent stresses (see also Parchure and Mehta, 1985). In Sec. 3.3.1, Fig. 3.16 we indicated that the bed shear stress exhibits a skewed probability distribution in time. Qualitatively, such a temporal distribution is sketched in Fig. 6.10. Moreover, the bed strength is also not constant, but varies over space. Here we discuss the spatial distribution in a horizontal plane (ignoring possible vertical variations), as a result of variations in sediment composition, bed level/structure, bioturbation, etc. As we have no data, we presume a Gaussian distribution of the critical shear stress for erosion, as sketched in Fig. 6.10. This approach was first proposed by Van Prooijen and Winterwerp (2010).

In the example of Fig. 6.10, bed shear stress and critical shear stress are such that the turbulence mean bed shear stress $\bar{\tau}_b$ is everywhere smaller than the critical shear stress for erosion τ_e of the bed. However, the instantaneous bed shear stress τ_b may exceed the critical shear stress $\tau_e(x_1, y_1)$ temporally at location (x_1, y_1). Then conditions for surface erosion are met (Eq. (6.15)), as indicated by the gray area in Fig. 6.10.

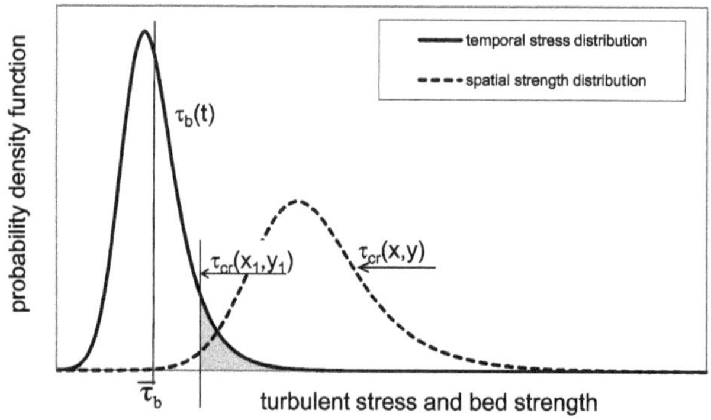

Fig. 6.10. Schematic of spatial distribution in critical shear stress and temporal distribution of bed shear stress at location (x_1, y_1). The shaded area represents the distribution of erosion events, where surface erosion conditions are met.

Van Prooijen and Winterwerp (2010) quantified this behavior. A skewed pdf of the bed shear stress was proposed, matching Obi's data (Fig. 3.16). The erodability of the bed was modeled with a Gaussian distribution, the parameters of which were obtained from calibration against data by Partheniades (1965). A parameterization of these pdf's yields three regimes, linking floc and surface erosion:

$$\frac{E}{M\tau_e} = \begin{cases} 0 & \text{for } \frac{\bar{\tau}_b}{\tau_e} < 0.52 \\ 0.204 - 0.823\left(\frac{\bar{\tau}_b}{\tau_e}\right) + 0.904\left(\frac{\bar{\tau}_b}{\tau_e}\right)^2 - 0.144\left(\frac{\bar{\tau}_b}{\tau_e}\right)^3 & \\ \frac{\bar{\tau}_b}{\tau_e} - 1 & \text{for } \frac{\bar{\tau}_b}{\tau_e} > 1.7 \end{cases} \tag{6.16}$$

in which the coefficients of the parameterization have been substituted. Equation (6.16) yields a smooth transition from a stable bed, via floc erosion to surface erosion, as shown in Fig. 6.11.

Thus floc erosion is defined by the stochastic behavior of turbulence and the stochastic (spatially varying) properties of the bed. Likely, spatial inhomogeneities in turbulence and bed properties dominate this so-called floc erosion. This model may therefore be interpreted in the sense that occasionally flocs are disrupted from the bed, which is the traditional definition of floc erosion.

From a practical, large-scale modeling point of view, the contribution of floc erosion to the overall erosion rates in open water can generally be ignored. In engineering studies, uncertainties are generally too large and

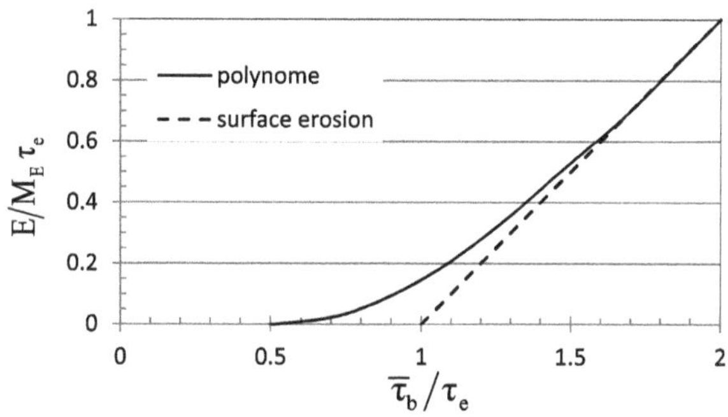

Fig. 6.11. Non-dimensional erosion rate as a function of normalized turbulence-mean bed shear stress — Eq. (6.16), after Van Prooijen and Winterwerp (2010).

proper data too scarce for legitimating the use of floc erosion models. However, understanding the role of such spatial inhomogeneities on the transport and fate of fine sediment transport is often an important step in performing a sound modeling study.

6.4 Mass erosion from a cohesive bed

We define mass erosion as an undrained erosion process, thus at large Péclet numbers. Mass erosion occurs when internal stresses induced by external loading exceed the undrained strength of the sediment bed c_u. Entrainment of fluid mud can be interpreted as a form of mass erosion, e.g. Chapter 7. However, in general, we refer to mass erosion when lumps of material are ruptured from a consolidated bed. Planes of failure follow inhomogeneities (biota), cracks, etc. in the bed. As these are stochastic by nature, the rate of mass erosion is stochastic as well, and therefore hardly predictable. However, the onset of mass erosion can be assessed to first order. It is noted that mass erosion generally induces irregular bed surfaces, locally enhancing hydrodynamic stresses on the bed. Mass erosion may therefore destabilize sedimentary beds at large scale, setting in motion a kind of snowball effect.

Mass erosion can occur by ductile failure (Fig. 6.12) or shear failure (Fig. 6.13). Ductile failure is induced by compression of the soil under confined conditions, for instance induced by the normal (turbulent) stresses of a flow. Failure occurs when the deviatoric stress q exceeds $2c_u$, e.g. Jacobs (2011); Winterwerp and Van Kesteren (2004).

Prandtl (1920) elaborated on shear failure by normal stresses, as sketched in Fig. 6.13. Later his theory was refined by Terzaghi (1943) and

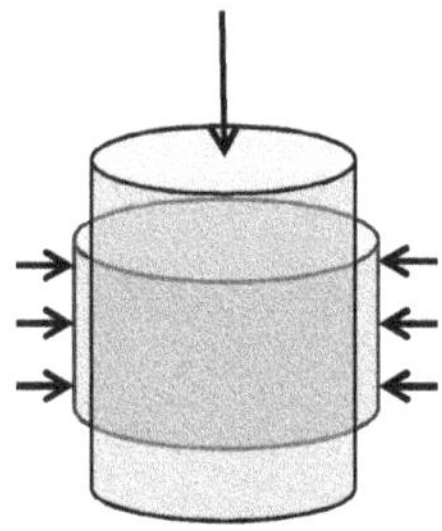

Fig. 6.12. Ductile failure of a soil by normal loads, see also Winterwerp and Van Kesteren (2004).

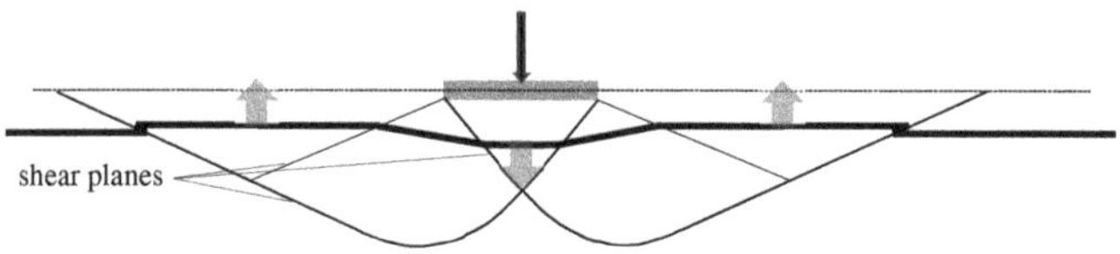

Fig. 6.13. Shear failure of a soil along Prandtl shear planes.

Michalowski (1997) to name a few. Prandtl suggested that shear failure would occur when $q > (2+\pi)c_u \approx 5c_u$.

This analysis leads to a criterion for the onset of mass erosion (e.g. Winterwerp and Van Kesteren, 2004):

$$q > (2-5)\, c_u \tag{6.17}$$

where the higher value ($5c_u$) would represent onset of mass erosion of a flat bed (cf. Fig. 6.13) and the lower value ($2c_u$) would be relevant for irregular beds.

The deviatoric stresses q within a bed mainly relate to the normal stresses induced by hydrodynamic loading, i.e., not to bed shear stresses (Sumer and Fredsøe, 2002). These stresses therefore scale with the dynamic pressure, so that the threshold for mass erosion equation (6.17) becomes:

$$1/2\rho U^2 > (2-5)\, c_u \tag{6.18}$$

where U is a characteristic free flow velocity. Figure 6.14 presents some data on the onset of mass erosion by turbulent flow in a straight flume as measured by Winterwerp and Van Kesteren (2004), in conjunction with Eq. (6.18), showing that a 3.6:1 fit captures the experimental data reasonably.

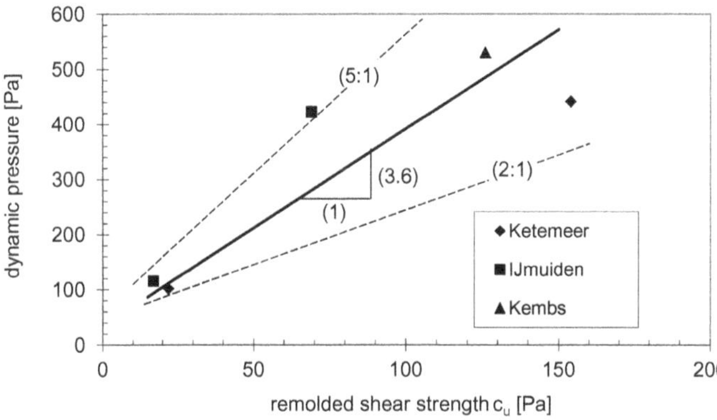

Fig. 6.14. Onset of mass erosion by turbulent flow — comparison with Eq. (6.18), see also Winterwerp *et al.*, 2012.

In an experimental study on mass erosion of very soft organic sediment in fresh water lakes in Florida (Mehta *et al.*, 2015) related the onset of mass erosion to the floc size of the sediment, applying fractal theory.

Mass erosion by waves

Mass erosion is in particular observed on intertidal areas, in the form of cliff erosion. The wave-induced stresses now relate to the velocity V of the approaching water mass in the crest of the waves, which equals the wave celerity c (Winterwerp and Van Kesteren, 2004). Thus the onset of mass (cliff) erosion under non-breaking waves reads:

$$1/2\rho V^2 = 1/2\rho c^2 = \frac{\rho g}{2k}\tanh\{kh\} > (2-5)\,c_u \tag{6.19}$$

where $k \equiv 2\pi/L$ with L = wavelength.

In case of breaking waves, the following empirical relation was proposed Wood (1997):

$$1/2\rho V^2 = 1/2\alpha\rho g\,(h + H_{rms}/2) > (2-5)\,c_u \tag{6.20}$$

where the empirical parameter measures $1 < \alpha < 3$, h is water depth and H_{rms} is the significant wave height.

6.5 Entrainment of cohesive sediment from the fluff layer

The previous section treats erosion of alluvial muddy beds, as sketched in Fig. 1.5(c) and 1.5(d). These beds may be stratified or not. Strength/density variations within these beds tend to be gradual. However, in many cases,

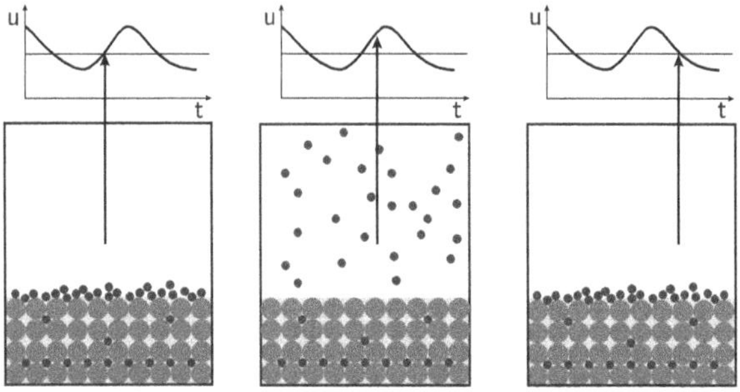

Fig. 6.15. Schematic explaining equilibrium between fluff layer and water column under tidal conditions.

the parent bed is fairly solid, containing coarse sediment or consolidated deposits. Fine sediment, depositing on such beds form a loose layer of fine cohesive sediment, referred to as a "fluff layer" (e.g. Sec. 6.2). Figure 6.15 shows a schematic on the formation of such a fluff layer during slack water, and the subsequent re-entrainment into the water column when flow velocities increase again. These beds are typical for settling-limited conditions (e.g. Fig. 1.4).

The existence/development of fluff layers has been reported before. Sanford *et al.* (1991) discussed the dynamics of fluff layers in the northern part of Chesapeake Bay (USA), Bedford *et al.* (1987), Wang *et al.* (2000), Wang (2003) found fluff layers in Long Island Sound (USA), while El Ganaoui *et al.* (2004) discussed the behavior of fluff layers in the Rhone River and the Mediterranean continental shelf in France. Recently, Mathew and Winterwerp (2017) analyzed fluffy sediment dynamics in the Passaic River (USA). Figure 6.16 presents a photo of a 2–3 mm thick fluff layer sitting on a bed of coarser sediment.

The density of these fluff layers has not been measured. However, these layers are under-consolidated and their density therefore should be around the gelling concentration, or lower, i.e. around 100 g/L. At a water depth of 6 m and a fluff layer thickness of 3 mm, full entrainment of such a fluff layer and subsequent mixing over the water column would be equivalent to a depth-averaged SPM-concentration of 50 mg/L, a typical value in the lower reaches of the Passaic River (see Mathew and Winterwerp, 2017).

A variety of data on the onset of entrainment of fluff layers has been published. Sanford *et al.* (1991) and Maa *et al.* (1993) carried out field measurements in Chesapeake Bay, USA (from an anchored station and with

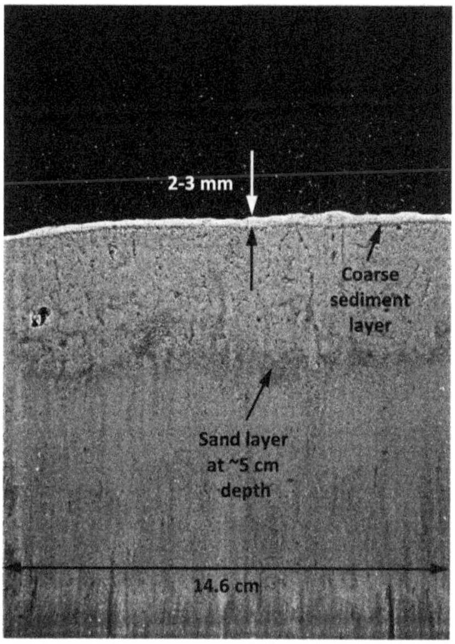

Fig. 6.16. Photo of fluff layer on otherwise solid bed as observed in the Passaic River (from Mathew and Winterwerp, 2017).

a sea carousel, respectively), reporting τ_e-values of 0.02–0.03 Pa. From field measurements in Long Island Sound, USA, Bedford *et al.* (1987) and Wang (2003) found τ_e-values of about 0.05 Pa, and El Ganoui *et al.* (2004) found τ_e-values of about 0.02–0.05 Pa, also from field measurements in the Rhone River and continental shelf, France.

Mathew and Winterwerp (2017) discuss the results of erosion experiments in the laboratory with a U-GEMS microcosm (Gust and Mueller, 1997) on ten undisturbed cores from the Lower Passaic River, USA. The critical shear stress for erosion measured about 0.04 Pa (except for one outlier at 0.075 Pa), while from the analysis of an extensive set of field data, $\tau_e = 0.03$ Pa was found.

Note that the power law in Fig. 6.2 would also give a drained strength (thus critical shear stress for erosion) of a few 0.01 Pa, assuming a gelling concentration of 100 g/L.

These U-GEMS experiments could not provide values for the erosion rate parameter M. However, from field measurements, the following values were reported (Bedford *et al.*, 1987): $M = 2 \cdot 10^{-6}$ kg/m^2/s (Wang, 2003): $M = 5 \cdot 10^{-6}$ kg/m^2/s (El Ganaoui, 2004): $M = 10^{-8}$–$6 \cdot 10^{-5}$ kg/m^2/s, and

Mathew and Winterwerp (2017): $M = 10^{-5} - 10^{-4}\,\mathrm{kg/m^2/s}$. These values are low to very low in comparison with M-values given in Sec. 6.1 obtained for alluvial muddy beds in the laboratory or in the field, which is likely related to the small values of τ_e. Further, we anticipate that these low values do not represent local erosion rates, but reflect horizontal inhomogeneities.

Such spatial inhomogeneities are particularly important in the real world, where sediment beds are characterized by a variety of bed forms (see Fig. 4.11) and other irregularities. Not only the sediment composition (for instances sand–mud distribution) is likely to vary spatially, but also the bed shear stresses are likely to vary largely over short distances. This has of course large implications for setting up numerical sediment transport models and the interpretation of their results. Thus two effects play a role:

(1) Fluff layers are expected not to be formed over the entire bed, and
(2) A fluff layer will entrain progressively over space, i.e. rapidly at one location, then at a second location, and then at a third, etc.

The net transport of fine sediment in estuaries and coastal seas cannot be understood without presuming the formation of a fluff layer during slack water. These net transports and the role of the fluff layer are subject of Sec. 9.2.2.

The effects of fluff layers on the fine sediment dynamics in estuaries and coastal seas can be modeled with for instance a two-layer approach, referred to as the buffer model. This type of modeling is discussed in Sec. 12.2.

6.6 Erosion of sand–mud mixtures

Sections 6.3, 6.4 and 6.5 dealt with cohesive sediment beds (see also Figs. 1.5(c) and 1.5(d)), or with "inert" beds with a loose layer of fine, fluffy sediment on top. Or, in terms of Fig. 1.4, these sections describe erosion under erosion-limited and settling-limited conditions.

In this section, we discuss the mobilization of fines from/in a granular (sandy) bed, sketched in Fig. 1.5(b). As the fines reside within the granular skeleton, i.e. within the pores of the sand, mobilization of fines is governed by the erosion of the sandy bed, and we refer to supply-limited conditions (Fig. 1.4). Whether a bed is cohesive or non-cohesive can be established from the activity plot, as presented in Sec. 2.2, e.g. Fig. 2.18 — see also Van Ledden *et al.* (2004).

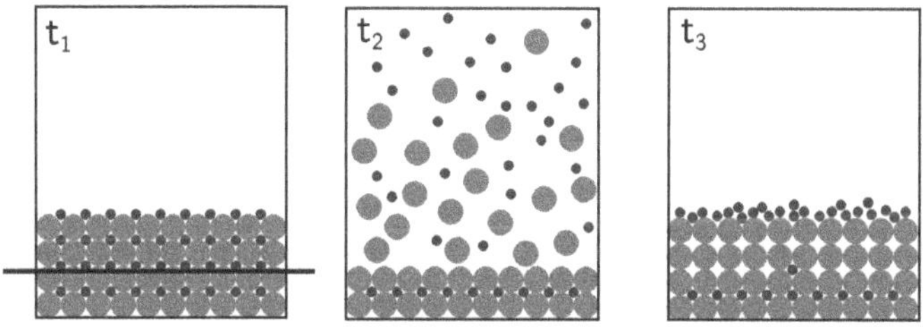

Fig. 6.17. Schematic of sorting from a sand–mud bed upon erosion.

As discussed at the beginning of this chapter, natural beds are in general either cohesive (possibly with some sand), or fully granular (possibly with a few % fines) because of hydrodynamics-induced sorting. This implies that the clay content of these beds are well beyond or well below its critical value ξ_0. Therefore the exact value of ξ_0 seems not too important. Note that spatially (for instance over a cross-section) a bed may exhibit both cohesive and non-cohesive features.

The schematic of Fig. 6.17 depicts an initially well-mixed bed, in which fines occupy the pores in between the coarser sand grains. When these sand grains are mobilized by hydrodynamic forces, for instance during accelerating tide, fines are washed out from the bed, and mixed over the water column. Owing to the smaller settling velocity of the fines, sorting occurs, and we retrieve the fluffy sediment system discussed in Sec. 6.5. During the next event (accelerating tide after the deposition during slack water), the fines are first picked up and mixed over the water column, and the coarser sand grains follow independently (and often later). Thus starting from supply-limited conditions, sorting induces settling-limited conditions. However, these fines may be re-buried into the sandy skeleton (Chapter 8), and supply-limited and settling-limited conditions may alternate.

When the hydrodynamic forcing increases, for instance towards spring tide or under storm conditions, deeper layers of sediment are mobilized, bringing more fines into the water column. This response is well illustrated by the SPM dynamics in the Dutch coastal zone of the North Sea, see Fig. 6.18. The right panel of Fig. 6.18 shows mean near-surface SPM-concentrations of a few 10 mg/L during summer, averaged over a number of spring-neap cycles. After a storm (left panel of Fig. 6.18) augmented SPM values up to a few 100 mg/L are encountered. As the horizontal scale in the Dutch coastal zone measures around 100 km, it is impossible that this increase in SPM is the result of horizontal advection: the fines must

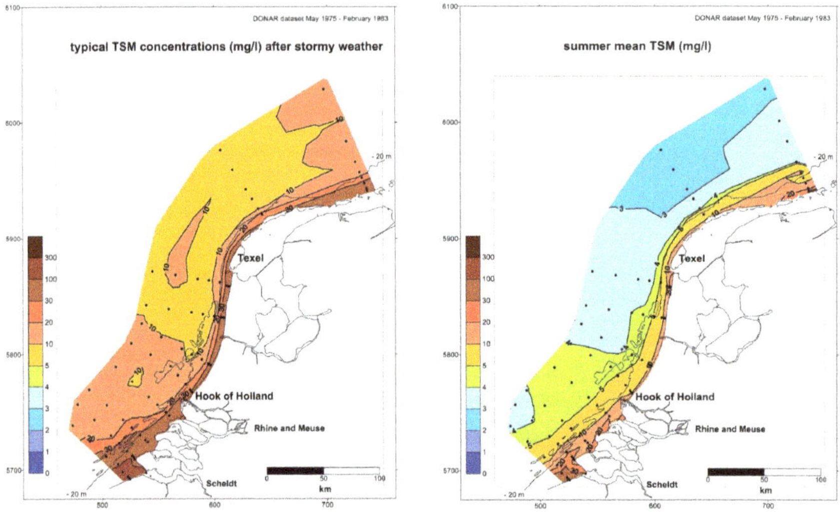

Fig. 6.18. Effect of storm conditions on SPM in the Dutch coastal zone (North Sea) as a function of wave conditions (from Suijlen and Duin, 2001).

have been entrained from within the bed (e.g. Van Kessel *et al.*, 2011a and Sec. 10.2.2).

During storm more sand is mobilized, thus entraining more fines into the water column. Unfortunately, this process is not well understood from a quantitative point of view, as studies on sand transport under currents and waves have focused on net effects, and hardly dealt with the thickness of the mobilized sand layer. However, given the amount of SPM in the water column after such a storm, it is likely that at least 10–20 cm sand is mobilized during storms, i.e. fines are washed out from this layer.

Furthermore, spatial variation in mud fraction along with bed erodability may play a role as well. With increasing hydrodynamic forcing, bed sediment will be mobilized over larger areas (see also Fig. 4.11 showing large irregularities in bed structure). Under the calm conditions following a storm, these fines must somehow remix with the coarser bed sediment as the patterns of Fig. 6.18 are observed frequently. Presently, three mechanisms have been identified (Fig. 6.19) to explain the re-entrainment of fines into the bed:

(1) Pore pressure gradients: Because of accelerating and decelerating flow over bed forms, pressure gradients within the bed are induced. Underpressures in the bed forms may then entrain fines into the bed. This

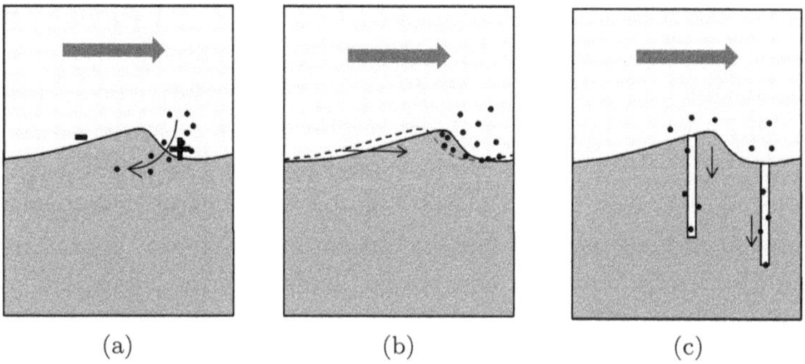

Fig. 6.19. Schematic of entrainment of fines into a granular, sandy bed: (a) pressure gradients over bed forms, (b) bed form migration, (c) bioturbation.

process was studied e.g. by Huettel *et al.* (1996) and Huettel and Rusch (2000). Also large scale pore pressure gradients, induced by waves and/or tidal variations, may entrain fines into a granular bed.

(2) Migration of bed forms: Fines accumulate in the troughs between bed forms (ripples). While migrating, such fines are buried into the bed, as proposed by Jennes and Duineveld (1985). Later, Martinius and Van den Berg (2011) showed that migrating bed forms bury large amounts of fines within the sediment bed of the Eastern Scheldt, The Netherlands.

(3) Bioturbation: Burying organisms are known to mix sediment over the upper 2–3 dm of a bed, a process referred to as bioturbation — see e.g. Graf and Rosenberg (1997) and Le Hir *et al.* (2007). Bioturbation is discussed in further detail in Sec. 8.3.

Webb and Theodor (1972), Van der Loeff (1981) and Harrison *et al.* (1983) argue that the contribution by pore water flows is the smaller one: entrainment of particulate matter appears effective mainly in coarse sediment under the influence of large pressure gradients. Moreover, advection by pore water flow is a slow process (Kleinhans *et al.*, 2005). The current view is therefore that the migration of bed forms is the more efficient burying process entraining fines into the upper part (cm's) of the bed, whereas bioturbation mixes these fines slowly to larger depths, as bioturbation rates are small (see Sec. 8.3).

Erosion rates of sand–mud mixtures

Van Ledden (2002, 2003) presented an analysis on the maximum mud content within in a granular bed. This analysis is based on the mass balance of mud in the water column, including the exchange with the bed, based on the Krone–Partheniades erosion/deposition formula. The equilibrium mud content ξ_∞^m within the bed under tidal conditions can be shown to depend on the relative settling flux F_{set} $(= W_s \xi_\infty^m / M)$ and the relative bed shear stress τ_b/τ_e. This relation indeed envelopes mud contents observed on mudflats in the Western Scheldt, as shown in Winterwerp and Van Kesteren (2004). However, as this relation does not contain mechanistic processes on the entrainment of fines into the bed, we do not elaborate on this relation.

Van Ledden proposed a conceptual model for the transport of fines in a mixed-sediment environment (see Van Ledden, 2002, 2003 and Van Ledden *et al.*, 2004). His model is based on the bed composition concept (Armanini, 1995). Van Ledden distinguishes between cohesive and non-cohesive beds, based on the critical clay content ξ_0, as discussed in Sec. 2.2. Note that Van Ledden refers to a critical mud content, referred to as ξ_0^m, thus including the effects of silt (i.e. $D_{50} < 64\,\mu$m).

For $\xi^m < \xi_0^m$, the total horizontal transport of sand F^s consists of bed load and suspended load, and is modeled through a modification of Van Rijn (1993) transport formula:

$$F^s = \nabla q_b + W_s^s \left(c_a - c^s\right) \tag{6.21}$$

in which q_b is the bed load, W_s^s is the settling velocity of sand, c^s is the suspended sand concentration, and c_a is a reference concentration, defined below. Van Rijn defines the following dimensionless parameters:

$$D_* = D_{50}\left(\frac{\Delta g}{\nu^2}\right)^{1/3} \quad \text{and} \quad T = \frac{\tau_b - \tau_{e,*}}{\tau_{e,*}} \tag{6.22}$$

in which D_{50} is the median diameter of the sand, Δ is the relative sand density $(\Delta = (\rho_s - \rho_s)/\rho_s)$, g is acceleration of gravity, ν is the dynamic viscosity of water, τ_b is the bed shear stress, and $\tau_{e,*}$ is the critical bed shear stress for sandy beds, as affected by the presence of mud. Van Ledden proposed a heuristic formulation for $\tau_{e,*}$. Later, Jacobs (2011) systematically investigated the influence of fines on the erodability of sand, and we prefer these latter formulations, which are summarized below.

The bed load q_b reads:

$$q_b = \frac{\alpha_{b1}\sqrt{\Delta g D_{50}^3}\, T_{nc}^{\alpha_{b2}}}{D_*^{0.3}} \tag{6.23}$$

The values for the coefficients α_{b1} and α_{b2} are given as follows:

- For $T < 3$: $\alpha_{b1} = 0.053$ and $\alpha_{b2} = 2.1$.
- For $T > 3$: $\alpha_{b1} = 0.1$ and $\alpha_{b2} = 1.5$.

The reference concentration c_a in Eq. (6.21) at height a from the bed reads:

$$c_a = 0.015 \frac{D_{50}}{a} \frac{T_{nc}^{1.5}}{D_*^{0.5}} \tag{6.24}$$

where T_{nc} represents the non-cohesive form of Eq. (6.22). Both q_b, c_a and a are identical to Van Rijn's original model. However, the value of $\tau_{b,*}$, thus T differs.

Van Ledden assumes that mud is washed out of the bed in conjunction with the mobilization of sand grains. As Eq. (6.23) yields the net flux of sand (the net bed load), Van Ledden used Van Rijn's relation between the gross erosion rate of sand E^s and q_b: $q_b = \lambda_s \times E^s$. Substituting from Van Rijn's model for the saltation length λ_s, Van Ledden proposes:

$$E^m = \frac{\alpha_{m1}}{3} \frac{\xi^m}{1 - \xi^m} \frac{\sqrt{\Delta g D_{50}}}{(D_* T_{nc})^{0.9}} T_{nc}^{\alpha_{m2}} \tag{6.25}$$

for the erosion of mud from a non-cohesive sand–mud bed, in which ξ^m is the mud content in the bed. Van Ledden proposed the following values for the coefficients in Eq. (6.23): $\alpha_{m1} = 0.075$ and $\alpha_{m2} = 1.9$.

Note that ξ^m does not appear in Eq. (6.23) or (6.24), implicitly assuming that the mud particles are few and small, residing only within the pores of the sand skeleton.

The latter is not the case in a cohesive bed ($\xi^m > \xi_0^m$), and the erosion rate for mud has to be corrected for the sand content ξ^s ($= 1 - \xi^m$). Furthermore, Van Ledden explicitly assumes that sand grains are washed out of the bed with the eroding mud bed, in proportion to their content. The erosion rate for sand E^s from a cohesive bed then becomes:

$$E^s = (1 - \xi^m)\, M_E\, (\tau_b - \tau_e) \quad \text{for } \tau_b > \tau_e \tag{6.26}$$

and the erosion rate for mud E_m from a cohesive bed reads:

$$E^m = \xi^m M_E\, (\tau_b - \tau_e) \quad \text{for } \tau_b > \tau_e \tag{6.27}$$

Role of fines on permeability

Jacobs (2011) developed erosion formulas for non-cohesive sand–mud mixtures (i.e. PI $< 7\%$), comparing the results with laboratory experiments. By adding fines to a sand skeleton, the soils permeability decreases, and pseudo-cohesive behavior may occur. Jacobs was able to quantify this pseudo-cohesive behavior on the undrained strength:

$$c_u = B_1 W_{rel}^{B_2} e^{\alpha \lambda_B} \tag{6.28}$$

The empirical parameters followed from calibration against laboratory experiments: $\alpha = 0.12$, $B_1 = 2.8\,\text{kPa}$ and $B_2 = -2.5$. Bagnold's (1954) linear concentration λ_B is defined as

$$\lambda_B = \frac{1}{(\phi_s/\phi_{s,\max})^{-1/3} - 1} \tag{6.29}$$

in which $\phi_{s,max}$ is the maximal volume concentration (packing) of a sand–silt skeleton, accounting for the sediment composition. The relative water content W_{rel} is defined as

$$W_{rel} = W/\text{PI}^* \tag{6.30}$$

in which PI* is the generalized plasticity index, i.e. obtained from the soils activity (Eq. (2.3)), as direct measuring PI is not possible for these non-cohesive soils.

Jacobs did not derive a relation for the effects of pseudo-cohesion on the critical shear stress for erosion τ_e of sand–mud mixtures. His experiments, however, show that in the range of $D^* = 1$–2, τ_e may increase by a factor three, compared to Shields original diagram for the onset of erosion of granular materials.

6.7 Biological effects

The stability of sedimentary beds, their composition, and even the hydrodynamic stresses on the bed may be affected largely by biota. Phenomenologically, these biological processes are discussed in Chapter 8. In this section, we elaborate on how erosion and bed composition can be influenced by biological and ecological effects and subsequently how such effects can be modeled.

Previously, a distinction was made between surface erosion and mass erosion. The rate of mass erosion is governed by inhomogeneities within the bed. Biotic activity has a major effect on these (in)homogeneities. This includes roots of vegetation, and burrows and trails by benthic species. For example, *Arenicula marina* (lugworm) digs burrows of 10s of centimeters long, (Volkenborn *et al.*, 2007). Cockels (*Cerastoderma edule*) can have an abundance of 100s–1000s of individuals per m^2. Beds with high abundance of macrobenthos are therefore very inhomogeneous and sensitive to mass erosion. Vegetation roots generally armor the soil, see Gedan *et al.* (2011). This would reduce mass erosion.

The effects of biota on surface erosion is analyzed by elaborating on the erosion rate formula:

$$E = M_E \left(\tau_b - \tau_e\right) \tag{6.31}$$

with erosion rate E, erosion rate coefficient M_E, bed shear stress τ_b and critical shear bed stress for erosion τ_e. All parameters and variables on the right-hand side of the equation can be affected by biota. We discuss these effects below.

Biological effects on the critical bed shear stress for erosion

One of the most studied effects of biota on erosion of muddy beds is related to the modification of the critical bed shear stress, or resistance to erosion. The formation of biofilms by algae or microphytobenthos leads to an increase of the critical bed shear stress τ_e, see Widdows *et al.* (1998); Austen *et al.* (1999); Tolhursf *et al.* (2002). It can increase by a factor 3–5, as elaborated by Le Hir *et al.* (2007), compared to beds without microphytobenthos — see Sec. 8.1 for a phenomenological description. Correlations have been made between chlorophyll-*a* contents and τ_e, using *in-situ* devices such as the cohesive strength meter, or in laboratory set-ups such as the annular flume (see Widdows *et al.* (2007) for an inter-comparison). Only non-damaged (intact) biofilms can increase τ_e. When disrupted, the biofilm is torn apart and τ_e drops again.

Biofilms are seasonal phenomenona, they come and go, depending on light, temperature, nutrients, etc. Furthermore, microphytobenthos sits at the base of the food chain for various higher trophic species. Weerman *et al.* (2011) studied the pattern formation by microphytobenthos on a mudflat in spring. Collapse of these spatial patterns is induced by an increasingly number of benthic herbivores, grazing on the biofilms. An example is presented in Fig. 6.20 showing the variation over time of SPM-concentrations measured in the Ems-Dollart, on a tidal flat ("Heringplaat") and in an adjacent channel ("Groote Gat"), an estuary in the northern of The Netherlands. This figure shows that SPM-concentrations are low when microphytobenthos is abundant, i.e. during spring season, with a second dip at the end of the summer (De Deckere *et al.*, 2002). It was observed that the abundance of meiobenthos on the mudflats increased in June, accompanied by the

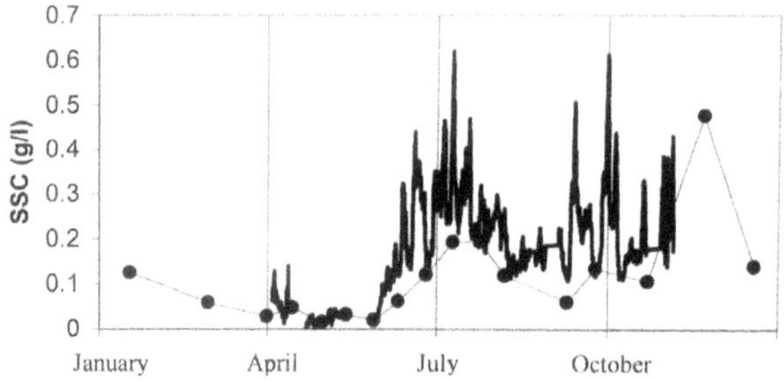

Fig. 6.20. Seasonal variation in suspended sediment concentration in the Dollart estuary ("Heringplaat" and "Groote Gat"), The Netherlands measured in 1996 (after De Deckere *et al.*, 2002).

decay of the microphytobenthos mat and the reduction of sediment stability. This is ascribed to the grazing of meiofauna on the microphytobenthos, as explained above. To the end of September, temperature and light climate become unfavorable and storm-induced waves more frequently. As a result the biofilms are no longer maintained by the microphytobenthos, while disrupted mechanically by the waves. Thus SPM-concentrations increase again in autumn and winter. These studies illustrate a strong interaction between sediment dynamics, influence of biota on sediment properties and interaction between species (see also Orvain *et al.*, 2004).

Some benthic species disturb the bed mechanically, affecting the erosion threshold. Species such as the *Hydrobia ulvae* (a small mud snail) crawl over and through the upper centimeters of the bed, loosening its structure and bonds (Orvain and Sauriau, 2002). Other species, such as the bivalve *Macoma balthica* search with their siphons for food on the bed surface, loosening the top layer as well, increasing the availability of erodible sediment (see Van Prooijen *et al.*, 2011). This "bulldozering" produces a fluff layer atop an undisturbed bed, likely lowering the critical bed shear stress for erosion. Quantifying this effect is however not straightforward, as it depends on the variety of species, the number of individuals, their activity, etc., which may vary over seasons as well. From a physical point of view, these biological effects at system scale depend on the amount of sediment that is affected. Note that these activities may have an impact on the erosion rate coefficient M_E as well (see below).

Reef structures, such as oyster reefs and mussel beds modify the composition of the seafloor largely (see also Chapter 8). It can be argued that the critical bed shear stress for erosion is modified within and around such reefs. However, at larger scale, reefs affect flow velocities, wave impact and sediment transport paths, thereby affecting the water-bed exchange processes as well. It is therefore difficult to isolate the effect of a reef on τ_e.

Biological effects on hydrodynamic forcing

The microphytobenthos mats addressed above form a smooth film atop the sediment bed, reducing skin friction values locally, thereby affecting the bed shear stress τ_b and vertical mixing. Reefs, mussel beds and other hard biota on the bed substrate generally enhance local bed shear stresses as well — a practical example is provided in Sec. 17.4.3.2. Reefs can have an impact on the larger scale of flow and wave patterns over and around reefs (see e.g. Walles *et al.*, 2015).

Vegetation has various effects on the flow, as elaborated by Nepf (2011, 2012). Generally speaking, vegetation induces extra drag, reducing flow velocity and/or increasing the water level slope locally. The plants themselves generate small-scale turbulence. In channels with vegetation, a significant increase in turbulent kinetic energy and bed shear stress is observed in comparison to channels without vegetation, at the same velocities and water depths (Yang *et al.*, 2015). It therefore depends on the local conditions whether the decrease in mean flow velocity or the increase in small-scale turbulence wins, resulting in reduced or enhanced bed shear stresses. Le Hir *et al.* (2007) summarized observations from literature and concluded that Manning's roughness coefficient n may vary from 0.018 to 0.200 for emerging vegetation — compare with the typical values given in Table 3.3.

As an example, consider salt marshes, characteristic vegetated areas along the higher fringes of estuaries. These marshes are cut through by creeks. The drag by vegetation reduces the flow velocity over the marshes, thereby enhancing the flow through the creeks (Temmerman *et al.*, 2007). Erosion of the vegetated areas is small, while enhanced in the creeks, see also Sec. 8.5.

Summarizing, various biotic "structures" induce drag to the flow, reducing flow velocities. Such structures do however also generate extra turbulence, which can enhance the local bed shear stress and therefore erosion and vertical mixing.

Biological effects on the erosion rate coefficient

The erosion flux parameter M_E determines how fast material erodes from the bed after the erosion threshold is exceeded. Various studies have been carried out to determine the effect of benthic species on M_E. For example, Andersen *et al.* (2002) has shown that *Hydrobia ulvae* (mudsnail) increases the erosion rate by loosening large amounts of fines. A process-based relation between biological activity and the erosion rate coefficient has, however, not yet been made. Conceptually, the surface erosion rate equation (6.15) may provide such a process-based formulation:

$$M_E = \frac{c_v \rho_{dry} \phi_s}{\alpha_\delta D_{p,50} c_u} \quad \text{for} \quad \tau_b > \tau_e \tag{6.32}$$

with dry density ρ_{dry}, volumetric solids concentration ϕ_s, coefficient α_δ, particle diameter $D_{p,50}$ and undrained shear strength c_u. It is likely that benthic species may change several parameters, i.e. ρ_{dry} and c_u by ingressing or egressing pore water, c_u and c_v by loosening the bed structure or cementing particles within the bed, and $D_{p,50}$ by breaking down or formation (pelletization) of flocs. Yet, even if successful, practical application in engineering models is difficult as these biological effect vary over time (tidal scales to seasons and beyond) and space. Moreover, mutual interaction between species may play a crucial role, as for instance the grazing of benthic herbivores on microphytobenthos.

Accounting for biological effects on erosion

The effects of biology on erosion rate and transport of (fine) sediment are ubiquitous. Accounting for them in numerical models is not straightforward however, because of the large spatial variations and sharp gradients in abundance of species, and temporal variations in behavior and presence (tidal scales to seasonal and longer). Furthermore, the variety of species and their interaction forbids a deterministic approach. Attempts have been made to model the effects of some biology on sediment transport (Orvain *et al.*, 2004; Van Prooijen *et al.*, 2011). Such models do however require a range of parameters for calibration.

More promising is the identification of groups of communities instead of elaborating on individual species. For instance, concepts/algorithms can be developed to study possible positive or negative feedback loops between the effects of living species on the transport and fate of fine sediment. In other words, a parameterization is required.

This implies that a conceptual understanding should be developed on the large impact of biology on erosion and its modeling, as well as on the potential temporal and spatial variations therein. Simulation of sediment dynamics therefore requires at least a qualitative overview of the type of species and abundance that is present in the area of interest.

Table 6.1. Occurrence of floods and bathymetrical surveys in Lower Passaic River.

Major flood	Survey
March 2007	September 2007 November 2008
February 2010	June 2010
September 2011 (Irene)	October 2011 September 2012

6.8 The effects of navigation

Ships in shallow water may mobilize large amounts of sediment, in particular when mooring and maneuvering. The major source of stirring is the ship propeller, but also return flow along and under a ship may stir up sediment from the bed.

Here we present some qualitative data to show the large potential impact of navigation on fine sediment dynamics, analyzing bathymetrical observations in the Lower Passaic River, USA. Five bathymetric surveys were carried out, before and after major floods in the river, as summarized in Table 6.1. The Lower Passaic River is relatively short (17 miles), debouching into Newark Bay, still under tidal influence, with a tidal range of 0.9–2.1 m. Mean river flows measure 34 m^3/s, allowing saline water to penetrate halfway the river. Erosion/scour of the river bed occurs at flow rates beyond 220 m^3/s. At lower flow rates, the river imports sediment from Newark Bay — for further details, see Mathew and Winterwerp (2020).

By comparing subsequent bathymetries, changes in the volume of the river can be determined. Further to Table 6.1, bathymetrical changes between 2007 and 2008, and between 2011 and 2012 would represent the effects of a prolonged period of low river flow, whereas the changes between 2008 and 2010, and between 2010 and 2011 would reveal the river's response to peak flood conditions.

The cumulative changes in bathymetry are presented in Fig. 6.21. The 2010–2008 and 2011–2010 curves represent erosion by the river flow (in combination with tide–induced velocities). The eroded fine sediment accumulates in the lower reaches of the river, below RM 2. This picture can be understood from the trumpet shape of the river, becoming more pronounced in the river mouth (Fig. 6.22). Here flow velocities, thus bed shear stresses, become so low that the eroded sediment can settle and accumulate. This

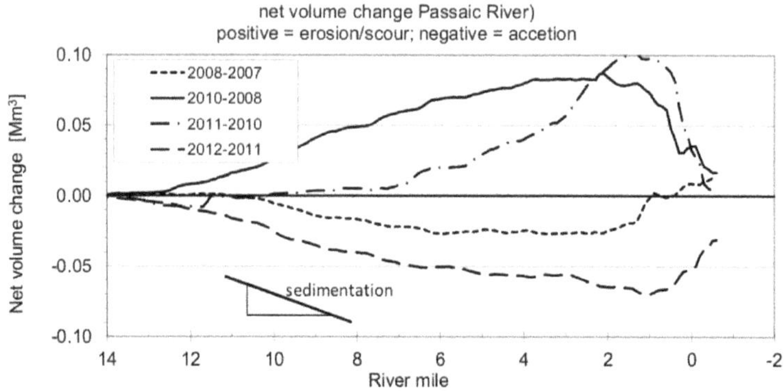

Fig. 6.21. Cumulative change in water volume of the Lower Passaic River — a positive slope of the curves reflects local erosion, while a negative slope sedimentation (milage measures positive up-river).

behavior was well simulated with a numerical sediment transport model, using the traditional erosion/deposition formulas discussed previously.

On the other hand, the 2008–2007 and 2012–2011 curves reflect sedimentation in the major part of the river. However, erosion is observed in the lower two miles. This erosion cannot be understood from the river/tide-induced flow velocities and bed shear stresses, as these are (much) lower than those during the floods of 2007, 2008 and 2011.

The only reasonable explanation for the erosion of sediment from the river mouth is navigation. To sustain this argument, 15 days of navigation were recorded, as presented in Fig. 6.22, together with the measured changes in bathymetry of 2011–2010 and 2012–2011. Figure 6.22 shows in particular intense mooring activities along the key walls in this area.

Only by adding a heuristic navigation-related erosion term to the numerical model, it was possible to simulate the curves in Fig. 6.21. Unfortunately, no generic formulation exists today, in spite of its large effect (in this river). However, a few guidelines can be given:

(1) If shipping is very frequent, a constant local source term can be added, simulating local scour by ships. If not this source term should be synchronized with the tide.
(2) Distinguish between areas where ships just sail, and where they maneuver, e.g. moor. Maneuvering ships stir up large amounts of sediment.

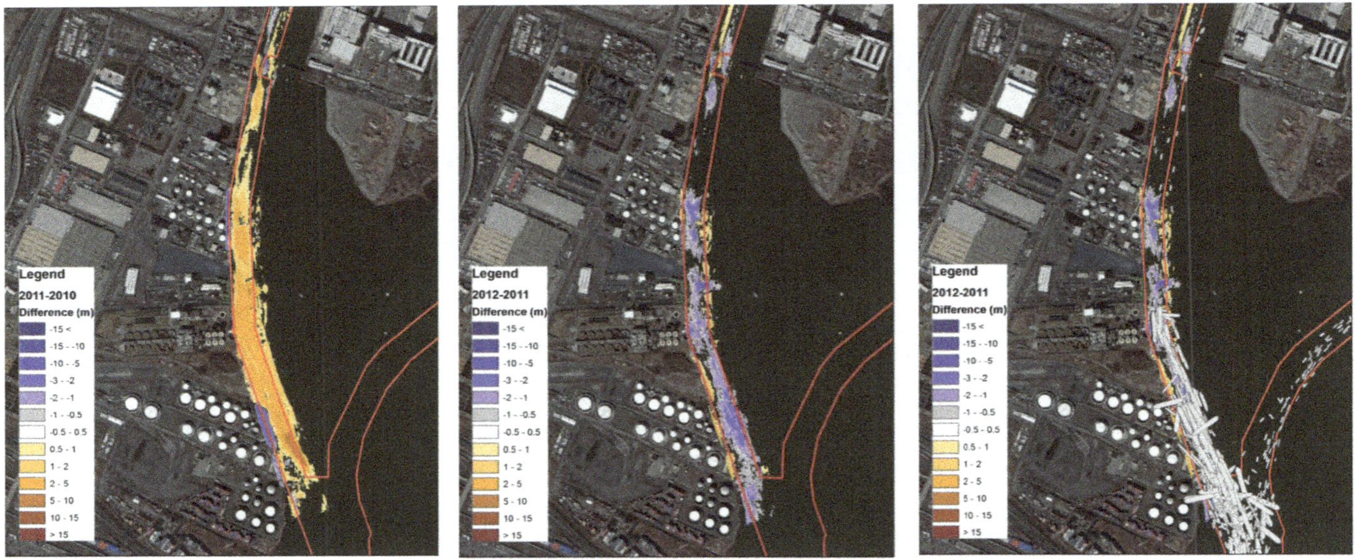

Fig. 6.22. Changes in bed level in the mouth of Lower Passaic River (left panel: 2011–2010, after erosive river flood; middle panel: 2011–2012, low river flow and river infill) and 15 days of ship movements (right panel).

(3) When scour holes become deeper, the efficiency of ship-induced scour reduces till an equilibrium is attained. A zero-order estimate of an equilibrium scour hole can be obtained by computing the bed shear stresses in these holes, modifying their shape and depth till the stresses no longer exceed the critical bed shear stress for erosion (e.g. Taylor *et al.*, 2007).

Chapter 7

Fluid Mud

The term "fluid mud" was probably first introduced by Inglis and Allen (1957) in their work on the sediment dynamics in the Thames estuary (UK). Since then, sediment suspended in water over a wide range of concentrations has been referred to as fluid mud. In literature, also the terms "creme de vase" (Migniot, 1968) and "sling mud" (NEDECO, 1968) have been used. Kirby and Parker (1983) have tried to bring order in the terminology, distinguishing between:

(1) a settled bed with a density of 1300 to 1700 g/L;
(2) a stationary suspension with a sediment concentration between about 0.1 and 200 g/L;
(3) a mobile suspension with a sediment concentration between about 1 and 150 g/L.

These definitions are not unambiguous because of the large overlap in concentration range. Later, Parker *et al.* (1994) advocated the term Hyper-Concentrated Benthic Layers instead of fluid mud, as the latter is generally related to its detection with double-frequency echo-sounders.

McAnally *et al.* (2007) proposed the following definition: "Fluid mud is a high concentration aqueous suspension of fine-grained sediment in which settling is substantially hindered by the proximity of sediment grains and flocs, but which has not formed an interconnected matrix of bonds strong enough to eliminate the potential for mobility."

In this book, we distinguish between three regimes of cohesive sediment suspensions, introducing a slightly different definition of fluid mud (see also Winterwerp, 1999):

(1) Fluid mud is a suspension/mixture of cohesive sediment at a concentration around (or a bit larger than) the gelling point (e.g. Eq. (4.12)),

i.e. of the order of several 10 to 100 g/L. In terms of the CSM, fluid mud is in a consolidating state. Effective stresses are negligible or small, but fluid mud exhibits highly non-Newtonian behavior, and it is either stationary or mobile. When mobile, fluid mud flow may be laminar or turbulent, its dynamics more or less disjunct from the flow in the water column above. When turbulent, one generally refers to turbidity currents, or debris flow when larger particles (up to boulders!) are carried by the flow.

(2) A high-concentration mud suspension (HCMS) is a suspension of cohesive sediment at concentrations of a few 100 to a few 1000 mg/L. This suspension has Newtonian properties, though possibly with a somewhat elevated viscosity, and is transported with the main flow. An important feature of HCMS is its interaction with the turbulent flow field (HCMS is treated in Sec. 7.1).

(3) A low-concentration mud suspension (LCMS) is a suspension of cohesive sediment at a concentration of the order of several 10 to maybe a few 100 mg/L. This suspension has Newtonian properties and does not (significantly) affect the turbulent flow field.

These phases may occur simultaneously, or alternatively at one location, depending on the driving forces in the aquatic system.

Fluid mud can only be stable under the influence of an external energy source, such as waves, horizontal pressure gradients and/or liquefaction by pumping water into the pores. If not, fluid mud is in a transient state, consolidating over time. As a result, dry density (and density gradients) and mechanical properties vary over the thickness of the fluid mud layer, and over time. For instance, Fig. 4.19 suggests that lower within the fluid mud, effective stresses already develop, while higher in the mud layer, the suspension is still in a hindered settling phase. These non-equilibrium properties of fluid mud are likely one reason for the different definitions/observations encountered in literature.

For practical purposes (e.g. dredging, navigation), dual frequency echo-sounders are used for detecting fluid mud layers. An echo-sounder emits an acoustic pulse through the water column, which reflects from layers with strong density contrast, such as the seabed in sandy environments. The distance from the acoustic source to the reflecting surface is proportional to the travel time of acoustic waves in the water column. In muddy environments, different frequencies for the emitted signal are deployed. Standard high frequency (about 210 kHz) and low frequency (about 33 kHz) signals are used to provide information about the water-fluid mud interface (lutocline) and the bed-fluid mud interface, respectively. Traveling with small wavelength, the high frequency reflection typically defines the surface of the fluid mud. However, the low frequency response is often irregular, inconsistent and exhibits many secondary echo's, owing to weak density gradients deeper within fluid mud layers (Kirby *et al.*, 1980).

Today other techniques exist for detecting and quantifying fluid mud appearances, e.g. Kirichek *et al.* (2018); McAnally *et al.* (2007). Detailed discussions on these techniques is however beyond the scope of this book.

7.1 High-concentration mud suspensions

The formation and properties of fluid mud can best be explained starting from a more dilute high-concentration mud suspension (HCMS). At SPM-concentrations beyond several 100 mg/L, SPM-induced density gradients become so large that turbulent mixing is suppressed, e.g. Secs. 3.2.3 and 3.4. This is illustrated in Fig. 7.1, which compares the measurements and 1DV-simulations (see Winterwerp (2001) for details on this model) of a 6 g/L sand suspension at 0.17 m water depth in a tilting flume with clear water conditions. The left panel shows a reduction in eddy viscosity by about 40%, whereas the right panel shows acceleration of the flow, as discussed in Sec. 3.2.3. Note that these sand concentrations are considerably larger than the few 100 mg/L mentioned above, because of the large settling velocity of the sand (≈1 cm/s), and the small water depth — see also Winterwerp (1999) and Winterwerp and Van Kesteren (2004).

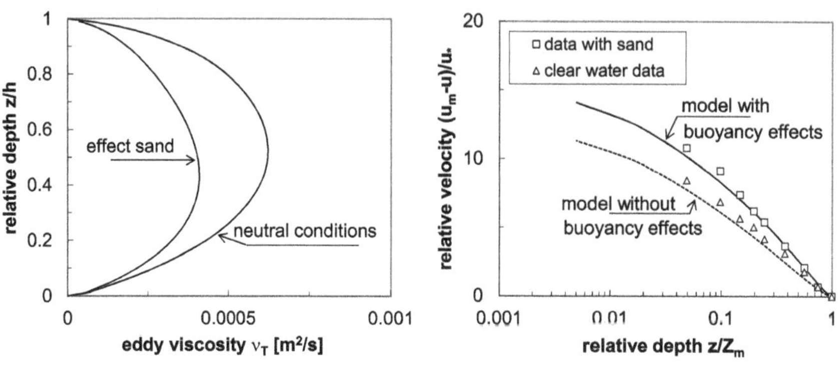

Fig. 7.1. Effect of 105 μm sand in 0.17 m tilting flume, computed with a 1DV-model, and compared with data by Coleman (1981).

The degree of (sediment-related) buoyancy-induced turbulence damping is measured by the flux Richardson number Ri_f, which is the ratio between the (potential) energy necessary for mixing a suspension over the water column, and the (turbulent kinetic) energy available for that mixing. Ri_f can therefore be considered as an efficiency parameter for vertical mixing:

$$Ri_f = -\frac{g\overline{w'\rho'}}{\rho\overline{u'w'}\partial u/\partial z} \tag{7.1}$$

where u' and w' represent the horizontal and vertical turbulent velocity fluctuations, ρ' the turbulent (SPM-induced) density fluctuations and $\partial u/\partial z$ the vertical mean velocity gradient (e.g. Sec. 3.2.3).

By assuming a logarithmic velocity profile ($\partial u/\partial z = u_*/\kappa z$), the corresponding parabolic viscosity profile ($\nu_T = \kappa u_* z\,(1 - z/h)$), local equilibrium between settling and mixing $\left(\overline{w'c'} = -w_s c\right)$, and taking into account hindered settling effects (the Richardson-Zaki formula, Eq. (4.36)), Eq. (7.1) can be written as:

$$Ri_f \propto \frac{\rho_s - \rho_w}{\rho_s}\frac{ghw_{w,0}c_{gel}}{\rho u_*^3}\phi_f(1-\phi_f)^5 \tag{7.2}$$

where $w_{s,0}$ is the reference settling velocity of a single sediment particle in still water, c_{gel} is the gelling concentration, and ϕ_f is the volumetric concentration of the mud flocs.

Experiments have shown that Ri_f attains a maximum value, referred to as $Ri_{f,cr}$, of about 0.15 to 0.2, beyond which turbulence collapses (Turner, 1973). This implies a maximum sediment load which can be carried by the flow. We refer to capacity conditions, and the accompanying SPM-concentration is indicated by the saturation concentration c_s. These capacity conditions are very similar to the capacity sediment transport discussed in Chapter 1. Substitution from Eq. (7.2) gives

$$c_s\,(z) = \frac{Ri_{f,cr}\rho}{\Delta g\kappa}\frac{u_*^3}{hw_s}\left(\frac{h}{z} - 1\right) \quad \text{with } w_s = w_{s,o}(1-\phi_f)^5 \tag{7.3}$$

where $\Delta = (\rho_s - \rho_w)/\rho_s$. Intuitively, it is to be expected that capacity conditions are met when Ri_f becomes critical over the entire water depth. This assumption is sustained by the analysis of data from the Amazon mouth by Trowbridge and Kineke (1994), showing a more or less constant

Richardson number at capacity conditions, and it was further elaborated by Toorman (2008). Thus a relation for the depth-averaged saturation concentration C_s can be derived (see Winterwerp, 2001):

$$C_s \equiv \frac{1}{h}\int_0^h c_s \mathrm{d}z = k_s \frac{\rho}{\Delta}\frac{u_*^3}{ghw_s} \approx K_s \frac{\rho}{\Delta}\frac{U^3}{ghw_s} \tag{7.4}$$

C_s can thus be interpreted as the sediment load that can be carried by a turbulent flow, i.e. the flow carrying capacity, thus being similar to the capacity concentration C_{cap} defined earlier (Chapter 1).

Numerical simulations with a 1DV-model, including the k–ϵ turbulence closure model with SPM-induced buoyancy destruction, give $K_s \approx 1.4\cdot10^{-4}$, (see Winterwerp, 2001). Note that Eq. (7.4) is very similar to Bagnold's auto-suspension criterion (Parker *et al.*, 1986).

Sediment stability diagram

Equation (7.4), with Richadson–Zaki's hindered settling formula, is plotted in the sediment stability diagram of Fig. 7.2. This diagram distinguishes between three regimes:

(1) Sub-saturated conditions for HCMS and LCMS along the left-lower branch of the curves: $Ri_f < Ri_{f,cr}$. The sediment-laden flow is below

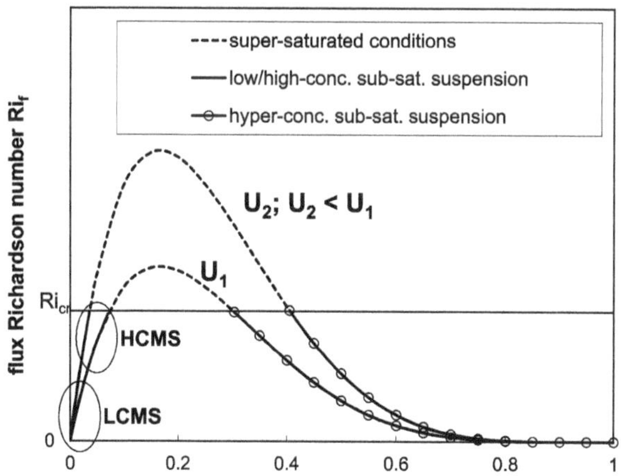

Fig. 7.2. Stability diagram for sediment-laden flow, showing stability curves for two different flow velocities U_1 and U_2 - for $Ri_f > Ri_{cr}$ turbulence collapses.

or at its carrying capacity and still Newtonian, possibly with some elevated viscosity.
(2) Sub-saturated hyper-concentration conditions along the right-lower branch: $Ri_f < Ri_{f,cr}$. The sediment-laden flow is below or at its carrying capacity, but highly non-Newtonian.
(3) Super-saturated conditions: $Ri_f > Ri_{f,cr}$. The sediment-laden flow is beyond its carrying capacity. The turbulent flow field collapses, and fluid mud can be formed (see next section). This is a "forbidden" region in the stability diagram.

It is illustrative to analyze the response of the sediment-laden flow in Fig. 7.2 to changes (1) in flow velocity, and (2) in SPM-concentration:

(1) Suppose that at HCMS-conditions, the flow velocity decreases from U_1 to U_2, while SPM-concentrations remain constant. When passing $Ri_{f,cr}$, the flow becomes supersaturated and fluid mud is (potentially) formed. As an example, such conditions are encountered when (high-concentration) SPM suspensions enter a harbor or navigation channel.
(2) Increasing the SPM-concentration along the left-lower branch of the stability diagram, at constant flow velocity (f.i. U_1), critical conditions are eventually met and the flow becomes supersaturated. SPM-values may increase by for instance wave-induced erosion (referred to as "auto-saturation" by Winterwerp (2001) or mechanical stirring of the bed.
(3) However, with increasing SPM-concentration along the right-lower branch of the stability diagram, at constant flow velocity, the sediment-laden flow becomes more stable. Supersaturated conditions are only met when the SPM-load decreases. These hyper-concentration conditions are not further discussed in this book — the reader is referred to Winterwerp (2001), Winterwerp *et al.* (2006a) and Toorman (2008).

Though this scheme does not distinguish between cohesive and granular material, Winterwerp (2001) argues that it is relevant in particular for cohesive sediment. When exceeding capacity conditions, a (thin) layer of fluid mud is formed, reducing turbulence production at the bed considerably, further reducing vertical mixing. This is not the case for sandy sediment, as beyond-capacity conditions, depositing grains form a rigid bed immediately, where full turbulence production remains possible.

7.2 Fluid mud formation

Fluid mud can be formed by deposition from the water column (as in the example above) and/or liquefaction/fluidization of a sediment bed. We distinguish between:

(1) Liquefaction yields the collapse of floc structure in the soil by external stresses, at constant water content. Liquefaction is therefore an undrained process, generating pore water over pressures.
(2) Fluidization occurs when water is pumped into the soil, either by natural processes, or mechanically (e.g. water injection dredging). Fluidization is a drained process, also generating pore water over pressures.

7.2.1 *Fluid mud formation from deposition*

A necessary condition for the formation of fluid mud by deposition is that the sediment's (initial) consolidation rate is smaller than its deposition rate. This condition can be formalized with Eq. (5.34), as

$$w_s \phi_{\text{SPM}} > \frac{\rho_s - \rho_w}{\rho_w} k \phi_{fm}^2 \tag{7.5}$$

in which ϕ_{SPM} is the volumetric concentration of the flocs in the water column, and ϕ_{fm} the volumetric concentration of the solids in the fluid mud layer, w_s is the (effective) settling velocity of the mud flocs, and k is the hydraulic conductivity within the fluid mud layer. ϕ_{fm} should be close to the gelling concentration, and in the following we assume $\phi_{fm} = 1$. Substitution from Eq. (4.36) and (5.10) gives the necessary condition for fluid mud formation:

$$\frac{w_{s,0}}{K_k} > \Delta \frac{(c_{fm}/\rho_s)^{\frac{4-2n_f}{3-n_f}}}{c_{\text{SPM}}/c_{gel}(1 - c_{\text{SPM}}/c_{gel})^5} \tag{7.6}$$

Equation (7.6) is plotted in Fig. 7.3 for $n_f = 2.6,\ 2.7$ and 2.8, and $c_{gel} =$ 100 and 150 g/L. Typical values for $w_{s,0}$ measure 0.1–1 mm/s, whereas K_k varies between 10^{-15} and 10^{-12} m/s, so that $10^{-12} < W_{s,0}/K_k < 10^{-8}$. This range is highlighted by a gray area in Fig. 7.3, showing that necessary conditions for fluid mud formation are generally met, unless consolidation rates are very high (i.e. large n_f, thus sand!).

Whether fluid mud is actually formed depends also on the local hydrodynamic (turbulent) conditions, which determine whether flocs can settle

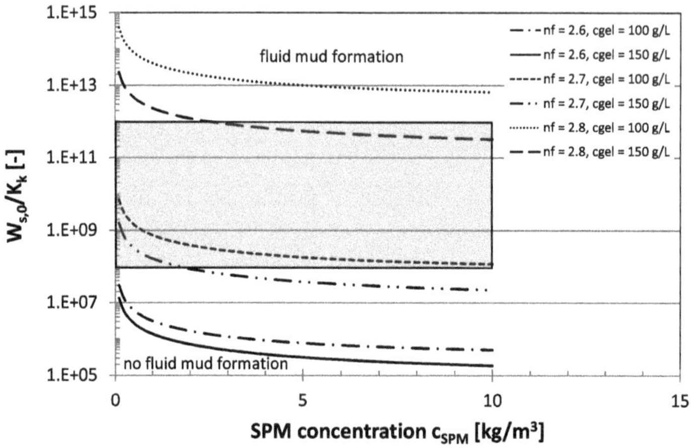

Fig. 7.3. Potential conditions for fluid mud formation for a variety of n_f and c_{gel}-values. The gray area represents common values for $w_{s,0}/K_k$.

on the bed, or are being picked up immediately. Figure 7.3 should therefore be interpreted as the potential for fluid mud formation.

Note that the formation of fluff layers, as discussed in Sec. 6.5, can also be considered as fluid mud formation, but at low n_f and low c_{gel}.

Fluid mud formation from sedimentation is a common feature in harbor basins and fairways, where flow velocities are (much) smaller than in the ambient waters, thus generating super-saturation conditions. Also in open sea, fluid mud may be formed from deposition, for instance after storm events, as discussed in the next section.

7.2.2 *Fluid mud formation from liquefaction*

Many laboratory experiments have shown liquefaction of mud layers by waves, e.g. Sakakiyama and Byker (1989); Maa and Mehta (1990); Maa *et al.* (1993); Foda and Tzang (1994); De Wit and Kranenburg (1993); Soltanpour *et al.* (2003); Vinzon and Mehta (1998); Van Kessel and Kranenburg (1998). While liquefying, pore water pressures increase, and at full liquefaction, pore water stresses equal the local hydrostatic pressure of the full suspension, i.e. effective stresses become zero (Foda and Tzang, 1994).

The stresses within a bed, induced by cyclic waves, follow butterfly patterns in the p–q diagram of the critical state model, at constant void ratio e, as liquefaction is an undrained failure process. Initially, the response of the bed is elastic, i.e. fully reversible, as sketched in Fig. 7.4. Over time,

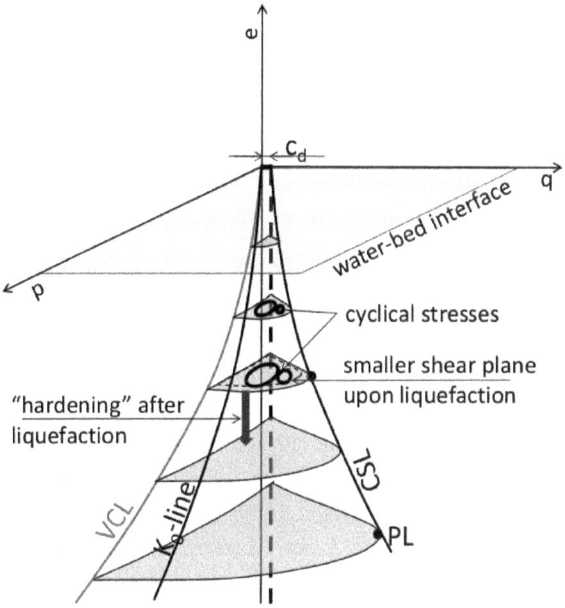

Fig. 7.4. Conceptual stress paths for liquefaction in the CSM at two depths, exhibiting elastic butterfly patterns induced by cyclic wave stresses. With time, the yield curves decrease by fatigue processes, and the soil may yield (fail), after which it hardens — see also Winterwerp and Van Kesteren (2004).

the yield curves may decrease in size by fatigue breakup of the soil structure (floc breakup), upon which the sediment structure yields (collapses, see also Winterwerp and Van Kesteren, 2004). Such breakup is a plastic response of the soil, hence irreversible. Failure takes place at the wet side of the diagram, i.e. the soil liquefies. Pore water can therefore escape from the soil, driven by the pore water over-pressures induced during wave activity. The soil densifies and becomes stronger — the yield curves shift towards lower e-values in the p–q–e diagram of Fig. 7.4. Thus, during liquefaction, the soil first weakens, and then becomes stronger ("hardens"), i.e. gets into an over-consolidated state (Sec. 5.2).

Winterwerp (2011) evaluated the potential for liquefaction of muddy soils from the Wadden Sea in The Netherlands in response to wave action. This study was carried out with triaxial tests on a number of unremolded samples from the Wadden Sea, grabbed from the intertidal, hence in an over-consolidated state (Sec. 5.2). The cyclical stresses were determined on the basis of wave measurements and numerical modeling, assessing the largest stresses to be expected. None of the samples underwent full liquefaction, i.e. full dissipation of effective stresses, even at wave loading well beyond the most extreme to be expected in the Wadden Sea. However, some increase in pore water pressure was found, and some energy losses by internal friction were measured.

As these results differ considerably from the results of laboratory experiments reported in the literature, a re-analysis of the literature was carried out:

(1) As far as reported by the authors, all yield strengths of the sediment used in the laboratory experiments measured a few tens to maximal 50 Pa. Mud beds in the flumes were obtained from normal consolidation or made from a slurry (Sec. 5.3). In contrast, the samples from the Wadden Sea were around the liquid limit, i.e. of the order of 1 kPa, thus at least more than one order of magnitude stronger.
(2) Lindenberg *et al.* (1989) carried out cyclical tests on kaolinite and unremolded soil samples from Lake Markermeer. The kaolinite samples could be liquefied easily, whereas the natural mud sample could not be liquefied. Next, the natural mud was remolded thoroughly at constant water content, while gas was allowed to escape — again no liquefaction could be attained. Finally, the sample was oven-dried, removing all organic material and gas, rewetted and remolded, upon which the mud liquefied as easily as kaolinite. The authors did not conclude which of the treatments was more important. However, the presence of gas can of course postpone liquefaction considerably.
(3) As discussed above, recovery after wave-induced liquefaction would increase soil strength, while pore water escapes, inducing overconsolidation. Full liquefaction is not required for this strengthening process, as pore water can already escape at small over-pressures. This may explain the large resistance of in-situ soils against wave-induced liquefaction — such soils likely have experienced many wave events in their lifetime. This may also explain why the Rotterdam mud became sensitive to liquefaction after rewetting and remolding.

Winterwerp (2011) also analyzed the literature on field experiments over muddy beds. The studies by Traykovski *et al.* (2015) on Louisiana mud coast and by Meirelles and Vinzon (2016) in Cassino Beach, Brazil, suggest both that during storm conditions part of the mud bed is eroded, and not liquefied. This is corroborated by Rogers and Holland (2009), showing significant wave damping throughout the Cassino Beach experiment, except during periods with high wave events. This suggests that the soft mud deposits are eroded and/or become mobile, traveling away form the measuring site.

Jaramillo *et al.* (2009) observed the formation of a soft mud layer during storm conditions off Atchafalaya coast (Louisiana, USA), reflected by a heightening of the lutocline. By definition, this observation excludes the occurrence of liquefaction, as liquefaction is an undrained process, characterized by the breakup of cohesive bonds within the bed, at constant water content.

Also measurements on wave damping (see Sec. 7.4) off the coast of Suriname by Wells and Kemp (1986) do not reflect wave-induced liquefaction. In fact the fluid mud in the coastal zone, dissipating the long ocean waves, is formed by erosion of the mud banks migrating along the coast, as discussed by e.g. Augustinus (1978) and Allison and Lee (2004).

Bed shear stresses by waves can be very large (e.g. Sec. 3.2.4), eroding large amounts of sediment from the mud bed. When these amounts exceed the carrying capacity of the turbulent tidal flow, saturation occurs, and fluid mud is formed upon deposition (Sec. 7.1). Winterwerp (2001) refers to auto-saturation. Thus fluid mud would be formed after a storm by deposition and not during storms by liquefaction.

Hence, wave-induced liquefaction of natural muddy beds is not likely to occur — all *in-situ* observations summarized above can be explained with the concept of auto-saturation. However, if *in-situ* soil strengths are low, of the order of those found in the laboratory experiments (i.e. several 10s Pa), *in-situ* liquefaction may occur. This is for instance the case with fresh mud deposits under normal consolidation, or organic-rich sediment, such as in the large lakes of Florida (Gowland *et al.*, 2007).

The behavior of the thick mud deposits off the mouth of the Amazon River is another example. Large tide-induced strains keep these deposits fluid and mobile (Vinzon and Mehta, 1998), preventing strength development within the mud and its immobilization.

Of course, fluid mud can be formed by mechanically pumping in water into the bed, such as during water injection dredging, which is thus a fluidization process.

7.3 Fluid mud properties

As discussed in Secs. 2.2 and 5.1, dense and/or densifying/consolidating mud suspensions build up a granular skeleton, developing effective stresses.

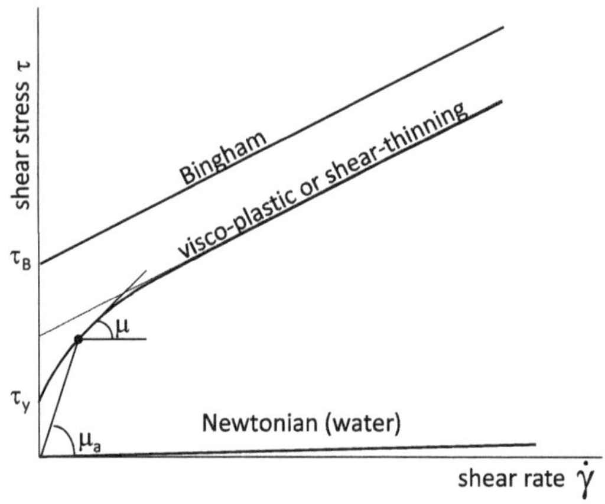

Fig. 7.5. Schematic of flow curves with shear-thinning and Bingham behavior. Also indicated are the yield strength τ_y, the local viscosity μ and the apparent viscosity μ_a.

Thus yield strength develops within the soil, while also the soil viscosity increases. Then the stress–strain relations of the soil become nonlinear, i.e. non-Newtonian.

As cohesive soils are formed from flocs (Sec. 4.3), the soil granular skeleton may break down upon shear. The viscosity of such soils then decreases with increasing shear rate, a property referred to as shear-thinning, see also Sec. 2.2.

The rheological behavior of cohesive soils is commonly represented by a flow curve, the stress (τ) — strain ($\dot{\gamma}$) relation obtained in a roto-viscometer (Mehta, 2013; Migniot, 1968). Such flow curves are sketched in Fig. 7.5 showing three lines: for water, for a so-called Bingham response and for a shear-thinning response. The local slope of a flow curve represents the viscosity of the fluid μ at a particular shear rate $\dot{\gamma}(=\mathrm{d}u/\mathrm{d}t)$ (see also Chapter 11). The ratio between shear stress τ and shear rate $\dot{\gamma}$ represents the so-called apparent viscosity μ_a (used in numerical models):

$$\mu_a = \frac{\tau}{\dot{\gamma}} \tag{7.7}$$

The Bingham model assumes constant viscosity μ (but NOT constant μ_a), while its extrapolation to the τ-axis defines the Bingham strength τ_B. The crossing of the shear-thinning curve with the τ-axis represents the yield strength τ_y (see also Fig. 2.20), while μ and μ_a decrease from initially high values.

One of the first shear-thinning rheological models is the Herschel–Bulkley model (Herschel, 1926):

$$\tau = \tau_y + K\dot{\gamma}^n \tag{7.8}$$

in which K is the so-called consistency index, which yields $K = \mu$ for $n = 1$, retrieving the Bingham model Eq. (2.4).

Thixotropy

Shear thinning in response to loading occurs because of a breakup in floc structure. This loss in strength is also referred to as the sensitivity of a soil, defined as the ratio of the undisturbed shear strength and remolded shear strength of a clay (e.g. Lambe and Whitman, 1979). When the shear strength recovers upon unloading, one refers to thixotropy, i.e. the reversibility of soil structure upon cycles of loading/unloading. Generally, the soft soils treated in this book ($c_u \ll$ LL) can recover completely, given enough time. Moore (1959) recognized that such thixotropic behavior can be quantified by modeling the (floc) structure within a soil. He therefore proposed a structure parameter λ_s, with $\lambda_s = 1$ for fully flocculated mud and $\lambda_s = 0$ at the complete loss of floc structure. The dynamics of λ_s are given by a rate equation:

$$\frac{\mathrm{d}\lambda_s}{\mathrm{d}t} = a\,(1 - \lambda_s) - b\lambda_s\dot{\gamma} \tag{7.9}$$

in which a is a flocculation parameter (in fact a flocculation time scale), and b a floc breakup parameter. Note the similarity to the flocculation equation in Sec. 4.3, describing floc size with time. The model parameters can be obtained from non-stationary roto-viscometer experiments. The general solution of Eq. (7.9) for constant $\dot{\gamma}$ reads:

$$\lambda_s = \frac{a}{a + b\dot{\gamma}} + C\exp\{-(a + b\dot{\gamma})\,t\} \tag{7.10}$$

in which C is an integration constant. At equilibrium, we find:

$$\lambda_{s,e} = \frac{a}{a + b\dot{\gamma}} \tag{7.11}$$

Note that in dynamic rheological experiments, $\dot{\gamma}$ is a function of time. For a simple linear function (i.e. $\mathrm{d}\dot{\gamma}/\mathrm{d}t$ = const), the solution contains an inconvenient error function. Therefore, it is more convenient to solve Eq. (7.9) numerically.

Worrall and Tuliani (1964) combined this approach with the following rheological model:

$$\tau = \lambda_s \tau_0 + (\mu_\infty + c\lambda_s)\,\dot{\gamma} \tag{7.12}$$

in which τ_0 is the equilibrium stress at zero deformation, and μ_∞ the local viscosity at large $\dot{\gamma}$. The equilibrium flow curve then reads:

$$\tau_e = \frac{a}{a + b\dot{\gamma}}\tau_y + \left(\mu_\infty + \frac{a}{a + b\dot{\gamma}}c\right)\dot{\gamma} \tag{7.13}$$

Toorman (1994, 1996) showed that Eq. (7.13) is identical to the shear-thinning equilibrium model by Worrall and Tuliani (1964):

$$\tau = \tau_0 + \left(\mu_\infty + \frac{\mu_0 - \mu_\infty}{1 + \frac{\mu_0 - \mu_\infty}{\tau_B - \tau_0}\dot{\gamma}}\right)\dot{\gamma} \tag{7.14}$$

provided that

$$c = \mu_0 - \mu_\infty, \quad \frac{b}{a} = \frac{\mu_0 - \mu_\infty}{\tau_B - \tau_0} \tag{7.15}$$

The Moore model was extended further by Cross (1965):

$$\frac{\mathrm{d}\lambda_s}{\mathrm{d}t} = a\,(1 - \lambda_s) - b\lambda_s\dot{\gamma}^m \tag{7.16}$$

which is not further elaborated upon.

Sensitivity to model parameters

In the following, we refer to the WT-model, which is the dynamic version of the equilibrium Worrall–Tuliani model. It is illustrative to study the behavior of the WT-model for various parameter settings by analyzing a virtual roto-viscometer experiment. We follow the numerical solution of the Moore model (Eq. (7.9), Toorman, 1996). The various model parameters are summarized in Table 7.1 and the resulting flow curves are shown in Figs. 7.6–7.8. Obviously, a variety of flow curves can be obtained, as in Fig. 7.6. Curve 1 reflects a classical shear-thinning behavior with small thixotropic effects, whereas decreasing c (curve 2) results in more pronounced thixotropic effects, with small μ_∞. When $b \ll a$, a classical Bingham curve (curve 3) is obtained. Curve 4 shows a strong reduction in yield at small viscosity.

The thixotropic effects (i.e. difference between accelerated and decelerated response) are illustrated in Fig. 7.7 by the difference between the

Table 7.1. Parameter settings (Eq. (7.14)) for virtual rheological experiments shown in Figs. 7.6–7.8.

Value	Curve 1	Curve 2	Curve 3	Curve 4
a [s^{-1}]	1.1	1.1	1000.0	1.0
b [-]	0.1	0.1	1.0	1.0
c [Pa.s]	5.0	1.0	0.5	2.0
τ_0 [Pa]	10.0	10.0	10.0	10.0
τ_B [Pa]	≈45	10.8	10.0	2.0
μ_∞ [mPa.s]	26.0	1.5	500.0	0.5

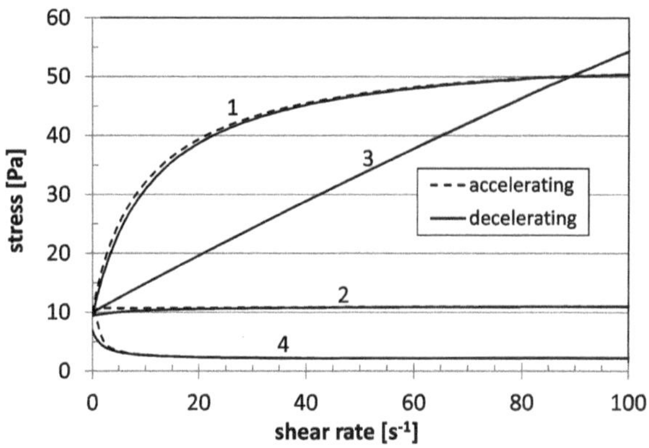

Fig. 7.6. Virtual rheometer experiment with accelerating and decelerating shear rate from 0 to 100 s^{-1} in 2 minutes, and parameter settings in Table 7.1 — shear stress τ as function of $\dot{\gamma}$.

accelerating and decelerating curves. These are particular manifest at low shear rates — note that in open waters shear rates of a few s^{-1} are common, i.e. well below 10 s^{-1}. Hence, thixotropy (hysteresis) is expected to be common under natural conditions. For some flow curves, the slope is negative, thus the local viscosity is non-defined. For that reason, the apparent viscosity μ_a has been introduced (Eq. (7.7), see also Fig. 7.5).

The apparent viscosity for these virtual experiments is plotted in Fig. 7.8, again showing thixotropic effects at low shear rates.

These examples show that the WT-model approach is very versatile and can produce a large variety of flow curves, depending on the parameter settings. This versatility, of course, is also a drawback when applied in numerical models, as these stress–strain relations can have a large effects on the outcomes of such models.

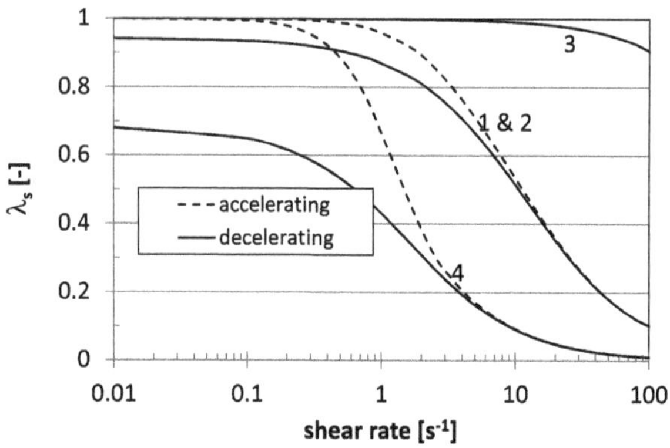

Fig. 7.7. Virtual rheometer experiment with accelerating and decelerating shear rate from 0 to $100\,\mathrm{s}^{-1}$ in 2 minutes, and parameter settings in Table 7.1 — structure parameter λ_s as function of $\dot{\gamma}$.

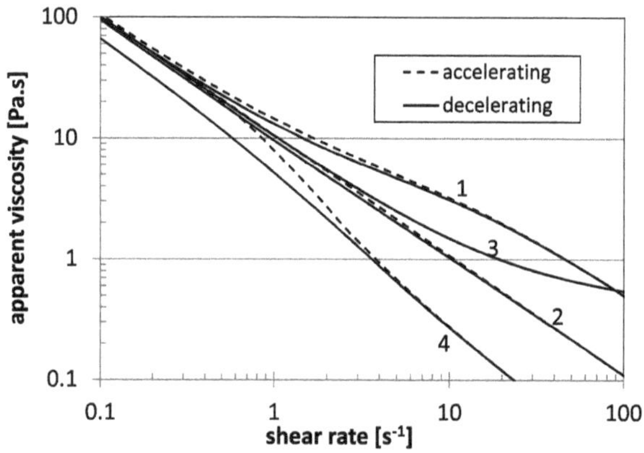

Fig. 7.8. Virtual rheometer experiment with accelerating and decelerating shear rate from 0 to $100\,\mathrm{s}^{-1}$ in 2 minutes, and parameter settings in Table 7.1 — apparent viscosity μ_a as function of $\dot{\gamma}$.

The reader is referred to for instance Coussot (1997) for the effect of various stress–strain relations on the vertical velocity profiles of mud flows.

Note that it is also possible to combine Moore's approach with models other than that of Worrall–Tuliani, for instance with the fractal approach introduced by Kranenburg (1994), e.g. Eqs. (5.31) and (5.32). Further, fluid mud is likely to consolidate, in which case one needs to combine the WT-approach with a consolidation model. In general, this is not really a problem, as the time scales for thixotropy and consolidation are fairly different, thus the models can be decoupled. However, by shearing fluid mud, pores and/or drainage channels may be opened in the mud structure, affecting the rate of consolidation through alteration of the effective hydraulic conductivity k.

Other rheological models

The flow curves discussed above are typical for so-called visco-plastic materials. As introduced in Sec. 7.2, soils can also exhibit elastic response to cyclic loadings, as under waves. When the hydraulic conductivity of a soil increases, for instance through the presence of sand, porosity effects can also play a role and soil deformation is no longer undrained.

The literature contains an overwhelming amount of constitutive models accounting for such porous and elastic effects in combination with viscous and/or plastic behavior. An extensive overview for fine sediment can be found in Mehta (2013). Here we introduce only the basic concept of the effects of visco-elasticity, yielding the butterfly stress paths in Fig. 7.4. The cyclic stresses and strains (in one dimension only) are formulated as

$$\gamma = \hat{\gamma} e^{i\omega x} \quad \text{and} \quad \tau = \hat{\tau} e^{i(\omega x + \theta)} \tag{7.17}$$

in which $\hat{\gamma}$ is the strain amplitude for deformations in x-direction at frequency ω, and $\hat{\tau}$ is the amplitude of the induced stress with phase lag θ. The ratio of τ and γ is defined as the complex shear modulus:

$$G^* \equiv \frac{\tau}{\gamma} = G' + iG'' = |G^*| \, e^{i\theta} \tag{7.18}$$

in which G' is the storage modulus (the elastic effects) and G'' the loss modulus (the viscous effects). A simple elasto-viscous model is known as the Kelvin–Voight model:

$$\tau = G' + \frac{G''}{\omega} \dot{\gamma} \tag{7.19}$$

The storage and loss moduli can be combined differently (a.o. the Maxwell model), and also the effects of plasticity can be added. A full elaboration of all these rheological models is beyond the scope of this book. For further information on the various rheological models applied for cohesive soils, the reader is referred to e.g. Mehta (2013).

Finally, in the above we discuss shearing in one plane only, for didactic reasons. For the full three-dimensional equations, the reader is referred to e.g. Malvern (1969).

7.4 Wave damping by fluid mud

Because of their large viscosity, soft/fluid mud layers can dissipate hydrodynamic energy effectively. For instance, damping of surface waves by viscous dissipation over muddy foreshores has been reported frequently in literature (e.g. Kurup, 1972; Meirelles and Vinzon, 2016; Sheremet *et al.*, 2011; Tubman and Suhayda, 1976). In particular wave damping in Suriname's foreshore is spectacular, as shown in Fig. 7.9, showing about 98% damping

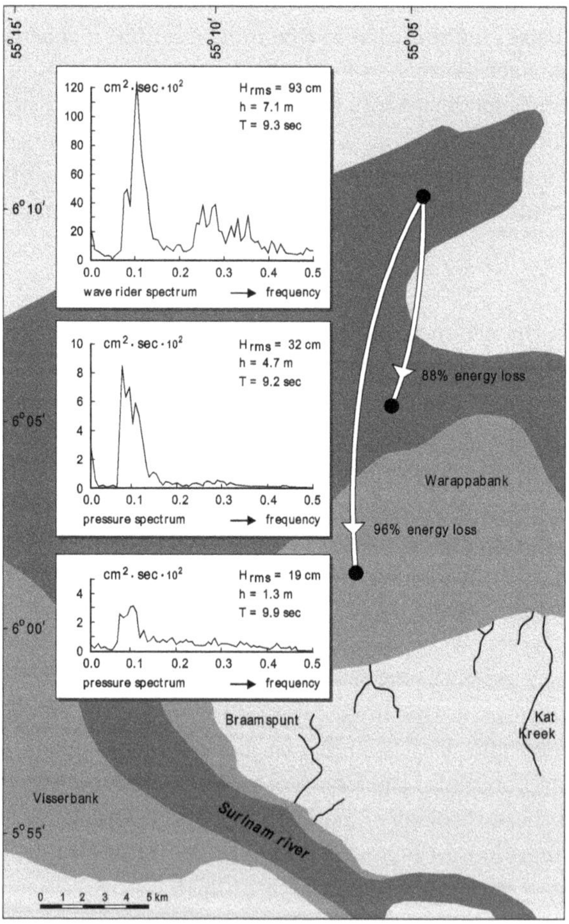

Fig. 7.9. Wave attenuation measured along the mud banks in Suriname coastal system (redrawn from Wells and Kemp, 1986).

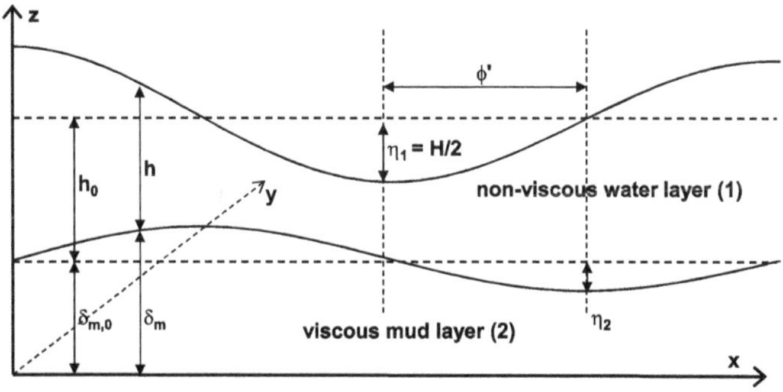

Fig. 7.10. The two-layer (2L) fluid mud model for analyzing viscous wave damping.

of the total wave energy over a 16 km trajectory (i.e. about 60 wavelengths for 13 s waves).

Gade (1958) was the first to analyze wave damping by viscous dissipation using a linearized wave model with experimental validation. In the following, we follow Gade, who treated the water column as a two-layer system, as sketched in Fig. 7.10. We refer to the upper water layer with subscript $\cdot_1$ and to the lower fluid mud layer with subscript $\cdot_2$. Neglecting the effects of advection and earth rotation, the continuity and momentum equations in a vertical x–z-plane read:

$$\frac{\partial u_i}{\partial x} + \frac{\partial w_i}{\partial z} = 0 \tag{7.20}$$

$$\frac{\partial u_i}{\partial t} + \frac{1}{\rho_i}\frac{\partial p_i}{\partial x} - \frac{\partial}{\partial x}\nu_{x,i}\frac{\partial u_i}{\partial x} - \frac{\partial}{\partial z}\nu_{z,i}\frac{\partial u_i}{\partial z} = 0 \tag{7.21}$$

$$\frac{\partial w_i}{\partial t} + \frac{1}{\rho_i}\frac{\partial p_i}{\partial z} - \frac{\partial}{\partial x}\nu_{x,i}\frac{\partial w_i}{\partial x} - \frac{\partial}{\partial z}\nu_{z,i}\frac{\partial w_i}{\partial z} = -g \tag{7.22}$$

In these equations, u is the velocity in horizontal x-direction, w is the velocity in vertical z-direction, p is pressure, ν_2 is the kinematic viscosity of the mud layer and ν_1 the (eddy) viscosity of the water layer above, and g is the acceleration of gravity. For the upper layer, $i = 1$, whereas for the lower layer $i = 2$.

This set of equations is easily extended with the second horizontal y-coordinate, as indicated in Fig. 7.10, but this is not elaborated in here. In solving these equations, no slip at the rigid bed is assumed, and normal velocities at the interfaces of the two-layer system are zero. This set of equations is linearized by assuming constant viscosity and small wave amplitudes ($\eta \ll h$), in which η is the wave amplitude. Moreover, no consolidation or other processes within the mud layer are assumed to take place. With these assumptions, Eq. (7.20)–(7.22) with their boundary conditions can be solved, assuming harmonic solutions:

$$\psi\,(x, z, t) = \Psi\,(z)\, e^{(i(kx-\omega t))} \tag{7.23}$$

in which ψ and Ψ are complex variables, representing wave height, velocity or pressure. Gade (1958) assumed depth-averaged flow velocities in the upper layer and neglected viscosity effects in x-direction, arriving at a dispersion equation for the propagation and damping of surface waves over a viscous fluid mud layer:

$$k = \omega \left(\frac{\left(1 + \Gamma \frac{\delta_{m0}}{h_0}\right) - \sqrt{\left(1 - \Gamma \frac{\delta_{m0}}{h_0}\right)^2 - 4\gamma\Gamma \frac{\delta_{m0}}{h_0}}}{2\gamma g \Gamma \delta_{m0}} \right)^{1/2} \tag{7.24}$$

in which $k \equiv k_r + ik_i$ is the complex wave number, and k_r (representing wavelength) and k_i (representing wave damping) are the real and imaginary wave numbers, respectively. Further, the following parameters have been introduced:

$$\Gamma = 1 - \frac{\tanh\,(m\delta_{m0})}{m\delta_{m0}}, \quad m = (1-i)\sqrt{\frac{\omega}{2\nu_2}} \quad \text{and} \quad \gamma = \frac{\rho_2 - \rho_1}{\rho_2} \tag{7.25}$$

where ω is the wave frequency and ν_2 is the constant mud viscosity. Eq. (7.24) contains two scaling numbers, i.e. the Stokes' boundary layer thickness $\sqrt{\omega/2\nu_2}$, and the ratio between fluid mud layer thickness and water depth δ_{mo}/h_0. Combining these yields the relative fluid mud layer thickness $\delta_{m0}/\sqrt{2\nu_2/\omega}$. Figure 7.11 presents the variation of k_i as function of $\delta_{m0}/\sqrt{2\nu_2/\omega}$, as given by Eq. (7.24), showing non-monotonic damping behavior. This behavior can be understood by realizing that for very large ν_2, the fluid mud behaves as a solid soil, whereas for very small ν_2, we have a homogeneous column of water. The influence of δ_m and ω is not directly obvious, as this influence is related to the phase difference between the two layers, determining the energy transfer between them (e.g. Gade, 1958). Also the real wave number k_r changes, however, only decreasing slightly with $\delta_{m0}/\sqrt{2\nu_2/\omega}$ (results not shown). For natural conditions, the peak in damping occurs at fluid mud thicknesses of a few 0.1 m.

The frequency-dependency in wave damping implies changes in the spectrum of waves traveling over soft/fluid mud layers. As these conditions are

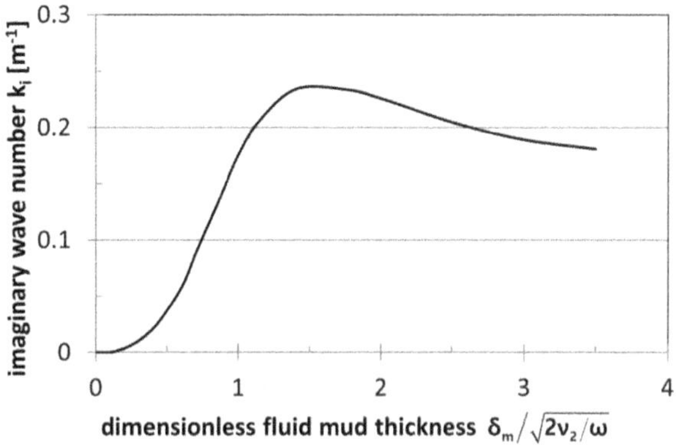

Fig. 7.11. Wave damping by viscous dissipation as function of the relative fluid mud layer thickness (after Gade, 1958).

generally met in shallow water conditions, energy transfer between the various frequencies may occur as well, modifying the spectrum further (e.g. Sheremet *et al.*, 2011).

Gade (1958), and later Dalrymple and Liu (1978); Jiang and Zhao (1989); de Wit and Kranenburg (1993); Kranenburg *et al.* (2011); Ng (2000) analyzed wave damping by viscous dissipation only. The various models differ by including viscosity and/or velocity gradients in the upper layer, or other linearizations (Ng, 2000) — see Kranenburg *et al.* (2011) for an overview.

However, energy losses may also occur by percolation and plastic effects, whereas under elastic conditions, energy is stored within the mud, released during other phases of the wave cycle. The viscous and elastic components of such models (see also Sec. 7.3) can be coupled in series (Kelvin–Voigt model, e.g. Macpherson, 1980; Chou *et al.*, 1993) or in parallel (Maxwell model, e.g. Maa and Mehta, 1990; Maa *et al.*, 1993; Rodriguez, 2000; Rodriguez *et al.*, 2001).

Jiang and Zhao (1989) and Mehta (2013) even proposed to combine the serial and parallel coupling through a three-element approach.

Another series of models was proposed by Mei and Ko-Fei Liu (1987) and Liu and Mei (1989) in which the effects of plastic deformations of the muddy seabed are accounted for in a so-called visco-plastic approach. The Bingham model is one of the most simple visco-plastic models used, but a number of other models have also been elaborated upon, such as the general-plastic Herschel–Bulkley model. Liu (1973), Yamamoto *et al.*

(1978), Yamamoto and Takahashi (1985) and Spierenburg (1987) included the effects of porosity of the seabed through a poro-elastic description.

An excellent overview of the various models is given in Mehta (2013); see also Jain and Mehta (2009). In practice, the lower fluid mud layer is thin, while its viscosity and thickness are generally poorly known. Therefore, from a practical/engineering point of view, we advocate the approach by De Wit and Kranenburg (1997) and Kranenburg *et al.* (2011) in which mud concentrations and viscosity are averaged over the lower layer thickness, whereas gradients in velocity in the upper layer are taken into account — the simple boundary layer model by Ng (2000) yields a good approximation for these conditions.

Wave dissipation

The rate of wave dissipation ε_m equals the work done on the interface between the two layers. Assuming $\eta \ll h$, the work done P [W] per unit area and unit time reads (Gade, 1958):

$$\varepsilon_m \left(\equiv P\right) = -\frac{1}{T}\int_T p\frac{\mathrm{d}\eta_2}{\mathrm{d}t}\mathrm{d}t \tag{7.26}$$

in which T is wave period, and where pressure p and wave height η follow from the dispersion relation and are given as harmonic functions, such as Eq. (7.23). The reader is referred to Gade (1958) or Kranenburg (2008) for a full analytical assessment of P.

The rate of wave dissipation ε_m by fluid mud also follows the decay in wave energy:

$$\varepsilon_m = -\frac{\mathrm{d}E_w}{\mathrm{d}t} = -C_g\frac{\mathrm{d}E_w}{\mathrm{d}x} \tag{7.27}$$

in which C_g is the group velocity of the waves, and E_w is the wave energy, defined as

$$E_w = \frac{1}{2}\rho g\eta_1^2 = \frac{1}{8}\rho g H_{1/3}^2 \tag{7.28}$$

for monochromatic waves with wave amplitude η_1, whereas the second term represents irregular waves with with wave height $H_{1/3}$. In case of a flat bed, Eq. (7.27) reduces to:

$$\varepsilon_m = 2C_g E_w k_i \tag{7.29}$$

relating energy dissipation directly to the imaginary wave number (see also Mehta, 2013). This flat bed assumption is used in Sec. 7.5.2 for assessing wave-induced mud transport.

The effects of wave dissipation by fluid mud are explicitly modeled by ϵ_m (Eq. (7.27)) in the right-hand term of the wave action density balance equation (Eq. (11.19)) implemented in the open source wave model SWAN, e.g. Kranenburg (2008) and Kranenburg *et al.* (2011). In this so-called SWAN-mud model, the asymptotic model by Ng (2000) is implemented, as this model converges well to the full solution by De Wit and Kranenburg (1997), while its numerical solution is much faster.

7.5 Fluid mud transport

By definition, fluid mud can behave as a fluid, and may therefore be mobile. Whether or not fluid mud is mobile depends on the ratio of the external stresses on the mud layer and the strength and viscosity within that mud layer. These external stresses may consist of (see also McAnally *et al.*, 2007):

(1) **Gravity:** If lying on a slope, the gravitational component along that slope induces a body force, which may propel the mud layer. Also gradients in the slope of the fluid mud interface itself can induce fluid mud flow.
(2) **Tide:** Pressure gradients from incoming tides induce pressure gradients within a mud layer and may induce fluid mud movement.
(3) **Drag:** (Turbulent) flow in the water column above a fluid mud layer may drag the mud in the direction of that flow.
(4) **Streaming:** Wave-induced fluid mud transport is induced in the direction of wave dissipation by so-called radiation stresses.

Mobile fluid mud is generally in a laminar hydrodynamic state. When turbulent, one commonly refers to turbidity currents or debris flow, the latter when larger sediment particles (up to boulders) are carried by the flow (see also Coussot, 1997). Mixing between fluid mud layers and the overlying water column occurs because of velocity differences between these layers.

Fluid mud flow by gravity, pressure gradients and drag, and subsequent mixing is discussed in Sec. 7.5.1, whereas wave-induced fluid mud transport (streaming) is treated in Sec. 7.5.2.

As discussed in Sec. 7.1, fluid mud is in a transient state, prone to consolidation. As time scales for consolidation are generally much larger than those for fluid mud flow, we may presume stationary fluid mud properties in the following.

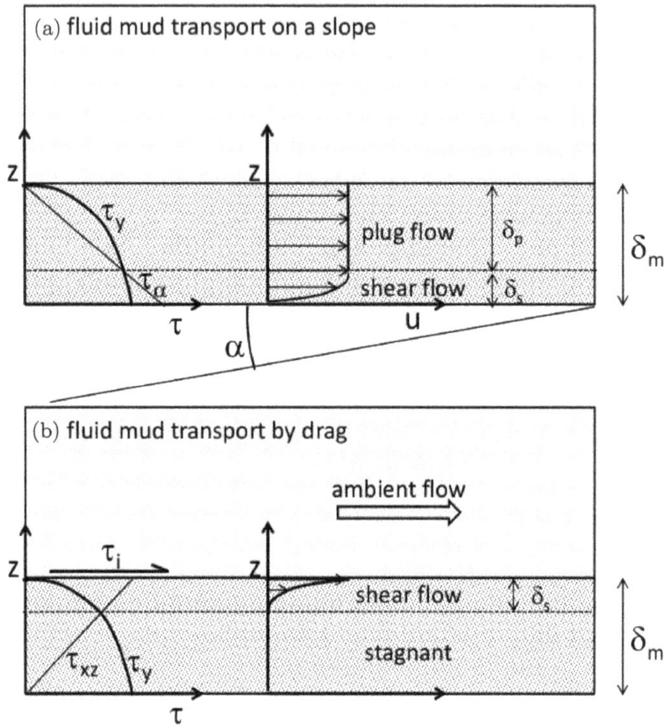

Fig. 7.12. Diagrams showing fluid mud flow (similar in case of a sloping interface): (a) plugflow of fluid mud on a slope; (b) drag-induced fluid mud flow — note that the stress distribution τ_{xz} within the fluid mud layer is only linear when the fluid mud bulk density is constant.

7.5.1 *Fluid mud transport by gravity and drag*

Let us analyze the behavior of fluid mud residing on a slope α in otherwise still water, residing on a flat bed with an ambient current of characteristic flow velocity U, or a combination, as sketched in Fig. 7.12. In this figure, the stress τ_{xz} within the fluid mud is assumed linear for didactic reasons, which is only true if the fluid mud density ρ_m is constant. Thus:

$$\tau_{xz} - \tau_i = g\left(\int_0^z (\rho_m(z) - \rho_w)\mathrm{d}z'\right)\sin\alpha \tag{7.30}$$

where τ_i is the interfacial shear, defined as $\tau_i = c_D\rho_w U^2$, with a typical value for the drag coefficient $c_D \approx 0.001$, and U is a characteristic ambient flow velocity.

Fluid mud movement is possible when the shear stress of Eq. (7.30) exceeds the yield strength of the mud, i.e. if $\tau_{xz} > \tau_y$. As neither τ_{xz} nor τ_y are constant over the fluid mud layer, only a part of the mud layer may become mobile. Assuming a simple Bingham model (Eq. (2.4)), the conditions for fluid mud mobility read (see also Kessel, 1997; Van Kessel and Kranenburg, 1998):

$$\tau_{xz} - \tau_i = \tau_y \mathrm{sgn}\left\{\frac{\partial u}{\partial z}\right\} + \mu_m \frac{\partial u}{\partial z} \qquad \text{for } |\tau_{xz} - \tau_i| > \tau_y$$

$$\frac{\partial u}{\partial z} = 0 \qquad \text{for } |\tau_{xz} - \tau_i| < \tau_y \tag{7.31}$$

Thus, when lying on a slope, the entire fluid mud layer may become mobile, with a thin shear flow layer (δ_s) near the bed, and plug flow ($\partial u/\partial z = 0$) higher in the water column (δ_p), as sketched in Fig. 7.12(a). Similar results may be obtained in case of fluid mud on a horizontal bed with a sloping interface.

In case of drag by an ambient current, it is likely that only the upper part of the fluid mud layer becomes mobile, indicated by (δ_s) in Fig. 7.12(b).

Of course, any other rheological model can be used for determining the strength of the fluid mud layer, including fractal descriptions for τ_y and μ.

Laminar or turbulent fluid mud flow

Generally, fluid mud flow is laminar, owing to its large viscosity. However, fluid mud may accelerate and may become turbulent, as is often the case on the continental slope. Then one generally refers to turbidity currents. Whether or not fluid mud is turbulent is reflected by an effective Reynolds number Re_e, which reads for Bingham material:

$$\frac{1}{\mathrm{Re}_e} = \frac{1}{\mathrm{Re}_f} + \frac{1}{\mathrm{Re}_y}, \quad \text{with } \mathrm{Re}_f = \frac{4U_m\delta_m}{\nu_m} \quad \text{and} \quad \mathrm{Re}_y = \frac{8\rho_m U_m^2}{\tau_B} \tag{7.32}$$

where U_m is the mean velocity in a fluid mud layer with thickness δ_m, τ_B is the Bingham strength of fluid mud, and ν_m and ρ_m are the viscosity and density of the fluid mud. According to Liu and Mei (1989), the critical, effective Reynolds number amounts to about $\mathrm{Re}_{e,cr} \approx (2\text{–}3) \cdot 10^3$, beyond which a mobile fluid mud layer becomes turbulent.

Low-turbulence

Even when turbulent, this turbulence may not be well developed, a flow condition generally described with a low-turbulence model, e.g. Pope (2001).

Many of these so-called low-turbulence models are quite complex. Here we present a simple approach by Chien (1982):

$$\mu_{LR} = \mu_T f_{LR} = \rho c_k f_{LR} \frac{k_T^2}{\varepsilon_T} \tag{7.33}$$

in which μ_{LR} is the low Reynolds number eddy viscosity, μ_T is the clear water eddy viscosity, as follows from a standard turbulence model (e.g. Chapter 11), and f_{LR} is a low-Reynolds damping function. In the literature, a number of models for f_{LR} are presented. Three of the more simple ones are by Hwang and Lin (1998):

$$f_{LR} = 1 - \exp\left\{-0.01\frac{z}{\sqrt{\nu k/\varepsilon_T}} - 0.008\left(\frac{z}{\sqrt{\nu k/\varepsilon_T}}\right)^3\right\} \tag{7.34}$$

by Georgiadis *et al.* (1994):

$$f_{LR} = \left(1 + \frac{3.45}{\sqrt{\mathrm{Re}_T}}\right)\tanh\left\{\frac{z}{70\sqrt{\nu k/\varepsilon_T}}\right\} \tag{7.35}$$

and by Uittenbogaard (1994), as implemented in the Delft3D software):

$$f_{LR} = 1 - e^{-p^2}, \quad \text{where } p = \frac{Re_T}{100} \tag{7.36}$$

The turbulent Reynolds number Re_T is defined as (see also Sec. 3.2.2):

$$\mathrm{Re}_T \equiv \frac{u'\ell}{\nu} = \frac{k_T^2}{\nu\varepsilon_T} \tag{7.37}$$

with u' is the characteristic turbulent velocity in the turbulent fluid mud layer, ℓ the characteristic length scale (mixing length, see also Sec. 3.2.2) and ν the fluid (mud) viscosity. Re_T can also be written in terms of the turbulent kinetic energy k_T and dissipation rate ε_T, as in Eq. (7.37) — see e.g. Georgiadis *et al.* (1994). These low-Reynolds formulas can be used in conjunction with for instance the classical k–ε turbulence model (see Chapter 11).

Friction and mixing

Friction at the interface of a fluid mud layer with the overlying water column may occur when one or both layers are mobile and turbulent. This interfacial shear stress, τ_i, can be modeled as

$$\tau_i = k_i \rho_1 |u_1 - u_2| (u_1 - u_2) \tag{7.38}$$

Table 7.2. Friction coefficients for two-layer stratified flows (from Abraham *et al.*, 1979).

k_i	Flow conditions
$4 \cdot 10^{-4}$	Upper layer flows, lower layer still
$15 \cdot 10^{-4}$	Lower layer flows, upper layer still
$7 \cdot 10^{-4}$	Both layers flow, but in opposite direction

where subscript $\cdot_1$ refers to the less dense upper layer and subscript $\cdot_2$ to the denser lower layer. Between the 1950s and 1970s much (experimental) research has been carried out on this interfacial shear stress of fresh-saline water stratified flows (e.g. Lofquist, 1960), resulting in a number of estimates for k_i. These are summarized in Table 7.2.

Mixing over the interface of a fluid mud layer with the overlying water column may occur when one or both layers are mobile. When one of the two layers is more turbulent, that layer "eats" from the other, as sketched in Fig. 7.13 — we refer to turbulent entrainment. The middle row of Fig. 7.13 represents "normal" mixing in a homogeneous, fully turbulent flow. The viscosity profile is parabolic (e.g. Eq. (3.22)), and the interface of an initially stratified passive tracer thickens symmetrically, decreasing the concentration gradient over that interface with time.

The upper row of Fig. 7.13 reflects stratified flow with the lower, denser layer the more turbulent one. This diagram reflects conditions for turbidity currents (debris flow), where turbulence is mainly generated at the rigid bed. Because of buoyancy destruction over the stratified interface, eddy viscosity in the upper layer is small (e.g. Sec. 3.2.3). The turbulent lower layer entrains less dense water (Bruens *et al.*, 2012). As a result, SPM-concentrations in the lower layer decrease, and the interface rises (conditions in the thinning upper layer do not alter).

The lower row of Fig. 7.13 reflects stratified flow with the upper, less dense layer the more turbulent one. This diagram depicts turbulent flow over a fluid mud layer in e.g. a fairway. Again, because of buoyancy destruction, eddy viscosity in the (stagnant) lower layer is small, or zero. The upper layer entrains denser water from the fluid mud layer, its SPM-concentration increases, while the interface lowers (conditions in the thinning fluid mud layer do not alter).

The turbulent entrainment rate $E_* (= w_e/u_*)$, where w_e is the entrainment velocity and u_* is the shear velocity, has been elaborated by Kranenburg and Winterwerp (1997) and Winterwerp and Kranenburg

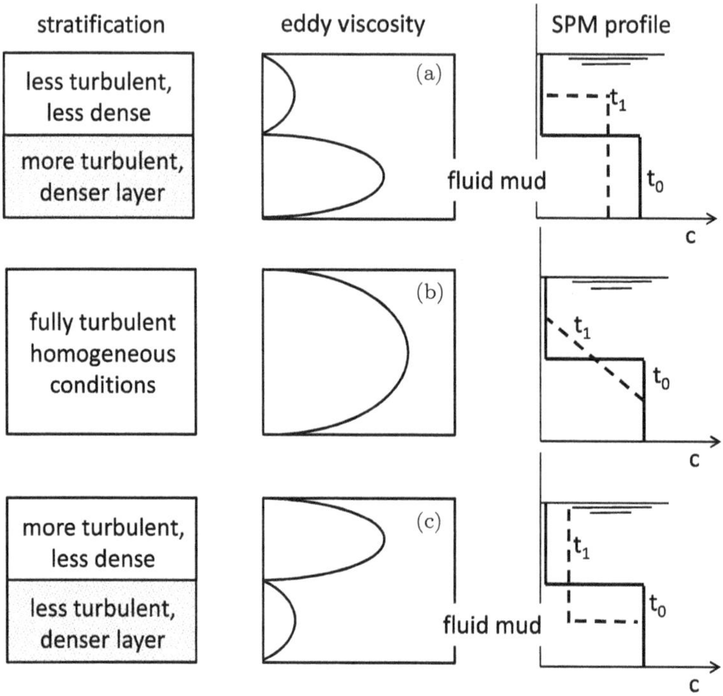

Fig. 7.13. Schematics showing mixing in stratified flows. Row (a) reflects a turbulent fluid mud layer (turbidity current), while in row (c) the upper, less dense layer is turbulent. Row (b) represents mixing of an initially stratified passive, non-buoyant tracer.

(1997), see also Mehta and Srinivas (1993):

$$E_* = \frac{w_e}{u_{*,i}} = \frac{c_\sigma}{\mathrm{Ri}_{*,e}} = \frac{0.42}{\mathrm{Ri}_{*,e}} \quad \text{with } Ri_{*,e} = \frac{\Delta\rho g h_i}{\rho_i u_{*,i}^2} \tag{7.39}$$

in which $Ri_{*,e}$ is the bulk Richardson number and the 0.42-value for c_σ follows from laboratory experiments and literature.

The subscript $\cdot_i$ reflects either of the cases of Fig. 7.13. When the lower layer is the more turbulent one, shear velocity at the bed has to be used. If the upper layer is the turbulent one, turbulence can only be produced at the interface (where production is low), or is advected from elsewhere in the system, e.g. off the fluid mud patch, in estuaries for instance from the banks and other irregularities, such as bed forms and jetties. However, this advected turbulence decays when traveling over the fluid mud, further to Eq. (3.31). For instance at 5 m water depth and a flow velocity of 1 m/s (hence $u_* \approx 3\,\text{cm/s}$), the advected turbulence would be largely dissipated after about 130 m (ca. 2 minutes). In combination with buoyancy destruction over the interface, we may conclude that fluid mud layers formed underneath an ambient current are generally very stable, in particular in open water.

7.5.2 *Fluid mud transport by waves*

Starting from a simple momentum balance equation, Longuet-Higgins (1953, 1958) showed that momentum losses by bed friction in the wave boundary layer are compensated by a tractive force, which has the form of the well-known turbulent Reynolds stress. These stresses are commonly referred to as radiation stresses, and the resulting tractive forces are known as streaming.

It can be shown that this radiation stress is proportional to the wave energy in the wave boundary layer. Therefore, any change in wave energy, for instance due to shoaling, bed friction, viscous damping, etc., must induce a change in radiation stress, hence a tractive force. In the cross-shore direction this accounts for the well-known wave setup in coastal waters with subsequent rip currents, and in longshore direction this yields the well-known longshore wave-induced currents, which can be quite large, in particular in the surf zone. Viscous wave damping over fluid mud induces setup of the overlying water as well. This water will flow off laterally, as a result of which mean water levels under waves over fluid mud are generally flat (analogy with rip currents).

In the following, we neglect any effects of ambient currents and discuss streaming in wave direction only — for the more general case, where tides and waves propagate in different direction, the reader is referred to e.g. Mei (1989) or Dingemans (1997). The rate of momentum transfer per unit width of/within the water layer in wave direction is referred to as the radiation stress S_{xx} [N/m], which is related to the mean wave energy E (Eq. (7.28)):

$$S_{xx} = \left(2n - \frac{1}{2}\right) E \quad \text{with } n = \frac{C_g}{c} = \frac{k_r}{\omega}\frac{\mathrm{d}\omega}{\mathrm{d}k_r} \tag{7.40}$$

in which c is the wave celerity (phase velocity) and C_g is the wave group velocity. The parameter $n(\equiv 1/2 + k_r h/\sinh 2k_r h)$ is 0.5 in deep water, and unity in shallow water. The Lagrangian momentum equation for the fluid mud layer (see Fig. 7.10) becomes:

$$\begin{aligned}\frac{\partial u_m^L}{\partial t} + u_m^L \frac{\partial u_m^L}{\partial x} + g\frac{\partial Z_s}{\partial x} + \frac{g\Delta\rho}{\rho_m}\frac{\partial}{\partial x}(Z_i - Z_s) \\ - \nu_m \frac{\partial^2 u_m^L}{\partial x^2} - \frac{\tau_{m,x}}{\rho_m\delta_m} - \left.\frac{\tau_{m,Z_i,x}}{\rho_m\delta_m}\right|_{z=Z_i} + \left.\frac{\tau_{m,Z_b,x}}{\rho_m\delta_m}\right|_{z=Z_b} = 0\end{aligned} \tag{7.41}$$

in which u_m^L is the Lagrangian velocity of the fluid mud layer (its mud particles), and $\partial Z_s/\partial x$ and $\partial Z_m/\partial x$ are the slopes of the water surface and fluid mud–water interface, respectively. Bed friction and interfacial shear are represented by $\tau_{m,Z_b,x}$ and $\tau_{m,Z_i,x}$ respectively. The other parameters have been defined before, except $\tau_{m,x}$, which is the tractive stress on the fluid mud by the dissipating waves. It can be assessed from the stress τ_x in the water column, which follows from the gradient of the radiation stress in the water column S_{xx}, e.g. Eq. (7.40):

$$\tau_x = -\rho_w h \frac{\mathrm{d}S_{xx}}{\mathrm{d}x} = -\left(2n - \frac{1}{2}\right)\frac{\partial E_w}{\mathrm{d}x} - E_w \frac{\mathrm{d}\,(2n - 1/2)}{\partial x} \tag{7.42}$$

Thus in shallow water ($n = 1$), $\tau_x \approx 1.5\mathrm{d}E_w/\mathrm{d}x$, ignoring other contributions (e.g. S_{xy} and S_{yy}), and E_w follows from Eq. (7.28). The radiation stress in the mud layer follows from the radiation stress in the water column τ_x:

$$E_m = \frac{\rho_m \hat{\eta}_2^2}{\rho_1 \hat{\eta}_1^2} E_w, \quad \text{hence } \tau_{m,x} = \frac{\rho_m \hat{\eta}_2^2}{\rho_1 \hat{\eta}_1^2} \tau_x \tag{7.43}$$

where $\hat{\eta}_1$ and $\hat{\eta}_2$ are the wave amplitudes of the water surface and water–mud interface (obtained from the dispersion equation in Sec. 7.4). Thus streaming is directly related to viscous dissipation of surface waves in the fluid mud layer (Eq. (7.27)), and for a flat seabed, to the imaginary wave number k_i (Sec. 7.4).

The actual fluid mud velocity u_m in an Eulerian frame of reference then becomes:

$$u_{m,x} \equiv u_{m,x}^E = u_{m,x}^L - U_{m,x}^S = u_{m,x}^L - \frac{E_m}{\rho_m \delta_m}\frac{k_r}{\omega} \tag{7.44}$$

in which $U_{m,x}^S$ is the Stokes drift (see also Sec. 9.2.4).

It is illustrative to elaborate on the conditions of Fig. 7.9 showing wave damping over a trajectory of about 9 km from $H_{1/3} = 0.93$ m to 0.32 m, implying an imaginary wave number of $k_i = 1.2 \cdot 10^{-4}\ \mathrm{m}^{-1}$. Using Eq. (7.42) with $n = 1$, we find $\tau_x \approx 0.16\,\text{Pa}$. The wave-induced stress on the fluid mud layer follows from Eq. (7.43), yielding $\tau_{x,m} \approx 0.01$–$0.02\,\text{Pa}$, where we assumed $\hat{\eta}_2/\hat{\eta}_1 \approx 0.3$ (see Gade, 1958). Such stresses would be insufficient to propel the fluid mud of Fig. 7.9 (towards the coast).

These values are very small, as follows from the analytical solutions by Gade (1958), who found a maximal value of $(k_{i,max}/\omega)^2\, hg \approx 0.02$, depending on fluid mud thickness and viscosity. Hence for the conditions of Fig. 7.9 ($h = 5\,\text{m}$; $T = 9.2\,\text{s}$), a maximal value for the damping rate of $k_{i,max} = 0.004$ would be found, i.e. almost a factor 30 larger than measured. The corresponding wave-induced stresses follow from Eq. (7.42), assuming shallow water and flat bed conditions:

$$\tau_x = \frac{\varepsilon_m}{c} \approx 2E_w k_i = \frac{1}{4} k_i \rho_w H_{1/3}^2 \tag{7.45}$$

which would yield maximal values for the stresses in the water column of several Pa and in the mud layer of several 0.1 Pa, respectively, i.e. more than a magnitude magnitude larger than above. Thus even for fluid mud conditions with less-than-maximal damping, streaming is expected to become an important mechanism is the coastal zone and on intertidal flats. The setup generated in the water column by τ_x would induce significant lateral currents along the coastline, to be compensated by shore-directed to-flow of sediment-laden water. Thus horizontal circulations of sediment-laden water are induced, similar to the well-known patterns of rip currents on sandy beaches. The (isolated) patches of fluid mud themselves are likely to be pushed towards the shore.

7.6 Fluid mud and navigation[1]

In this section, we discuss the navigability of fluid mud, as this is a major reason for studying fluid mud.

Safe navigation in ports and waterways should be guaranteed by port authorities, and a good safety record is of utmost importance for the competitive position of harbors. An important factor for safe navigation is the available space under a ship's keel (see Fig. 7.14), known as the under-keel-clearance (UKC). Sufficient UKC is achieved by setting restrictions to the vessel draft and/or by maintaining the desired navigable depth by dredging operations. The first option may be undesirable from an economic point of view if this leads to restrictions with respect to the access of large vessels. Then the second option becomes a necessity, although operational costs and

[1]This section was written by Dr. O. Kirichek.

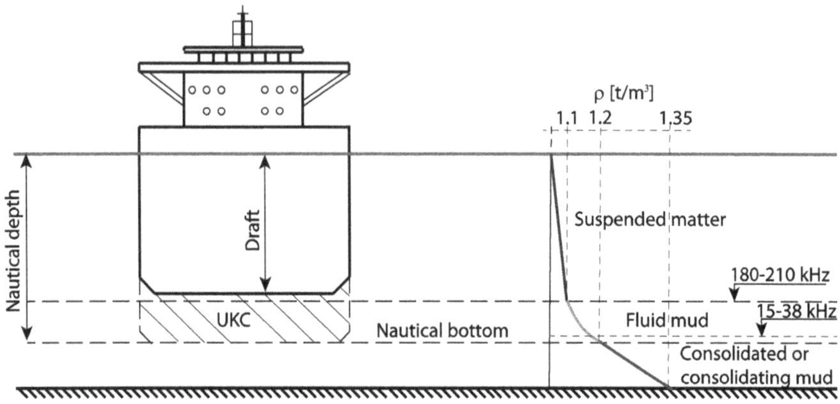

Fig. 7.14. The nautical bottom concept, which was developed at a critical density of 1.2 T/m^3. Modified after Nederlof (1978).

environmental impact of disposing dredged sediments may become substantial as well — this is in particular true in the case of (contaminated) fine sediment (see a.o. Sec. 3.5).

However, there is third option in case fine sediment deposits are very soft, i.e. when fluid mud layers are formed in navigation channels and/or harbor basins (Sec. 7.2). This third option involves sailing through the fluid mud, i.e. the ships keel is beneath the water-fluid mud interface, which is referred to as the nautical depth concept, yielding a "negative" UKC.

In this section, we explain how sailing over and through fluid mud affects the navigability of ships in a qualitative way.

Navigation in ports and waterways with fluid mud layers can be challenging McBride *et al.* (2014) as fluid mud layers can influence the navigability and maneuverability of vessels. This influence depends on the rheological properties of fluid mud, yielding additional drag on the vessel's hull, rudder and propeller.

Fluid mud layers can also affect the navigability of a ship if it sails over the fluid mud, and not through. The vessel motion may trigger internal waves on the water–mud interface, which can influence control of the vessel, affecting its maneuverability (Sellmeijer and Oortmerssen, 1984; Van Craenenbroeck and Vantorre, 1991). The celerity and amplitudes of such internal waves depend on the rheological properties of fluid mud, as studied by Delefortrie and Vantorre (2016).

Furthermore, traditional acoustic surveying provides ambiguous positioning of the "bed" yielding misleading nautical charts in muddy areas (see introduction to this chapter). Being effective in mud-free environments with a sharp density contrast at the water-bed interface, echo-soundings are often ambiguous in channels with fluid mud layers, where acoustic reflections may be difficult to interpret due to density gradients, and sometimes gas in the deposited sediment.

When sailing in fluid mud-free areas, the drag on a ship increases with decreasing UKC. The vessel experiences a sharp increase in drag when touching the rigid bed (such as consolidated mud). In case of fluid mud, the increase in drag is more gradual when the vessel penetrates a fluid mud layer.

In mud-free environments, a decrease in UKC limits a ship's maneuverability because of an increase in hydrodynamic inertia, in particular in yaw and sway damping forces. Fluid mud layers cause similar effects, especially when a vessel would navigate in close proximity to the mud-water interface. Hydrodynamic inertia for sway and yaw increase when the keel penetrates a fluid mud layer. When sailing through fluid mud, the viscous mud sticks to the ships hull, increasing the apparent mass of a ship. This added mass may amount up to seven times the ships clear water mass (Delefortrie and Vantorre, 2006). Further penetration of the ships hull into the mud would increase unfavorable effects on the ship's controllability because of heterogeneities within fluid mud layers. For instance, a ship may be pulled into a direction where fluid mud induces less drag on the hull.

Apart from additional drag on the hull, fluid mud layers can affect the propulsion of a vessel. The presence of fluid mud may cause significant losses in propulsion efficiency, especially when the vessel gets in contact with fluid mud.

Internal waves

The flow around a rudder is unfavorably affected by the generation of internal waves and enhanced drag when the ship penetrates fluid mud layers. However, the flow around a rudder improves when the vessel penetrates further into more concentrated mud layers, restoring the ships response to rudder maneuvers (Vantorre, 1991).

The controllability and maneuverability of a vessel can also be hindered by internal waves when navigating over fluid mud. Such waves can typically be detected on the interface of two fluids with different densities. One of the first effects of internal waves on navigability was reported by Ekman (1904) during a polar expedition in Norwegian fjords. In these fjords, the water column is stratified as fresh river water is not mixed with the denser seawater. Ekman noticed that a slow sailing vessel was impacted by an unknown drag. The vessel seemed to be affected by unexpected mysterious powers and could not be controlled properly. This phenomenon was referred to as "dead-water" by the vessels crew.

This magic drag was later attributed to internal waves generated on the interface between fresh and salt water. Next, this phenomenon was confirmed by a series of laboratory experiments, while Lamb (1916) derived the underlying mathematical theory.

As discussed above, such internal waves can also be generated at a fluid mud-water interface, affecting the efficiency of rudder and propeller, as was observed during both full-scale and laboratory experiments with a real vessel and a scaled model, respectively (Delefortrie and Vantorre, 2006). It was concluded, that a vessel navigating just above a fluid mud layer (with a small positive UKC related to the water depth detected by high-frequency echo-sounder) can experience more problems with maneuvering than vessels navigating through fluid mud (with a negative UKC).

The amplitude and celerity of internal waves induced by ships navigating over fluid mud increase with increasing sailing speed. However, when the ship speed increases beyond a critical velocity U_c, the unfavorable effects of internal waves on the ships maneuverability decrease again, as these internal waves detach from the ship Vantorre (2001). An empirical value for this critical speed in clear-water stratified systems (salt–fresh water, warm–cold water) is given by the following relationship:

$$U_c = \sqrt{\frac{8}{27} g h_1 \left(1 - \frac{\rho_w}{\rho_m}\right)} \tag{7.46}$$

where g is the gravitational acceleration, h_1 is the water depth above the fluid mud, ρ_w and ρ_m are the bulk water and fluid mud densities, respectively.

Fluid mud monitoring

Currently, there exists no universal surveying method in waterways with fluid mud layers. Port authorities use various measuring methods to secure navigation in muddy areas. For instance, the majority of ports across the world use dual frequency echo-sounders, as discussed above.

Another approach deployed in ports with fluid mud layers is known as the "nautical depth" concept. Currently, the following definition for the nautical depth is used: "The nautical bottom is the level where the physical characteristics of the bottom reach a critical limit beyond which contact with a ship's keel causes either damage or unacceptable effects on controllability and maneuverability of the ship" (PIANC, 2014). Traditionally, the density of fluid mud is used as a criterion to define the nautical depth in ports and fairways. For instance, in the Port of Rotterdam, fluid mud is considered to be navigable if its density does not exceed 1.2 T/m^3. However, some ports adapted other density values. For instance, the Port of Zeebrugge uses density values between 1.15 and 1.35 T/m^3 varying over the port area. In China, the ports of Lianyungang and Tianjin use densities between 1.2 and 1.3 T/m^3 as a criterion for nautical depth.

Though elaborate to measure, the density criterion is fairly successful, as can be understood from Sec. 5.1, which explains that the mean density of a consolidating mud layer is not very sensitive to its total mass — *in-situ* densities do not change too much over time. However, yield strength and, to a lesser extend, viscosity scale highly nonlinear with the mud (dry) density, and may vary largely in space and over time (e.g. Sec. 5.1), which is therefore subject of ongoing research.

In other ports, rheological parameters are used to define nautical depth. For instance, a dynamic viscosity of 10 Pa.s is used as a criterion in the muddy channels of several German ports (McBride *et al.*, 2014). On the other hand, since 2005, the Port of Emden uses a yield stress for defining the nautical depth (Wurpts and Torn, 2005). Together with a special dredging method (so-called conditioning dredging), the fluid mud in this port is kept navigable at a yield stress of 100 Pa.

Due to the heterogeneity of fluid mud, its rheological properties exhibit a nonlinear relationship with density as shown in Fig. 7.15. The complexity can be triggered by different properties of fluid mud, for instance by

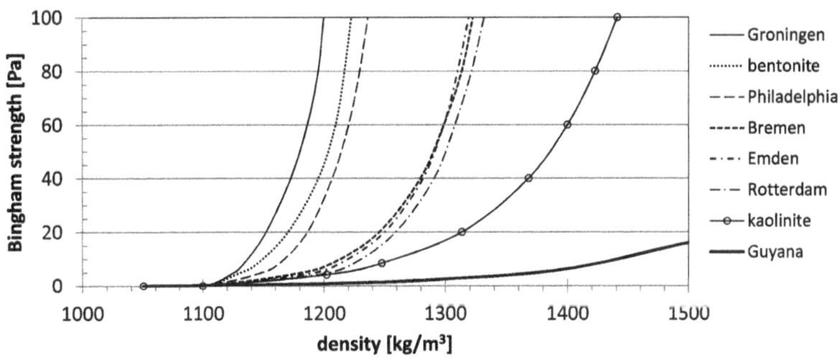

Fig. 7.15. Variation of the Bingham shear stress with bulk density of mud for different ports. Redrawn after Fontein and Byrd (2007).

its composition and thixotropic behavior. Also organic matter affects the rheology properties of fluid mud, as discussed elsewhere in this book. Thus, careful studies of the organic matter have to be performed to improve our understanding of fluid mud rheology.

Mud properties (such as the thickness of fluid mud layers, rheological properties and density) change in time over the water column while the sediment settles and consolidates. Fluid mud is therefore in a transient condition. In addition, fluid mud layers can be mobile, thus substantial amounts of mud can be relocated to deeper port areas. Even though up-to-date surveying in muddy areas is conducted on a regular basis, the current surveying strategies do not capture the time-dependency relevant for mud layers in ports and waterways.

Chapter 8

Biological Effects*

This chapter presents examples on the role of biota in fine sediment dynamics.

Organic matter is the basis of life and trophic interactions in ecosystems. It is characterized by biomass (mass of organic matter) and dead organic matter ("detritus"). When transferred through the food web, organic matter is partly respired (oxidized to CO_2 and various minerals), and partly transformed into the tissues of organism. However, the latter is not an efficient process. As a rule of thumb, only 10% of the organic matter consumed is transformed stably into tissues. The rest is lost, either by respiration, or as faeces (a kind of detritus, which also forms a food source for other organisms).

In shallow marine and estuarine waters, where light can reach the bed, various primary producers (i.e. organisms producing elementary organic matter) are found. These organisms use minerals and an energy source to synthesize organic matter. These so-called photo-autotrophic producers generate chemical energy from light by photosynthesis. They are the most important primary producers in the majority of ecosystems. These producers need mineral nutrients as well. Essential macromolecules for life, such as amino and nucleic acids, enzymes and membranes contain elements other than carbon, hydrogen and oxygen. Quantitatively, nitrogen (N) and phosphorus (P) are the most important, while metals are used also as cofactors in enzymes.

A second class includes chemo-autotrophic producers, mainly bacteria and archaea. They gain energy from oxidation–reduction reactions, primarily by re-oxidation of reduced substances (e.g. ammonium NH_4^+ or sulfide S^{2-}), which are produced by anoxic oxidation of organic matter through anaerobic metabolism.

*This chapter is largely based on the course notes of and comments by Prof. P. Herman.

In deeper waters, light cannot reach the bed and photoautotrophy is not possible. Sedimentation, resuspension and/or bed load are then important sources of organic matter. Some organisms feed directly on organic matter (such as phytoplankton) suspended in the water column, others consume organic matter settling on the bed. By their activities, these organisms mix organic matter actively into the sediment, a process called bioturbation. This changes the nature and availability of organic matter for consumption, thereby affecting the mechanical properties of the sediment bed as well.

Apart from uptake by organisms, organic matter can also be lost by resuspension and subsequent advection. Finally, burial can take place, storing organic matter in accumulating sediment. This organic matter escapes oxidation at biological time scales. Thus it can become incorporated within depositional rock, oxidized only at geological time scales upon erosion of these rocks. In the biological carbon budget, burial has a small contribution, i.e. less than about 1% of the total flux of organic matter towards the bed. However, at geological time scales, burial has great significance. It extracts carbon from the biosphere over longer periods, influencing the global CO_2 balance. This provides an essential feedback in the climate system, contributing to its regulation. Burial is enhanced in warm, productive, CO_2-rich periods that often coincide with anoxic near-bed waters and sediment.

Taxonomy

From a taxonomy point of view, benthic organisms can be classified as follows:

(1) bacteria and archaea,
(2) fauna, and
(3) flora.

In pelagic systems, the number of individuals and species is exponentially distributed with respect to size (ranging from the smallest bacteria to whales). In contrast, the biomass and species in the benthos are clustered in three size groups. These roughly correspond to the traditional categories of micro-, meio- and macro-fauna, see Fig. 8.1 and express the different life modes in the benthos, in particular the relation between the size of individuals and of sediment particles. Microbenthos lives on individual particles and in their pores. This benthos is so small that it can sit on microsites rich in organic resources, or move through the pores of the sediment. Meiofauna

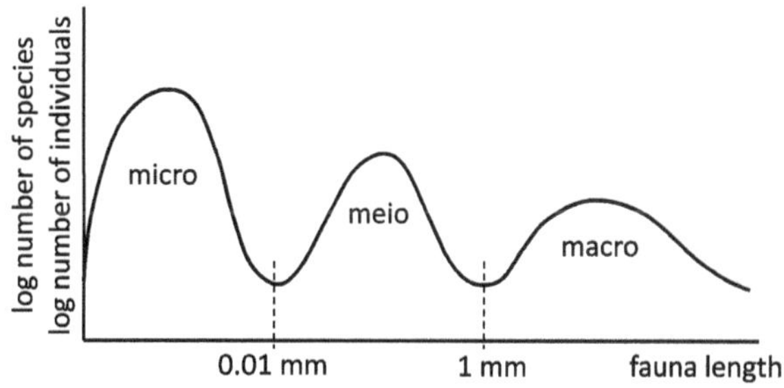

Fig. 8.1. Size spectra benthic organisms (redrawn from Warwick and Clarke, 1984).

moves in between particles, distinguishing individual particles and grazing on food items associated with these particles. Macrofauna is much larger than the individual particles and move in burrows through the sediment forming patches of about 1 mm or larger. This fauna reacts to (chemical) gradients.

This size spectrum suggests that fauna at in-between sizes are hampered in their movements, and are therefore largely absent. Some of the more important species are summarized in Table 8.1.

In the following sections, the role of biota on the mechanical properties of fine sediment is discussed. These biota may stabilize or destabilize the bed, affect bed porosity, modify settling velocity and deposition efficiency, etc., thus modifying the morphology of aquatic systems at a variety of scales, ranging from ripple formation to entire estuaries. The reader is referred to Le Hir *et al.* (2007) for a discussion on the modeling of biological effects.

By modifying their physical environment, biota may engineer their own living conditions in terms of habitat structure, physical stress, food or water availability and other characteristics. This may also change living conditions for other species — thus one species may facilitate conditions for other species. This process is referred to as "ecosystem engineering" (Jones *et al.*, 1994). Ecosystem engineering is an important class of ecological interactions, next to e.g. predation and competition for resources. Together, these ecological interactions shape the communities of organisms occurring in a particular place and at a particular time, and the fate of these communities over longer times, in conjunction with the abiotic system. Though the abiotic system is often a boundary condition for predation and competition, direct interaction between biota and the abiotic system is essential for the ecosystem engineering concept.

Table 8.1. Organisms and vegetation in and on the sediment bed.

	Taxonomy	Importance	Species
Fauna			
Microbenthos	*Bacteria*		
	Archaea		
	Protista		
	Microphytobenthos	$\sim$100 gC/m^2/yr	Diatoms Green algae Cyanobacteria
Meiobenthos	*Nematoda* (roundworm)	millions #/m^2	
	Copepoda	10000–100000 #/m^2	
	Small phyla		
Macrobenthos	*Annelida*	$\sim$100–1000 #/m^2	*Arenicola marina* (lugworm) *Hediste diversicolor* (ragworm) *Oligochaeta* (small worms)
	Crustacea		*Callianassa* (sand ghost shrimp): *Corophium* (mud shrimp) *Balanus* (barnacle)
	Echinodermata	$\sim$1–100 #/m^2	*Echinocardium* (sea potato) *Amphiura filiformis* (brittle star) *Asterias rubens* (sea star)
	Mollusca	$\sim$10–10000 #/m^2	*Mytilus edulis* (mussel) *Macoma balthica* (clam) *Crassostrea* (Japanese oyster) *Cerastoderma edule* (cockle)
Flora			
Macroalgae	*Laminaria* (kelp)	$\sim$1000 gC/m^2/yr	
	Ulva lactuca (sea lettuce)	$\sim$1000 gC/m^2/yr	
Sea grass	*Zostera marina* (eelgrass)	$\sim$500 gC/m^2/yr	
Saltmarsh plants	*Spartina anglica* (cord-grass)	$\sim$1000 gC/m^2/yr	
	Puccinellia maritima (salt grass)	$\sim$1000 gC/m^2/yr	
Mangroves	Div. species	$\sim$1000 gC/m^2/yr	

8.1 Bio-stabilization

Microbenthos is composed of unicellular groups, such as flagellates, ciliates and small *Foraminifera* (large *Foraminifera* are part of the meiobenthos), bacteria and archaea. Species diversity in the microbenthos can be extremely high, even without counting the (yet poorly known) "species" diversity of bacteria and archaea. It is quite common to find hundreds of species of protists in a single sample of a few cm^3. Identification of species is difficult, and must be done on live samples for a number of groups. In general, the biodiversity of this group is not well-known.

The role of bacteria in the breakdown of organic matter is well-studied, but the interaction between bacteria and other groups in the microbenthos remains largely unresolved. Possibly, tight predator–prey relations exist, comparable to those in the pelagic. In the benthos also viruses are active, and it is likely that these are responsible for high turnover rates of bacteria.

The best-documented direct influence of organisms on sediment characteristics is the effect of microphytobenthos on sediment erodability (see also Secs. 2.1 and 6.7). Microphytobenthos excretes EPS (extracellular polymeric substances), a kind of mucus mostly composed of polysaccharides, for two different reasons. The first is locomotion. Algal cells migrate up and down within the sediment bed at the rhythm of the tide, most probably to prevent resuspension during immersion. They excrete mucus as a propulsion mechanism. This mucus remains in the sediment and "glues" to the sediment particles. A large fraction of mucus is water-soluble, but during low tide it can stick very effectively to sediment particles, and some is kept upon immersion. Secondly, EPS is excreted as an overflow mechanism during photosynthesis (Blanchard *et al.*, 2000; Ruddy *et al.*, 1998). When cells have insufficient nutrients to process the carbohydrates produced by photosynthesis, they cannot stop photosynthesis. Therefore, they excrete the excess carbohydrates to prevent osmotic problems. Also this EPS binds to sediment particles at the sediment surface.

This would imply that nutrient limitations might stimulate sediment stability by microphytobenthos, and *vice versa*, that in eutrophic waters, sediment deposits would be less stable. However, the factors determining the production of EPS, its nature and binding capacity are yet largely unknown (Decho, 2000).

Note that polysaccharides consist of organic polymers, as introduced in Sec. 2.1. The binding mechanisms discussed above are therefore induced by electrostatic and molecular forces (Van der Waals forces), and therefore sensitive to the chemical conditions of the aquatic environment, such as salinity and acidity (pH), especially, the latter may be of importance. Both primary production by microphytobenthos and oxidation–reduction reactions in a bed modify the pH. Extreme values as low as 6 and as high as 9 have been observed in marine sediment, often with strong gradients over a few mm. The consequences for sediment stability have, however, not yet been quantified.

Bed erosion and organic matter

The effect of EPS on sediment stability has been studied by erosion experiments. These have mainly been carried out in laboratory flumes on sediment samples from the field. The bed is eroded by gradually increasing the bed shear stress exerted by the flow, while monitoring the concentration of suspended sediment (SPM) in the flume. The two important parameters determined in these experiments are the erosion threshold and the erosion flux.

Organisms can influence either or both of these two parameters (see Sec. 6.7). Most experiments with microphytobenthos show primarily an influence on erosion threshold. EPS enhances the cohesion (stickiness) of sediment particles and a kind of "skin" is formed on the sediment surface (this is not the crust of Sec. 5.2). This skin is not easily broken by the flow because of its enhanced strength, and because the bed becomes very smooth, reducing the erosive stresses themselves as well. However, once it is broken, further erosion depends on the intrinsic (physical) properties of the underlying sediment, which is less influenced by EPS and less over-consolidated, and erosion rates may increase rapidly.

An example of the influence of benthic algae on the erosion threshold is given in Fig. 8.2, showing experimental results by Austen *et al.* (1999). Note that chlorophyll-*a* is used as a proxy for the biomass of benthic algae, while other studies suggest that the EPS concentration itself is the variable most directly related to the erosion threshold. Smaller effects were found by Tolhursf *et al.* (2002), reporting an increase in τ_e by a factor 3–5, while De Brouwer *et al.* (2000) and Christie *et al.* (2000) found factors of 5 and 4, respectively.

A consequence of the increase in erosion threshold by microphytobenthos is that sediment may start to accumulate once microphytobenthos mats develop in spring. This positive feedback can have system-wide consequences. Figure 8.3, for instance, shows that in the central (the most quiet) part of the Molenplaat (a small intertidal mudflat in the Western

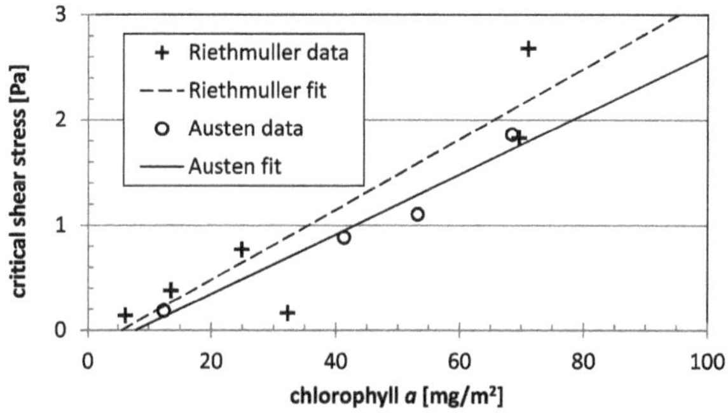

Fig. 8.2. Erosion threshold as function of chlorophyll-a content — circles reflect data for sediment from the Sylt-Romo tidal flats, whereas crosses represent the Lister Dyb tidal flats (redrawn from Austen *et al.*, 1999).

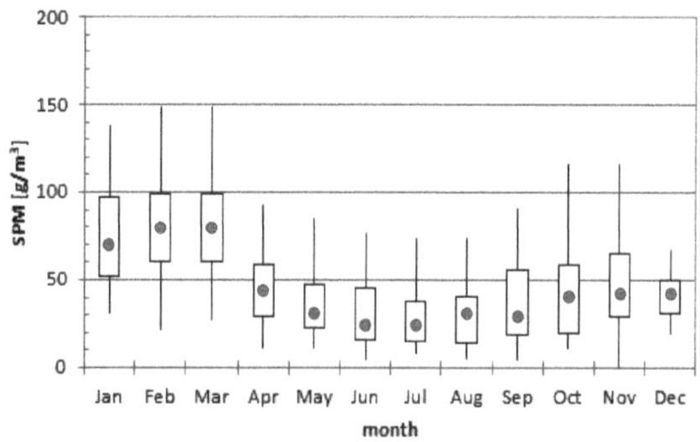

Fig. 8.3. Seasonal variation in SPM in Western Scheldt with mean and 25% and 75% percentiles.

Scheldt, The Netherlands), mud accumulates in late spring and summer, while eroding and then transported away from the flat during the following winter.

Also epifauna, i.e. fauna living on the sediment, affects hydraulic processes. As an example, the effect of mussel density is discussed. Individual mussels disturb/block the flow, inducing local scour, thus enhancing erosion. However, when mussels form compact beds, they stabilize the underlying sediment. This behavior was quantified by Widdows *et al.* (2002), as shown in Fig. 8.4.

Note that the effective roughness induced by mussel beds can become very large with roughness length k_s of the order of the mussel dimensions, i.e. many centimeters (Van Leeuwen, 2008); see also Sec. 17.4.3.2.

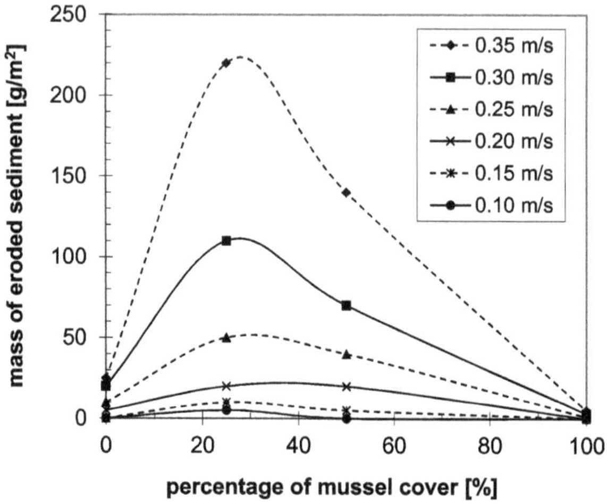

Fig. 8.4. Erodability as a function of mussel cover (redrawn from Widdows *et al.* (2002).

8.2 Bio-deposition

The binding effect of polysaccharides not only affects the stability of bed sediment. Dissolved polysaccharides also affect flocculation. As discussed in Sec. 4.3, flocs become denser when the sticking efficiency between particles increases. On the other hand, organic filaments themselves may result in very large, open flocs, as shown in Fig. 4.4 — see also marine snow, below.

Generally, two effects of (dissolved) organic material on the depositional behavior of cohesive sediment in the water column are distinguished:

(1) the effects of organic polymers on flocculation processes, and
(2) the effects of pelletization on particle aggregation.

We discuss both effects below.

8.2.1 *Bio-flocculation*

From an extensive literature survey, Van Leussen (1988, 1994) concluded that the effects of organic matter on aggregation rates is more or less maximal at organic matter concentrations of 1 to 2 mg/L. As such conditions are always met in estuaries and coastal waters, aggregation by polyelectrolites (the polymers in organic matter — see Sec. 2.1) always plays a role.

For instance, Uiterwijk Winkel (1975) observed that floc sizes in the Rhine-Meuse estuary decrease in the highly dynamic salinity intrusion zone in the estuary mouth. Van Leussen (1994) concluded that the cohesive sediment was already highly flocculated in the fresh water regions of the Rhine, owing to the effects of organic matter (and dissolved metals). The higher shear stresses in the estuarine zone of the river would breakup these flocs, rather than enhancing flocculation by salinity effects (Sec. 2.1)).

At large organic matter concentrations, aggregation efficiencies are high (Secs. 2.1 and 4.3) yielding large, porous flocs with high water content (Vicsek, 1992). This is observed in the deep sea, where residence times are large enough to attain equilibrium at small SPM-concentrations (Sec. 4.3). Wells (1987) , for instance, studied flocculation dynamics around Cape Lookout Bight (NC, USA). He found high sedimentation rates of "marine snow". Marine snow contains very open and fragile mud flocs, often highly elongated, measuring many centimeters, formed under conditions of low suspended sediment concentrations and flow velocities.

8.2.2 *Bio-pelletization*

Fauna living on the bed (one refers to epifauna) deposits pelletized fine sediment on that bed. Mussels (*Mytilus edulis*), for instance, are suspension feeders, filtering water at a rate of about 1 m^3/hr per 10 kg (flesh weight) of mussels (Dankers and Koelemaij, 1989). During a mussel-rich year, the entire Wadden Sea (The Netherlands) water mass can be filtered in a little over one week, whereas during lean mussel years, filtering would take about one month.

Mussels filter water and suspended particles over their gills. They retain most particles larger than approximately 1 μm. The filtered material is subsequently processed by the labial palps (small organs at the base of the gills) part of which is ingested in the animal gut. The remainder is packaged with mucus into "pseudo-faeces" and ejected by the mussel. Particulate matter taken up by mussels is excreted as faeces, after passage through their digestive system. Also this faeces is packaged in mucus ("faecal pellets"). These two processes are referred to as pelletization.

Pseudo-faeces and faeces form particles much larger and denser than the natural flocs in the ambient water and consequently have much higher settling velocities. Unless the organic mucus is broken down by bacteria, these pellets are quite stable — they even have been found in geological deposits. In comparison, the pellets from oysters and clams are less stable, thus their impact on the sediment dynamics is more temporary.

Faecal pellets consist for 70–90% (by weight) of phylo-silicates, mostly clay minerals and silt, and for 4–8% of organic carbon. Thus pelletization arrests large amounts of physils and organic matter, being deposited in areas where normally the hydrodynamic conditions do not allow deposition of fine sediment. Faecal pellets can therefore largely affect the overall sediment dynamics in marine waters. Local sedimentation rates by pelletization in Delaware Bay were found to be about 200 times larger than the fluvial sediment input (Biggs and Howell, 1984). Verwey (1954) estimated deposition rate in the Dutch Wadden Sea by pelletization at about 25–175 kton/yr. During summer, Flemming and Delafontaine (1994) found deposition rates in a mussel bed of up to 0.5 mm/day. However, much of this material was lost again by erosion in winter. The yearly net effect on the sediment dynamics was therefore difficult to assess.

It was found that mussels affect the mud content (and consequently the bio-geochemistry and ecology) in the seabed of the Wadden Sea over areas which are about 100 m larger than the mussel beds themselves. The deposition and subsequent incorporation of fines into the sediment has a large effect on net sedimentation rates, with accumulation rates up to 1 m/yr. Within the bed, mussels also shield sediment particles from erosion, leading to accumulation of considerable amounts of fine sediment, thereby forming mussel reefs.

Also beds of Japanese oyster increase in height as young oysters settle on top of older individuals, while the space between the older shells fills up with (muddy) sediment. Vertical accretion rate can be several, up to tens of centimeters per year, up to a level above which tidal inundation times limit further oyster growth. In the Eastern Scheldt (The Netherlands), Japanese oyster beds have been growing for a few decades. However, today all mature oyster beds have reached their maximum height, which was found relate to high water levels only, i.e. independent of the initial water depth.

Thus mussels and oysters can also be considered as eco-engineers, affecting sediment stability by their presence alone. Dankers (personal communication) distinguishes between different forms of mussel occurrences:

(1) Naturally, mussels adhere to each other and to the seabed, forming beds, thus stabilizing the seabed, while also pelletizing the sediment.

As mussel beds can become decades old, mud contents within these beds remain permanently high.

(2) In the short term, mussels cultivated on the bed (common in The Netherlands) have similar effects on the mud dynamics as natural mussel beds. However, these mussels are harvested after 1–2 years, after which the seabed is usually "cleaned", i.e. mud is removed, before new mussels are placed on the bed. The long-term effects of these mussel cultures therefore differ from natural mussel beds.

(3) Mussel cultivations suspended in the water column do pelletize sediment, but do not protect deposited material from erosion.

8.3 Bioturbation

"Bioturbation" is a somewhat loosely defined term referring to mixing by animals within the bed. In its generic meaning, bioturbation also includes the effects of fauna on the pore water within the sediment, as discussed in Sec. 8.4. Here, we focus on mixing of particles within the bed, which is a consequence of feeding and burrowing, as sketched in Fig. 8.5. Bioturbation may change the sediment composition and structure up to several decimeters depth. It has bio-geochemical consequences for the distribution of organic matter and oxidants within the sediment. Bioturbation also modifies the distribution of fine particles in a sandy bed, and influences the compaction and erodability of a bed.

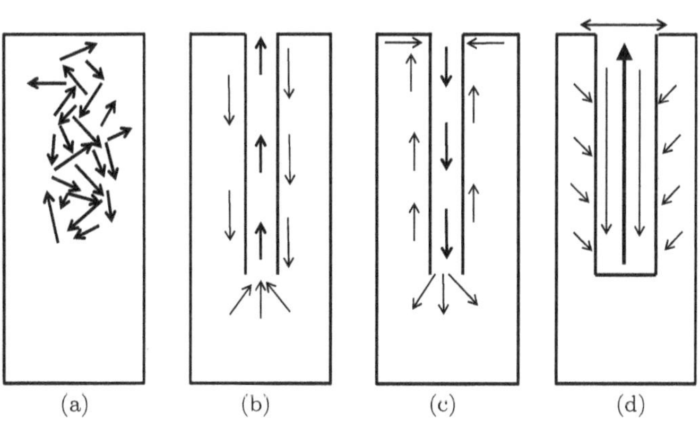

Fig. 8.5. Schematics of bioturbation and sediment reworking by (a) biodiffusors, (b) upward conveyers, (c) downward conveyers, and (d) regenerators (redrawn from Kristensen *et al.*, 2012).

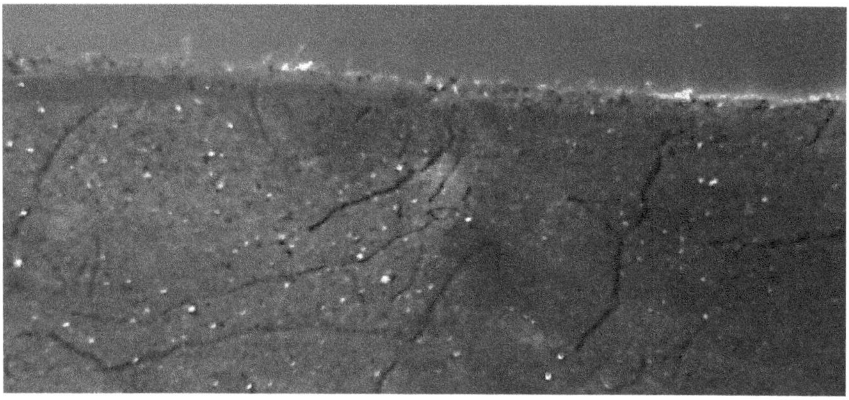

Fig. 8.6. Photo of burrows by *Tubifex* in sediment from Lake Markermeer with oxidized patches of sediment; width of photo represents 1 cm of soil (courtesy M. de Lucas Pardo).

Bioturbational mixing is the consequence of a variety of processes, depending on how animals feed, how they dig and use their burrows for ventilation, and where faecal material is deposited (e.g. Fig. 8.6). Animals may feed deep within the bed, at the bed surface or in the water column. Faeces may be deposited within the bed, at the bed surface or in the water column. Depending on these processes, biotic and abiotic material fluxes may decrease or increase. Mathematically, these fluxes can be characterized as non-local advective transports, often adequately represented as a diffusion process. Either way, there are considerable effects on the capacity to store fines within a bed, on sediment biogeochemistry, and on sediment surface topology. Below, a few examples are presented.

The lugworm (*Arenicola marina*) creates a feeding funnel (burrow), efficiently transporting surface material towards a feeding pocket deeper within the bed. It deposits its faeces on the bed surface. For oxygenation, the lugworm creates a current through its burrow from the faecal cast towards the feeding pocket. Burial was studied experimentally in the Western Scheldt using luminophores (fluorescent particles of the sand size) placed onto the sediment surface and retrieved 24 hr later. Luminophores were found throughout the feeding funnels (Montserrat *et al.*, 2011).

In another field experiment at Sylt (German Wadden Sea), lugworms were removed from the bed. This caused large changes in the structure of the sediment surface, in the mud content of the sediment, in sediment bio-geochemistry and in the associated sediment fauna (Volkenborn *et al.*,

2007). It is likely that the lugworm plays a pivotal role in maintaining the composition and ultimately also the morphology of tidal flats where they are abundant (Volkenborn *et al.*, 2007).

Callianassa (small shrimp-like fauna) is probably world champion bioturbation. Species of this genus form galleries up to 2.5–3 m depth (Pemberton *et al.*, 1976). In the Mediterranean, *Callianassa truncata* collects seagrass leaf material and other refractory detritus into storage chambers, where the material is partly subject to microbial breakdown and then used as food. The animals ventilates its burrow systems, and also modifies the sediment surface substantially (Ziebis *et al.*, 1996).

A good example of bioturbation is the bed composition of Lake Markermeer, The Netherlands (see also Chapter 17). This fresh water lake has a surface of about 760 km^2 and mean water depth of about 3.5 m. Its bed consists of a 1–2 decimeter thick layer of soft mud atop stiff Holocene deposits. Currently, the majority of this soft mud is in an anoxic state, whereas the upper few mm are oxic, as sketched in Fig. 8.7 (see also Fig. 8.6). The oxic sediment is more or less in equilibrium with SPM in the water column, causing high turbidity levels, and poor water quality conditions in the lake as a consequence. The anoxic sediment is mobilized only under storm conditions, yielding SPM-concentrations in the lower meter of the water column up to a few g/L (e.g. Vijverberg *et al.*, 2011).

Lake Markermeer was formed in 1976, when a closure dike (the Houtribdijk) isolated the lake from the large fresh water Lake IJsselmeer (closed off from the Wadden Sea/North Sea in 1932). Since then, fine sediment eroded

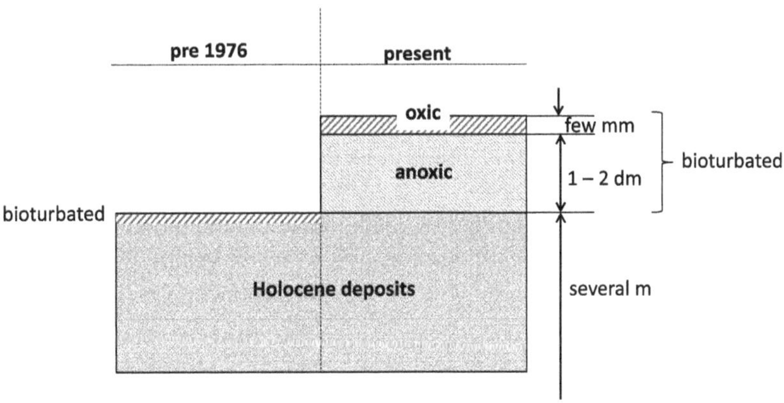

Fig. 8.7. Schematic of sediment composition of Markermeer bed before (pre 1976) and after construction of the Houtribdijk.

from the Lake Markermeer bed is trapped behind the dike, accumulating within the lake, see Fig. 8.7.

De Lucas Pardo (2014) and De Lucas Pardo *et al.* (2013) studied the erodability of the Holocene deposits, and concluded that under the hydrodynamic conditions prevailing in Lake Markermeer (mainly wave action), these deposits cannot be eroded. However, De Lucas Pardo also found a profound effect of bioturbating meiofauna (in particular *nematodes* and *tubifex*), loosening the upper mm's of these deposits. Extrapolating laboratory experiments, De Lucas Pardo concluded that this bioturbation had produced the entire stock of soft sediment in Lake Markermeer in a period of about five years after the erection of the Houtribdijk.

Bioturbation continues till today, but only in the soft sediment layer atop the Holocene deposits. The Holocene deposits themselves are protected from further bioturbation/erosion by the anoxic soft mud layer above, as meiofauna needs oxygen to live.

Thus meiofauna initially initiated erosion of the Holocene Lake Markermeer bed, whereas later the eroded sediment stabilized the bed, while bioturbation keeps the sediment soft.

Bioturbation as a diffusion process

Mathematically, bioturbation is often modeled as a diffusion process (see also Sec. 12.2). Meysman *et al.* (2003, 2008) have rigorously derived a diffusion-like model, discussing the underlying assumptions. Even though the criteria for a continuous local diffusion approach are not always fulfilled (Meysman *et al.*, 2003), its application is generally successful. This can be explained by the fact that a diffusion-approach yields an asymptotic solution for a wide variety of conditions (Meysman *et al.*, 2008). The intrinsic variability of rates and processes caused by a mixed community of bioturbating animals is the basis for this generalization. Nevertheless, particular cases may require the use of non-local models in order to represent field observations.

The fundamental parameter of a bio-diffusion model is the bioturbation coefficient D_b (diffusion coefficient). Various values for D_b can be found in literature. Cournane *et al.* (2010) studied bioturbation by the ragworm (*Nereis diversicolor*) and found values of D_b = 2–14 cm^2/yr (i.e. (6–44)$\cdot 10^{-12}\,\text{m}^2/\text{s}$). Lecroart *et al.* (2007) used isotopes and found values of D_b = 0.1–400 cm^2/yr (i.e. (0.03–1270)$\cdot 10^{-12}\,\text{m}^2/\text{s}$), while Middelburg *et al.* (1997) reported values of $D_b = 55 \pm 9\,\text{cm}^2/\text{yr}$ (i.e. $(170 \pm 30) \cdot 10^{-12}\,\text{m}^2/\text{s}$). These studies illustrate the variability of observed biodiffusion coefficients, but also confirm the generally accepted range from $0.1\,\text{cm}^2/\text{yr}$ for deep-sea sediment to $10\,\text{cm}^2/\text{yr}$ for coastal sediment.

It can be deduced from these D_b-values that bioturbation is a slow process. It induces considerable mixing on a time scale of years, but not on time scales of days or weeks. In sediment that is subject to strong physical forces, e.g. tidal currents or waves, the latter will probably dominate biological mixing (see also Sec. 6.6). In deeper or less exposed sediment, biological mixing takes over this dominant role. Deep-sea sediments, for instance, are almost exclusively mixed by bioturbation.

8.4 Bio-irrigation

Most metazoan animals need oxygen for their metabolism. This is a problem in fine sediment, where pore water is usually anoxic, even at shallow depths. Animals therefore construct burrows and use various types of pumping techniques, to get the oxygen needed (Fig. 8.8). These burrows change the physical structure of the bed, providing e.g. drainage channels along which water can escape enhancing consolidation rates of a bed. Burrowing and pumping also lead to pelletization of the sediment. Pumping induces diffusive and/or advective input of oxygen-rich water into the soil. This primarily affects the sediment biogeochemistry, but can also have a secondary

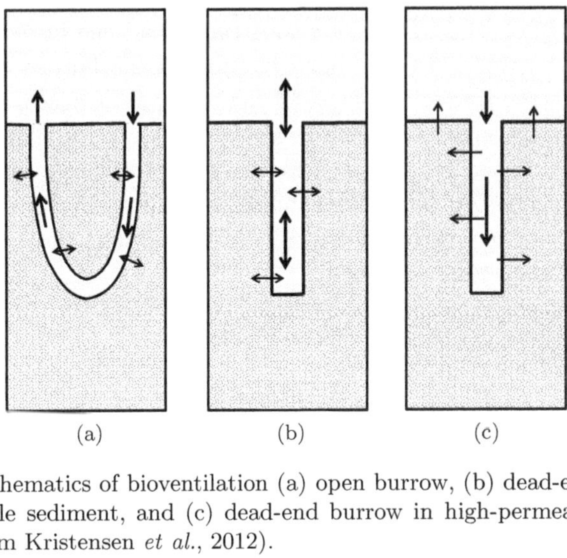

(a) (b) (c)

Fig. 8.8. Schematics of bioventilation (a) open burrow, (b) dead-end burrow in low-permeable sediment, and (c) dead-end burrow in high-permeable sediment (redrawn from Kristensen *et al.*, 2012).

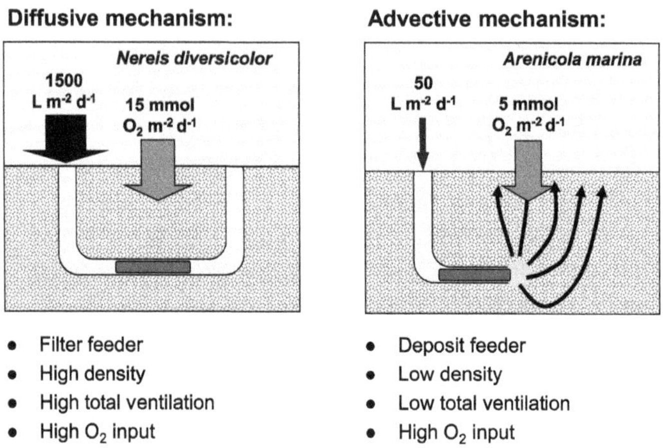

Fig. 8.9. Schematic of burrow ventilation by *Nereis* and by *Arenicola* (redrawn from Meysman *et al.*, 2006).

influence on sediment stability against erosion through the oxidation of iron precipitates.

As an example, *Nereis diverscolor* and *Arenicola marina* use different strategies for burrow ventilation (e.g. Meysman *et al.*, 2006), as sketched in Fig. 8.9. *Nereis* makes two-sided open burrows through which oxygen-rich water is pumped. However, water does not enter the interstitial pores of the sediment and oxygen only penetrates from the burrow into the soil by diffusion. In contrast, *Arenicola* constructs dead-end burrows. It pumps water from the tail end of its burrow, through the burrow over its gills, and then into the interstitial space of the sediment. Consequently, oxygen is transported by advection (with burrow water) into the sediment matrix. The volume pumped by *Nereis* is almost two orders of magnitude larger than that of *Arenicola*, while the rate of oxygen entrainment into the sediment is of the same order of magnitude. Note that both animals burrow up to about 20 cm deep into the bed.

8.5 The role of vegetation

Submerged or intertidal vegetation, in seagrass meadows, salt marshes or mangrove forests, has profound effects on the morphological development of estuaries and coasts. Patches of vegetation (called "tussocks") induce very different hydrodynamic conditions inside and outside tussocks. First, a tussock diverts ambient flow. Part of the flow is diverted around a tussock,

Fig. 8.10. Tussock of English cordgrass (*Spartina anglica*) on a mud flat.

another part is concentrated higher in the water column over the plants, while the flow velocity is small in between the plants. As a result, sedimentation is generally enhanced within a tussock. However, there are stagnation effects where the flow hits a tussock, locally enhancing turbulence generated by the stem–flow interaction, thus affecting the local sediment dynamics. This generally results in scour in front of a tussock, while erosion around a tussock often occurs as well. Figure 8.10 illustrates such an eco-engineering pattern, showing a photograph of a tussock of English cordgrass (*Spartina anglica*) on a mud flat with elevated bed level, while surrounded by a scour moat.

The main effect of vegetation on the flow is the drag caused by the stems. This transforms part of the kinetic energy into turbulent kinetic energy, which is subsequently dissipated in a turbulent energy cascade (see Sec. 3.2.2). As a result, considerable shear is induced over the water depth, as sketched in Fig. 8.11.

In between stems, the turbulent length scale is set by the diameter of the stems, as these are the main agents generating drag. This length scale is usually small, but hydraulic drag is considerable, and flow velocities are largely reduced. Above, around the top of the canopy, the turbulent length scale is set by the distance between stems, as the stems operate as a kind of filter to turbulent eddies protruding higher into the water column. In this layer, water velocities are large and the turbulent kinetic energy reaches a maximum over the vertical. Finally, above the canopy, the turbulent length scale is set by the difference between the water surface and the height of

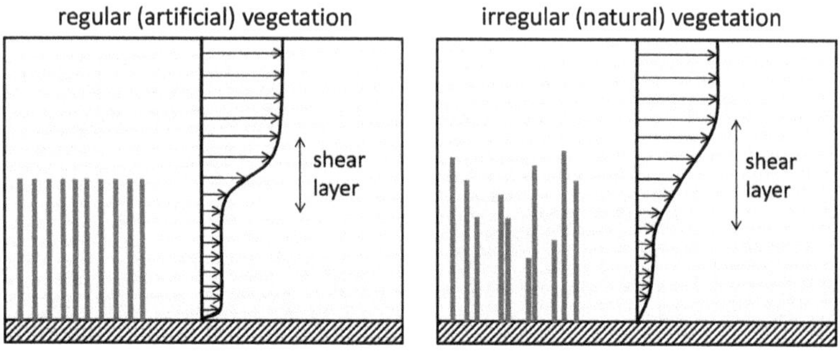

Fig. 8.11. Diagram showing flow over/through regular (left panel) and irregular (right panel) vegetation (after Poggi *et al.*, 2004 and Horstman *et al.*, 2018).

the canopy. Flow velocities are highest, often exceeding the ambient velocity (i.e. without canopy).

Quantitative information and computational rules for assessing the role of vegetation tussocks on hydrodynamics and sediment dynamics stem from laboratory experiments, in which vegetation is modeled as regular sticks, or plastic plants. Horstman *et al.* (2018) shows that vertical velocity gradients in natural, irregular canopies are much smaller than over regular configurations (Fig. 8.11). Bed shear stresses within the irregular canopy were about half of those in a regular configuration. It is likely that also wave damping by irregular canopies is higher, as flow velocities within a canopy are larger — this effect has not yet been studied, however.

The principles underlying these different controlling elements on turbulence have been implemented in turbulence models, representing the effects of vegetation on flow (and waves) in numerical models. It is stressed that the effects of vegetation on hydrodynamics can be modeled often by increasing the bed shear stress (locally). However, this would prevent sediment to settle in between the vegetation, as bed shear stresses would become unrealistically high. Therefore, Baptist *et al.* (2009) proposed to model energy losses by the vegetation in stead, see also Tanino and Nepf (2008).

Wave damping

The effect of vegetation on waves is reasonably well understood, both through laboratory experiments and from field observations (Mendez and Losada, 2004). Wave damping by vegetation can be considerable. The distance over which significant wave damping is observed varies with the characteristics of the waves and density of the vegetation. Usually, the first tens

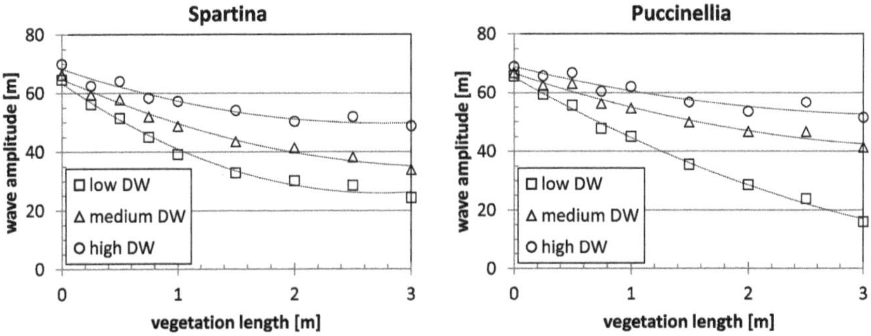

Fig. 8.12. Wave damping by *Spartina* and *Puccinellia* as a function of tussock dimension (from Bouma *et al.*, 2010).

of meters from the vegetation leading edge are the most important, but for swell and infra-gravity waves the effect can extend over hundreds of meters.

The effect of a single stem on wave damping changes considerably with the stem stiffness. It has been observed that wave damping by a single stem of the stiff *Spartina* (Fig. 8.12) is much larger than that of a single stem of the flexible grass *Puccinellia*. However, per unit surface *Puccinellia* contains many more stems than *Spartina*. As a consequence the effects per unit biomass are very comparable (DW, for dry weight, Fig. 8.12). It is not yet known how general this result is, but it holds promise for a generalization of vegetation effects beyond the specificities of single species (Bouma *et al.*, 2010).

Bio-morphology

The interaction between vegetation, hydrodynamics and sediment transport may affect the morphology of aquatic systems considerably. When habitat conditions are favorable, vegetation begins colonizing bare mudflats (and tidal flats). The initial plants/tussocks catch sediment and stabilize the deposits (thus eco-engineering their own environment/habitat). In between these initial plants/tussocks, the flow accelerates and scours the bed, as in Fig. 8.10. Thus a pattern of elevated, vegetated areas with non-vegetated channels/creeks develops, forming the well-known morphology of vegetated tidal flats.

This development was recorded by Temmerman *et al.* (2007), as shown in Fig. 8.13. Accounting for some fairly simple interactions between vegetation, hydrodynamics and sediment transport, these patterns can

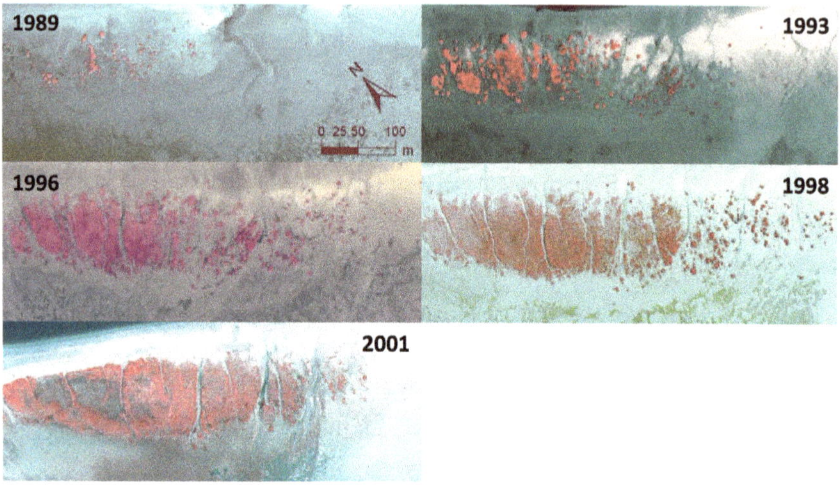

Fig. 8.13. Aerial photographs of colonization by English cordgrass (*Spartina anglica*) of a tidal flat in the Scheldt estuary and subsequent channel formation (after Temmerman *et al.*, 2007).

qualitatively be simulated with numerical models (Temmerman *et al.*, 2007), and we refer to the discipline of bio-morphology. However, the discipline of morphology is beyond the scope of this book, and not further addressed.

Mangroves

Mangroves are amongst the most important eco-engineers in tropical and sub-tropical coastal waters, i.e. between roughly 25°N and 25°S. These species are very robust, and when conditions are favorable (Balke *et al.*, 2013), they can colonize the intertidal rapidly. Young mangroves grow fast; *Avecennia* for instance can grow 1 m/yr in height. The mangrove habitat is typically found between high-high water spring (HHWS) and mean sea level (MSL) or mean high water (MHW), depending on the dynamics of the wave climate. Around 2010, mangroves covered about 150,000 km^2 of the worlds coastline, its habitat shrinking rapidly, though, at a rate of 1–1.5% per year (Spalding, 2010).

There are about 110 different mangrove species across the world. They are salt-tolerant (but need some fresh water) and therefore out-compete other species in salty and brackish water. Pristine mangrove fringes and forests are 100s m to a few km wide, and generally very dense.

Fig. 8.14. Aerial photographs of young, pristine mangrove forest along Suriname coast.

Mangroves provide important ecosystem services (Russi *et al.*, 2013). They play an important role in coastal protection, as the mangrove habitat can grow with (moderate) sea level rise, and the dense forest attenuates waves efficiently (see below). Mangroves offer nurseries and shelter to many fish — it is estimated that about 70% of all commercial fisheries in the tropics depends on mangroves. Mangrove forests purify water, provide timber and other forest products, such as honey, and sequester large amounts of carbon.

Further to the discussion above on saltmarsh vegetation, also mangroves engineer their environment largely, even much more than saltmarsh vegetation, as mangroves are much larger, and present throughout the year. While growing, mangrove juveniles affect the hydrodynamics by their drag, and induce sedimentation by inducing still/calm water conditions. Net siltation rates in mangrove forests can be high, up to many cm's per year, while gross rates can be an order of magnitude larger (Cuong and Brown, 2012). In between groups of mangroves, the flow accelerates, forming channels/creeks, as shown in Fig. 8.14.

Depending on the species, but in particular on its density, a 100 m wide mangrove forests can damp between 13% and 66% of incoming wave energy (McIvor *et al.*, 2012, 2013). In particular, young, dense forests are very effective in damping waves. Also, the dense mats of pneumatophores (*Avecennia*) can damp small waves very effectively.

Chapter 9

Transport and Fate of Mud in Estuaries and Tidal Basins

In this and the following chapter, the mechanisms inducing net transport of fine sediment (and sand, to some extend) in marine waters (hence tidal conditions) are discussed. Net SPM transport is defined as the flux F of fine sediment in horizontal direction averaged over a specific time window (e.g. tidal cycle, or longer) through a cross-section. Except for Sec. 9.1.3, SPM is treated as a passive tracer, though may affect vertical mixing through buoyancy effects.

Net SPM transport is induced by asymmetries in hydrodynamics and/or bathymetry in combination with nonlinearities in the sediment transport processes and/or their relevant time scales (see also Sec. 3.4.3). Table 9.1 summarizes these nonlinearities and relevant time scales; asymmetries in hydrodynamics and bathymetry are the subject of the sections below.

The scaling of bed and suspended load (for sand) follows from e.g. Van Rijn (1993), or other classical textbooks on sediment transport. The scaling for the transport of SPM for capacity conditions (i.e. HCMS-conditions) follows from Sec. 7.1, assuming $F_{mud} \propto UC_{sat}$.

It is illustrative to compare T_s and T_m for fines under typical estuarine conditions with $h = 10\,\text{m}$, $W_s = 0.5\,\text{mm/s}$ and $U = 1\,\text{m/s}$ (thus $u_* \approx 0.05\text{m/s}$) (see also Sec. 3.4.3). This yields $T_s \approx 5.5\,\text{hr}$ and $T_m \approx 0.5\text{hr}$, implying pronounced asymmetry between mixing and settling, thus between the conditions during accelerating and decelerating tide. The implications of this asymmetry on the tidal variations of vertical SPM profiles were discussed in Sec. 3.4.3, showing profound deviations from the equilibrium Rouse profile.

Residual horizontal SPM transport is the result of the difference between two gross transport fluxes in opposite directions, for instances, up- and

Table 9.1. Nonlinearities in sediment processes and the relevant time scales.

	Process	Scales with	Time scale
Non-equ. transp.	Erosion	$E \propto (\tau_b - \tau_{cr}) \sim U^2$	Not defined†
	Settling and deposition		$T_s \sim T_D \sim h/W_s$
	Vertical mixing	$\Gamma_z \propto U$	$T_m \sim h^2/\Gamma_z \approx 10h/u_*$
	Tidal period		$T/4$ or $T/2$
Equ. transport	Bed load	$F_{bl} \propto U^3$	
Equ. transport	Suspended load	$F_{sl} \propto U^{4-5}$	
Equ. transport	Mud load (SPM)	$F_{mud} \propto U^4$	

Note: †Not defined for fine sediment, see Sec. 3.4.3.

down-estuary. As gross transport rates are generally much (orders of magnitude) larger than their difference, small variations in gross transport have large effects on the residual transport. This observation is further elaborated in Sec. 9.4, and has large consequences for the use of numerical models (see Sec. 13.6).

The net sediment transport discussed in this and the following chapter is the result of non-zero correlations between the various tidally varying state parameters, induced by nonlinearities. Tidal flow is driven by pressure gradients. In case of temperature-, salinity-, and/or SPM-induced density gradients (Sec. 3.2.3), the horizontal pressure gradient becomes:

$$p = p_{atm} + g\int_z^{\zeta} \rho(x,z)\mathrm{d}z \quad \text{thus} \quad \frac{\partial p}{\partial x} = \rho g \frac{\partial \zeta}{\partial x} + g\int_z^{\zeta} \frac{\partial \rho}{\partial x}\mathrm{d}z \tag{9.1}$$

where ζ represents the water level. The second-last term in Eq. (9.1) is known as the barotropic pressure gradient, whereas the last term is referred to as the baroclinic pressure gradient.

Net sediment transport can be induced by asymmetries in velocity and pressure gradients, and in water density. Further to the definition above, one refers to barotropic and baroclinic effects, respectively. The first are treated in Sec. 9.2, whereas the latter in Sec. 9.1. These effects have been elaborated in detail by Burchard *et al.* (2018) in their review on sediment trapping in estuaries.

Finally, it is noted that an interpretation of the net and gross SPM fluxes also requires understanding of the sediment sources and sinks, which may be sea-borne, riverine, or originate from local sources (i.e. within the model domain), such as bank or bed erosion. Identification of sources and sinks is part of the task of system understanding, which is addressed further in Chapter 13.

9.1 Baroclinic effects

This chapter analyses (potential) net SPM horizontal transport by asymmetries (i.e. gradients) in salinity, as encountered in estuaries, tidal lagoons and coastal waters (the latter are discussed in detail in Chapter 10). Such (horizontal) salinity gradients induce gradients in the density of water. Density gradients can also be induced by temperature and SPM. Turbidity currents and debris flows are manifestations of the latter (see Sec. 7.5.1). When the density of a fluid depends on temperature, salinity and/or SPM, it is called baroclinic, hence the term baroclinic effects, and the induced currents are referred to as density currents.

In general, density gradients induced by salinity, temperature and/or SPM are quite small. Yet even small density gradients may have profound effects on the net transport of SPM, as will become clear from the examples given below and in Sec. 9.5 and Chapter 18.

9.1.1 *Gravitational circulation*

Figure 9.1 sketches the generation of a net circulation by a horizontal salinity gradient. For illustrative purposes, we assume that the sketched stretch of the estuary is at rest. Then the total forces, amounting to the vertical integration of the hydrostatic pressure distribution, on either side of this stretch should be at equilibrium. However, as the density of saline water is larger than that of fresh water (Eq. (3.26)), the vertical pressure distributions at the sea side and river side of this stretch differ, inducing

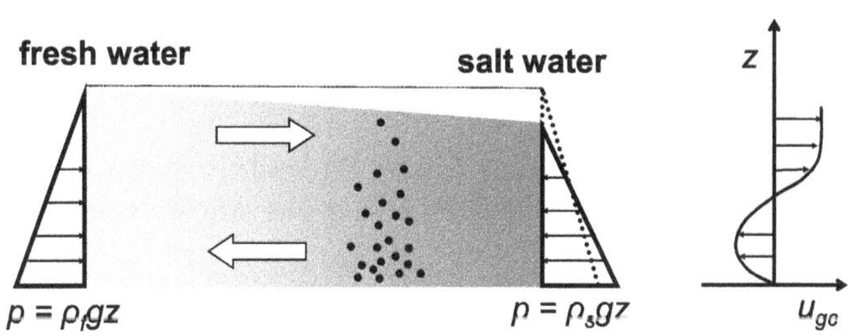

Fig. 9.1. Diagram explaining gravitational circulation by longitudinal salinity gradients, also showing the induced gravitational circulation velocity U_{gc}.

a longitudinal near-bed pressure gradient in the direction of the salinity gradient. The force balance whence requires a slope in the water surface to balance that salinity gradient.

These opposite near-bed and near-surface pressure gradients induce a residual circulation, being zero when integrated over the water depth. This residual circulation is known as the estuarine or gravitational circulation. We prefer the latter term, as this circulation may also occur in tidal basins and coastal zones. It is the longitudinal density gradient which induces this circulation. Vertical salinity gradients cannot induce a longitudinal current — however, vertical stratification can enhance gravitational circulation significantly, as discussed below.

The impact of horizontal density gradients can be assessed easily from the 2DV momentum balance, assuming vertically homogeneous saline water and fresh water conditions at either side of the stretch of Figure 9.1. The hydrostatic pressure gradient reads, assuming constant density over the water depth (no stratification):

$$\frac{\partial p}{\partial z} = \rho g \quad \text{thus} \quad p = \int_z^h \rho g \,\mathrm{d}z = \rho g(h - z) \tag{9.2}$$

The horizontal pressure gradient thus becomes:

$$\frac{\partial p}{\partial x} = \frac{\partial \rho g h}{\partial x} = \rho g \frac{\partial h}{\partial x} + g(h - z)\frac{\partial \rho}{\partial x} \tag{9.3}$$

which should balance the vertical stress distribution:

$$\frac{\partial p}{\partial x} - \frac{\partial \tau}{\partial z} = 0 \tag{9.4}$$

To first order, we assume a parabolic viscosity profile (thus no or little stratification):

$$\tau = \mu_z \frac{\partial u}{\partial z} = u_* \kappa \rho \frac{z}{h}(h - z)\frac{\partial u}{\partial z} \tag{9.5}$$

Further to Sec. 3.2.2, Eq. (3.22), we substitute Eqs. (9.3) and (9.4) into Eq. (9.5), and integrate Eq. (9.5) over the water depth, which gives the logarithmic velocity distribution (following from the barotropic effects), modified by a term induced by the longitudinal salinity gradient (i.e. the baroclinic effects):

$$u(z) = -\frac{u_*}{\kappa} \ln\left\{\frac{z}{z_0}\right\} + \frac{1}{2}\frac{gh}{\kappa u_*}(z - z_0)\frac{\alpha}{\rho}\frac{\partial S}{\partial x} = u_{btr}(z) + u_{bcl}(z) \tag{9.6}$$

with the barotropic u_{btr} and baroclinic u_{bcl} velocity components and where $\alpha \approx 0.75$ (see equation of state Eq. (3.26)). Integration of Eq. (9.6) over the water depth h gives:

$$U = -\frac{u_*}{\kappa}\left[\ln\left\{\frac{h}{z_0}\right\} - 1\right] + \frac{1}{4}\frac{\alpha g h^2}{\kappa u_*}\frac{1}{\rho}\frac{\partial S}{\partial x} = U_{btr} + U_{bcl} \tag{9.7}$$

Equations (9.6) and (9.7) show that the barotropic and baroclinic pressure gradients induce a residual flow in an estuary, which includes the gravitational circulation introduced above. The gravitational velocity U_{gc} is defined as

$$U_{gc}(z) = u_{bcl}(z) - U_{bcl} \tag{9.8}$$

For given depth-averaged velocity U, u_* can be computed from Eq. (9.7), which is then substituted into Eq. (9.6), yielding the vertical velocity profile for given U and $\partial S/\partial x$. Figure 9.2 shows this modification of the logarithmic velocity profile for $\partial S/\partial x = 0.3\,\mathrm{ppt/km}$, typical for the Sea Scheldt around Antwerp (Belgium) (Winterwerp *et al.*, 2006b). This picture is characteristic for many other estuaries in the world. Figure 9.2 shows that the effect is quite large, but that Eq. (9.6) does not fully capture the measured velocity pattern — this difference will be explained in Sec. 9.1.2.

The subject of gravitational circulation has been (and is still being) studied extensively, resulting in an overwhelming body of literature. The reader is referred for instance to the classical work by Hansen and Rattray

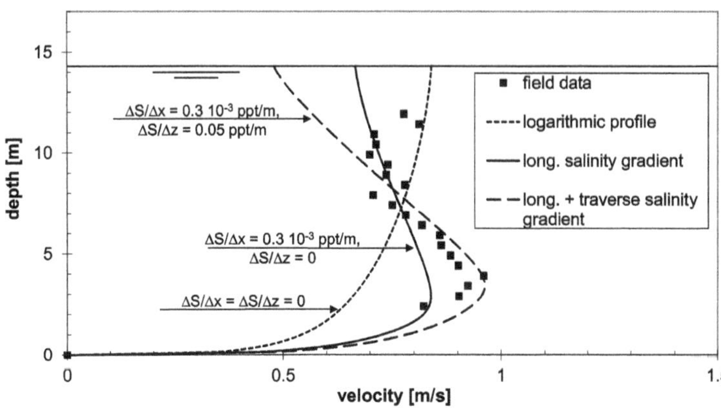

Fig. 9.2. Modification of logarithmic velocity profile by longitudinal (Eq. (9.6)) and transverse salinity gradient (Eq. (9.17)).

(1965); or to Valle-Levinson (2010), Ralston *et al.* (2008), MacCready and Geyer (2010) and Wolanski and McLusky (2011).

Ralston *et al.* (2008, following on Hansen and Rattray) assessed the magnitude of the residual velocity U_{gc} (Fig. 9.1) induced from the one-dimensional momentum balance, assuming non-stratified conditions and constant, characteristic viscosity ν_z:

$$U_{gc} = \frac{gh^3 \partial\rho/\partial x}{48\rho\nu_z} = \frac{g\alpha h^3 \partial S/\partial x}{48\rho\nu_z} \tag{9.9}$$

in which $\partial\rho/\partial x$ is the baroclinic pressure gradient, α follows from the equation of state Eq. (3.26), e.g. $\rho = \rho_0 + \alpha S$, h is water depth, and $\partial S/\partial x$ is the longitudinal salinity gradient. At 10 m water depth, a typical value for $\partial S/\partial x = 0.3\,\text{ppt/km}$ and $\nu_z = 0.02\,\text{m}^2/\text{s}$ gives $U_{gc} \approx 23\,\text{cm/s}$, which largely agrees with the analysis above (Fig. 9.2).

Salinity intrusion

These self-induced up-estuary baroclinic pressure gradients are a major mechanism for the intrusion of salt water into estuaries (e.g. Dyer, 1997). Depending on tidal conditions (spring/neap) and river flow, salt water can intrude estuaries by several 10s of kms. The actual penetration is governed by the balance between the baroclinic and barotropic pressure gradients affected by internal friction.

Note that the net up-estuary salt flux, induced by U_{bcl} is compensated by down-estuary river-induced flushing of salty water mixed higher into the water column by turbulence.

Currents induced by internal pressure gradients resulting from gradients in water density by temperature, salinity and/or SPM are referred to as density currents — this is the barotropic term in Eqs. (9.6) and (9.7). Also the transport of fluid mud (Sec. 7.5.1) can be a form of a density current.

Salinity intrusion in estuaries was studied extensively in the 1960s, see for instance the work by Keulegan (1966) and Ippen and Harleman (1961). The literature contains various algebraic formulas quantifying salt water intrusion in estuaries (e.g. Hansen and Rattray, 1965; Prandle, 1981; Savenije, 2005). Here we follow Kuijper and Van Rijn (2011) in determining the maximum salinity intrusion length L_{max} and the corresponding longitudinal cross-sectionally averaged salinity distribution in exponentially converging estuaries, i.e. the flow-carrying cross-section of the river is given by $A(x) = A_0 \exp(-x/L_a)$ with L_a the so-called convergence length of the estuary, and A_0 the cross-sectional area at the estuary mouth:

$$L_{max} = \zeta L_a \ln\left\{1 + \frac{2A_0 D_S}{\zeta L_a |Q_r|}\right\} \tag{9.10}$$

in which the parameter ζ is defined by

$$\zeta = \frac{1}{1 - \Phi L_a} \quad \text{with} \quad 2\Phi = k_i + \frac{3}{L_b} - \frac{1}{L_a} \tag{9.11}$$

where L_b is the width-related convergence length of the flow-carrying cross-section (in a prismatic channel, $L_b = L_a$). Further, k_i is the imaginary part of the tidal wave number (i.e. the wave damping: $U(x) = U_0 \exp\{k_i x\}$), U_0 is the amplitude of the tidal velocity, A_0 is the cross-section at the mouth, and Q_r the river discharge,while the river flow induced velocity in the mouth is denoted by $U_{r,0}$.

The longitudinal salinity-induced dispersion coefficient D_S (Sec. 12.1) is given by

$$D_S = \alpha_c 60\sqrt{\pi} h_0 c_i \sqrt{\frac{U_{r,0}}{U_0}\frac{C}{\sqrt{g}}\frac{L_{t,e}}{L_a}} \tag{9.12}$$

in which the internal celerity c_i equals $\sqrt{\Delta\rho g h_0/\rho}$, C is the Chézy coefficient, $L_{t,e}$ is the tidal excursion (which varies with spring-neap variation of U_0), and the calibration coefficient $\alpha_c = 1.0 \pm 0.3$.

The tidal mean longitudinal salinity distribution $S(x)/S_0$, where S_0 is the tidal mean salinity at the mouth ($x = 0$) is given by

$$\frac{S(x)}{S_0} = \left[1 - \frac{1}{2}\frac{|Q_r|\,\zeta L_a}{A_0 D_S}\left(\exp\left(\frac{x}{\zeta L_a}\right) - 1\right)\right]^2 \tag{9.13}$$

Equation (9.13) was compared with data from the Scheldt River, using the following parameters: $L_a = 26.26$ km, $L_b = 50.28$ km, $A_0 = 168 \cdot 10^3$ m^2, $U_0 = 0.73$ m/s, $k_i = 0.0036$ km^{-1}, $D_S = 290$ m^2/s, $\alpha_c = 1.3$, and $S_0 = 31$ ppt. This comparison is presented in Fig. 9.3, showing excellent agreement with the measured maximal salinity distribution over 100 km in the Scheldt. Also, comparison with 12 other rivers was favorable (results not presented — see Kuijper and Van Rijn, 2011). Today salinity intrusion is generally assessed with the help of three-dimensional numerical models. Yet, the algebraic relation by Kuijper and Van Rijn is still useful to obtain a first assessment.

In case the river flow is too large, the internal flow becomes super-critical and salt water cannot intrude the estuary. Then a salinity front is formed

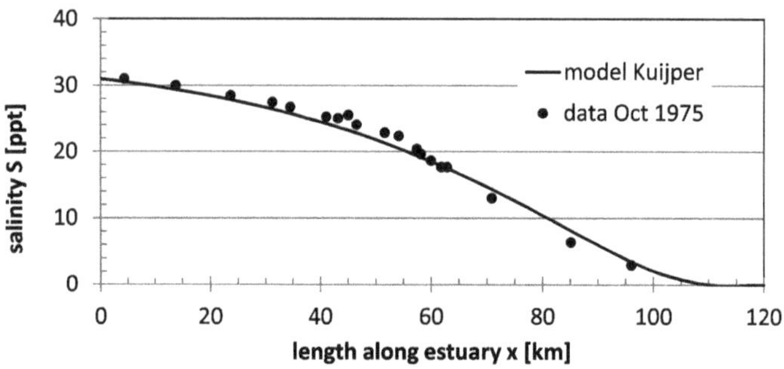

Fig. 9.3. Comparison of salinity distribution according to Eq. (9.13) and data of the Scheldt River (October, 1975), modified from Kuijper and Van Rijn (2011).

well outside the river mouth. This is the case for instance in the Amazon River, and in the Yangtze River at high river discharges.

Estuarine turbidity maximum (ETM)

Gravitational circulation (baroclinic effects) and subsequent salinity intrusion are very important for the transport of fine suspended matter in estuaries. There are three effects:

(1) The baroclinic pressure gradient induces a net transport of saline water up-estuary, as in Eq. (9.7) — if this saline water contains SPM, this pressure gradient also induces net up-estuary transport of SPM.
(2) The gravitational circulation itself (Eq. (9.8)) — as SPM-concentrations near the bed are larger than higher in the water column (Sec. 3.4), gravitational circulation induces net SPM transport in up-estuarine direction, as sketched in Fig. 9.1.
(3) Near-bed velocity gradients during flood are larger than during ebb (internal asymmetry — Fig. 9.6), thus the larger bed shear stresses mobilize more SPM during flood, or mobilize this sediment earlier after HWS than after LWS (Sec. 9.2.2).

At the head of the salinity intrusion, baroclinic effects become zero. This is therefore a zone of SPM convergence, as sketched in Fig. 9.4 — further up-estuary, down-estuary SPM transport prevails (possibly apart from the effects of tidal asymmetry, as treated in Sec. 9.2).

This zone of SPM convergence is known as the estuarine turbidity maximum (ETM), and SPM-concentrations in this ETM can easily exceed SPM-values elsewhere in the river by many factors — concentrations up to and over 500 mg/L are no exception.

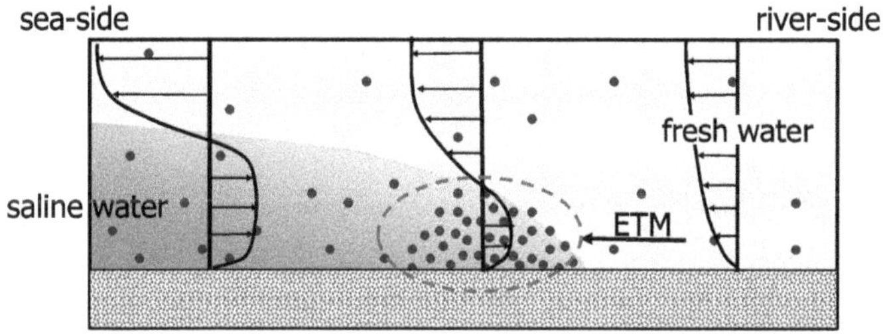

Fig. 9.4. Schematic on the formation of an estuarine turbidity maximum (ETM) at the head of salinity intrusion and a few suspended particles by baroclinic effects only (not to scale).

Likely, the maximum SPM-concentration in an ETM is close to the local carrying capacity of the flow if sufficient sediment is available. However, SPM-concentrations in an ETM are not easily established, as they depend on at least the following parameters:

(1) The availability of fine suspended sediment, including history effects.
(2) The phase within the tide and within the spring-neap cycle, as ETM dynamics are highly variable.
(3) Local mixing, affected by stratification due to vertical salinity and SPM gradients.

On the other hand, the location of ETMs is known fairly well, as they are closely related to the head of salinity intrusion. As mentioned, in the Scheldt estuary, the ETM is found about 100 km from the estuary mouth, while about 50% of the sediment in that ETM originates from the North Sea, thus 100 km down-estuary (Verlaan *et al.*, 1998) — see also Chapter 15.

9.1.2 *Secondary effects*

In the previous section, we emphasized that gravitational circulation is induced by longitudinal salinity gradients. However, density gradients can also occur over depth when the water column is stratified. These stratification effects may be significant. However, as the primary driving force is the horizontal salinity gradient, we refer to stratification as a secondary effect.

If the supply of fresh water is large, stratification is enhanced, while tide-induced turbulence mixes fresh and saline water masses over the vertical. Thus, estuaries in for instance the microtidal Mediterranean are quite stratified. These effects can be quantified to first order with the flood number

α_F (Ippen and Harleman, 1961; Simmons, 1955):

$$\alpha_F = \frac{Q_r}{Q_t} = \frac{Q_r T}{P_t} \tag{9.14}$$

where Q_r and Q_t are the river and tidal discharge, and P_T is the tidal prism with T the tidal period. Vertically well-mixed conditions would prevail if $\alpha_F < 0.1$, whereas for $\alpha_F > 1$, the estuary is highly stratified.

Prandle (2009) and Geyer (2010) developed a more quantitative formula for stratification, defining ΔS as the difference in salinity near the bed and water surface:

$$\frac{\Delta S}{\bar{S}} = \frac{10 U_r^{4/5}}{U_0^{2/5}(gh\alpha\bar{S}/\rho)} \tag{9.15}$$

in which $\bar{S}$ is the depth-mean (cross-sectional mean) salinity, and U_r, U_0 and α were defined in Sec. 9.1.1. The importance of stratification on vertical mixing was discussed in detail in Sec. 3.2.3, arguing that the vertical eddy viscosity ν_z and the equivalent eddy diffusivity ε_z can decrease by a factor 10–100 (e.g. Fig. 3.11).

Tidal straining

The stratification parameters above describe some mean characteristics of salinity distribution in estuaries. However, Simpson *et al.* (1990) and Jay and Musiak (1994) have shown that even in vertically well-mixed estuaries, vertical stratification may occur temporarily by tidal straining. They refer to internal asymmetry. It is best explained graphically — see Fig. 9.5.

Let us start at HWS with a vertically mixed water column, thus with a longitudinal salinity gradient only. When ebb tide commences, fresh water is advected over the denser saline water, as the flow velocity higher in the water column is larger than near the bed. Thus stratification is induced (middle panel of Fig. 9.5). In this case, the stratification is stable, reducing vertical mixing (e.g. Sec. 3.2.3).

If we start our argument at LWS, denser sea water would be advected over fresh water (lower panel of Fig. 9.5), and again vertical stratification is induced, but with the denser water higher in the water column. Then stratification becomes unstable, and the denser saline water sinks, owing to the difference in buoyancy, enhancing vertical mixing.

The stable/unstable stratification and their opposite effects on vertical mixing have a strong influence on the vertical velocity- and SPM-profiles, as sketched in Fig. 9.6 (free after Jay and Musiak, 1994). As turbulence is

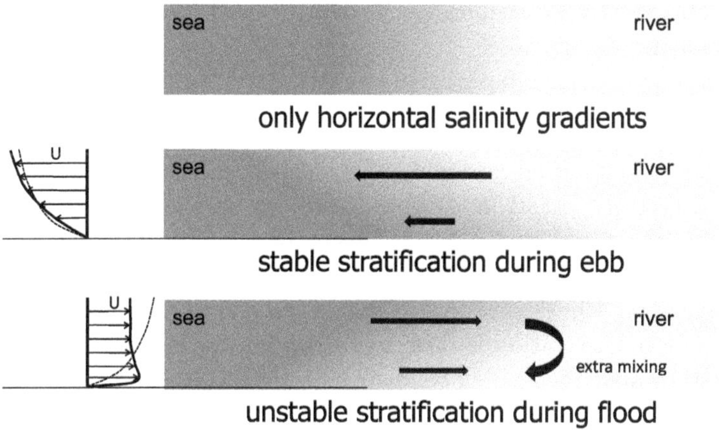

Fig. 9.5. Stable and unstable stratification induced by tidal straining. The dotted lines in the velocity diagrams represent the logarithmic velocity profile for buoyancy-neutral conditions.

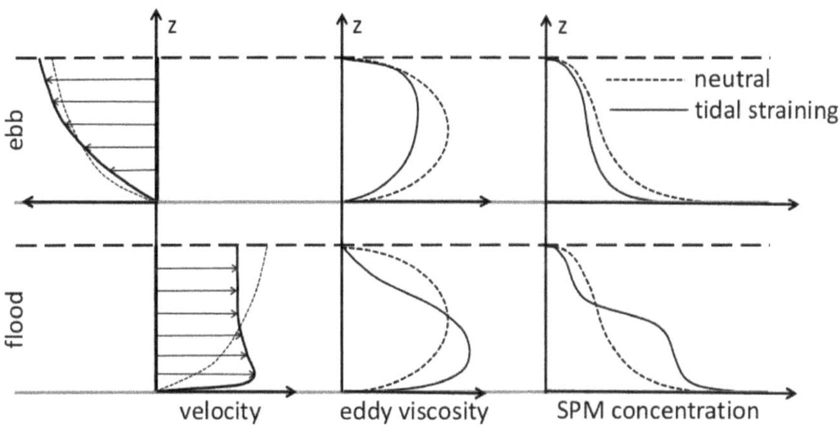

Fig. 9.6. The effect of internal asymmetry (asymmetry in vertical mixing) induced by tidal straining on vertical SPM profiles.

mainly produced near the bed, where velocity gradients are the larger, tidal straining induces asymmetry in longitudinal SPM-transport:

(1) Velocity gradients near the bed during flood are substantially larger than during ebb, thus more turbulence production takes place during flood, and bed shear stresses are larger.

(2) Stable stratification during ebb reduces vertical mixing in the water column.
(3) The degree of stratification varies continuously over the tidal cycle, thus vertical mixing of SPM varies with time.

As a result, tidal straining (internal tidal asymmetry) enhances up-estuary transport of SPM, which has been shown to be an important mechanism in establishing fine sediment balances in estuaries and tidal inlets (Jay and Musiak, 1994).

In Chapter 10, we explain how tidal straining in combination with the effects of Earth's rotation can play an important role in the SPM distribution in coastal waters, depending on their latitude.

Bends and differential advection

In the above, we implicitly assume straight channels with a rectangular cross-section. However, in general, estuaries and tidal inlets have meandering channels, sometimes multiple ebb and flood channels, e.g. Sec. 9.3. Because flow velocities at the outer bend are larger than at the inner bend, lateral gradients in salinity are induced by so-called differential advection.

The magnitude and direction of secondary currents in a river bend are computed from the momentum balance equation in radial direction r (r positive towards inner bend), e.g. De Vriend (1981) and Verbeek *et al.* (1999):

$$-\frac{u^2}{R} = -g\frac{\partial h}{\partial r} + \frac{\partial}{\partial z}\left(\nu_z \frac{\partial v}{\partial z}\right) - \alpha g \left(h - z\right) \frac{1}{\rho}\frac{\partial S}{\partial r} \tag{9.16}$$

where u is tangential velocity in the river, v is the lateral velocity, being zero integrated over the water depth, while R is the radius of the bend. Integrating Eq. (9.16) over the water depth, and subtracting its result from Eq. (9.16), eliminates the transverse water slope $\partial h/\partial r$ (see Winterwerp *et al.* (2006c) for details):

$$-\frac{u^2 - \overline{u^2}}{R} = \frac{\partial}{\partial z}\left(\nu_z \frac{\partial v}{\partial z}\right) + \frac{\tau_r}{\rho h} - \alpha g \left(\frac{h}{2} - z\right) \frac{1}{\rho}\frac{\partial S}{\partial r} \tag{9.17}$$

Assuming a parabolic viscosity profile (Eq. (3.22)), the vertical velocity profile (Eq. (9.6)) is modified with Eq. (9.17). If we assume the measured $\partial S/\partial x = 5$ ppt/km of Fig. 9.2, the resulting profile shows a further deviation from the initial logarithmic profile, becoming close to the observations. As the velocities near the bed increase due to gravitational circulation, saline water penetrates further and stratification is further enhanced, reducing vertical mixing further. Winterwerp *et al.* (2006c) implemented Eq. (9.17) into a 1DV-model, resolving the time dependency of $u(z,t)$, including a k–ε turbulence closure model with buoyancy destruction, explicitly modeling the effects of stratification.

Large deviations from the logarithmic velocity profile, and their spatial variability, have important consequences for the net transport of SPM.

With respect to numerical modeling, simulation of the salinity field is therefore a necessity for the accurate modeling of SPM. Whether or not salinity is correctly modeled often cannot be established from field data alone, as these generally consists of (a few) point measurements only. The salinity field to be expected is therefore important in the system understanding described in Chapter 13. It is not possible to describe all possible (and observed) salinity patterns. The reader is referred to the summaries in literature, e.g. Burchard *et al.* (2018); Valle-Levinson (2010); Wolanski and McLusky (2011).

9.1.3 *SPM-induced density currents*

In the previous sections, only salinity-induced baroclinic effects are discussed: SPM follows the water movement passively, though can settle of course, and may affect vertical mixing by buoyancy destruction. However, horizontal gradients in SPM may also induce density currents, owing to their contribution to the bulk density of the fluid (equation of state (3.26)). This behavior is best understood from the famous lock exchange experiment, as sketched in Fig. 9.7. At $t < t_0$, two fluids of different density are separated by a door in a channel of depth h. At $t = t_0$, the door is removed, and the denser fluid flows underneath the less dense fluid, and *vice versa.*

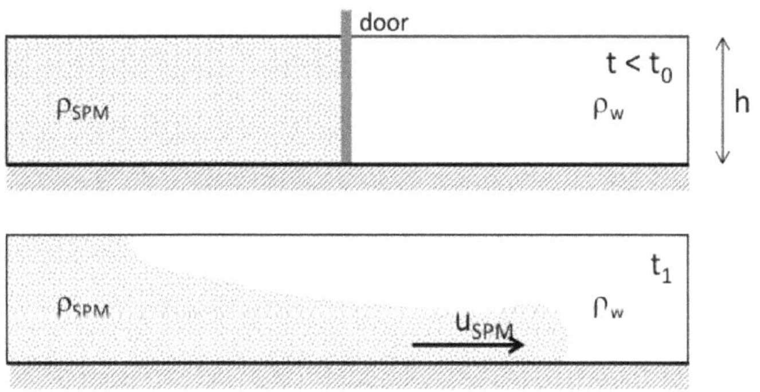

Fig. 9.7. Schematic of lock exchange experiment.

This configuration has been studied in detail in the 1950s and 1960s. For instance, Abraham and Vreugdenhil (1971) and Barr (1967) derived at the celerity of the denser fluid u_{SPM}:

$$u_{\mathrm{SPM}} \approx (0.4 - 0.6)\sqrt{\frac{(\rho_{\mathrm{SPM}} - \rho_w)}{\rho_w} gh} \tag{9.18}$$

where ρ_{SPM} and ρ_w are the density of the sediment-laden denser fluid, and of clear water, while the factor (0.4–0.6) reflects the effects of bed friction.

A sediment-laden flow with SPM-concentrations of 100 mg/L in a 10 m deep channel would induce density currents with a velocity of about 4–6 cm/s. Depending on the prevailing conditions, this can be significant, for instance with respect to lateral spreading of fine sediment, and siltation into harbor basins.

9.2 Barotropic effects

The effects of asymmetries in tidal velocity on net SPM transport are known as barotropic effects (e.g. asymmetry in pressure gradients at constant density). It is convenient to include a discussion on Stokes drift in this section as well.

Figure 9.8 sketches the deformation of a simple sinusoidal tidal wave in a rectangular channel at relative small water depth. The celerity of the wave crest amounts to $c_c = \sqrt{g(h+a)}$, where h is the mean water depth and a is the wave amplitude. Similarly, at the wave trough $c_t = \sqrt{g(h-a)}$. Thus $c_c > c_t$, and the wave crest overtakes its trough. In extremo, a tidal bore is formed. In an Eulerian framework (thus at a fixed tidal station), the water level rises fast after the turn of tide, and the duration of rising water becomes smaller than that of falling water, even though the original, incoming wave is symmetric.

To first order, the tidal velocity scales with celerity: $u \approx ac/h$. The tidal velocity therefore also becomes deformed, and asymmetries in peak velocity and duration of slack water occur. As an example, the flow velocity and tidal elevation measured in the Ems River in Fig. 9.9 show peak flood velocities (>1.5 m/s) about 50% larger than peak ebb velocity (<1.0 m/s), while rising water amounts to about 5.5 hrs in contrast to the 7 hrs of falling water. The co-occurrence of low water (LW) and low water slack (LWS), and HW and HWS is a peculiarity of the Ems River (owing to its very low effective

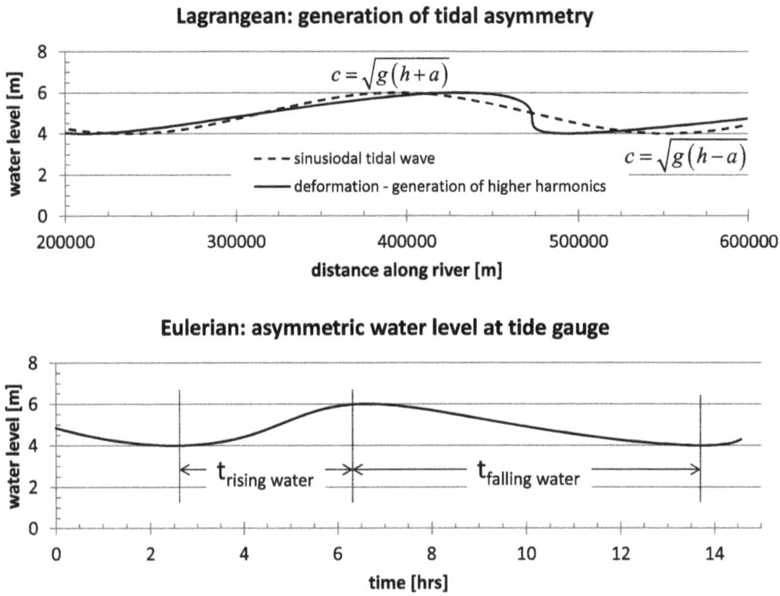

Fig. 9.8. Deformation of sinusoidal tidal wave, generating higher harmonics (overtides) because the celerity at the wave crest is larger than at its trough. The upper panel explains deformation of the tidal wave, whereas the lower panel shows the Eulerian equivalent at a fixed measuring station.

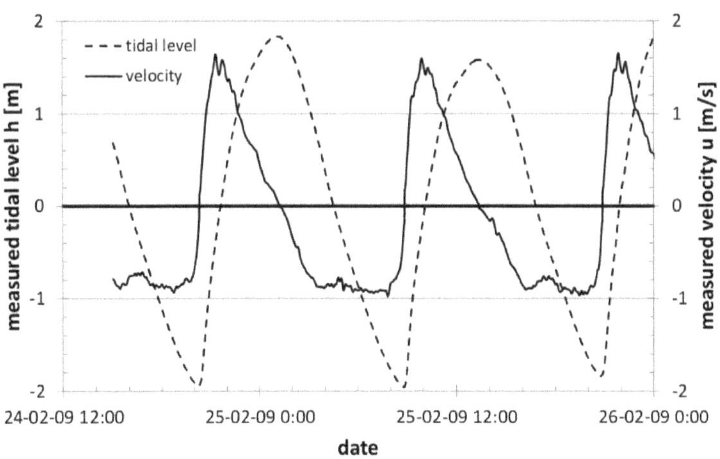

Fig. 9.9. Measurements of flow velocity and tidal level in the Ems River around spring tide, close to Leerort (after Wang (2010); acknowledgment to M. Becker).

hydrodynamic drag), and not a common feature in estuaries (e.g. Friedrichs, 2010, 2012; Winterwerp and Wang, 2013).

The deformation of a tidal wave can also be understood from the non-linear terms in the hydrodynamic equations (see Chapter 11), for instance in the momentum equation:

$$\frac{\partial u}{\partial t} + \frac{\partial uu}{\partial x} + \cdots = 0 \tag{9.19}$$

If we assume a sinusoidal tide, with $u = U\cos(\omega t)$, where U is the velocity amplitude and ω is the wave frequency, double-frequency terms are generated:

$$u^2 = U^2\cos^2(\omega t) = \frac{U^2}{2}(1 + \cos(2\omega t)) \tag{9.20}$$

As the primary tidal component is commonly the semi-diurnal lunar component M_2, the first higher harmonic (or overtide) is the M_4 component. The most simple asymmetric tide is thus given by

$$u\,(t) = U_2\cos\,(\omega t - \phi_2) + U_4\cos\,(2\omega t - \phi_4) \tag{9.21}$$

where subscript $\cdot_2$ refers to the M_2 component, subscript $\cdot_4$ to the M4 component, while ϕ represents the phase of the tide. As shown below, the actual shape of $u(t)$ depends on the phase difference between the M_2 and M_4 component, which is given by $2\phi_2 - \phi_4$, shown in Eq. (9.22), which is obtained after some rearrangement of Eq. (9.21):

$$u\left(t + \frac{\phi_2}{\omega}\right) = U_2\cos\,(\omega t) + U_4\cos\,(2\omega t + 2\phi_2 - \phi_4) \tag{9.22}$$

As an example, Fig. 9.10 shows two tidal velocity curves for phase differences of $2\phi_2 - \phi_4 = 45°$ and $90°$ (using $U_2 = 1\,\text{m/s}$, $U_4 = 0.2\,\text{m/s}$ and $\omega = 2\pi/T$ with $T = 12.5\,\text{hrs}$). For 90°, peak ebb and peak flood velocity are equal, but the period of low water slack (LWS) amounts to about 45 min, whereas the HWS-period is almost twice as long, i.e. 83 min. Here, t_{LWS} and t_{HWS} are defined as the time period during which $u < u_{crit}$, where u_{crit} represents a critical velocity for erosion.

On the other hand, for 45°, the peak flood velocity is larger than the peak ebb velocity, i.e. about 1.2 m/s in comparison to about 0.9 m/s, while the ebb period of about 6:48 hrs is an hour longer than the flood period of 5:42 hrs. However, slack water periods are now identical. The durations of the ebb and flood periods are often used as proxies for distinguishing between ebb- and flood-dominance, which is correct as long as fresh water river discharges are small.

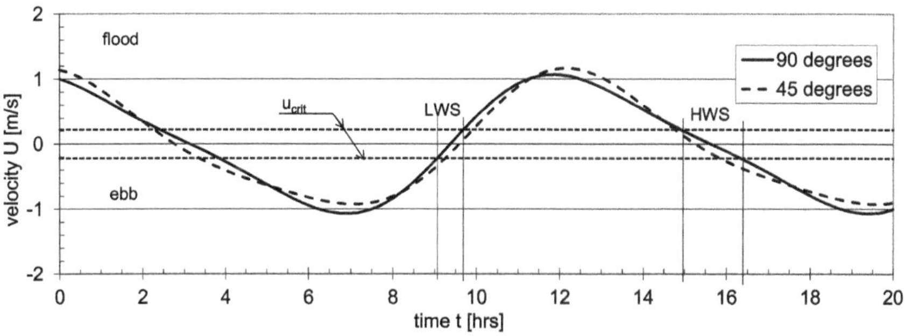

Fig. 9.10. Example of asymmetric tidal velocity with $2\phi_2 - \phi_4 = 45°$ and $2\phi_2 - \phi_4 = 90°$.

Table 9.2. Net SPM transport by tidal asymmetry under capacity and below-capacity conditions.

Asymmetry in:		Capacity	Below-capacity	Section
Peak velocity	Hor. transport	X	–	9.2.1
	Vertical mixing†	X	X	9.2.1
Slack water duration		x‡	X	9.2.2
Stokes drift		X	X	9.2.4

Notes: †Most likely very small under non-stratified conditions.
‡Generally negligible.

Note that for both cases, the tide-integrated water discharge is zero, i.e. ebb- and flood-discharge are equal.

Thus we distinguish between tidal asymmetry in:

(1) peak velocity,
(2) slack water duration.

Of course, these may occur simultaneously, depending on the phasing between the M_2 and M_4 tides. In the following sections, the role of these asymmetries for capacity and below-capacity transport conditions is elaborated, as summarized in Table 9.2. The special role of intertidal areas is elaborated in Sec. 9.2.3.

We focus on the effects induced by the semi-diurnal tidal component, and its first overtide. However, in addition to asymmetries generated by tidal distortion, the interaction between diurnal and semi-diurnal constituents, such as M_2, K_1, and O_1, can also lead to tidal asymmetry and residual sediment transport, as elaborated by Hoitink *et al.* (2003).

Open- and closed-bed conditions

Finally, we stress that barotropic asymmetries only induce net sediment transport if sediment can (temporarily) reside on/in the bed. This can be illustrated with a numerical simulation of the SPM distribution in a schematic estuary. It is 104 km long, with a rectangular cross-section of constant depth $h = 6$ m and an exponentially converging width $b = b_0 \exp(-x/L_b)$ with the width at the estuary mouth $b_0 = 3$ km, $L_b = 30$ km, and $b_{104\,\mathrm{km}} = 90$ m. At the open sea boundary (kmr 0) we prescribe a diurnal tide ($M_2 = 2.13$ m, $\phi_{M2} = 91°$) with its first overtide ($M_4 = 0.12$ m, $\phi_{M4} = 171°$), and at the up-estuary closed boundary a constant fresh water river discharge of 30 m^3/s, containing no sediment, is prescribed. Salinity and SPM-concentration at the open sea boundary are set at 20 ppt and 100 mg/L, respectively. Computed maximum salinity intrusion is about 20 km.

Two simulations are carried out, both starting without any sediment in/on the bed ("empty-bed" conditions, see Sec. 12.3). In simulation 1, sediment may settle, but no water-bed exchange processes are modeled ("closed bed"), while in simulation 2, water-bed exchange processes are included, i.e. sediment can settle in/on the bed, and stay there for some time, until the critical shear stress for erosion is exceeded again ("open bed", fluff layer dynamics — Sec. 6.5). The relevant sediment model parameters read: $W_s = 0.5$ mm/s (continuous deposition, e.g. Sec. 4.2), $c_{gel} = 100$ g/L (thus including the effects of hindered settling), and erosion parameters of the Partheniades model $M = 10^{-4}$ kg/m^2/s and $\tau_e = 0.5$ Pa.

As SPM affects the hydrodynamics via buoyancy destruction in the k–ε turbulence model, the water movement for simulation 1 and 2 are a bit different, and maximal SPM values occur at different times within the tidal cycle. We focus on these maximal values only. The computed along-estuary SPM distributions after 18 months of simulation (dynamic equilibrium conditions) are presented in Fig. 9.11. The left panel shows the SPM distribution for "closed-bed" conditions. SPM enters the estuary by baroclinic effects only, and cannot penetrate beyond the head of salinity intrusion. SPM-values do not exceed those at the open model boundary. In contrast, the right panel with "open-bed" conditions shows that sediment can penetrate the entire estuary, and that SPM-concentrations may well exceed open boundary values.

We may draw four generic conclusions from these simulations:

(1) Baroclinic effects can transport fine sediment into an estuary, but not beyond the head of salinity intrusion.

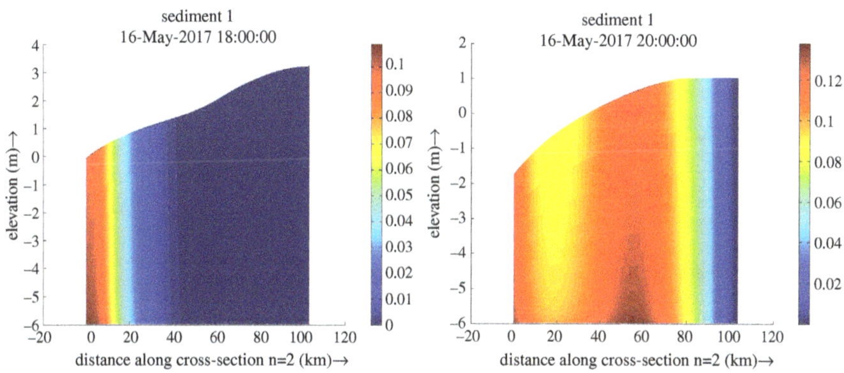

Fig. 9.11. Computed SPM [g/L] distribution along a hypothetical, exponentially converging estuary for "closed-bed" (left panel) and "open-bed" (right panel) conditions — note that the computed maximum SPM values are shown, at different times though because of phase shift in these maximums (courtesy M. de Lucas Pardo).

(2) Barotropic effects can transport fine sediment into an estuary beyond the head of salinity intrusion — these effects can only be modeled if water-bed exchange processes are accounted for.

(3) SPM-concentrations induced by baroclinic effects cannot exceed those at the open sea boundary, while accumulation, and therefore higher than sea boundary values of SPM-concentrations are possible.

(4) The actual values of SPM in an estuary, induced by barotropic effects are sensitive to the model parameters of the water-bed exchange formulations (i.e. settling velocity and erodability).

9.2.1 *Asymmetry in peak tidal velocity*

In this section, the role of tidal asymmetry under capacity conditions is discussed.

The large role of asymmetries in peak velocity with respect to capacity transport can be illustrated by further analyzing Fig. 9.10 (the 45°-case). The tidal water flux follows from integration of the tidal curve over the ebb and flood periods, yielding 16.1 m per unit cross-section, equal for ebb and flood, thus no net flow over the full tidal period, as it should be.

Next, we assume that SPM transport scales as $F_{\text{SPM}} \propto u^4$. The tidal variation in F_{SPM} is shown in Fig. 9.12. Integrating over the flood and ebb period yields $F_{\text{SPM},flood} = 14.6\,\text{m}^4/\text{s}^3$, while $F_{\text{SPM},ebb} = -7.2\,\text{m}^4/\text{s}^3$, i.e. 50% smaller. The larger flood transport is referred to as flood-dominance.

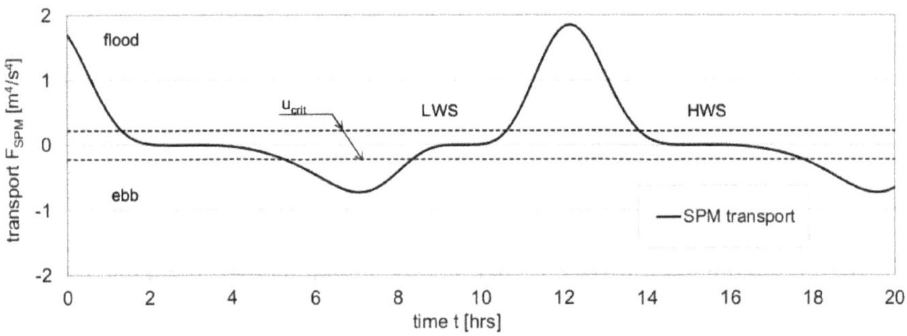

Fig. 9.12. Example of asymmetric SPM-transport, assuming $F_{\text{SPM}} \propto u^4$ and $2\phi_2 - \phi_4 = 45°$, at mean water depth $h = 5\,\text{m}$.

Alternatively, one refers to ebb-dominance when ebb-directed transport exceeds flood-directed transport.

Further to the graphical analysis above, we consider the case of bed load $F_{bl}(= \alpha_b h u^3)$, assuming that the depth-averaged (characteristic) flow velocity is given by:

$$u = U_0 + U_2 \cos(\omega t - \phi_2) + U_4 \cos(2\omega t - \phi_4) \tag{9.23}$$

where U_0 represents the contribution of river flow. Integration over the tidal period yields:

$$\langle F_{bl} \rangle = \frac{3}{2}\alpha_b h \left[u_0 u_2^2 + \frac{1}{2} u_2^2 u_4 \cos(2\phi_2 - \phi_4) \right] \tag{9.24}$$

in which it is implicitly assumed that $U_0 \ll U_2$. Equation (9.24) shows that F_{bl} depends on the phase difference $2\phi_2 - \phi_4$ between the primary tide and its first overtide, and that the contribution of river flow on F_{bl} is not linear: sediment is not just washed out of the river by the (stationary) river flow. This nonlinear effect of river flow is further elaborated below.

9.2.1.1 *Effect of river flow and second* ETM

In general, one may conclude that flood-dominance occurs when the peak flood velocity exceeds the peak ebb velocity. However, this inference may be modified by the effects of river flow, as not only flood velocities decrease and ebb velocities increase with increasing river flow, but also the durations of ebb and flood may change. Of course, river flow also affects the baroclinic effects and stratification, as elaborated upon in Sec. 9.1.

The barotropic effects of river flow were quantified for the Ems River by Winterwerp *et al.* (2017), defining the asymmetry in peak velocity as A_p and in the SPM flux T_{SPM} as A_f:

$$A_p = \frac{\hat{u}_{flood}}{\hat{u}_{ebb}} \tag{9.25}$$

$$A_f = \int_{\text{LWS}}^{\text{HWS}} u_{flood}^4 \text{d}t \Big/ \int_{\text{HWS}}^{\text{LWS}} u_{ebb}^4 \text{d}t \tag{9.26}$$

where $\hat{u}$ reflects peak velocities. The variation in ebb and flood velocities, and the durations of ebb and flood tide as function of river flow were computed with numerical simulations for a variety of river flows, using a calibrated hydrodynamic numerical model of the river. The cross-sectionally averaged velocities obtained from these simulations are substituted in Eqs. (9.25) and (9.26), the results of which are presented in Fig. 9.13 along the river thalweg, measured from the weir at Herbrum (i.e. tidal limit). Flood-dominance is defined as conditions for which A_p and $A_f > 1$ (and < 1 for ebb-dominance). As the river has an exponentially converging cross-section, the transition between flood- and ebb-dominance migrates down-estuary with increasing river flows. Figure 9.13 shows that the A_f-criterion is stricter than the A_p-criterion, mainly because in the A_f-criterion the different durations of ebb and flood periods are accounted for.

The curves in Fig. 9.13 can also be interpreted as the locations of converging sediment transport. These lines thus reflect the location of a

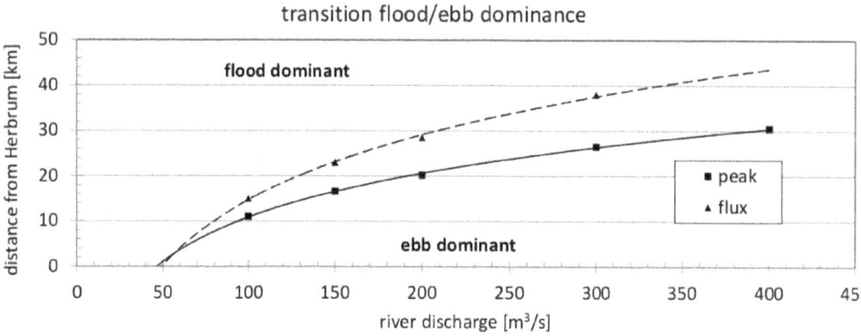

Fig. 9.13. Transition between flood and ebb dominance along the Ems River as a function of river discharge (after Winterwerp *et al.*, 2017). This graph can also be interpreted as the variation in the location of a possible fresh water turbidity maximum (ETM) as function of river flow.

second turbidity maximum (ETM) for alluvial bed conditions — see also Burchard *et al.* (2018). This second ETM moves up- and down-estuary with the river flow and over the spring-neap cycle, and may be located up-estuary of the salt water-induced ETM. Whether or not such a secondary ETM will actually be found in a tidal river depends further on the local availability of fines.

It is emphasized that this second ETM is not related to flushing by fresh water river flow (as for a passive tracer, such as salt), but is formed by the deformation of the tidal velocity, reducing flood-dominant conditions, in conjunction with water-bed exchange processes.

We started Sec. 9.2 with explaining the deformation of the tide through Fig. 9.8, showing that for flood-dominant conditions, flood duration is shorter than ebb duration. Thus the ratio between flood and ebb duration is often used as a proxy for flood-/ebb-dominance. However, this proxy may be misleading, as the effects of river flow on the tide itself is generally less pronounced than on the tidal velocities. If possible/available, we therefore recommend using velocity data for analyzing flood-/ebb-dominance in tidal rivers, in stead of water level (alone).

9.2.1.2 *Asymmetry in vertical mixing*

As explained in Sec. 3.4, vertical mixing scales linearly with velocity, i.e. $\varepsilon_z \propto u$. Yet, residual transport may occur when the velocity profile is deformed, owing to self-stratification of suspended sediment, reflected by the non-homogeneity of the Rouse profile (see Sec. 3.4). This effect is sketched in Fig. 9.14: the larger SPM-concentrations higher in the water

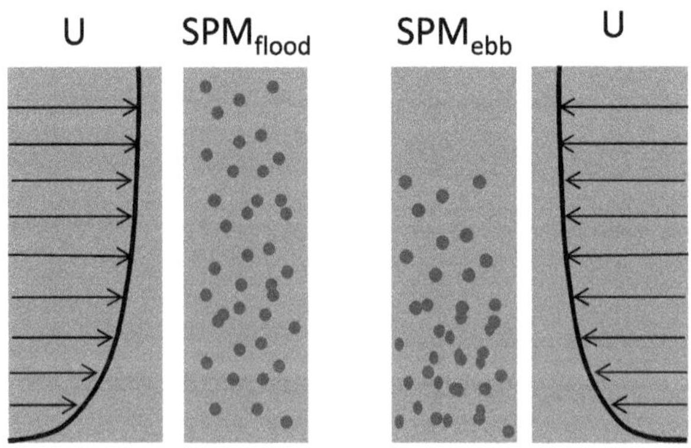

Fig. 9.14. Schematic of enhanced SPM-transport with increasing vertical mixing.

column, where velocities are larger, occur during flood. If the degree of stratification (Sec. 9.1.2) would become asymmetric (more stratified during flood, for instance), further asymmetries develop and pronounced flood- or ebb-dominant conditions may occur. These asymmetries are often referred to as internal asymmetry (Jay and Musiak, 1996). As explained above (Sec. 9.1.2, tidal straining) internal asymmetry may induce large residual SPM fluxes.

On the other hand, the ebb period becomes longer than the flood period, for the asymmetry of Fig. 9.10, thus affecting the integrands of Eq. (9.26).

In the following, the effect of asymmetry in vertical mixing is quantified, using the velocity variation of Fig. 9.10. Stratification effects are not included in the analysis below, these have been discussed above. However, these may be present in field data, and would be accounted for implicitly in 3D numerical modeling, assuming $h = 10\,\mathrm{m}$, $u_* = 0.01\,\mathrm{m/s}$ and $W_s = 2\,\mathrm{mm/s}$ ($\beta = 0:34$), and determine the vertical profiles of velocity and SPM from Eqs. (3.21) and (3.64) (thus assuming constant water depth), determining the Rouse number from Eq. (3.61). Furthermore, we assume $z_0 = 1\,\mathrm{mm}$. The flux is computed as $F_{\mathrm{SPM}} = \frac{1}{h}\int_h uc\,\mathrm{d}z$, and shown in Fig. 9.15. Integration over the ebb and flood period shows that the ebb flux is about 0.2% larger than the flood-directed SPM flux. Apparently, the effect of an approximately 1-hour longer ebb period is almost fully compensated by the more efficient mixing during flood, as sketched in Fig. 9.14. Simulations with a full 1DV-model, accounting for SPM-induced stratification effects show similar results (in fact an ebb-directed residual flux of 0.1% of the gross flux). For this example, the effects of asymmetry in vertical mixing appear to be negligible.

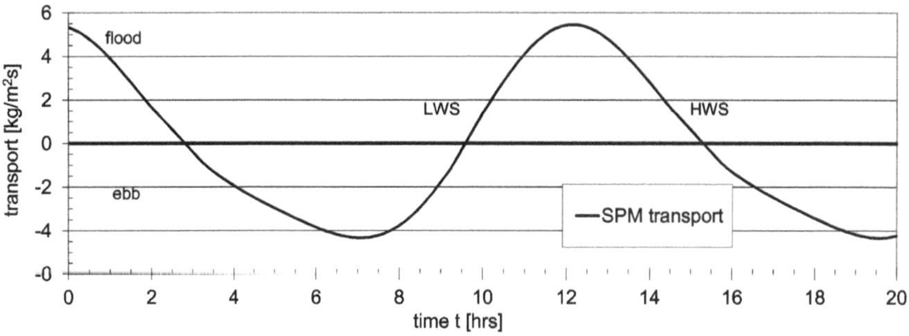

Fig. 9.15. Computed SPM flux $= \frac{1}{h}\int_h uc\,\mathrm{d}z$ for the asymmetric tide of Fig. 9.10, assuming a logarithmic velocity profile and an equilibrium Rouse SPM-profile.

Note that we assumed a constant sediment mass throughout the tidal cycle, and equilibrium Rouse profiles. This analysis fails when water-bed exchange processes were to be included, i.e. the Partheniades–Krone formulations (see also Sec. 3.4.3). In that case, SPM-concentrations would increase till saturation conditions, or would become entirely depleted (Sec. 12.2). Exploratory simulations with a 1DV model showed that either flood-dominant or ebb-dominant conditions can occur, depending on the initial conditions of the simulations (initial SPM-concentration and ebb- or flood-flow). This is obviously not a realistic result, implying that such 1DV simulations are unsuitable to establish residual transports.

9.2.2 *Asymmetry in slack water duration*

Net SPM transport by asymmetries in peak velocity occurs because the total amount of sediment mobilized, brought into suspension and carried by the flow is larger during one phase of the tide (for instance flood) than during the next phase. This asymmetry therefore plays a role in particular under capacity conditions, with unlimited sediment availability. For below-capacity conditions, the amount of sediment that can be mobilized, brought into suspension and carried by the flow, is however limited. For these conditions, it is important *when* that limited amount of sediment is mobilized and mixed over the water column, and how long it stays in suspension. This is mainly governed by the hydrodynamic conditions around slack water. We refer to slack water asymmetry.

It is noted that, once mobilized, mixing and the subsequent transport by the main flow are affected by internal asymmetry, i.e. the asymmetry in vertical mixing, which is a function of the asymmetry in peak velocity and vertical stratification (tidal straining), as discussed in previous sections.

We distinguish between scour lag and settling lag. Scour lag occurs because of asymmetries in the time that sediment stays on/in the bed around HWS and LWS. Settling lag occurs when sediment remains longer in the water column around HWS than LWS, or vice versa. These asymmetries may be induced by asymmetries in tidal velocities (local, or Eulerian asymmetry), or because of spatial gradients (asymmetries) in tidal velocity or water depth (spatial, or Lagrangian asymmetry). The reader is referred to Gatto *et al.* (2017) for further details.

The effects of slack water asymmetry may also occur under capacity conditions, but are then generally overshadowed by the asymmetries in peak velocity.

As discussed in the next sections, flood-dominance for below-capacity conditions prevails when the duration of HWS is longer than of LWS. However, the duration of HWS and LWS can be affected by river flow. Then, up-estuary SPM transport by slack water asymmetry is nullified where HWS and LWS periods become equal and a secondary fresh water ETM may be formed (Burchard *et al.*, 2018), the location of which would vary with varying river flow.

To get a feeling for the river flows needed to change this asymmetry, we refer to Fig. 9.10, showing that for $2\phi_2 - \pi_4 = 90^\circ$ the flood and ebb velocity curves around slack water are more or less straight for a considerable period. This means that large river flows are required to modify the duration of HWS- and LWS-periods. As the formation and location of this secondary ETM for below-capacity conditions is not easy to visualize in general terms, we do not elaborate further on it, focusing on the net transports instead.

9.2.2.1 *Eulerian scour lag*

In this section, we discuss net SPM transport by asymmetry in erosion. We presume uniform flow conditions in a straight, prismatic channel, in which the tide is deformed as in Fig. 9.10 with $2\phi_2 - \phi_4 = 90^\circ$.

Figure 9.16 sketches the variations in transport direction over a tidal cycle. It is assumed that at the beginning of HWS, sediment settles directly on the bed, and stays there for 83 min (see Fig. 9.10). Then, when the ebb flow increases beyond $u_{ebb} > u_{crit}$, particles are picked up from the bed

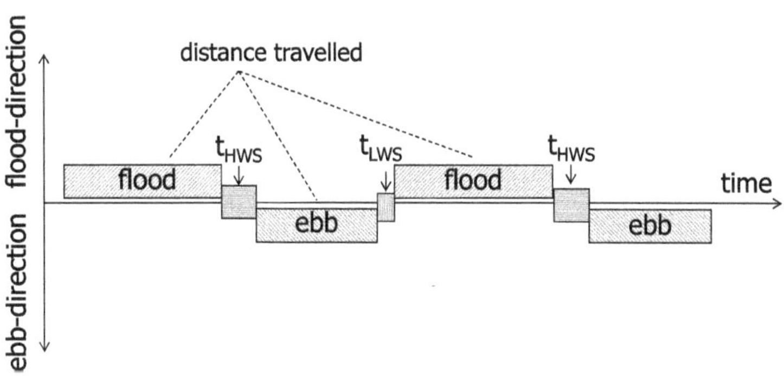

Fig. 9.16. Schematic of Eulerian (local) scour lag for the situation of Fig. 9.10 with $2\phi_2 - \phi_4 = 90^\circ$, i.e. $t_{\mathrm{LWS}} < t_{\mathrm{HWS}}$.

and travel down-estuary, settling on the bed immediately at the beginning of LWS, staying there for 45 min. Thus, around HWS particles settle earlier and are resuspended later (there is a time-lag in scour, thus the name scour lag) than around LWS, thus remaining longer on the bed. This asymmetry induces a net up-estuary SPM-transport.

It is noted that we follow the particles along the river. This implies that the terminology used, i.e. "local or Eulerian scour lag" is somewhat misleading. However, because of consistency with literature, we stay with it.

9.2.2.2 *Lagrangian scour lag*

Lagrangian scour lag reflects net SPM transport because of gradients in (mean) flow velocity, for instance in a tidal inlet, over an intertidal flat, or in a tidal basin. The latter was analyzed by Van Straaten and Kuenen (1957) and Postma (1961). Van Straaten and Kuenen were likely the first introducing the terms scour and settling lag.

Let us analyze a closed, 10 km long rectangular basin with a flat bed (water depth is not relevant) with a simple sinusoidal tide (no overtides). The tidal velocity amplitude at the mouth measures $U = 1\,\mathrm{m/s}$, linearly reducing along the basin to zero at its head (e.g. Eq. (9.29)). Furthermore, we assume an erosion threshold of $u_{crit} = 0.2\,\mathrm{m/s}$. Thus when $u < u_{crit}$, the sediment particles remain on the bed. Again, we assume that the particles settle immediately on the bed when $u < u_{crit}$.

Starting at $x = 0$, a particle remains on the bed until the accelerating flood velocity exceeds u_{crit} (after about 24 min), upon which it is transported with the flood current to about $x = 5.6\,\mathrm{km}$, as depicted in Fig. 9.17. There, the particle stays on the bed for about 108 min, owing to the smaller velocity at $x = 5.6\,\mathrm{km}$. Thus, after being picked up by the ebb current, there is too little time for it to reach its original position, and the particle settles at about $x = 600\,\mathrm{m}$ up-basin from its initial position. Here it stays for about 72 min before being picked up again by the accelerating flood current, traveling to about $x = 5.7\,\mathrm{km}$, gaining a bit more than 100 m in up-basin direction. The resulting path is plotted in Fig. 9.17

In this way, a particle slowly migrates up-basin up to the position where the velocity amplitude $U < u_{crit}$. Then the particle stays on the bed permanently. This condition was referred to as "threshold lag" by Gatto *et al.* (2017) and "final settling lag" by Van Straaten and Kuenen (1957).

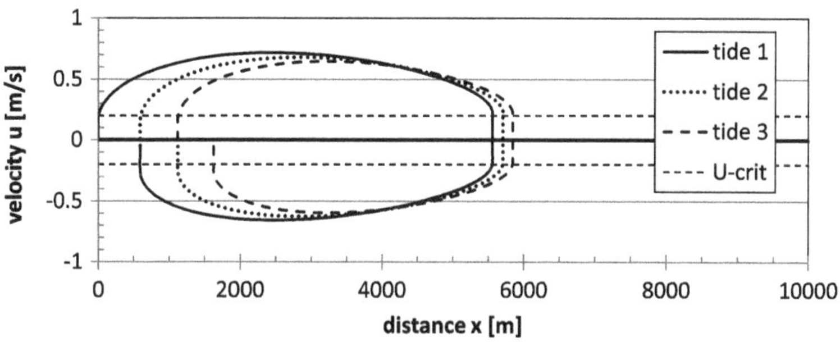

Fig. 9.17. Particle trajectory x in $x - u$ diagram during three successive tides for below-capacity conditions, assuming a simple sinusoidal tide with linearly decreasing amplitude, and $u_{crit} = 0.2\,\mathrm{m/s}$.

In general, the effects by Lagrangian scour lag are larger than those by Eulerian scour lag.

9.2.2.3 *Eulerian settling lag*

The analyses in the two sections above assume that during decelerating tide, particles settle immediately when $u < u_{crit}$. This, of course, is not realistic as settling takes time (see also the settling time scale T_s in Table 9.1, and the analysis in Section 3.4) — we referred to settling-limited conditions. This settling time induces another asymmetry, known as settling lag (see again Van Straaten and Kuenen, 1957; and Groen, 1967). Again, we distinguish between Eulerian and Lagrangian effects.

De Swart and Zimmerman (2009) summarized net SPM transport ($F_{Esl} = huc$) by Eulerian settling lag, accounting for erosion of the bed (but neglecting scour lag). We follow their derivation. The simple asymmetric tide of Eq. (9.21) is assumed, with an M2 and M4 component only. The depth-averaged, one-dimensional sediment mass balance is given by

$$h\frac{\partial C}{\partial t} = hM^*u^2 - W_sC \tag{9.27}$$

in which the threshold velocity for erosion (u_{crit}) is ignored, which is acceptable, as we follow particles after they are eroded from the bed (recall that slack water asymmetry is about "the when" of mobilizing a limited amount of sediment). The erosion parameter M^* in Eq. (9.27) is labeled with a $\cdot^*$ to distinguish from the erosion parameter M in Chapter 6. Substitution of Eq. (9.21) into Eq. (9.27), and integration over the tidal period gives

(De Swart and Zimmerman, 2009):

$$\langle F_{Esl}\rangle = h\frac{M^*}{W_s}\left[\frac{1}{4}U_2^2U_4\cos(2\phi_2-\phi_4)\left[\frac{2}{1+(\omega h/W_s)^2}+\frac{1}{1+(2\omega h/W_s)^2}\right]\right]$$
$$+h\frac{M^*}{W_s}\left[\frac{1}{2}\left(\frac{\omega h}{W_s}\right)U_2^2U_4\sin(2\phi_2-\phi_4)\right.$$
$$\left.\times\left[\frac{1}{1+(\omega h/W_s)^2}+\frac{1}{1+(2\omega h/W_s)^2}\right]\right] \tag{9.28}$$

Equation (9.28) shows that the net/residual SPM transport $\langle F_{Esl}\rangle$ depends on:

(1) the time scale for settling with respect to the tidal period, i.e. $\omega h/W_s$ and $2\omega h/W_s$;
(2) the ratio between erosion and settling, i.e. hM^*/W_s; large M^* means that the available sediment is mobilized rapidly, thus enhancing net transport. Note that below-capacity transport is not explicitly assumed in the derivation of Eq. (9.28);
(3) the phase difference between the semi-diurnal tide M2 and its first overtide M4, $(2\phi_2 - \phi_4)$.

Equation (9.28) has a maximum when $2\phi_2 - \phi_4 = \pm 90°$, with maximal flood dominance for $2\phi_2 - \phi_4 = 90°$ (thus maximal HWS period). Furthermore, F_{Esl} becomes small for small W_s, when the sediment behaves more and more as a passive tracer (see also the "closed" and "open" bed discussion above).

9.2.2.4 *Lagrangian settling lag*

Similar to Sec. 9.2.2.2, we assume a simple sinusoidal M2 tide, the amplitude of which decreases linearly with x:

$$u = U_2(x)\cos(\omega t) \tag{9.29}$$

which is substituted into the one-dimensional mass balance equation for the depth-averaged SPM-concentration:

$$h\frac{\partial C}{\partial t} - h\frac{\partial uC}{\partial x} = hM^*u^2 - W_sC \tag{9.30}$$

This approach forms the basis of the papers by Van Straaten and Kuenen (1957) and Postma (1961). De Swart and Zimmerman (2009) assumed that the advection term in Eq. (9.30) is small, allowing an analytical solution for the net transport by Lagrangian settling lag F_{Lsl}:

$$\langle F_{Lsl} \rangle = -\frac{3}{4}\frac{h}{\omega}\frac{M^*}{W_s}U_2^3\frac{\mathrm{d}U_2}{\mathrm{d}x}\left[\frac{(\omega h/W_s)^2}{1+(\omega h/W_s)^2}\left[1+\frac{1}{2}\frac{1-2(\omega h/W_s)^2}{1+4(\omega h/W_s)^2}\right]\right] \tag{9.31}$$

As the sign of the right-hand side of Eq. (9.31) depends solely on the sign of $\mathrm{d}U_2/\mathrm{d}x$, $\langle F_{Lsl} \rangle$ is always in the direction of the lower velocity (thus up-basin), similar to the graphical presentation of Fig. 9.17.

Lagrangian scour lag and settling lag explain why fine sediment is always found at the head of tidal inlets, where velocities are small. However, surface waves may disturb this picture, as explained in Sec. 9.3.2, responsible for seasonal variations in mud occurrences in temperate climates.

9.2.3 *Asymmetry in compound channels*

In the above, we have assumed prismatic tidal channels with rectangular cross section, i.e. without intertidal area. However, natural estuaries and tidal basins are characterized by channels and intertidal flats/shoals. Their cross sections are described by so-called hypsometric curves, which reflect the width of a cross-section as a function of depth. As an example, Fig. 9.18 sketches a typical cross-section of a tidal channel in an estuary or tidal basin with mean depth h and tidal amplitude a. It is composed of a flow-carrying section (the part below low water (LW)) with width b_c, and intertidal areas, with combined width b_s. Thus, the total width of a cross-section is b_{tot} $(= b_c + b_s)$. In the linear, one-dimensional approach below, it is assumed that longitudinal flow velocities in the intertidal areas are zero, i.e. these store water only.

Analytical solutions for the flow velocities through such tidal channels are possible by assuming a rectangular (or otherwise) channel below

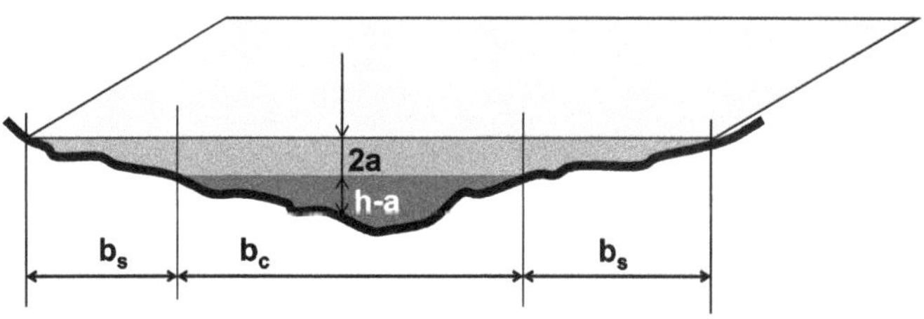

Fig. 9.18. Sketch of typical tidal channel with intertidal areas.

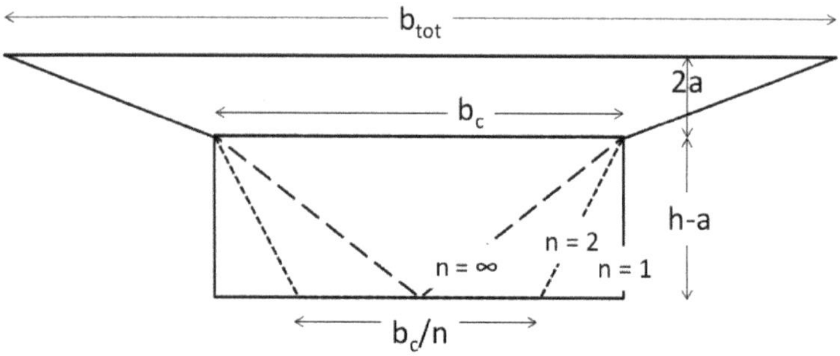

Fig. 9.19. Schematic cross section of a tidal channel with various flow-carrying cross-sections, i.e. rectangular: $n = 1$, trapezoid: $2 < n < \infty$ and triangular: $n = \infty$.

low water, and a trapezoidal intertidal area between low and high water (HW). The behavior of such compound channels has been analyzed e.g. by Dronkers (2016) and Friedrichs (2012). Here we follow Wang *et al.* (1999, 2002). The cross-section of the tidal channel is schematized as in Fig. 9.19. The flow-carrying cross-section can have various shapes, depending on the shape parameter n. The intertidal area is sloping linearly over its width. The mean water depth is h and the tidal amplitude is a.

The celerity c in a rectangular channel is $c = \sqrt{gh}$. However, in a compound channel with rectangular cross-sections, it becomes:

$$c = \sqrt{g \frac{V_c}{b_{tot}}} \tag{9.32}$$

where V_c is the area of the flow-carrying cross-section. In the first paragraphs of this chapter, we showed that tidal deformation (thus asymmetry) is induced by a difference in tidal propagation at high and low water. Hence, a convenient way to analyze tidal asymmetry is by defining a local tidal deformation parameter γ as the ratio between the local celerity at high water (c_{HW}) and at low water (c_{LW}). For a rectangular channel without intertidal area:

$$\gamma_r = \frac{c_{\mathrm{HW}}}{c_{\mathrm{LW}}} = \frac{\sqrt{g\,(h+a)}}{\sqrt{g\,(h-a)}} \tag{9.33}$$

which is always ≥ 1, thus flood-dominant. In a compound channel, γ_c becomes:

$$\gamma_c = \frac{c_{\mathrm{HW}}}{c_{\mathrm{LW}}} = \frac{\sqrt{g\,(h+a)\,b_c/b_{tot}}}{\sqrt{g\,(h-a)}} \tag{9.34}$$

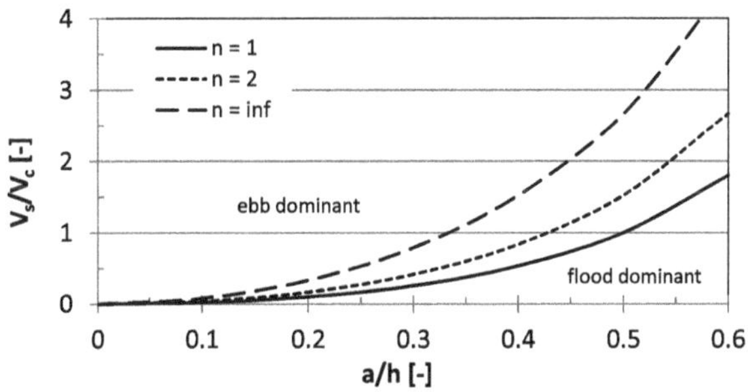

Fig. 9.20. Transition between flood and ebb dominance as a function of the relative intertidal area and relative tidal amplitude for the three configurations of Fig. 9.19.

The tidal wave does not deform if $\gamma = 1$, i.e. when the nominator and denominator of Eq. (9.34) are identical. To compare with Friedrichs and Aubrey (1988), Wang *et al.* (1999, 2002) generalized Eq. (9.34) using the surface area of the flow-carrying cross-section V_c and of the intertidal area V_s. Some geometrical analysis for the cross-sections of Fig. 9.19 then gives

$$\gamma_c = 1 \quad \text{if} \quad \frac{V_s}{V_c} = \frac{\left(\frac{4n}{n+1}\right)\left(\frac{a}{h}\right)^2 \Big/ \left(1 - \frac{a}{h}\right)}{\left(\frac{n+1}{2n}\right) + \left(\frac{n-1}{2n}\right)\frac{a}{h}} \tag{9.35}$$

This relation is plotted in Fig. 9.20. For $V_s = 0$ (i.e. $b_s = 0$), the tide is always flood-dominant (if $a > 0$), as mentioned above. Figure 9.20 further shows that ebb-dominance prevails when the intertidal area becomes large enough, while flood-dominance prevails at large amplitudes (of course, if the amplitude becomes too large, the present linear analysis fails). It is further noteworthy that this simple analysis predicts that the transition between ebb- and flood-dominance depends on the shape of the flow-carrying cross-section at identical volume.

Qualitatively, Fig. 9.20 is very similar to the results by Friedrichs and Aubrey (1988), which were based on a numerical analysis of a 7 km long semi-enclosed tidal basin, whereas Fig. 9.20 is identical to analytical results by Friedrichs (2012) and Dronkers (2016).

9.2.4 *Stokes drift*

It can be shown that the phase difference between tidal velocity and water level in a converging estuary with a rough, sandy bed is about $-45°$,

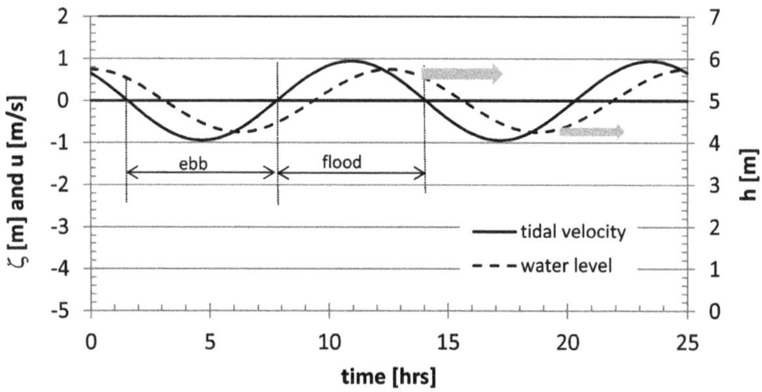

Fig. 9.21. Mean water level during flood and ebb for a phase difference between velocity and water level of $-45°$; gray arrows indicating celerity of the wave crest and trough.

e.g. Friedrichs (2012); Dronkers (2016); Winterwerp and Wang (2013), implying that HW precedes HWS by about 2 hrs. As a result, the mean water level during flood is higher than that during ebb, as shown in Fig. 9.21. This induces a net water flux, known as Stokes drift. In short wave theory, Stokes drift is explained by the observation that the orbital circles/ellipses are not closed.

At first order, Stokes drift can be assessed from the continuity equation, i.e. the nonlinear term hu. Assume:

$$h = h_0 + \zeta(t) = h_0 + a\cos(\omega t) \quad \text{and} \quad u = U\cos(\omega t - \phi) \tag{9.36}$$

The Stokes drift then follows from the water flux F_S, integrated over a tidal period:

$$\begin{aligned} F_S &= \int_T hu\mathrm{d}t = \int_T [(h_0 + a\cos(\omega t))\, U\cos(\omega t - \phi)]\mathrm{d}t \\ &= aU\frac{T\cos\phi}{2} \end{aligned} \tag{9.37}$$

Thus, for a progressive wave, we find $F_S = aUT/2$, whereas for a standing wave $F_S = 0$ (as $\phi = 0$). The latter follows also from Fig. 9.21, as the mean water level during ebb and flood are then identical.

Let us next examine the tidal propagation in a semi-enclosed bay, assuming zero river flow. Continuity requires that the tide-integrated tidal discharge at any cross-section along the river should be zero. This implies that the Stokes drift has to be compensated. This compensation is called "rectification of the Stokes drift", inducing a residual current, referred to as the rectification velocity U_r. U_r can only be generated through an extra energy slope in the water level $\Delta(\mathrm{d}h/\mathrm{d}x)$. Let us estimate this slope for the conditions of Fig. 9.21, i.e. $\phi = -45°$ (e.g. exponentially converging estuary with rough, sandy bed), so that $U_r \approx 0.35\,aU/h_0$. Then:

$$\frac{\mathrm{d}\tau}{\mathrm{d}z} = \frac{\mathrm{d}p}{\mathrm{d}x} = g\rho\frac{\mathrm{d}h}{\mathrm{d}x} \quad \text{i.e.} \quad \frac{\mathrm{d}h}{\mathrm{d}x} = \frac{(U+U_r)^2}{hC^2} \quad \text{or} \quad \Delta \approx 0.7\frac{aU^2}{h_0^2C^2} \tag{9.38}$$

For $a = 0.75\,\mathrm{m}$, $h = 5\,\mathrm{m}$, $U = 0.9\,\mathrm{m/s}$ and $C = 60\,\mathrm{m}^{1/2}/\mathrm{s}$, we find a rectification velocity $U_r \approx 5\,\mathrm{cm/s}$, and an extra slope of about $\Delta \approx 5 \cdot 10^{-6}$ (i.e. 5 mm per km, thus difficult to measure).

However, in small, exponentially converging estuaries, river-induced velocity is often not too large either, and U_r may substantially contribute to the net sediment fluxes.

9.3 Other asymmetry effects

The previous sections explain how asymmetries in hydrodynamics, in conjunction with the nonlinear behavior of sediment, can induce (large) residual sediment transports, even against the direction of the mean flow.

However, asymmetries in bathymetry, forcing, etc. may also induce residual sediment transports, against the direction of mean net flow. The following sections give some examples.

9.3.1 *Effects of wind*

Wind generates waves, flow and turbulence. Wind therefore modifies all the transport mechanisms mentioned above, e.g. tidal asymmetry (e.g. storm during flood and mild conditions during ebb); scour lag (wave induced bed shear is more important on shallow areas than in deeper channels); baroclinic processes (wind induces additional vertical mixing), etc. The effects of wind therefore cannot be treated in an isolated manner, as it co-occurs and interacts with other driving forces. In many cases, the effects by wind enhance other processes. This is well illustrated in De Vet *et al.* (2018). Simulations show that the isolated effects of waves and wind have a minor influence on the sediment transport around an intertidal flat in the Eastern Scheldt (The Netherlands), while their combination induces significant sediment transport. In these simulations, waves stir up sediment from the bed, being advected by wind-induced currents.

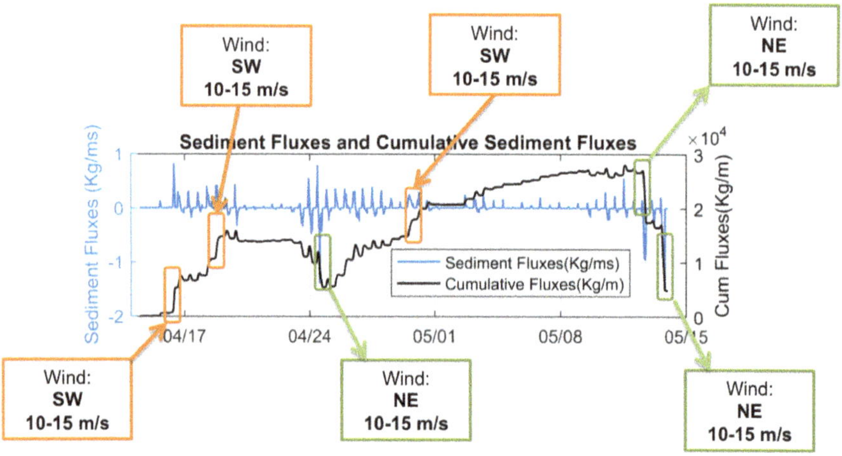

Fig. 9.22. Time series of the instantaneous (blue) and cumulative (black) sediment flux on a tidal flat in the Wadden Sea. The green (NE-wind) and orange (SW-wind) boxes indicate the time windows of wind events (courtesy Colosimo).

To illustrate the effects of wind on the net sediment flux, we analyze the measured sediment flux on a tidal flat in the Wadden Sea (The Netherlands), see Colosimo *et al.* (2020). Several wind events occurred during this month, leading to relatively large fluxes in SW- or NE-direction. Although the no-wind conditions do contribute, the net flux is highly influenced by wind events exceeding a threshold of 10–15 m/s — note that this part of the Wadden Sea is characterized by considerable tidal velocities and salinity-induced gravitational circulation. The effect of wind is illustrated in Fig. 9.22, showing large instantaneous sediment fluxes in NE-direction (positive) when the SW-wind exceeds the threshold, while fluxes in SW-direction become large (negative) when NE-winds exceed this threshold. As a result, the cumulative fluxes depict strong gradients when wind speeds exceed 10–15 m/s.

Accounting for wind in numerical models requires imposing a realistic or characteristic wind climate. Such real time simulations are computationally expensive. Yet, there is ample evidence that wind can have a decisive influence on the transport and fate of fine sediment in tidal basins, as shown by Duran-Matute *et al.* (2016) and Sassi *et al.* (2015). Using real-time simulations, they show the significant influence of wind on the yearly mean fluxes of fine sediment in the Dutch Wadden Sea.

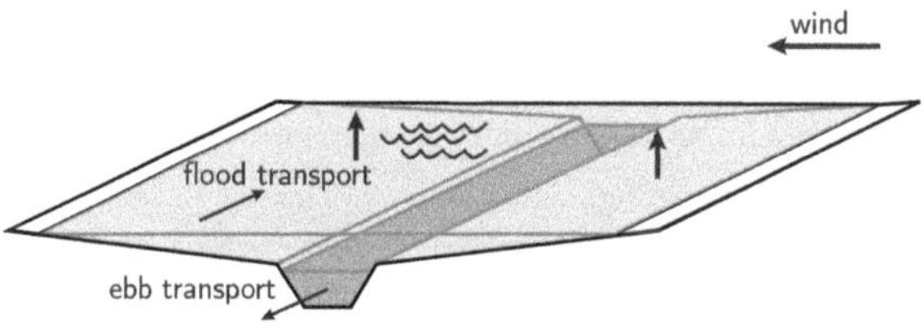

Fig. 9.23. Sketch of wave action in tidal basin with intertidal area.

9.3.2 *Effects of waves*

Figure 9.23 sketches a tidal channel with large intertidal area. From Sec. 9.2.3 we know that such a system would be ebb-dominant. In this configuration, fine sediment can accumulate on intertidal areas, as flow velocities over the intertidal tend to be small, in particular under calm weather.

However, when waves are present, this picture changes. As the intertidal area is very shallow, small (and short) waves may (re-)mobilize large amounts of fine sediment, whereas wave effects in the tidal channel remain small. These effects can be illustrated with the linear wave theory, assuming locally generated waves.

The bed shear stress induced by waves relates to the orbital velocity near the bed u_{orb} (Sec. 3.3.2):

$$\tau_b \propto u_b^2 \left(\equiv u_{orb}^2\big|_{z=0}\right) \tag{9.39}$$

From linear wave theory, the amplitude of the orbital motion $\hat{u}_{orb}$ becomes:

$$\hat{u}_{orb} = \frac{\omega H}{2} \frac{\cosh\{k(h-z)\}}{\sinh\{kh\}} \tag{9.40}$$

where H is the wave height, ω the wave frequency and k the wave number. Thus the ratio between the bed shear stress in a tidal channel and its intertidal area becomes:

$$\frac{\hat{\tau}_{b,c}}{\hat{\tau}_{b,a}} = \frac{u_{b,c}^2}{u_{b,a}^2} = \left(\frac{\sinh\{kh_a\}}{\sinh\{kh_c\}}\right)^2 \approx \left(\frac{h_a}{h_c}\right)^2 \tag{9.41}$$

which reduces with the inverse ratio of the squares of the channel depth h_c and water depth on the surrounding intertidal h_a, as $\lambda \gg h$, thus $kh \ll 1$.

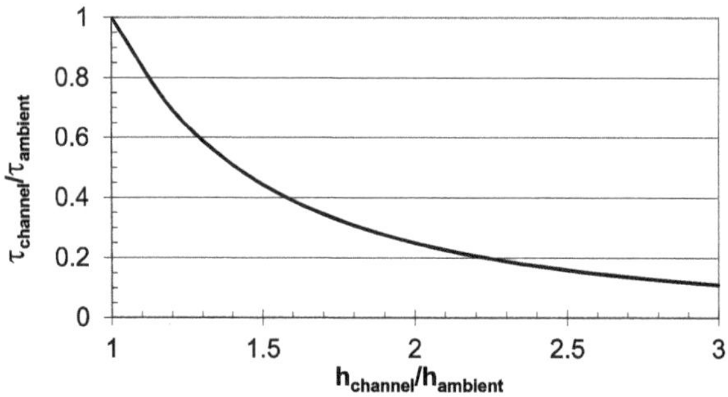

Fig. 9.24. Bed shear stresses in a tidal channel relative to the stresses on the intertidal area as a function of the relative channel depth.

This relation is plotted in Fig. 9.24, showing that at a channel depth twice the surrounding water depth, wave-induced bed shear stresses in the channel are at least 70% smaller than over the intertidal. This reduction increases rapidly with increasing channel depth. We have implicitly assumed that the waves are not modified when propagating over/through the tidal channel, which is reasonable, as the channels are generally small in comparison to the intertidal zone.

Waves on intertidal areas influence net SPM fluxes, depending on the phase of tide. As an example, we may discuss the conditions in a tidal basin where tidal level and velocity are $-45°$ out of phase, as in Fig. 9.21. During flood, the mean water depth is larger than during ebb. Thus during flood, a larger part of the intertidal area is flooded, and waves affect a larger part of the basin. In this situation, residual SPM transport in flood direction is induced.

Figure 9.25 shows a satellite image of the Dutch Wadden Sea, with muddy intertidal fringes at the head of the basin along the Dutch mainland. During storm conditions with higher water levels, (part of) this sediment is eroded and SPM concentrations in the basin are larger than at sea ($c_{basin} \gg c_{sea}$). Further to Sec. 9.4 and Fig. 9.27 fine sediment is then exported from the Wadden Sea into the North Sea (see also Nauw *et al.*, 2014 and Sassi *et al.*, 2015).

9.3.3 *Effects of floc size and settling velocity*

In Sec. 4.3, time scales for flocculation are derived, arguing that the time scale for floc breakup T_b is generally much smaller than that for flocculation (aggregation) T_f, Eqs. (4.27) and (4.28). Thus, flocs tend to adjust to the

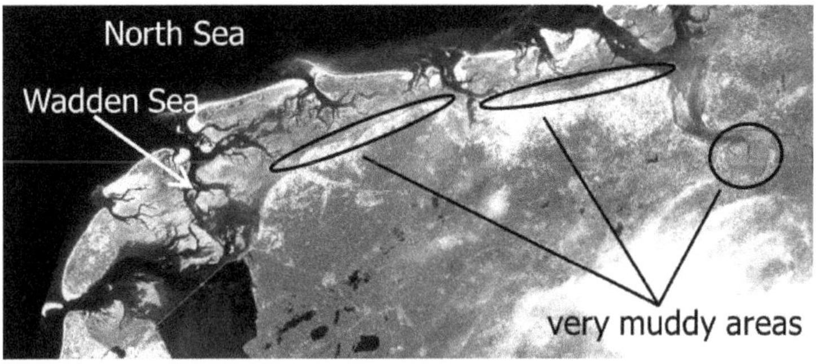

Fig. 9.25. The Dutch Wadden Sea with muddy areas at the head of the basin.

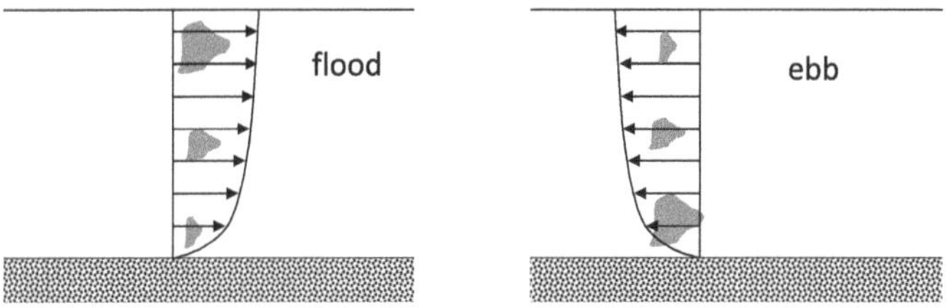

Fig. 9.26. Schematic of asymmetry in floc size over a tidal cycle.

higher shear zones rapidly, while (re-)aggregation takes much more time. This implies asymmetries in floc size over a tidal cycle.

These asymmetries in floc size were further elaborated in Winterwerp (2011) for the hyper-turbid Ems River. Because sediment-induced stratification during ebb is much larger than during flood, owing to the very strong asymmetry in peak velocity (e.g. Fig. 9.9), the larger flocs are found high in the water column during flood ("G wins"), whereas during ebb, the larger flocs are found near the bed ("c wins") — see Eq. (4.22).

These effects are sketched in Fig. 9.26. The larger flocs higher in the water column do not necessarily imply larger mass fluxes. However, as flocs are larger around HWS than around LWS, the trapping efficiency of the river for fine sediment increases.

9.4 Net and gross transport in estuaries

In the previous sections, we discussed net sediment transport mechanisms in open marine waters by asymmetries in hydrodynamics, in conjunction with nonlinearities in sediment dynamics. The actual net SPM fluxes, however,

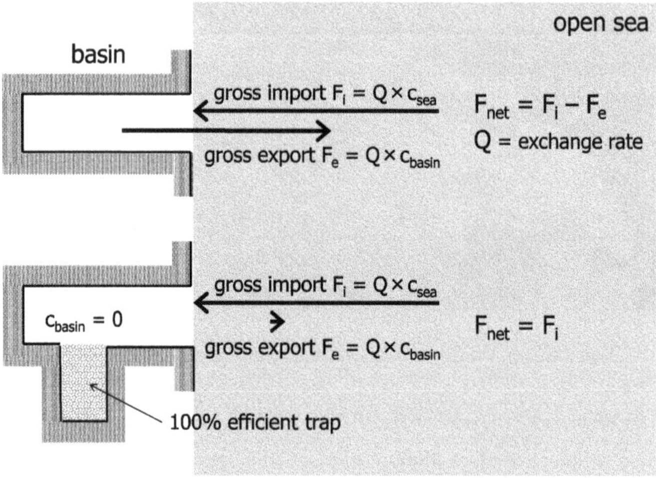

Fig. 9.27. Schematic showing the net transport (F_{net}) of SPM from the open sea into a basin, as the between the gross flux (F_i) into that basin and the gross flux (F_e) out of that basin.

also depend on the trapping in these systems, i.e. where can sediment accumulate, and at which rate. This is best understood from the residual between two large sediment fluxes in opposite direction.

Let us analyze the sediment flux through the mouth of a tidal basin, as in Fig. 9.27. The exchange of water is denoted by Q, which is assumed zero when integrated over a tidal cycle (no river input). The gross sediment import during flood is then $F_i(= Qc_{sea})$, while the gross export amounts to $F_e(= Qc_{basin})$. Hence, the net SPM flux into the basin is $F_{net} = F_i - F_e$.

If for some reason, all sediment entering the basin would be trapped within that basin, $F_e = 0$ and $F_{net} = F_i$. This somewhat trivial example has two important consequences:

(1) In general, gross effects are much larger than net effects, as discussed below. If that is the case, small alterations within an aquatic system may yield large effects. This has also far reaching implications on the use of process-based numerical models (see Sec. 13.6).
(2) Net SPM transport at any location is largely dependent on processes elsewhere in the aquatic system. Therefore, net transport by asymmetries should be interpreted as potential net transports. In particular, the morphodynamic state (in fact the off-equilibrium state) of an aquatic system determines the trapping rates in that system.

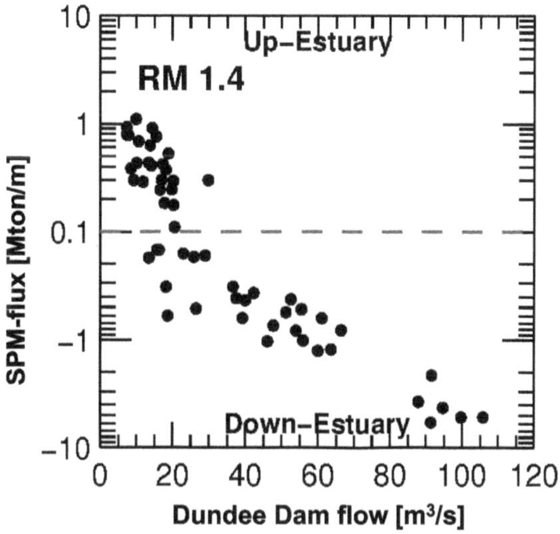

Fig. 9.28. Net SPM flux measured near the mouth of the Passaic River in 2009, integrated over water depth and tidal period.

As an example, consider the sediment dynamics in the Lower Passaic River Mathew and Winterwerp (2020), which is currently at morphodynamic equilibrium. During low river flow, the river imports fines, while during high river flows, the river bed is scoured, and sediment is exported, see Fig. 9.28. These fluxes are governed by gravitational circulation and tidal asymmetry, as affected by tide and river flow. The net import rates during low river flow depend on:

- The availability of fine sediment, in the case of the Lower Passaic River originating from river inflow and from Newark Bay, downstream of the river mouth, and
- The amount of sediment eroded during previous high river flow events — if much sediment was eroded, the river is far away from equilibrium and trapping efficiencies are high.

The diagram in Fig. 9.28 for 2009 therefore differs from that of 2010, owing to different river flow histories Mathew and Winterwerp (2020). This argument also holds in case of human interventions, when a river is locally deepened, for instance to remove legacy sediment. Such a scour hole will be filled up with sediment, inducing net sediment transport which may be

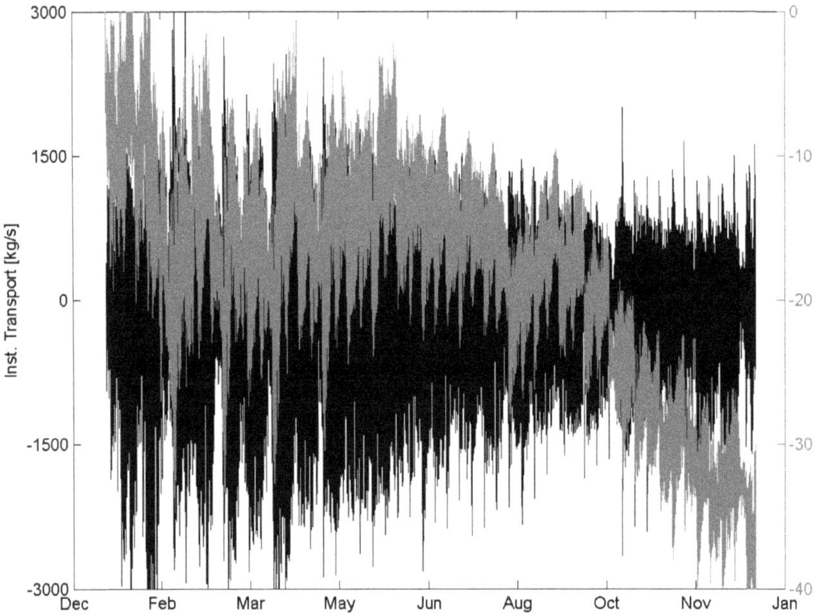

Fig. 9.29. Computed gross (black line) and net (gray line) SPM flux through the mouth of the Dutch–German Ems Estuary.

against the direction of density circulation and tidal asymmetry effects, bringing the estuary back to morphodynamic equilibrium.

Another example of large differences between gross and net SPM fluxes is presented in Fig. 9.29, showing the computed gross and net SPM fluxes through the mouth of the Ems Estuary. The calibrated model used reflects long-term equilibrium conditions. However, due to temporal variations in river flow, wave climate, etc., gross SPM fluxes vary largely over the year, and the net flux is never zero. These net effects are not only the result of the processes discussed above, but are also largely governed by temporal variations in trapping potential in the river:

(1) During low river flow and calm weather conditions, there are more calm water areas in the river where sediment can accumulate.
(2) During high river flow and rough weather conditions, the river bed is scoured — these locations are filled in again during the next low river flow and calm weather conditions.

Quantifying the computational results of Fig. 9.29 yields gross SPM fluxes of 1000–3000 kg/s, or 50–150 ton/tide. The net flux would be equivalent to maximal 40 ton/year, i.e. on average 0.06 ton/tide. Though these numbers cannot be compared directly, it is obvious that the gross fluxes are three orders of magnitude larger than the net fluxes.

9.5 Harbor siltation

A special, and very common configuration of Fig. 9.27 is the presence of a harbor basin along a (tidal) river. In general, such harbor basins trap sediment efficiently, requiring frequent maintenance dredging. Here we focus on fine sediment (SPM) entering a harbor basin in suspension. Maintenance dredging may involve large costs, in particular when the fines are contaminated. In the following, we analyze the various processes governing the siltation of harbor basins along tidal and non-tidal open waters. The mechanisms governing siltation in fairways is discussed in Sec. 10.5.

Ambient water is exchanged with sediment-laden water from a harbor basin through the three mechanisms described below. Because the basin is semi-enclosed, no net exchange of water occurs (over a long-enough period). However, there is a gross exchange of water (and therefore sediment): sediment-rich water from the basin surroundings is exchanged with sediment-poor water from within the basin itself. In the zeroth-order approach followed in this chapter, it is assumed that the sediment flux may be quantified as the product of water flux and SPM-concentration, whereas the trapping efficiency (the fraction of sediment remaining within a harbor after entering) determines the difference between the SPM flux into and out of the harbor basin, similar to the discussion in Sec. 9.4.

A harbor basin can be situated along a non-tidal river, a lake, a tidal river, a coast, an estuary, etc. The water body in front of a harbor entrance flows with a characteristic velocity U. In tidal rivers and along coasts, tides play a role as well, whereas in estuaries, and along some coasts, density currents are also important. In lakes, U is generally small, as often induced by wind.

The three relevant mechanisms exchanging sediment-laden water between the ambient water body and a harbor basin are (see Fig. 9.30):

(1) **Horizontal turbulent exchange:** Large-scale circulations in the harbor's mouth are driven by flow separation in the mouth, by entrainment and by stagnation effects at the downstream side of the basin

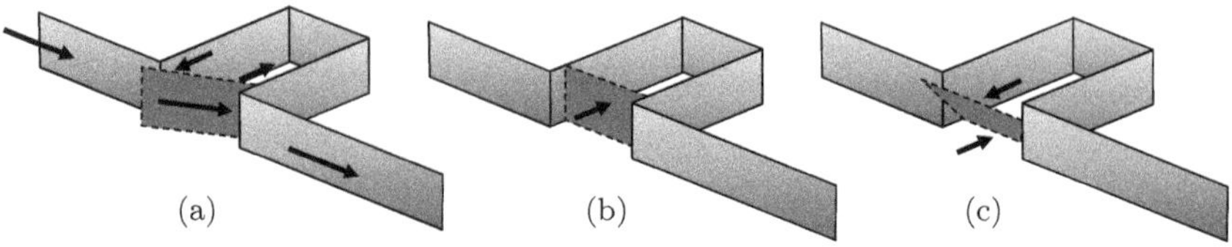

Fig. 9.30. Schematic of the three exchange processes: (a) turbulent mixing; (b) tidal filling; and (c) density currents.

entrance. This mechanism always plays a role in flowing ambient water. Net exchange of water is always zero, at all time scales.

(2) **Tidal filling:** During rising tide, sediment-laden water flows into the basin, while during falling water, the same amount of water flows out of the basin, generally containing less sediment, though. Over a tidal period, no net amount of water is exchanged. This mechanism plays a role in tidal rivers and along open coasts.

(3) **Density currents:** Gradients in salinity induce density currents with a near-bed current (which contains the majority of the sediment) against the direction of that gradient. Salinity-induced density currents play a role in estuaries and coastal systems with salinity gradients. Eysink (1989) emphasized the dominant role of this mechanism. Density currents can also be induced by gradients in SPM and temperature.

These three mechanisms interact. For instance, during tidal filling, the wake induced by flow separation in the harbor mouth is deflected into the basin, and thus horizontal exchange becomes less effective in exchanging water between the basin and the surrounding water body.

It is illustrative elaborating on these three mechanisms in an estuary, where salinity variations are more or less in phase with ebbing and flooding of the tide. In general, water level and tidal velocity are out of phase, and in the following we assume a phase difference of $-45°$. During flood, flow separation occurs at the down-estuary side of the harbor basin, forming a wake which is deflected into the basin during rising tide (Fig. 9.31). During falling tide, this wake is pushed out of the basin, and little water is exchanged by horizontal turbulent exchange, which is therefore neglected in this zeroth-order estimation. Tidal filling of course is in phase with rising and falling water. The salinity in the estuary in front of a harbor basin is more or less in phase with flood and ebb; during flood, salinity in the

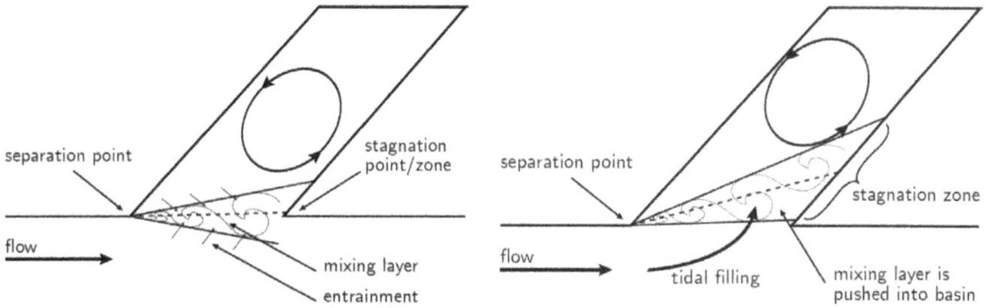

Fig. 9.31. Schematic of deflection of turbulent wake into basin without tide (left panel) and during rising tide (right panel).

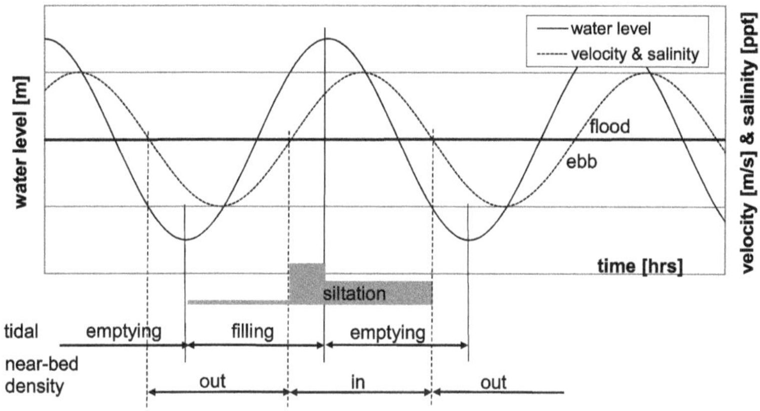

Fig. 9.32. Phasing of water exchange processes in harbor mouth over a tidal period, assuming tidal velocity and salinity are in phase.

estuary is larger than in the basin, while the opposite is the case during ebb. Hence, a near-bed sediment-rich density current flows into the basin during flood, while, owing to siltation within the basin, a near-bed sediment-poor density current flows out during ebb. This phasing is sketched in Fig. 9.32, showing that sediment import into a harbor basin is unevenly distributed over a tidal period. When all mechanisms play a role, density currents are generally the major contributor to harbor siltation, followed by the effects of tidal filling.

Eysink (1989) was the first to quantify the siltation rate in harbor basins, using a zeroth-order assessment (see also Van Rijn, 2005; Winterwerp and Van Kesteren, 2004). The basis for this assessment is the zero-order SPM

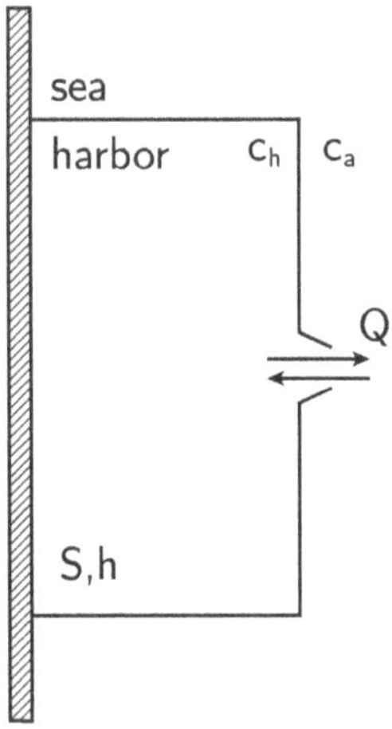

Fig. 9.33. Schematic of sediment exchange through harbor mouth.

balance, sketched in Fig. 9.33, which reads for this configuration reads:

$$\frac{\mathrm{d}Vc_h}{\mathrm{d}t} = Qc_a - Qc_h - F_s \quad \text{thus} \quad \frac{\mathrm{d}c_h}{\mathrm{d}t} = \frac{Q}{V}c_a - c_h\left(\frac{Q}{V} + \frac{\alpha W_s}{h}\right) \tag{9.42}$$

where V, S and h are the (constant) volume, surface and depth of the basin, Q is the exchange rate of water through the basin mouth, c_a and c_h are the (uniform) SPM-concentration in the ambient water and within the harbor basin, αW_s is the effective settling velocity of SPM (accounting for local turbulent mixing and non-uniform c_h-distribution). The siltation flux F_s is given by

$$F_s = \alpha S W_s c_h \tag{9.43}$$

The equilibrium solution $c_{h,e}$ to this simple differential equation reads:

$$c_{h,e} = \frac{Q}{Q + \alpha S W_s} c_a \tag{9.44}$$

Hence, we assume that the equilibrium SPM values in a harbor basin $c_{h,e}$ (and thus the siltation rate) are proportional to the ambient SPM value c_a. The exchange flow Q is determined by the three processes described above:

$$\frac{1}{T}\int_T Q\mathrm{d}t \equiv \langle Q \rangle = \langle Q_t \rangle + \langle Q_e \rangle + \langle Q_d \rangle \tag{9.45}$$

f_e	$f_{t,e}$	f_d	$f_{d,t}$
0.02 – 0.05	0.1 – 0.25	0.05 – 0.125	0.1 – 1.0

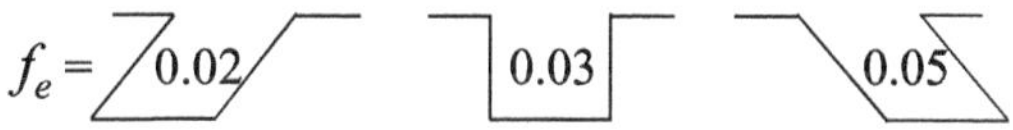

Fig. 9.34. Coefficients for Eq. (9.46).

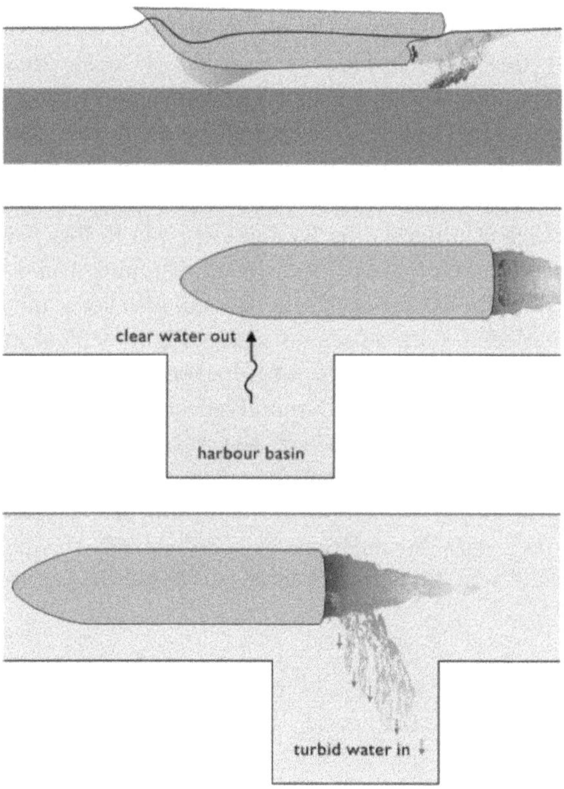

Fig. 9.35. Schematic of effect shipping on harbor siltation.

where $\langle Q_t \rangle$ is the gross water exchange by tidal filling (angular brackets implies averaging over the tidal period), $\langle Q_e \rangle$ is the gross water exchange by horizontal turbulent circulation (entrainment), and $\langle Q_d \rangle$ is the gross water exchange by density currents. Eysink (1989) proposes a number of coefficients to quantify these gross water exchange rates:

$$\begin{aligned}\langle Q_t \rangle &= V_t/T \\ \langle Q_e \rangle &= f_e A U - f_{t,e}\langle Q_t \rangle \\ \langle Q_d \rangle &= f_d A \sqrt{\frac{\Delta \rho s g h}{\rho}} - f_{d,t}\langle Q_t \rangle \\ \alpha &= 1 - u_h^2/u_{crit}^2\end{aligned} \tag{9.46}$$

where V_t is the tidal volume of the harbor basin, T is the tidal period, A is the cross-section of the harbor entrance, $\Delta\rho_s$ is a characteristic salinity-induced density difference in front of harbor entrance, the coefficients $f_{t,e}$ and $f_{d,t}$ reflect reductions in exchange efficiency during rising tide, u_h is a characteristic velocity in the harbor basin, and u_{cr} is a threshold velocity below which sediment can permanently settle on the basin bed. The empirical coefficient proposed by Eysink are given in Fig. 9.34.

Next to these three exchange mechanisms, shipping itself can also induce sediment import into harbor basins, in particular in lakes and canals, where the above-mentioned mechanisms are small. Figure 9.35 sketches this process. When a ship arrives in front of a harbor basin, water levels on the fairway decrease owing to the Bernoulli-effect, and sediment-poor water flows out of the basin. After passage, the water level is restored, and the basin is filled again, but now with sediment-lade water, fine sediment being mobilized by the ship's boundary layer and propeller. To the knowledge of the authors, no general formulations exist to quantify this process.

Chapter 10

Transport and Fate of Mud in Coastal Waters

Chapter 9 discusses the various mechanisms transporting fine sediment (SPM/mud) in semi-enclosed marine waters, such as estuaries and tidal basins/lagoons. In this chapter, we discuss the fate and transport of fine sediment in coastal waters. Coastal systems occur in a large variety, because of the wide range in driving mechanisms, sediment availability and geological settings (e.g. Bosboom and Stive, 2015). Moreover, the driving forces operate at scales much larger than in semi-enclosed basins; we refer in particular to large ocean circulations and the effects of Earth's rotation.

We do not discuss coastal systems in general (the reader is referred to e.g. Bosboom and Stive, 2015; Dean and Dalrymple, 2001; Komar, 1998), limiting ourselves to muddy coastal waters. Muddy sediment is very fine, thus easily erodible. Over time, mud can therefore only accumulate in quiescent coastal waters, over gentle slopes (1:1000 and smaller), and at relatively low wave activity. Wave-driven cross- and longshore transport, so important in sandy coastal systems, are much smaller or absent in muddy coastal waters. This does not mean that waves do not play an important role, as they stir up fine sediment from the seabed (see Sec. 3.2.4). Moreover, streaming (e.g. fluid mud transport by waves, Sec. 7.5.2) may occur, generally inducing onshore mud transport.

In this chapter, the role of tide, Earth's rotation, wind and waves on the transport and fate of SPM in coastal waters is elaborated upon, and we discuss a few oceanographic features, necessary for understanding coastal processes. This is done through the analysis of four different coastal systems, which cover a large part of the variety in coastal systems (with respect to fine sediment dynamics). These systems are summarized in Table 10.1, together with the relevant driving forces in these coastal water, and elaborated upon in the next sections.

Table 10.1. The four muddy coastal waters treated in this chapter, and the driving forces dominating fine sediment dynamics in these waters.

Mechanism	Patos Lagoon	North Sea	Amazon mouth	Java Sea
Climate zone	Temp	Temp	Trop	Trop
Ocean current	—	—	√	—
Wind-induced flow and setup	√	∼	—	√
Earth's rotation	√	√	—	—
Tide	√	√	√	√
Long waves (swell)	√	√	√	—
Short waves (sea)	√	√	—	√
Grav. circulation	√	√	√	√

Note: ∼ some effect, but not dominant.

10.1 Tidal propagation in coastal waters

In this section, we elaborate on large-scale tidal effects, summarizing the tidal dynamics relevant to the transport and fate of SPM in the coastal zone. More information on tides can be found in many textbooks, such as Gerkema (2019). Overtides, so important in estuaries and tidal basins, may occur as well in shallow coastal waters. Where relevant, we refer to Sec. 9.2.

Tidal ranges in the open ocean are small, a few decimeters at most. Amplification of the tide occurs through interaction with land masses, e.g. the continental shelf. For instance, the mouth of the Amazon River, Brazil and the west-facing coastal waters of France (Saint Malo) and the UK (Bristol Channel) are renowned for very large tidal ranges, over 10 m.

We distinguish between coasts along shallow, semi-enclosed seas, and those facing the open ocean:

(1) Tidal velocities in coastal waters fringing the open ocean are generally more or less perpendicular to the coast — for instance along the Guiana coasts (see below).
(2) Tidal velocities in coastal waters along shallow and semi-enclosed seas generally follow the coastline, whereas cross-shore velocities are small — this is the case for instance in the North Sea (see below).

First we discuss tides in coastal waters facing the open ocean. Though tidal ranges in the open ocean are small, tidal propagation is not, as celerity scales with $\sqrt{gh}$, where h is water depth. In shallow coastal waters, celerities are much smaller. Thus the tide in coastal waters, fringing a continental

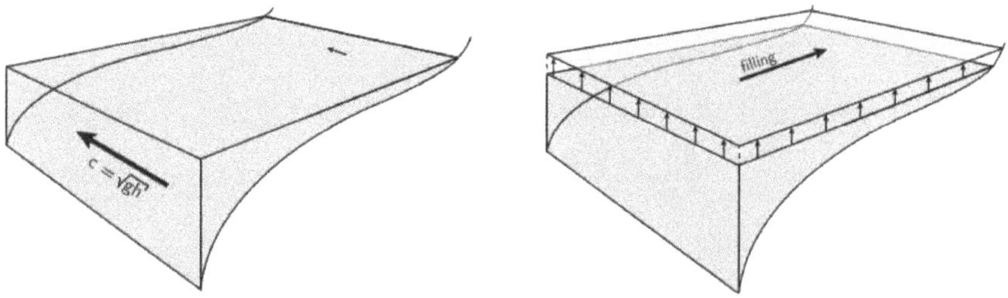

Fig. 10.1. Tidal propagation and filling along the continental shelf, resulting in predominant cross-shore tidal velocities in the adjacent shallow coastal waters, induced by tidal filling.

shelf lags well behind that in deeper water. In that case, tidal velocities are predominantly cross-shore, as sketched in Fig. 10.1.

Away from the equator, effects of Earth's rotation become important. While propagating, the tide is deflected by Earth's rotation (Coriolis effects) and the presence of land masses (continental shelves), forming so-called Kelvin waves. The tide therefore follows rotational paths, known as amphidromic systems, which are found everywhere from small semi-enclosed basins to the open ocean. As an example, Fig. 10.2 shows such a set of amphidromic systems in the North Sea for the M_2-tide. The center of an amphidromic system is known as the amphidromic point, characterized by small (ideally zero) tidal amplitude. Note that tidal velocities are not necessarily small at such points. The major amphidromic system yields a counter-clockwise Kelvin wave, whereas in the southern North Sea, a secondary, clockwise tidal rotation is found. As a result, the tide along the Dutch coast propagates from south to north (see also Sec. 10.2).

The location of an amphidromic point not only depends on size, shape and depth of a basin, but also on the wave frequency. Therefore, in theory there are as many amphidromic points as there are tidal components. However, in practice these are difficult to identify as the resulting Kelvin wave is a composition of all individual tidal components.

Next to these amphidromic systems, tidal velocities may also exhibit amphidromic behavior, in particular in semi-enclosed seas, such as the North Sea, where the tide forms a standing wave.

An extreme case is the tidal propagation in the Singapore Strait. Singapore is located at the transition between the dominantly semi-diurnal tide in the Indian Ocean and the mixed to diurnal tide in the South China Sea. The dominance of either component is defined by the tide form factor

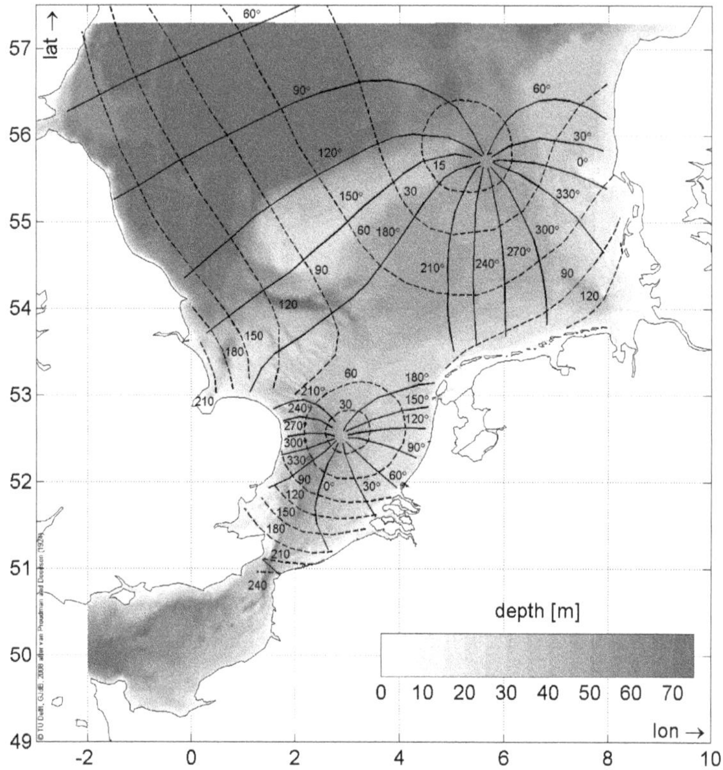

Fig. 10.2. Tidal propagation in the North Sea. Because of Earth's rotation, the primary tide propagates in anti-clockwise direction, known as a Kelvin wave (redrawn from Proudman and Doodson (1924) by De Boer).

$F(= (A_{K1} + A_{O1})/(A_{M2} + A_{S2}))$, where A is tidal amplitude. A diurnal tidal amphidromic point exists close to Singapore, and therefore the tides around Singapore are dominantly semi-diurnal ($F \approx 0.25$, see Fig. 10.3), forming a standing wave. In conjunction with an amphidromic system, strong diurnal currents are generated. The tidal currents are therefore classified as mixed, diurnal to locally even semi-diurnal (Fig. 10.3). These currents are large, and flow velocities by the main diurnal component are over 1 m/s (Fig. 10.4), more than twice the velocity of the M_2 constituent. This has important implications for sediment transport.

Thus in the Singapore Strait rapid changes occur over short distances from semi-diurnal to diurnal tidal regimes. Note that such diurnal tides are prominent in many tropical waters. The diurnal and semi-diurnal

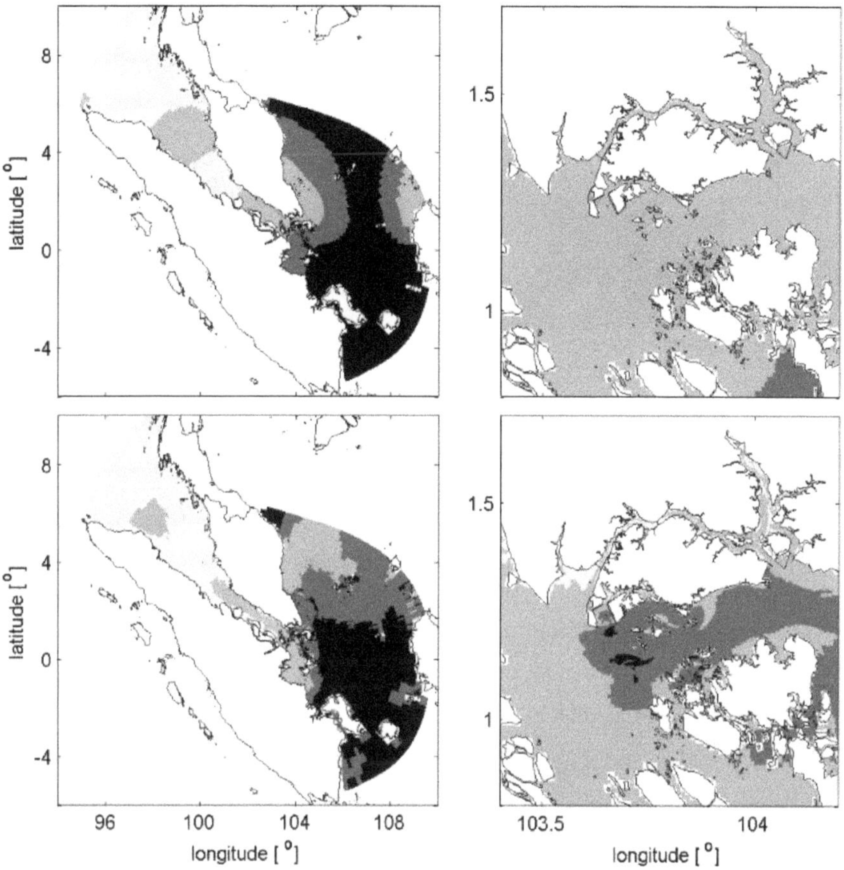

Fig. 10.3. Tidal regime around Malaysia and Sumatra (Indonesia), based on the tidal form factor F for water levels (top panels) and currents (bottom panels) at larger scale (left panels) and at Singapore regional scale (right panels). Black is diurnal ($F > 3$), light gray is semi-diurnal ($F < 0.25$), gray-shaded in-between are mixed, dominantly semi-diurnal ($0.25 < F < 1.5$) to diurnal ($1.5 < F < 3$) (after Van Maren and Gerritsen, 2012).

components may also induce pronounced tidal asymmetries and therefore residual sediment transport (especially the O_1, K_1, M_2 triad; see Hoitink *et al.*, 2003). As fine sediment is often abundant in tropical waters, this asymmetry may induce significant residual SPM transports. In open coastal waters, this type of asymmetry is probably more important than the better known M_2–M_4 tidal asymmetry (Song *et al.*, 2011). In the case of the Singapore Strait, the direction of residual transport by tidal asymmetry

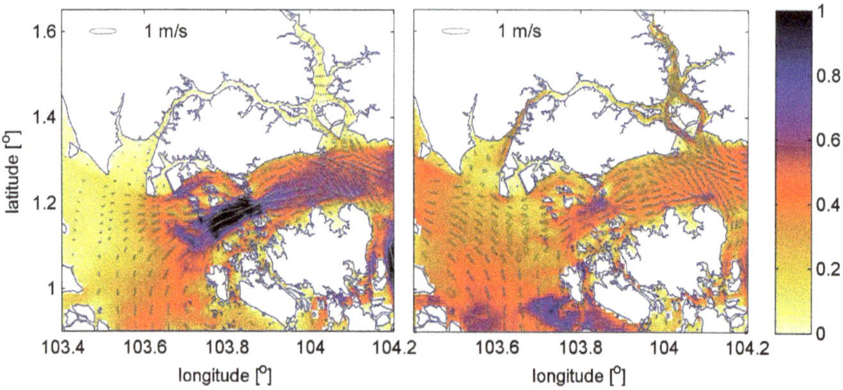

Fig. 10.4. Velocity amplitudes and tidal ellipses of K_1 (left panel) and M_2 (right panel) (after Van Maren and Gerritsen, 2012).

is opposite to the direction of monsoon-induced flows. Fine sediment is stronger influenced by residual flows than coarse sediment, and therefore the direction of residual sand transport is opposite to that of the residual SPM transport (Van Maren and Gerritsen, 2012).

10.2 Effects of Earth's rotation

Earth's rotation induces an apparent acceleration deflecting currents to the right in the northern hemisphere, and to the left in the southern (see Fig. 10.14). Around the equator, the effects of Earth's rotation are negligible. Also, at small spatial scales, i.e. for $Ro < 1$, the effects of Earth's rotation are small relative to other effects. Hence, in many estuaries and tidal basins, Earth's rotation may be neglected. The Rossby number Ro is defined as:

$$Ro \equiv \frac{\sqrt{gh}}{\ell f} \tag{10.1}$$

where h is local water depth, ℓ is a characteristic length scale, f ($\equiv 2\Omega \sin\varphi$) is the Coriolis frequency, Ω (=7.2921 $\cdot$ 10^{-5} rad/s) is the Earth's rotation, and φ is latitude. In the coastal zone, $Ro = 1$ typically yields ℓ-values of about 100–200 km. However, the effects of Earth's rotation on stratified flows occurs at much smaller spatial scales, as discussed in Sec. 10.2.2.

Earth's rotation, in conjunction with (tidal) currents may have a considerable effect on the large-scale SPM dynamics. Here we discuss two examples,

Fig. 10.5. The Patos Lagoon system in Southern Brazil (Google Earth).

i.e. plume formation around Patos Lagoon, Brazil, and the fine sediment dynamics in the Dutch coastal zone.

10.2.1 *Patos Lagoon*

Patos Lagoon in south Brazil (Fig. 10.5 has a length of about 250 km and a width of about 60 km, with a surface area of 10,400 km^2). It is shallow with a mean depth of 5 m (Möller *et al.*, 2001).

Various rivers drain a catchment of about 2,000,000 km^2, of which the Camaquã and Guaiba Rivers are the more important, with a mean discharge of 2000 m^3/s, peaking in winter at 25,000 m^3/s (Möller *et al.*, 2001). Low discharges occur during summer and autumn.

The tide is mainly semi-diurnal, with a daily inequality, a mean tidal range of about 1 m and a maximum of about 2.4 m. The lagoon is connected to the continental shelf by a 20 km long, 1–3 km wide channel of 15 m depth. Along the entrance fairway, 3 km long jetties have been constructed. Owing to the large surface area of the lagoon, tidal velocities well above 1 m/s can occur in this channel, even at smaller tidal amplitudes (Fernandes *et al.*, 2002).

Velocities in the lagoon itself are small, a few dm/s at most, and the salinity field is therefore stratified.

Patos Lagoon is oriented in the direction of the main winds in the area, i.e. NE–SW (Calliari *et al.*, 2009). The direction of these winds is governed by the passage of frontal systems with a period of 1–2 weeks (Möller *et al.*, 2001). Locally, i.e. at the scale of the lagoon, these winds induce setup and setdown in the direction of the winds, while also generating internal circulations. One refers to the "meteorological tide".

The major effect of these winds, however, is at the regional scale. NE-winds induce set-down by Ekman transport along the coast (southern hemisphere), with a lowering of the mean water level up to several dm. For the same reason, SW-winds induce setup, with a similar mean water level increase.

The combined effects of setdown on the shelf and setup in the lagoon by NE-winds induce outflow from the lagoon to the shelf, as indicated in Fig. 10.6. These effects are of course modified by the effects of tidal filling/emptying, which induces a modulation of the wind-induced effects. Owing to its shallowness, wind-induced (short) waves in the lagoon also mobilize fine sediment from the bed, which are thus exported with the outflow.

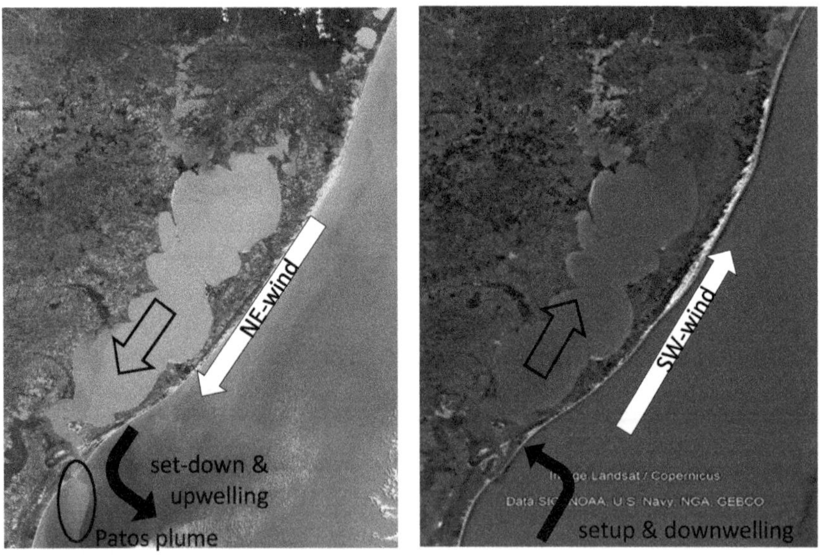

Fig. 10.6. Plume formation in Patos Lagoon coastal zone during NE-wind (left panel), and sediment import into the lagoon during SW-wind (right panel) (Google Earth).

These exported fines form a plume south of the lagoon mouth, clearly visible in Fig. 10.6 (left panel). Part of these fines accumulate in the foreshore. When the wind turns, and these deposits are stirred up from the bed by the longer ocean waves, the sediment is transported back again into the lagoon. This re-import is augmented by salinity-induced gravitational circulation in the entrance channel to the lagoon.

Because of its dominant effect, water level variations due to changing winds are referred to as the meteorological tide, which has a 1–2 week return period in this part of the world.

The astronomical tide alone would export fines from the lagoon to the shelf, owing to the higher SPM-concentrations within the lagoon. Also the local effect of the NE-winds would export fines from the lagoon. However, the effects of Earth's rotation at a larger scale greatly augment these exports. Vinzon *et al.* (2009) estimate, on the basis of numerical modeling, that inflow/outflow through the entrance channel by regional wind effects, in combination with the effects of Earth's rotation, is a factor eight larger than by the local wind only.

Because the entrance channel has been deepened well beyond its equilibrium depth, allowing access to larger ships to the Port of Rio Grande do Sul, siltation occurs in the channel, and occasionally fluid mud is formed.

10.2.2 *North Sea — The Dutch coastal zone*[1]

In this section, we analyze the interaction between river inflow, tide and Earth's rotation in shallow coastal waters. As an example, we discuss the fresh water and SPM distribution in the Dutch coastal zone (southern North Sea). Here, the yearly mean distribution of SPM (Fig. 10.8) shows a strong similarity with the yearly mean salinity field, presented in Fig. 10.7. The band of fresh water along the Dutch coast is referred to as a river plume or a region of freshwater influence (ROFI, Simpson *et al.*, 1990), or in more popular terms as the "coastal river". This ROFI is formed by fresh Rhine water. When the fresh water plume becomes large enough, Earth's rotation effects become important, as measured by the internal Rossby number Ro_i, defined as:

$$Ro_i \equiv \frac{\sqrt{\Delta\rho g h/\rho}}{\ell f} \tag{10.2}$$

[1]This section is based on the PhD-thesis by De Boer (2009).

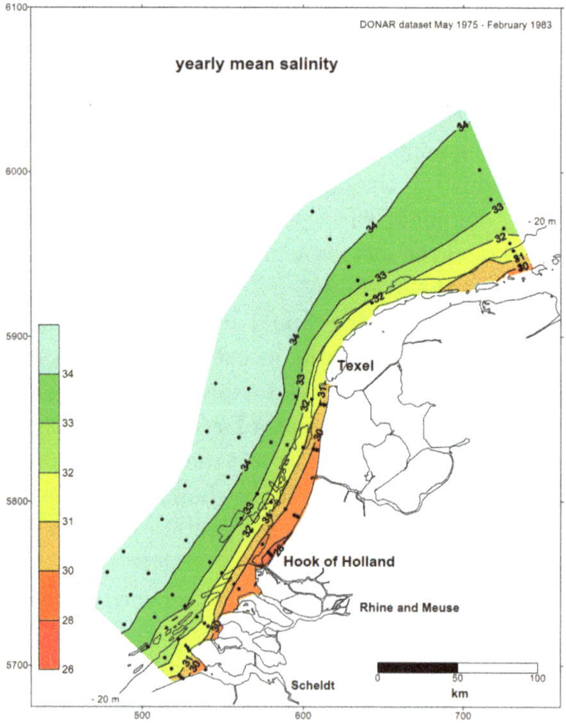

Fig. 10.7. Yearly-mean salinity distribution along Dutch coast.

where $\Delta\rho/\rho$ is the fresh water buoyancy, g is gravitational acceleration, h is water depth, f is the Coriolis frequency and ℓ is the horizontal dimension of the river plume (e.g. Garvine, 1987; Lie-Yauw Oey and Mellor, 1993). For the Rhine plume at latitude $\approx 52°$, ℓ would measure about 17 km for $Ro_i = 1$, which is also more or less the width of the Rhine ROFI. The river plume is deflected to the right by the effects of Earth's rotation, forming the coastal river. In the following, we discuss how tide, fresh water and Earth's rotation interact in keeping SPM close to the coast.

The majority of fine sediment within the Dutch coastal zone, e.g. within the Rhine plume, originates from further south, i.e. the English Channel, erosion of the French cliff coast, and likely erosion of Holocene clay deposits off the Belgian coast (Adriaens *et al.*, 2018; Eisma, 2009; Fettweis and Van den Eynde, 2003).

The Rhine itself contributes only a few % to the total SPM load in the coastal zone.

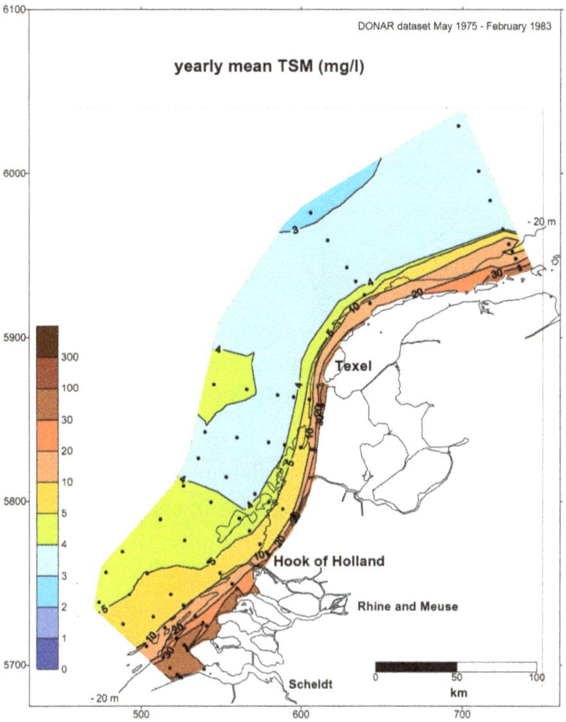

Fig. 10.8. Yearly-mean SPM distribution along Dutch coast.

Figure 10.2 shows that the tide in the Dutch coastal zone propagates counter-clockwise. Thus flood velocities are northward, whereas the (smaller) ebb velocities are due south. However, it is illustrative to discuss first the effects effects of Earth's rotation on the fresh water distribution without tidal influences.

River plume and Earth's rotation without tidal effects

In Sec. 9.1.1, we explained the generation of vertical circulations induced by horizontal density gradients, due to e.g. salinity. This so-called gravitational circulation was quantified to first order by the Eqs. (9.2)–(9.5).

In the coastal zone, fresh water is not confined by the banks of an estuary, and can therefore be deflected by Earth's rotation, to the right in the northern hemisphere, as sketched in Fig. 10.9. This is the basic mechanism forming river plumes (ROFI) along a coast. Thus cross-shore density gradients are formed, which induces a cross-shore gravitational circulation. However, Earth's rotation not only deflects the river plume, but also the

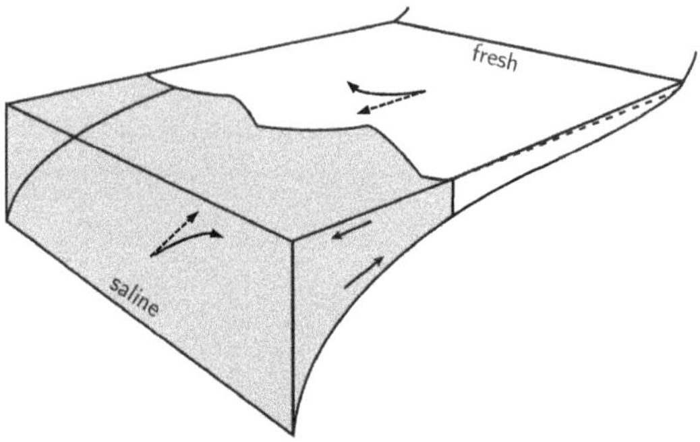

Fig. 10.9. Induction of vertically spiraling current by the effects of Earth's rotation on gravitational circulation (thermal wind).

cross-shore currents themselves. The offshore-directed near-surface current is deflected towards the north, while the onshore-directed near-bed current is deflected to the south. In oceanography, this effect is known as the thermal wind.

Combining these two effects induces a vertically spiraling flow, as sketched in Fig. 10.9. Heaps (1972) was the first to quantify this flow from a simple balance between the cross-shore pressure gradient, Earth's rotation and vertical mixing (see also Eq. (9.2):

$$\begin{aligned} &\frac{\partial \tau_x}{\partial z} + fv = 0 \quad \text{or} \quad && \nu \frac{\partial^2 u}{\partial z^2} + fv = 0 \\ -\frac{\partial p}{\partial x} + &\frac{\partial \tau_y}{\partial z} - fu = 0 \quad \text{or} \quad && -\frac{gz}{\rho}\frac{\partial \rho}{\partial y} + \nu \frac{\partial^2 v}{\partial z^2} - fu = 0 \end{aligned} \tag{10.3}$$

where x, y, z are the longshore, cross-shore and vertical coordinates, τ_x, τ_y are the longshore and cross-shore shear stresses, respectively, ν is the (constant) vertical (eddy) viscosity, and the Coriolis frequency f was defined above. The solution to Eq. (10.3) for constant viscosity, ranging from almost frictionless to highly viscous fluids, is presented in Fig. 10.10, showing strong vertical spiraling shear flow, with a near-bed velocity perpendicular to the shore, while near the water surface, a longshore current is induced. These longshore currents can be of the order of a few dm/s, while the cross-shore velocities are typically a few cm/s, i.e. an order of magnitude smaller.

Because of continuity (see also below), the cross-shore coast-directed near-bed current (upwelling) needs to be compensated by a near-surface

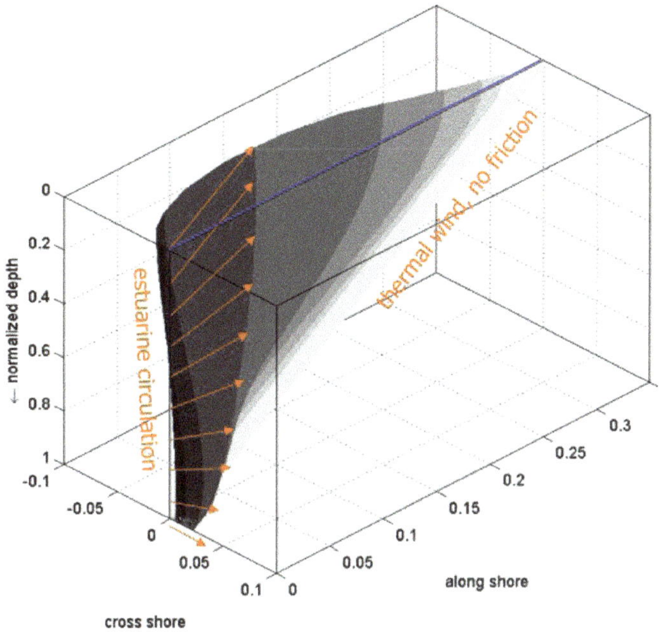

Fig. 10.10. Vertically spiraling current [m/s] profile for a range of constant eddy viscosities, ranging from highly viscous (black) to frictionless pure thermal wind (light gray). The arrows indicate the profile for typical North Sea conditions (De Boer, 2009 — after Heaps, 1972).

offshore directed current. Sediment particles in the water column therefore follow circular (ellipsoid) paths in cross-shore direction induced by the cross-shore gravitational circulation, while transported in the longshore/offshore direction, when advected higher into the water column. Thus, SPM follows vortex-like paths along the coast (e.g. Visser, 1993), coarser sediment cannot be advected higher into the water column — this "Heaps-transport" is therefore not relevant for coarser sediment.

River plume and Earth's rotation with tidal effects

Along the Dutch coast, flood velocities are due north, while ebb velocities are directed southwards. The tidal velocities are of course also affected by Earth's rotation: flood velocities tend to be deflected towards the coast, while ebb velocities in offshore direction. Because of continuity, flood velocities therefore induce down-welling, whereas ebb velocities induce up-welling, as flow velocities higher in the water surface are generally larger than near the bed.

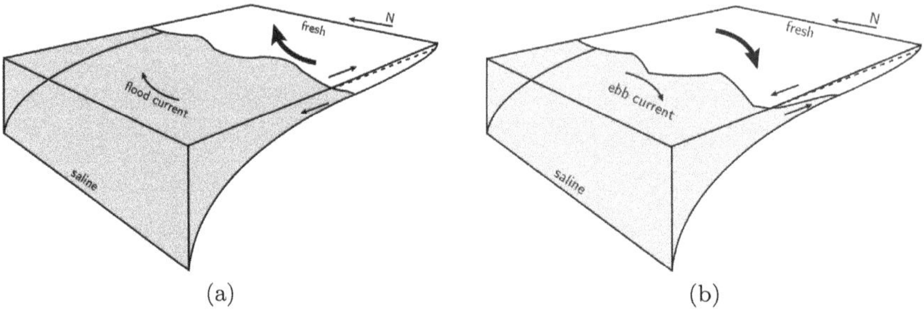

Fig. 10.11. Induction of near-bed residual currents by the interaction of tidal flow, stratification (neap tide) and Earth's rotation; (a) mixing (destratification) by down-welling during flood and (b) enhancement of stratification by up-welling during ebb.

Thus in case of the river plume along the Dutch coast, the ROFI is pushed against the coast during flood, see Fig. 10.11(a). In contrast, during ebb, the river plume is pushed away from the coast (Fig. 10.11(b)).

This behavior of the ROFI has important consequences for the vertical salinity distribution: the water column becomes less stratified during flood, opposing buoyancy spreading (Fig. 10.11(a)), and more stratified during ebb, enhancing buoyancy spreading (Fig. 10.11(b)). These variations in stratification with the ebb and flood tide were first recognized by Simpson *et al.* (1990), who refer to Strain Induced Periodic Stratification (SIPS). Later (De Boer, 2009; De Boer *et al.*, 2006) elaborated on the important effects of tide-induced advection on the fresh water dynamics of the ROFI, and extended Simpson's terminology to ASIPS (advection and strain induced stratification).

The effects of stratification in the Dutch coastal zone also vary over the spring-neap cycle. The tidal range at Hoek van Holland varies between about 1.5–2.1 m from neap to spring tide. This large variation in tidal range (thus tidal energy) explains why fresh Rhine water, brought into the coastal waters, is highly stratified during neap, whereas vertically well-mixed during spring tide.

The strong stratification during neap tide induces a two-layer system with an upper fresh water layer (the ROFI), decoupled from the more saline North Sea water underneath. Thus during ebb, the effects of SIPS become much larger, as elaborated in detail by De Boer (2009). Here we present his two diagrams on the hydrodynamics and salinity distribution during spring (Fig. 10.12) and neap (Fig. 10.13) tides.

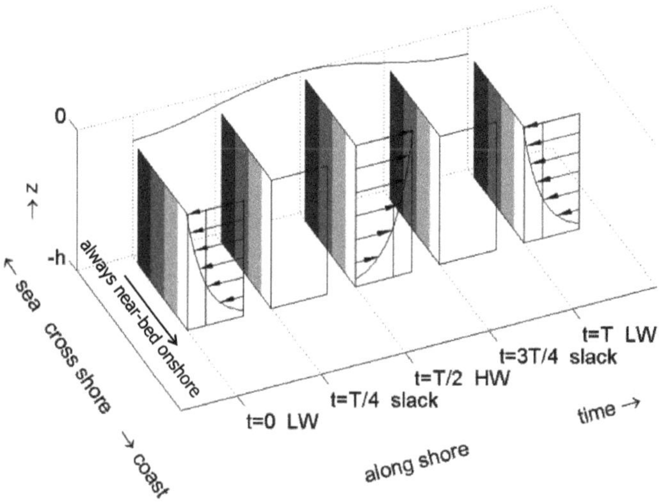

Fig. 10.12. Sketch of long-shore velocity profiles over a tidal cycle in the Dutch coastal zone under spring-tide conditions (vertically well-mixed) — after De Boer (2009).

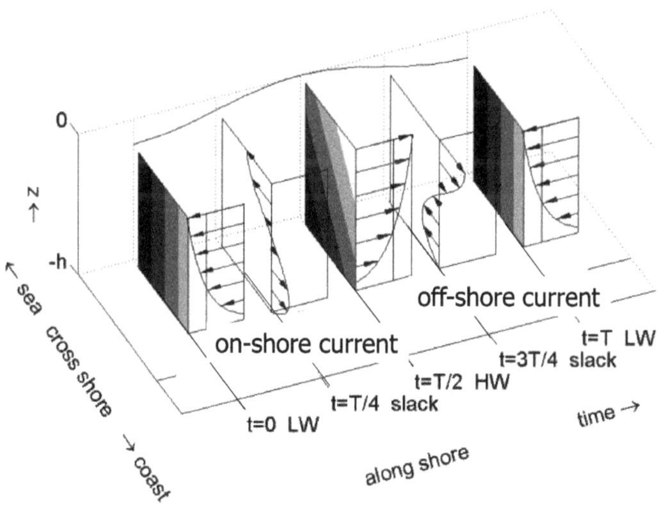

Fig. 10.13. Sketch of long-shore and cross-shore velocity profiles over a tidal cycle in the Dutch coastal zone under neap-tide conditions (vertically stratified) with alternating near-bed onshore and offshore currents — after De Boer (2009).

During spring tide, vertical mixing is so strong that it dominates the effects of SIPS. The tidal current remains more or less rectilinear along the coast, while the salinity field remains vertically well-mixed, e.g. Fig. 10.12.

During neap tide, the above-mentioned decoupling between the fresh Rhine water and the lower saline North Sea water is amplified by the effects of SIPS, as sketched in Fig. 10.13. Around slack water, a fairly strong undertow towards the coast (LWS) or off the coast (HWS) is induced, while the flow becomes stratified. During maximum ebb and flood velocities, the salinity field becomes more mixed over the vertical.

The effects of thermal wind are tide-independent, and only vary with the amount of fresh water. Flood velocities are therefore augmented by thermal winds, whereas ebb-velocities are reduced.

Though tidal and thermal wind effects are large in the Dutch coastal zone, wind itself has a considerable direct influence, in particular during ebb conditions. The prevailing SW winds enhance northern currents, thus contributing to down-welling during flood, while reducing upwelling during ebb. Wind also induces cross-shore currents, as discussed below.

These hydrodynamics and salinity patterns have strong implications for the SPM dynamics. As SPM-concentrations are larger near the bed, the cross-shore gravitation circulation sketched in Fig. 10.13, in combination with SIPS-induced undertows keeps SPM close to the coast, i.e. in within the ROFI, as shown in Fig. 10.8.

The up-welling and down-welling effects along the coast were illustrated by De Boer (2009), discussing the visibility of SPM from satellite images in the Dutch coastal zone, varying with the phase of the tide. Because of the origin of SPM in the Dutch coastal zone, it is not visible at the water surface during stratified conditions, as SPM remains in the lower, denser North Sea seawater.

These analyses show how the interaction of tide, Earth's rotation and stratification govern the transport and fate of fine sediment in the Dutch coastal zone. This picture becomes further complicated by the effects of waves, which can mobilize fine sediment from the North Sea seabed, e.g. Sec. 6.6.

10.3 Effects of winds

Winds are the result of gradients in atmospheric pressure and Earth's rotation. Pressure gradients are largely induced by temperature differences, affecting the density of air — winds are therefore density density currents (e.g. Sec. 9.1).

Winds occur at a variety of scales, from very local winds affected by irregularities in the landscape (cities, mountains, ...), to land-sea breezes induced by local temperature differences, to the global trade winds, and every type of wind in between.

Winds induce (surface) currents by drag, and generate waves. In this section, focus is on wind-induced currents, affecting the transport and fate of SPM in coastal waters. Section 10.3.1 discusses wind-induced currents/circulations at global (mega) scale, whereas in Sec. 10.3.2, we discuss wind-induced currents at the regional scale of coastal systems.

10.3.1 *Wind effects at global scale*

Hot, moist air at the equator rises because of its relative low density, inducing winds, i.e. the to-flow of air, from the so-called Intertropical Convergence Zone (ITCZ), which spans the tropics from more or less −30°S and +30°N. In the northern hemisphere, these winds are deflected to the right by the effects of Earth's rotation, whereas in the southern hemisphere, they are deflected to the left, as sketched in Fig. 10.14. These winds are known as the Trade Winds, and they blow to the west. More accurately, in the

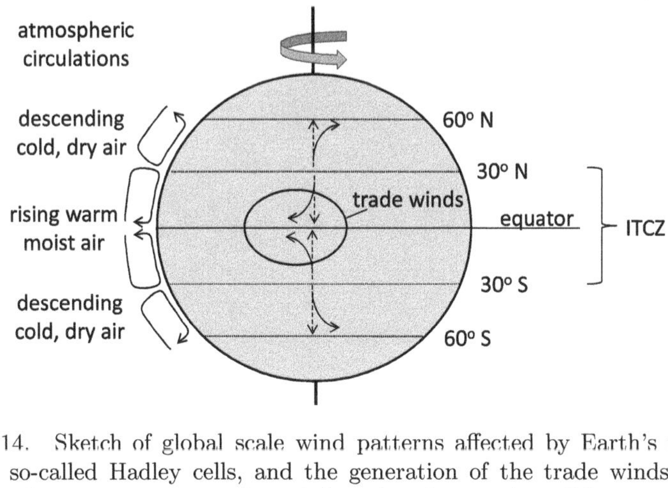

Fig. 10.14. Sketch of global scale wind patterns affected by Earth's rotation, forming so-called Hadley cells, and the generation of the trade winds — dotted, straight arrows represent intended path, whereas the solid, curved arrows represent the actual path.

northern hemisphere, Trade Winds are more or less from the NE, and in the southern hemisphere from the SE.

These winds induce currents/circulations in the upper 50–100 m of the ocean, which are relevant to coastal SPM dynamics, as for instance in the larger Amazon delta (see below).

The pattern of Trade Winds is significantly affected by the presence of the continents. They not only prohibit the undisturbed development of Trade Winds, but also induce a second large-scale wind system, known as the monsoon. Monsoon winds are generated by asymmetric heating of the ocean and the continent, owing to the larger heat capacity of seawater. The major monsoon systems are the western Africa monsoon and the Asia-Australian monsoon. These monsoons locally dominate the effects of the Trade Winds. The monsoon system in the Java Sea is an example of the latter (Sec. 10.3.2).

Figure 10.15 shows a satellite image of the Amazon plume and the Guiana Current along the larger Amazon delta. It consists of the Amazon delta itself, while the low lands of North-East Brazil, French Guiana, Suriname and Guyana can also be considered part of this delta, as the majority of the fine sediment, forming these low lands, originated from the Amazon River.

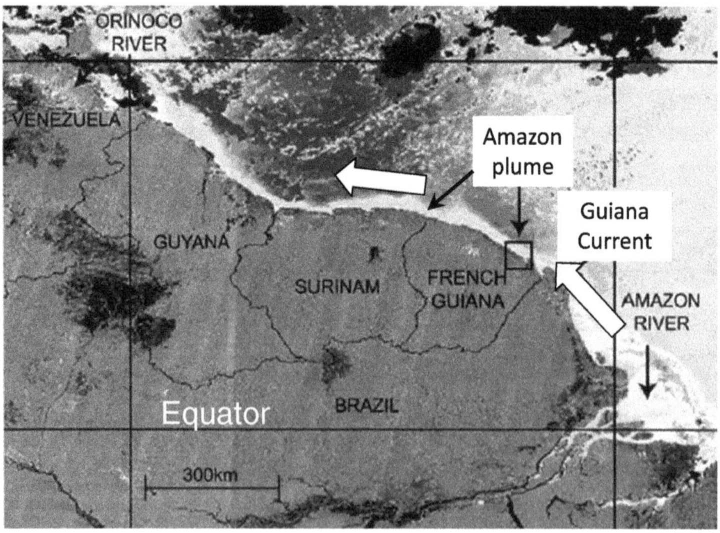

Fig. 10.15. Satellite image of Amazon plume (light band along the coast), with the Guiana Current along the greater Amazon delta.

The plume of fresh Amazon water deflects to the left, affected by the Trade Winds and the Guiana Current. The latter is driven by the Trade Winds themselves and fed by the Brazilian Current and the South Equatorial Current. (Earth's rotation does not play a role close to the equator.)

Trade Winds also generate waves on the Atlantic Ocean. Because of the persistent NE-direction of these winds, the coasts of the larger Amazon delta is affected by swell from that direction. Significant wave heights amount to 1–2.5 m, with peak periods of 8–12 s.

These wind-induced effects govern the dynamics of SPM in the Guiana coastal waters, in conjunction with tidal effects:

(1) Fine sediment is mobilized from the seabed by persistent swell, resulting in SPM-concentrations of many 100 mg/L, even up to a few g/L (NEDECO, 1968; Anthony *et al.*, 2010).
(2) The tide is significant with neap ranges of about 1.2 m, and spring ranges of about 2.5 m. However, tidal velocities are small, i.e. of the order of 0.1–0.2 m/s, and are more or less perpendicular to the coast, as explained by Fig. 10.1.
(3) The Guiana Current induces longshore currents of about 0.3–0.4 m/s in NW–W direction in the coastal zone. The interaction between the cross-shore tidal velocities and the longshore Guiana Current induces a zig-zag flow pattern in the coastal zone, resulting in net SPM transport to the west.
(4) The 20–50 km wide band of fresh Amazon water induces a near-bed onshore undertow through gravitational circulation keeping fine sediment in the coastal zone. Note that, contrary to the North Sea (Sec. 10.2.2), no thermal winds are induced owing to the absence of Coriolis effects.

The transport of fine sediment in the coastal waters of the greater Amazon delta takes place in the form of large, migrating mudbanks as well. These mudbanks measure 10–30 km in longshore direction and extend to about the −20 m isobath, migrating at about 1 km/year. During migration, large amounts of fluid mud are formed at the lee-side of their banks, inducing significant wave damping, as explained in Sec. 7.4.

For an elaboration of the mechanisms which form these mudbanks, the reader is referred to e.g. NEDECO (1968); Allison and Lee (2004); Anthony *et al.* (2010); Augustinus (2004).

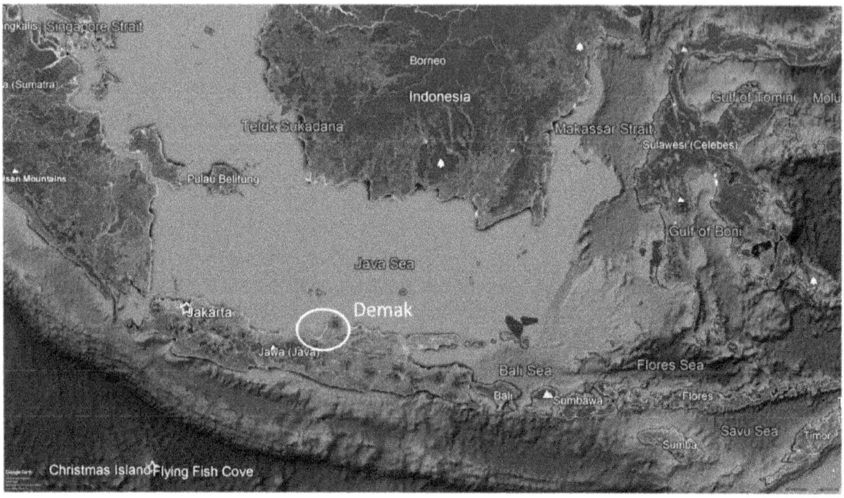

Fig. 10.16. The Java Sea with Demak, mid Java (Google Earth).

10.3.2 *Wind effects at regional scale*

An example of the importance of winds at the regional scale is the effects of the monsoon winds on SPM transport in the Java Sea (Fig. 10.16). Here we discuss SPM dynamics in the Demak coastal zone, mid-Java, NE of Semarang. The climate in the Java Sea is characterized by a SE-monsoon (May–September) and a NW-monsoon (October–April). The residual ocean currents along the north coast of Java follow these monsoons (Wrytki, 1961), i.e. a residual to the west during the SE-monsoon and to the east during the NW-monsoon.

The NW-monsoon along the Java coast is wet, with 30–50 cm rainfall per month, whereas the monthly rainfall during the SE-monsoon does not exceed 10 cm/month. As a result, large amounts of fresh water are brought into the Java coastal zone during the NW-monsoon, kept towards the shore by the wind stresses, as sketched in Fig. 10.17. Owing to the small tidal velocities (see below), the water column becomes easily stratified. The gravitational circulation induced in this wet period generates a near-bed current towards the coast, keeping fine sediment within the coastal zone. At present, these fines stem mainly from wave-induced erosion of the muddy seabed and coastline, though were originally brought into the coastal system by local rivers.

During the dry season, gravitational circulation is small or absent, and onshore winds induce an offshore near-bed current (Fig. 10.17).

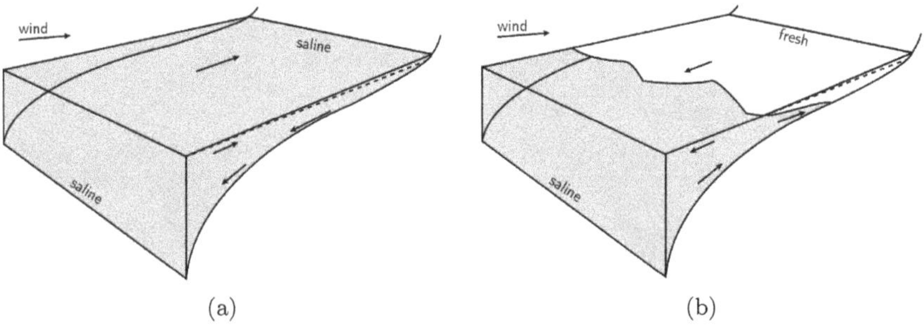

Fig. 10.17. Induction of near-bed currents by onshore wind in (a) dry period and (b) wet period.

Furthermore, as the fetch for SE-winds is small, the waves are also small, with heights not exceeding a few dm. On the other hand, waves during the NW-monsoon frequently exceed 1 m, eroding the coast and the seabed.

Thus, fine sediment eroded during the NW-monsoon may remain in the coastal zone, though with a small eastward residual flux along the coast. During the SE-monsoon, little sediment is mobilized, and offshore fluxes are small.

The tide in the Java Sea is complex. Along the Demak coastline, the tide is mainly diurnal, with a small semi-diurnal component. Neap and spring tidal range amount to about 40–60 cm, though can be a bit larger when the semi-diurnal components synchronize with the diurnal components, occurring more or less at the semi-diurnal spring-neap cycle. Tidal currents in the coastal waters measure around 0.1 m/s (Yusuf and Yanagi, 2013).

As wind- and tide-induced currents are of the same order of magnitude, SPM dynamics in Demak coastal zone are driven by a complex combination of multiple sediment transport processes, governed by tide, winds, waves and fresh water. Proper system understanding and subsequent modeling requires addressing all these aspects.

10.4 Net and gross transport rates in coastal waters

In Sec. 9.4, we discussed the relation between the gross and net SPM fluxes in estuaries, concluding that, generally, gross sediment fluxes are much larger than the net fluxes. In this section, we discuss the relation between gross and net SPM fluxes in coastal waters on the basis of an example from The Netherlands.

Van Kessel established the sediment budget in the Dutch Wadden Sea from simulations of SPM transport with a calibrated numerical model. Their

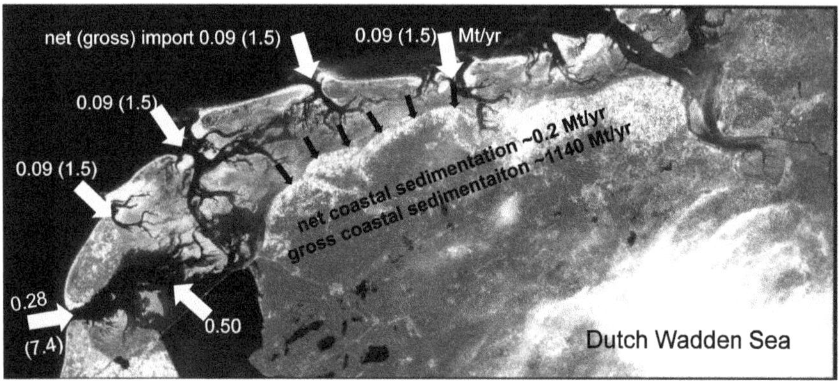

Fig. 10.18. Computed gross and net SPM transport in the Dutch Wadden Sea. Numbers between brackets represent gross sediment fluxes through the tidal inlet, summed over a year, while the non-bracketed numbers represent the net fluxes.

results are summarized in Fig. 10.18, showing the computed gross and net fluxes through the tidal inlets. For instance in the Marsdiep (inlet in the SW-corner of Wadden Sea), a gross import of 7.4 Mton/yr was computed, whereas 7.1 Mton/yr leaves the Wadden Sea again. Thus the net import amounts 0.28 Mton/yr. Gross deposition rates on the mainland coast were also computed, amounting to 1140 Mton/yr. This mainland coast is known to accrete, and indeed the simulations suggest a net sedimentation rate of 0.2 Mton/yr, i.e. almost four orders of magnitude smaller.

These very large differences in net and gross effects imply that small alterations in the gross fluxes may switch the long-term behavior of these coasts entirely from for instance accretion to erosion, while their assessment is subject to significant errors.

10.5 Fairway siltation

From a practical/engineering point of view, siltation in navigation channels is an important aspect of sediment dynamics in coastal waters. Siltation can occur with fine (mud/silts) and coarse (sands) sediment, which may be transported as suspended load and/or as bed load. We emphasize the adjective "open water", as fairways in rivers and estuaries behave differently. Here, full morphodynamic analyses are required to assess these siltation rates, as navigation channel and the ambient waters generally interact strongly. This is for instance the case in the Western Scheldt where the deepening of the fairway affects the morphodynamic development of that

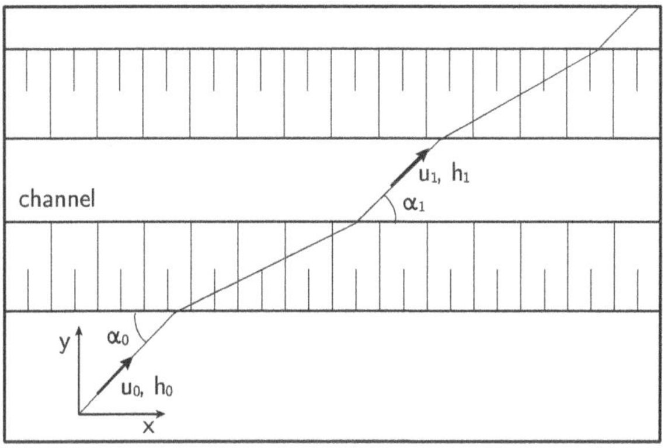

Fig. 10.19. Refraction of oblique current over a fairway.

estuary Jeuken and Wang (2010). Navigation channels in open waters are generally small compared to the spatial scales of relevant hydrodynamics, and therefore do not interact with the overall system dynamics.

In the following, we discuss siltation mechanisms and rates by currents oblique to a navigation channel, as sketched in Fig. 10.19. To understand the processes, the scales of the system have to be considered. A cross-current experiences a sudden increase in water depth, as the width of the channel is small compared to the systems dimensions. Thus the flow decelerates locally, as the water flux does not change. However, along the channel, the relevant scales are of the same order of magnitude as the system scales. Given a constant pressure gradient along a channel, the larger depth within the channel reduces the effective hydraulic drag, thus accelerating the current. As a result, depending on the angle of incidence, the flow in/over a channel accelerates of decelerates. We elaborate further on this configuration.

In the following, we use simple geometrical arguments to establish the flow velocity in a channel as a function of relative channel depth and angle of incidence α using a simplified set of continuity and momentum equations:

$$\begin{aligned} \frac{U_{y,1}}{U_{y,0}} &= \frac{h_0}{h_1} \\ \frac{U_{x,1}}{U_{x,0}} &= \frac{C_1}{C_0}\sqrt{\frac{h_1}{h_0}} \approx \sqrt{\frac{h_1}{h_0}} \quad (\text{as } C_1 \approx C_0) \end{aligned} \tag{10.4}$$

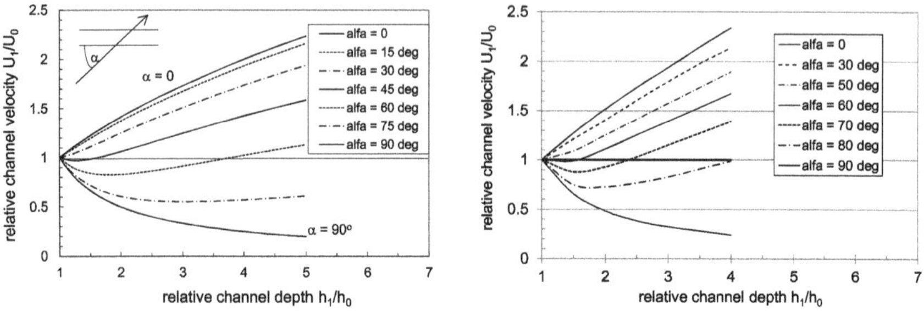

Fig. 10.20. Ratio flow velocity inside and outside channel according to Eq. (10.4) (left panel) and the three-dimensional approach (redrawn from Jensen *et al.* (1999), right panel).

Thus:

$$U_1 = U_0 \frac{h_0}{h_1} \sqrt{\sin^2 \alpha_0 + \left(\frac{h_1}{h_0}\right)^3 \cos^2 \alpha_0}, \quad \text{and}$$
$$\tan \alpha_1 = \sqrt{(h_0/h_1)^3} \tan \alpha_0 \tag{10.5}$$

where subscript $\cdot_0$ and $\cdot_1$ refer to ambient and in-channel conditions, x and y are the coordinates along and across the channel, U and h are velocity and depth and C is the Chézy coefficient. The ratio between the flow velocities inside and outside the navigation channel are plotted in Fig. 10.20 as a function of the angle of incidence α_0 of the ambient flow. Note that up to quite large α_0, the flow velocity within the channel is larger than outside the channel. Three-dimensional effects align the flow even more with the channel, as demonstrated by the numerical results by Jensen *et al.* (1999), Fig. 10.20, right panel. A larger flow velocity implies a larger transport capacity. Does this than mean that the channel becomes self-cleansing?

However, the larger velocity implies also a larger specific discharge. The "extra" water can only come from the surroundings of the channel. But this extra water carries also extra sediment. Let us examine whether this extra load balances the larger flow velocity (larger transport capacity) in the channel. For simplicity, only along-channel flow is discussed, referring to Fig. 10.21. Continuity and substitution from Eq. (10.4) yields a relation for an "effective channel width" b_0, representing the area from which the "extra" water and sediment is attracted (continuity equation):

$$\frac{b_0}{b_1} = \frac{U_1 h_1}{U_0 h_0} = \frac{C_1}{C_0}\left(\frac{h_1}{h_0}\right)^{3/2} \tag{10.6}$$

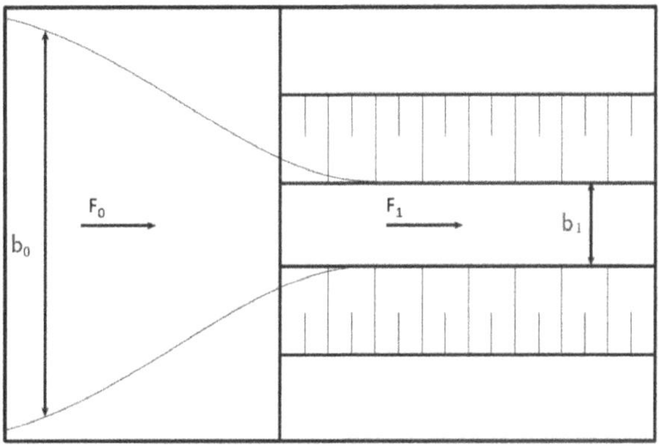

Fig. 10.21. Diagram explaining attraction of extra water by a fairway.

Next, we assume that capacity transport F_e scales with the local flow velocity: $F_e \propto u_*^n$, where u_* is the shear velocity, and in which $n = 3$ represents bed load and $n =$ 4–6 represents suspended load. Substitution from Eq. (10.4) and (10.6) yields a relation for the sediment transport capacity inside a channel in relation to that outside:

$$\frac{F_{e,1}}{F_{e,0}} = \left(\frac{u_{*,1}}{u_{*,0}}\right)^n = \left(\frac{U_1}{U_0}\frac{C_0}{C_1}\right)^n = \left(\frac{h_1}{h_0}\right)^{n/2} \tag{10.7}$$

The ratio of the sediment transport inside and next to the channel F_1/F_0 then becomes:

$$\frac{F_1}{F_0} = \frac{b_1 F_{e,1}}{b_0 F_{e,0}} = \frac{C_0}{C_1}\left(\frac{h_1}{h_0}\right)^{(n-3)/2} \approx \left(\frac{h_1}{h_0}\right)^{(n-3)/2} \tag{10.8}$$

As C_1 is (slightly) larger than C_0, the sediment transport capacity within the channel is smaller than outside in case of bed load ($n = 3$). Thus the channel will silt up, as one expects. However, for suspended load, when $n =$ 4–6, Eq. (10.7) predicts that the fairway would be self-cleansing, and possibly even eroding! Of course this is against one's experience. Note that one would find such results also, at least qualitatively, when using advanced three-dimensional numerical sediment transport models. Hence, something is missing in this approach.

This paradox is explained from omitting the effect of waves. These may stir up and transport large amounts of sediment in the shallows surroundings a fairway, whereas wave activity on the channel bed would be small. From the linear wave theory, we can derive a first-order estimation of the

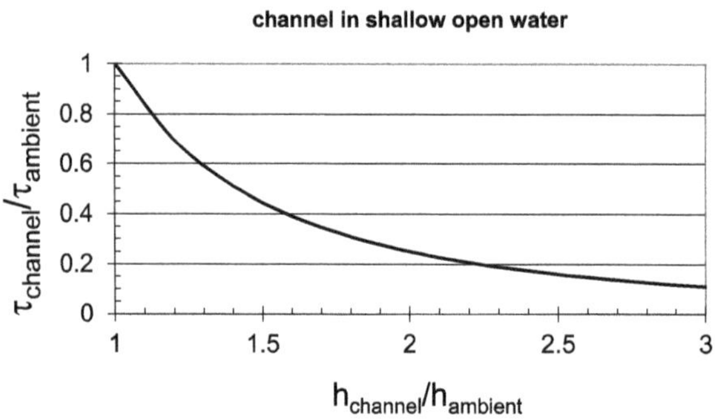

Fig. 10.22. Maximal wave-induced bed shear stresses on channel bed as a function of wave-induced bed shear stresses in the surrounding waters and relative channel depth.

ratio of maximal wave-induced bed shear stresses:

$$\frac{\hat{\tau}_{b,1}}{\hat{\tau}_{b,0}} = \frac{u_{b,1}^2}{u_{b,0}^2} = \left(\frac{\sinh\{kh_o\}}{\sinh\{kh_1\}}\right)^2 \approx \left(\frac{h_0}{h_1}\right)^2 \quad \text{as } kh \ll 1 \quad \text{and} \quad \lambda \gg h \tag{10.9}$$

This ratio is plotted in Fig. 10.22, showing that bed shear stresses on the channel bed decrease rapidly with channel depth. Moreover, wave-induced shear stresses are generally larger than flow-(tide) induced stresses. Thus neglecting the effects of waves on the siltation rates in navigation channels may give an entirely wrong picture of channel maintenance needs.

On the other hand, this analysis yields two practical observations:

(1) Channel siltation may vary strongly over the year, in particular if the local climate is characterized by seasonal variations in wave conditions.
(2) Channel orientation may influence siltation rates, thus maintenance costs.

The literature provides four simple engineering rules for assessing channel siltation $F_{s,c}$ per unit channel length and channel width b:

Mayor-Mora *et al.* (1977):

$$F_{s,c} = \frac{T_0}{\cos(\alpha_0 - \alpha_1)}\left[1 - \exp\left\{-\frac{A_m h_0 b}{h_1 \sin\alpha_1 \cos(\alpha_0 - \alpha_1)}\right\}\right] - T_1\left[1 - \exp\left\{-\frac{A_m b}{\sin\alpha_1}\right\}\right]\sin\alpha_1 \tag{10.10}$$

$$A_m = W_s^2/\varepsilon_1 U_1, \quad \varepsilon_1 = 0.085 h_1 u_{*,1}$$

Bijker (1980):

$$
\begin{aligned}
F_{s,c} &= \left(\frac{T_0}{\cos(\alpha_0 - \alpha_1)} - T_1\right)\left[1 - \exp\left\{-\frac{A_b b}{\sin\alpha_1}\right\}\right]\sin\alpha_1 \\
A_b &= \frac{F_B W_s}{h_1 U_1}; \quad F_B = \frac{a_1}{a_2 - a_3} \\
a_1 &= \frac{u_{*,1}}{u_{*,0} - u_{*,1}}; \quad a_2 = \frac{W_s}{0.085 u_{*,1}}\left[1 - \exp\left\{-\frac{W_s}{0.085 u_{*,0}}\right\}\right] \\
a_3 &= \left(\frac{u_{*,1}}{u_{*,0}}\right)^3 \frac{W_s}{0.085 u_{*,0}}\left[1 - \exp\left\{-\frac{W_s}{0.085 u_{*,1}}\right\}\right]
\end{aligned}
\tag{10.11}
$$

Eysink and Vermaas (1983):

$$
\begin{aligned}
F_{s,c} &= \left(\frac{T_0}{\cos(\alpha_0 - \alpha_1)} - T_1\right)\left[1 - \exp\left\{-\frac{A_{ev} b}{h_1 \sin\alpha_1}\right\}\right]\sin\alpha_1 \\
A_{ev} &= 0.03\frac{W_s}{u_{*,1}}\left[1 + \frac{2W_s}{u_{*,1}}\right]\left[1 + 4.1\left(\frac{k_s}{h_1}\right)^{0.25}\right]
\end{aligned}
\tag{10.12}
$$

Note that from our analysis above, the shear velocity has to be corrected for the effects of waves, otherwise misleading results will be obtained.

Allersma (1971):
The fourth model is by Allersma, known as the "volume-of-cut" method, which, however, has never been published:

$$
h(t) = h_e - (h_e - h_0)\exp\left\{-\frac{h_e}{h_0}\frac{t}{\tau}\right\} \tag{10.13}
$$

This simple model is particularly useful in analyzing the siltation rates upon deepening an existing channel — data on previous dredging volumes can be used for calibration of the time constant for siltation τ. t is an arbitrary time scale (one year, five years, ...), h_0 the initial channel depth, and h_e the channels equilibrium depth (which may be the local water depth in open water).

Chapter 11

The Governing Equations

11.1 General

This chapter presents the various sets of governing equations relevant for the transport of SPM in open water for Newtonian fluids (except for Sec. 11.4), i.e.:

(1) continuity equation for water,
(2) momentum equation,
(3) transport equation for dissolved and suspended matter,
(4) equation of state,
(5) k–ε turbulence closure model,
(6) a phase-averaged wave model.

We assume one-phase incompressible flow, i.e. flow in which suspended particles follow turbulent water movements, which is the case if $w_s \ll w'_{rms}$, where w_s is the settling velocity of a particle and w'_{rms} is the root mean square of the (vertical) turbulent velocity fluctuations (Uittenbogaard, 1994). As $w'_{rms} \approx u_*$, the Rouse number β $(= \sigma_T w_s/\kappa u_*)$ (see Sec. 3.4) yields a good proxy for this condition. In Sec. 3.4 it was shown that fine sediment mixes over the water column when $\beta \ll 1$. In other words, when fine sediment is well-mixed, a one-phase approach is feasible. Uittenbogaard (1994) showed theoretically that even sand particles with a diameter up to 200 μm can properly follow turbulent movements typically induced in tidal flows. This is supported by Muste and Patel (1997) through measurements of the fluctuating velocity components of suspended 250 μm sand particles in a turbulent flow. They found that RMS-values of the particle velocities are only about 15–20% smaller than the RMS-values of the fluctuating velocity components of the carrying fluid. In case of strong stratification, this condition may not be met. However, then often other effects, such as non-Newtonian behavior, likely start to play a role.

Further we apply the Boussinesq approximation, neglecting all effects of density variations, except in the pressure terms.

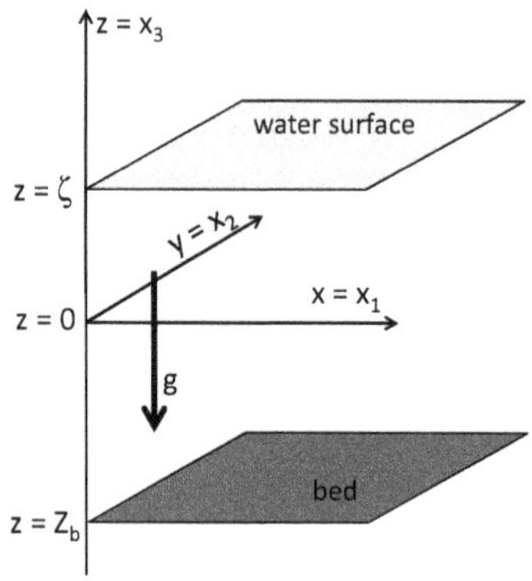

Fig. 11.1. Coordinate system with bed and surface levels.

We will start with the full three-dimensional form of the equations, and present the relevant shallow water formulations in the next four sections.

Continuity equation

The continuity equation reads:

$$\frac{\partial u_i}{\partial x_i} = 0 \tag{11.1}$$

in which u is velocity and x is the spatial coordinate, using the summation convention ($i = 1 \equiv x,\ = 2 \equiv y,\ = 3 \equiv z$), see Fig. 11.1, where also water level ζ and bed level Z_b are also indicated (thus Z_b attains negative values in this coordinate system).

Momentum equation

The momentum equation reads:

$$\frac{\partial u_i}{\partial t} + \frac{\partial u_i u_j}{\partial x_j} - e_{ijk} f_k u_j = \frac{1}{\rho}\frac{\partial \sigma_{ij}}{\partial x_j} - \delta_{i3} g \tag{11.2}$$

where $i, j = 1, 2, 3$ as above, f is the Coriolis frequency, g is the acceleration due to gravity, ρ is bulk density of water (see also Eq. (11.10)), σ is the stress tensor, δ_{ij} is the Kronecker delta ($\delta_{ij} = 1$ for $i = j$ and $\delta_{ij} = 0$ for $i \neq j$) and e_{ijk} the permutator ($e_{ijk} = 1$ for cyclical i, j, k and $e_{ijk} = -1$ for anti-cyclical i, j, k; otherwise $e_{ijk} = 0$). The first term in Eq. (11.2) represents flow acceleration, the second term represents advection, the third term apparent acceleration by Earth's rotation (Coriolis effects), the fourth term represents internal stresses within the fluid, and the last term represents a body force, i.e. acceleration by gravity.

The stress tensor consists of a molecular part (superscript $\cdot^m$) and a turbulent part (superscript $\cdot^T$), which are linearly superposed (Malvern, 1969). Non-Newtonian and soil mechanical effects (Sec. 11.4) are included in the molecular terms:

$$\sigma_{ij} = \sigma_{i,j}^m + \sigma_{i,j}^T = \tau_{i,j}^m - \delta_{ij}p + \tau_{i,j}^T - \frac{2}{3}\delta_{ij}\rho k_T \tag{11.3}$$

in which p is the normal stress (pressure)and k_T the turbulent kinetic energy (Eq. (3.4)). The molecular shear stress τ_{ij} reads:

$$\tau_{ij}^m = \mu\left(\frac{\partial u_i}{\partial x_j} + \frac{\partial u_j}{\partial x_i}\right) \tag{11.4}$$

and its turbulent counterpart:

$$\tau_{ij}^T = \rho\nu_T\left(\frac{\partial u_i}{\partial x_j} + \frac{\partial u_j}{\partial x_i}\right) \tag{11.5}$$

in which μ (or μ_a) is the (apparent) molecular viscosity and ν_T is the eddy viscosity ($\nu_T = \mu_T/\rho$; Eq. (3.22)).

Transport equation for salinity

The transport equations for dissolved matter and salt have the same form — we will only give the equation for salinity:

$$\frac{\partial S}{\partial t} + \frac{\partial u_i S}{\partial x_i} - \frac{\partial}{\partial x_i}\left((\varepsilon_m + \Gamma_T)\frac{\partial S}{\partial x_i}\right) = 0 \tag{11.6}$$

in which S is salinity, ε_m is the molecular diffusion coefficient and Γ_T is the eddy diffusivity (Eq. (3.56)). For other matter, S is replaced by another quantity, possibly with internal reactions, if not conservative. The first term in Eq. (11.6) represents temporal variation in salinity, the second advection and the latter the transport of salinity by molecular and turbulent diffusion.

Transport equation for SPM

The transport equation for suspended sediment (SPM) reads:

$$\frac{\partial c^{(i)}}{\partial t} + \frac{\partial}{\partial x_i}\left(\left(u_i - \delta_{i3}w_s^{(i)}\right)c^{(i)}\right) - \frac{\partial}{\partial x_i}\left((D_s + \Gamma_T)\frac{\partial c^{(i)}}{\partial x_i}\right) = 0 \tag{11.7}$$

where superscript $\cdot^{(i)}$ refers to the i-fraction of sediment. The first term in Eq. (11.7) represents temporal variations in SPM, the second advection, including the effects of settling, and the last term transport of SPM by molecular and turbulent diffusion. In general, molecular diffusion coefficients for dissolved and particulate matter are not identical. The latter is given by the Stokes–Einstein equation (Bird *et al.*, 1960):

$$D_s = \frac{kT}{6\pi\mu D} \tag{11.8}$$

where D is particle size, k is the Boltzmann constant ($= 1.38 \cdot 10^{-23}\,\mathrm{J/K}$), and T and μ are the absolute temperature and dynamic viscosity of water. Thus, D_s attains a value of about $2 \cdot 10^{-14}\,\mathrm{m^2/s}$ for 10 μm particles at 15°C, hence it is generally negligible in natural environments.

In general, the turbulent diffusion coefficients for dissolved matter and SPM are assumed to be equal, in regard to the one-phase approach. If not, correction coefficients can be used, see for instance Van Rijn (1993).

Equation of state

As discussed in Sec. 3.2.3, suspended and dissolved matter affect the effective (bulk) density of water ρ_b (in a one-phase approach):

$$\begin{aligned} \rho(c) &= \phi_s \rho_s + (1 - \phi_s)\rho_w = c + \left(1 - \frac{c}{\rho_s}\right)\rho_w \\ &= \rho_w + \left(1 - \frac{\rho_w}{\rho_s}\right) c = \rho_w + \alpha c \approx \rho_w + 0.6c \end{aligned} \tag{11.9}$$

where ρ_b, ρ_w and ρ_s are the bulk density of water and the specific densities of water (with dissolved substances) and sediment, respectively, and the coefficient α has been introduced before. The equation of state, summing over all sediment fractions $c^{(i)}$, thus becomes:

$$\rho_b(S,T,c) = \rho_w(S,T) + \sum_i \left\{ \left(1 - \frac{\rho_w(S,T)}{\rho_s^{(i)}}\right) c^{(i)} \right\} \tag{11.10}$$

where T is temperature. A good engineering approximation of Eq. (11.10) was given by Eq. (3.26).

Turbulence closure

The k–ε turbulence closure model is state-of-the-art for engineering applications, and is used in this book, unless stated otherwise. It assumes isotropy and is applicable if the turbulence Reynolds number $Re_T > 100$

(Nezu and Nakagawa, 1993; Nezu and Rodi, 1986; Rodi, 1984), where

$$Re_T \equiv \frac{u'\ell}{\nu} \approx \frac{\sqrt{k_T}\ell}{\nu} = \frac{k_T^2}{\nu\varepsilon_T} = \frac{\nu_T}{c_\mu \nu} \tag{11.11}$$

At lower Re_T, viscosity effects become important, and a low-Reynolds model or correction has to used, as described in Sec. 7.5.1, see also Patel *et al.* (1985) or Goldberg and Apsley (1997). Possible modifications for non-Newtonian conditions are discussed in Sec. 11.4.

The turbulent kinetic energy per unit mass k_T was defined in Eq. (3.4), but repeated here for convenience, together with the definition of the rate of turbulence dissipation per unit mass ε_T:

$$k_T \equiv \frac{1}{2}\overline{(u'_i u'_i)} \quad \text{and} \quad \varepsilon_T \equiv \frac{1}{2}\nu\overline{\left(\frac{\partial u'_i}{\partial x_j} + \frac{\partial u'_j}{\partial x_i}\right)^2} \tag{11.12}$$

where a prime $\cdot'$ denotes a turbulent quantity and ν is the kinematic viscosity ($\equiv \mu/\rho$). The eddy viscosity and eddy diffusivity are related to k_T and ε_T via:

$$\nu_T = c_\mu \frac{k_T^2}{\varepsilon_T} \quad \text{and} \quad \Gamma_T = \frac{\nu_T}{\sigma_T} \tag{11.13}$$

where σ_T is the turbulent Schmidt number (Eq. (3.56) and Sec. 3.4 for typical values, and effects of stratification) and c_μ a coefficient given below. Note that the subscripts $\cdot_T$ have been added to k and ε to distinguish between wave length and molecular diffusion.

The k–ε turbulence model reads:

$$\begin{aligned}\frac{\partial k_T}{\partial t} &+ \frac{\partial u_i k_T}{\partial x_i} - \frac{\partial}{\partial x_i}\left(\nu + \frac{\nu_T}{\sigma_k}\right)\frac{\partial k_T}{\partial x_i} \\ &= \nu_T\left(\frac{\partial u_i}{\partial x_j} + \frac{\partial u_j}{\partial x_i}\right)\frac{\partial u_i}{\partial x_j} + \delta_{i3}\frac{g}{\rho_b}\frac{\nu_T}{\sigma_T}\frac{\partial \rho_b}{\partial x_i} - \varepsilon_T\end{aligned} \tag{11.14}$$

and

$$\begin{aligned}\frac{\partial \varepsilon_T}{\partial t} &+ \frac{\partial u_i \varepsilon_T}{\partial x_i} - \frac{\partial}{\partial x_i}\left(\nu + \frac{\nu_T}{\sigma_\varepsilon}\right)\frac{\partial \varepsilon_T}{\partial x_i} \\ &= c_{1\varepsilon}\nu_T\frac{\varepsilon_T}{k_T}\left(\frac{\partial u_i}{\partial x_j} + \frac{\partial u_j}{\partial x_i}\right)\frac{\partial u_i}{\partial x_j} + \delta_{i3}\left(1 - c_{3\varepsilon}\right) \\ &\quad \times \frac{\varepsilon_T}{k_T}\frac{g}{\rho_b}\frac{\nu_T}{\sigma_T}\frac{\partial \rho_b}{\partial x_i} - c_{2\varepsilon}\frac{\varepsilon_T^2}{k_T}\end{aligned} \tag{11.15}$$

The first term in the turbulent energy equation (11.14) represents the temporal variation in turbulent kinetic energy k_T, the second and third terms represent variations in k_T by advection and molecular and turbulent diffusion respectively, the fourth term represents turbulence production by shear, the fifth term represents buoyancy destruction in case of stratification (see Sec. 3.2.3), and the last term represents dissipation of turbulent kinetic energy into heat.

The various terms in the dissipation equation (11.15)[1] can be understood only from their derivation (Launder and Spalding, 1972), for which the enstrophy Z is introduced (Tennekes *et al.*, 1972), defined as

$$Z \equiv \frac{1}{2}\omega^2 \equiv \frac{1}{2}(\nabla \times \mathbf{u})^2 \tag{11.16}$$

with vorticity ω as the curl of the velocity vector $\mathbf{u}$. For turbulent flows, the rate of enstrophy dissipation per unit mass relates to the turbulent vorticity fluctuation:

$$\varepsilon_T = \frac{1}{2}\nu\overline{(\omega')^2} \tag{11.17}$$

The ε_T-equation follows from the total rate of change of ω multiplied by its turbulent fluctuations (Tennekes and Lumley, 1994), i.e. $\omega' \mathrm{D}\omega/\mathrm{D}t$. In the turbulent energy cascade, about 60% of the turbulent energy is transferred to smaller scale eddies, and the other 40% to larger scale eddies. Thus, velocity gradients at smaller scale increase, which can be interpreted as a production of enstrophy. Finally, neighboring eddies deform the fluid in between, which requires work by these eddies, hence a loss in vorticity. This loss is described with a dissipation term.

With these considerations in mind, the terms in Eq. (11.15) can be interpreted in a more or less physical way. The first term represents the temporal variation in turbulent kinetic energy dissipation ε_T, the second and third terms represent advection and diffusion of ε_T respectively, the fourth term yields "production" of ε_T (production of enstrophy), the fifth term accounts for buoyancy effects in stratified flows, and the last term is a sink term (dissipation of Z).

The various coefficients are given e.g. by Launder and Spalding (1972) and Rodi (1984): $c_\mu = 0.09$, $c_{1\varepsilon} = 1.44$, $c_{2\varepsilon} = 1.92$, $\sigma_k = 1.0$, $\sigma_\varepsilon = 1.3$, $\sigma_T = 0.7$, $\kappa = 0.41$, and $c_{3\varepsilon} = 1.0$ or 0, in case of stable or unstable stratification, respectively. These coefficients were obtained from calibration

[1] We are grateful to Dr. R.E. Uittenbogaard for his guidance in this explanation.

of the k–ε model against turbulence measurements in a boundary layer (logarithmic velocity profile) and behind a grid (homogeneous turbulence, and its decay). The values for $c_{3\varepsilon}$ were recommended by Uittenbogaard (1994).

The turbulent mixing length (Sec. 3.2.2) in terms of k_T and ε_T reads:

$$\ell = c_\mu^{3/4} \frac{k_T^{3/2}}{\varepsilon_T} \tag{11.18}$$

Note that in this k–ε-model, possible effects of sediment particles themselves on the turbulence itself are not accounted for.

Wave model

Here, waves are described with a phase-averaged wave model (Booij *et al.*, 1999; Ris *et al.*, 1999), solving the wave action density balance $N(\sigma, \theta)$ (Whitman, 1974)

$$\frac{\partial N}{\partial t} + \frac{\partial c_{g,x} N}{\partial x} + \frac{\partial c_{g,y} N}{\partial y} + \frac{\partial c_\theta N}{\partial \theta} + \frac{\partial c_\sigma N}{\partial \sigma} = \frac{S_{tot}}{\sigma} \tag{11.19}$$

where σ is the relative radial frequency in a frame of reference moving with the mean flow velocity (Lagrangian framework), θ is the wave direction, $c_{g,x}$, $c_{g,y}$, c_θ and c_σ are propagation speeds of wave action in (x, y, θ, σ) space. The wave action density $N(\sigma, \theta)$ [Js/m^2] is defined as the energy density E_w [J/m^2] divided by the relative frequency: $N \equiv E_w(\sigma, \theta)/\sigma$. The first term in Eq. (11.19) represents the time evolution of the wave-action density, and the second and third terms represent the propagation of wave energy, including the effects of shoaling. The fourth term yields a shift in frequency in case of the presence of currents and variations in depth, while the fifth term describes current- and depth-induced refraction, respectively. The last term S_{tot} is a source and sink term representing energy generation by wind S_{wind}, nonlinear interactions between waves of various frequencies S_{nl}, energy dissipation by whitecapping S_{wc}, breaking S_{br} and friction over a rigid bottom S_{fr}. Kranenburg *et al.* (2011) also added a viscous dissipation term S_{mud} to S_{tot} to account for wave damping by/over fluid mud (Sec. 7.4).

In the absence of ambient currents, σ equals the radial wave frequency ω, and no current-induced shifting over frequencies occurs (Mei, 1989). Then Eq. (11.19) can be written as a normalized wave energy balance:

$$\frac{\partial E_w}{\partial t} + \frac{\partial c_{g,x} E_w}{\partial x} + \frac{\partial c_{g,y} E_w}{\partial y} + \frac{\partial c_\theta E_w}{\partial \theta} = S_{tot} \tag{11.20}$$

The significant wave height H_s is obtained from integration of E_w over the wave frequencies and directions:

$$H_s = 4\sqrt{\frac{\int\int E_w(\sigma,\theta)\mathrm{d}\sigma\mathrm{d}\theta}{g\rho}} \tag{11.21}$$

The influence of waves is important for:

(1) stirring up sediment from the bed (i.e. the erosion formulations),
(2) vertical mixing (see Sec. 3.2.4), and
(3) possibly streaming in case of fluid mud on the bed (Sec. 7.5.2).

Modeling the effects of wave-driven transport (streaming) on soft (fluid) mud is not included in state-of-the-art engineering models, and therefore not elaborated upon in the governing equations in this chapter (see also Sec. 7.5.2).

The boundary conditions for the various equations are discussed in the following sections.

11.2 The 3D shallow water equations

As introduced in Chapter 1, we treat transport and fate of SPM in shallow water only, in which case vertical accelerations are assumed to be small in comparison to the acceleration of gravity. The momentum equation in the vertical direction then simplifies to the hydrostatic pressure distribution:

$$\frac{\partial p}{\partial x_3} = -g\rho_b \tag{11.22}$$

Integration to the water surface ζ yields:

$$p\left(x_3\right) = p_{atm} + g\int_{x_3}^{\zeta} \rho_b \mathrm{d}x'_3 \tag{11.23}$$

where p_{atm} is the atmospheric pressure at the water surface $x_3 = \zeta$ — see also the introduction of barotropic and baroclinic effects in the beginning of Chapter 9. The continuity and momentum equations (11.1) and (11.2) in x, y, z coordinates (i.e. x_1, x_2, x_3) then read:

$$\frac{\partial u}{\partial x} + \frac{\partial v}{\partial y} + \frac{\partial w}{\partial z} = 0 \tag{11.24}$$

$$\frac{\partial u}{\partial t}+u\frac{\partial u}{\partial x}+v\frac{\partial u}{\partial y}+w\frac{\partial u}{\partial z}-fv-g\frac{\partial \zeta}{\partial x}-\frac{g}{\rho_b}\int_z \frac{\partial \rho}{\partial x}\mathrm{d}z'$$
$$=\frac{\partial}{\partial x}\left(2\nu_H\frac{\partial u}{\partial x}\right)+\frac{\partial}{\partial y}\left[\nu_H\left(\frac{\partial u}{\partial y}+\frac{\partial v}{\partial x}\right)\right]$$
$$+\frac{\partial}{\partial z}\left[(\mu+\rho_b\nu_T)\frac{\partial u}{\partial z}\right] \tag{11.25}$$

$$\frac{\partial v}{\partial t}+u\frac{\partial v}{\partial x}+v\frac{\partial v}{\partial y}+w\frac{\partial v}{\partial z}+fu-g\frac{\partial \zeta}{\partial y}-\frac{g}{\rho_b}\int_z \frac{\partial \rho}{\partial y}\mathrm{d}z'$$
$$=\frac{\partial}{\partial x}\left[\nu_H\left(\frac{\partial v}{\partial x}+\frac{\partial u}{\partial y}\right)\right]+\frac{\partial}{\partial y}\left(2\nu_H\frac{\partial v}{\partial y}\right)$$
$$+\frac{\partial}{\partial z}\left[(\mu+\rho_b\nu_T)\frac{\partial v}{\partial z}\right] \tag{11.26}$$

$$p(z)=p_{atm}+g\int_z^{\zeta}\rho_b\mathrm{d}z' \tag{11.27}$$

Here, the diffusion coefficient ν_H is introduced, modeling horizontal turbulence. In deriving these equations, the following approximations are applied:

(1) the τ_{xx}-term is much smaller than the other stress terms in x-direction;
(2) $\frac{\partial w}{\partial x} \ll \frac{\partial u}{\partial z}$ and $\frac{\partial w}{\partial y} \ll \frac{\partial v}{\partial z}$, which follow from the shallow water approximations.

The ν_H-term represents horizontal viscosity effects by large horizontal eddies (e.g. Smagorinsky, 1963). Apart from these effects, isotropic turbulence is assumed, and ν_T follows from Eq. (11.13) and the k–ε turbulence closure model. Possible non-Newtonian effects (Chapter 7) are accounted for in the molecular (apparent) viscosity term μ (replace μ by μ_a; see also Sec. 11.4).

The momentum equations in x- and y-direction are written in so-called non-conservative form, obtained by substitution of the continuity equation into the momentum equations. In the conservative form, the advection terms would read $\partial uu/\partial x$, etc. The same holds for the balance equations for salinity and SPM below.

Boundary conditions for hydrodynamics

Solving these equations requires boundary conditions. In general, we assume no-slip conditions at the rigid bed $z = Z_b$, i.e. $u|_{z=Z_b} = v|_{z=Z_b} = 0$. Further, $h = \zeta - Z_b$, where h is water depth. Flow velocities perpendicular to the free water surface and the rigid bed are also assumed to be zero, which gives the kinematic boundary conditions:

$$\begin{aligned} u\frac{\partial Z_b}{\partial x} + v\frac{\partial Z_b}{\partial y} + w = 0 \\ \frac{\partial \zeta}{\partial t} + u\frac{\partial \zeta}{\partial x} + v\frac{\partial \zeta}{\partial y} - w = 0 \end{aligned} \tag{11.28}$$

where derivatives in bed level Z_b reflect local bed slopes. The stresses at the bed yield the dynamic boundary conditions, and follow from:

$$\left[(\mu + \rho_b \nu_T)\frac{\partial u}{\partial z}\right]_{z=Z_b} = \tau_{b,x} \quad \text{and} \quad \left[(\mu + \rho_b \nu_T)\frac{\partial v}{\partial z}\right]_{z=Z_b} = \tau_{b,y} \tag{11.29}$$

where $\tau_{b,x}$ and $\tau_{b,y}$ are the flow-induced bed shear stresses in x, y direction, as follows from Sec. 3.3. Possible effects of wind-induced shear stresses at the water surface are similar to Eq. (11.29).

At the open boundaries of a hydrodynamic model, (time varying) flow velocities and water levels are prescribed. With respect to the flow velocities, in general a logarithmic profile is assumed over the vertical.

Equations and boundary conditions for salinity

The transport equations for dissolved and particulate matter follow from their 3D original, again introducing horizontal turbulent diffusion:

$$\begin{aligned} \frac{\partial S}{\partial t} + u\frac{\partial S}{\partial x} + v\frac{\partial S}{\partial y} + w\frac{\partial S}{\partial z} - \frac{\partial}{\partial x}\left((\varepsilon_m + \Gamma_H + \Gamma_T)\frac{\partial S}{\partial x}\right) \\ - \frac{\partial}{\partial y}\left((\varepsilon_m + \Gamma_H + \Gamma_T)\frac{\partial S}{\partial y}\right) - \frac{\partial}{\partial z}\left((\varepsilon_m + \Gamma_T)\frac{\partial S}{\partial z}\right) = 0 \end{aligned} \tag{11.30}$$

Here, the diffusion coefficient Γ_H is introduced, modeling horizontal turbulent mixing — see also Sec. 12.1.

At the rigid bed, bed-normal salinity transport is assumed to be zero, i.e. $[w - (\varepsilon_m + \Gamma_T)\,\partial S/\partial z]_{z=Z_b} = 0$. A similar condition is prescribed at the water surface.

At the open boundaries, often the so-called Thatcher and Harleman (1972) formulation is used, as sketched in Fig. 11.2. During most of the flood period, sea water with salinity S_{sea} enters a model domain. After HWS, salinity at the open boundary is determined by the transport processes within the model domain, reducing salinities in case of fresh water input into that domain. After LWS, salinity should return to its prescribed value S_{sea} (i.e. $\partial S/\partial n = 0$, with n = coordinate perpendicular to open boundary). To account for mixing processes outside the model domain,

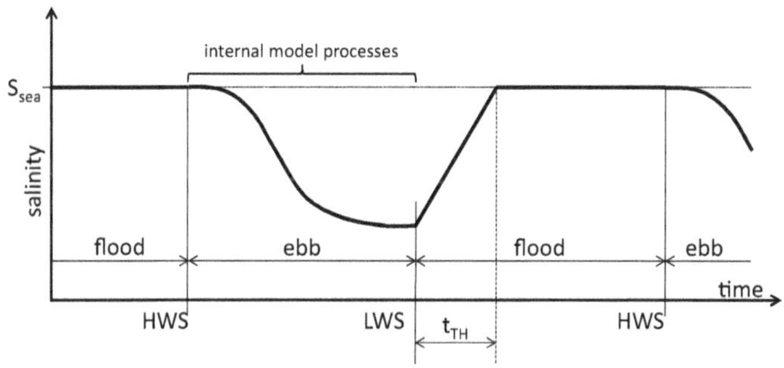

Fig. 11.2. Definition of the Thatcher–Harleman time scale t_{TH}.

and to prevent shocks in a numerical model, Thatcher and Harleman (1972) proposed a linear return of the salinity to S_{sea} in a finite time interval t_{TH}, referred to as the Thatcher–Harleman return time. Typical values for t_{TH} in model applications vary around 1 hr.

Over the vertical, a homogeneous salinity is generally prescribed.

Equations and boundary conditions for SPM

The 3D shallow water equation for SPM reads:

$$\begin{aligned}&\frac{\partial c^{(i)}}{\partial t}+u\frac{\partial c^{(i)}}{\partial x}+v\frac{\partial c^{(i)}}{\partial y}+w\frac{\partial c^{(i)}}{\partial z}-\frac{\partial w_s^{(i)}c^{(i)}}{\partial z}\\&\quad-\frac{\partial}{\partial x}\left((D_s+\Gamma_H+\Gamma_T)\frac{\partial c^{(i)}}{\partial x}\right)-\frac{\partial}{\partial y}\left((D_s+\Gamma_H+\Gamma_T)\frac{\partial c^{(i)}}{\partial y}\right)\\&\quad-\frac{\partial}{\partial z}\left((D_s+\Gamma_T)\frac{\partial c^{(i)}}{\partial z}\right)=0\end{aligned} \tag{11.31}$$

with the same horizontal diffusion coefficient Γ_H as above.

The boundary conditions at the bed are governed by erosion and deposition processes described in the Chapters 4 and 6. This gives

$$\left[\left(w-w_s^{(i)}\right)c^{(i)}-(D_s+\Gamma_T)\frac{\partial c^{(i)}}{\partial z}=\xi^{(i)}E_{m,s}-D^{(i)}\right]_{z=Z_b} \tag{11.32}$$

where $E_{m,s}$ is the erosion flux for mud or sand, and $\xi^{(i)}$ is the sediment content (mud and sand) of fraction (i) in the bed, as described in the Secs. 6.3, 6.4 and 6.6. Currently, it is not possible to account for the erosion fluxes of individual cohesive sediment fractions — the bed is eroded in bulk, its composition depending on sediment composition within the bed (bookkeeping). Thus distinction is made only in the erosion fluxes of mud

(E_m) and sand (E_s), taking into account whether the bed is cohesive or non-cohesive (Sec. 6.6). The deposition flux $D^{(i)}$ is resolved for each sediment fraction in the water column separately (Sec. 4.2), thus affecting the sediment composition within the bed.

At the open boundaries, a Thatcher–Harleman concept can also be applied, while vertically, SPM is generally set to be constant.

Boundary conditions for turbulence

The transport equations for the k–ε model for shallow water conditions are similar to Eqs. (11.14) and (11.15), apart from a horizontal diffusion term as in Eq. (11.30) and (11.31), and are not repeated.

The boundary conditions for the k–ε model at a rigid bed ($z = Z_b$) read:

$$k_T\big|_{z=Z_b} = \frac{u_*^2}{\sqrt{c_\mu}} \quad \text{and} \quad \varepsilon_T\big|_{z=Z_b} = \frac{u_*^3}{\kappa z_{wc}} \tag{11.33}$$

where u_* is the shear velocity (accounting for both waves and currents, e.g. Sec. 3.2.4) and z_{wc} is the roughness height for waves and currents (Sec. 3.2.4). At the water surface, k_T and ε_T are zero, unless there is turbulence input at the surface by for instance wind forcing:

$$k_T\big|_{z=\zeta} = \frac{u_{*,s}^2}{\sqrt{c_\mu}}, \quad \varepsilon_T\big|_{z=\zeta} = \frac{u_{*,s}^3}{\kappa z_{wc,s}} \tag{11.34}$$

where subscript $\cdot_s$ refers to surface (wind) conditions.

At the open model boundaries, generally no values for k_T and ε_T are prescribed, but these are computed from assuming equilibrium between production and dissipation for given velocity.

Wave model

The phase-averaged wave model Eq. (11.19) computes wave energy (height) and frequency. Orbital velocities and excursion are computed with linear wave theory to assess bed shear stresses relevant for the 3D SPM transport and k–ε equations above — see also Sec. 3.3.2.

Bed boundary conditions are implicit in this model, through the subterms of S_{tot}. At the open boundaries, a (measured) wave spectrum is generally prescribed.

11.3 The 2DH shallow water equations

In spite of impressive developments in three-dimensional modeling, depth-averaged (2DH) models are still being used in engineering studies. They are often used also in a first phase of setting up and calibration of 3D models (but see its pitfalls in Sec. 12.1).

In this section, we present the depth-averaged shallow water equations, obtained from averaging the 3D shallow water equations (Sec. 11.2) over the water depth, i.e. from $z = Z_b$ to $z = \zeta$. We do not present the integration steps themselves (see for instance Vreugdenhil, 1994).

The depth-averaged continuity equations reads:

$$\frac{\partial h}{\partial t} + \frac{\partial h\bar{u}}{\partial x} + \frac{\partial h\bar{v}}{\partial y} = \frac{\partial \zeta}{\partial t} + \frac{\partial h\bar{u}}{\partial x} + \frac{\partial h\bar{v}}{\partial y} = 0 \tag{11.35}$$

where $\bar{u}$ and $\bar{v}$ represent the depth-averaged flow velocities in x- and y-directions:

$$\bar{u} \equiv \frac{1}{h}\int_{Z_b}^{\zeta} u\mathrm{d}z, \text{ etc.} \tag{11.36}$$

The depth-averaged momentum equations in x- and y-directions read:

$$\frac{\partial \bar{u}}{\partial t} + \bar{u}\frac{\partial \bar{u}}{\partial x} + \bar{v}\frac{\partial \bar{u}}{\partial y} - \frac{\partial}{\partial x}\left[\nu_H\left(\frac{\partial \bar{v}}{\partial x} + \frac{\partial \bar{u}}{\partial y}\right)\right]$$

$$-f\bar{v} + g\frac{\partial \zeta}{\partial x} + \frac{gh}{\bar{\rho}_b}\frac{\partial \bar{\rho}_b}{\partial x} + \frac{\tau_{b,x}}{\rho_b h} - \frac{\tau_{s,x}}{\rho_b h} = 0 \tag{11.37}$$

$$\frac{\partial \bar{v}}{\partial t} + \bar{u}\frac{\partial \bar{v}}{\partial x} + \bar{v}\frac{\partial \bar{v}}{\partial y} - \frac{\partial}{\partial y}\left[\nu_H\left(\frac{\partial \bar{v}}{\partial x} + \frac{\partial \bar{u}}{\partial y}\right)\right]$$

$$+f\bar{u} + g\frac{\partial \zeta}{\partial y} + \frac{gh}{\bar{\rho}_b}\frac{\partial \bar{\rho}_b}{\partial y} + \frac{\tau_{b,y}}{\rho_b h} - \frac{\tau_{s,y}}{\rho_b h} = 0 \tag{11.38}$$

with ν_H and τ_b defined in Sec. 11.2, and $\bar{\rho}_b$ the depth-averaged (bulk) water density, which may be a function of temperature, salinity and/or SPM (Eq. (11.10)). The stresses at the bed τ_b and surface τ_s follow from quadratic frictions laws, as before.

The depth-averaged equation for the transport of salinity reads:

$$\frac{\partial \bar{S}}{\partial t} + \bar{u}\frac{\partial \bar{S}}{\partial x} + \bar{v}\frac{\partial \bar{S}}{\partial y} - \frac{\partial}{\partial x}\left(D_{S,x}\frac{\partial \bar{S}}{\partial x}\right) - \frac{\partial}{\partial y}\left(D_{S,y}\frac{\partial \bar{S}}{\partial y}\right) = 0 \tag{11.39}$$

with $\bar{S}$ the depth-averaged salinity. The dispersion terms $D_{S,x}$ and $D_{S,y}$ represent "dispersion", a kind of diffusion induced by spatial inhomogeneities (see Sec. 12.1) and self-induced gravitational circulation. Values for $D_{S,x}$ and $D_{S,y}$ are discussed in Sec. 12.1. In general, $D_{S,x} \gg D_x$ and $D_{S,y} \gg D_y$.

The Thatcher–Harleman boundary condition (Fig. 11.2) is also used in the depth-averaged mode.

The depth-averaged equation for the transport of SPM reads:

$$\frac{\partial \bar{c}^{(i)}}{\partial t} + \bar{u}\frac{\partial \bar{c}^{(i)}}{\partial x} + \bar{v}\frac{\partial \bar{c}^{(i)}}{\partial y} - \frac{\partial}{\partial x}\left(D_x \frac{\partial \bar{c}^{(i)}}{\partial x}\right) - \frac{\partial}{\partial y}\left(D_y \frac{\partial \bar{c}^{(i)}}{\partial y}\right)$$
$$= \xi^{(i)} E_{m,s} - D^{(i)} \tag{11.40}$$

with $\bar{c}$ the depth-averaged SPM concentrations. Obviously, the bed boundary conditions are now part of the transport equation. The dispersion coefficients D_x and D_y in the 2DH SPM transport equations are generally much larger than in a 3D model, as discussed in Sec. 12.1.

A vertical turbulence model is not necessary, but horizontal diffusion may still play a role. The wave model described in Sec. 11.1 is already in 2DH mode.

11.4 The continuous model concept

In the sections above, water-bed exchange of SPM is described explicitly as a bed-boundary condition to the hydro-sediment models. These conditions may be affected by the exchange rates themselves, which are accounted for with a (multi-layer) bed model. This is basically the state-of-the-art engineering approach, and straightforward in case of very rigid and/or sandy beds. However, in case of soft beds, as discussed in Chapter 7, this approach becomes problematic, because the transition between bed and water column becomes gradual and is generally poorly defined.

In that case, a so-called continuous approach becomes more feasible. It implies that all sediment mass is confined within the computational domain. Erosion and deposition thus become implicit processes, resolved within the hydro-sediment model itself — these exchange processes then are reflected by a kind of "internal mixing". The movement of the water-sediment interface in the computational domain, where SPM concentrations become high, requires the use of rheological models, as introduced in Sec. 7.3. Also, consolidation and subsequent strength development has to be accounted for, together with the role of sediment on the turbulent flow field (at least the effects of sediment-induced buoyancy destruction). This internal mixing can be further affected by processes such as over-consolidation, soil deformation by shear, etc., which, however, are currently not accounted for. Continuous models have been developed by for instance Le Hir and Cayocca

(2002); Toorman (2002); Winterwerp (2001); and Hsu *et al.* (2007) and his later work. These models should be considered as research tools, because of the uncertainty in underlying physics (see the various previous sections in this book) and the required computational times. The continuous approach refers in particular to processes in the vertical — the governing equations do not have to be modified in the horizontal direction.

Momentum equation

In the following we will follow Winterwerp (2001) for modeling the processes over the vertical, recognizing that formulations in other models differ in details only. We still assume that a one-phase approach is applicable. The momentum equations for the water-sediment mixture under shallow water conditions read:

$$\frac{\partial u}{\partial t}+u\frac{\partial u}{\partial x}+v\frac{\partial u}{\partial y}-fv+\frac{1}{\rho_b}\frac{\partial p}{\partial x}=T_{xy}+\frac{\partial}{\partial z}\left[\left(\frac{\mu}{\rho_b}+\nu_T\right)\frac{\partial u}{\partial z}+\frac{\tau_{xz}}{\rho_b}\right] \tag{11.41}$$

and

$$\frac{\partial v}{\partial t}+u\frac{\partial v}{\partial x}+v\frac{\partial v}{\partial y}+fu+\frac{1}{\rho_b}\frac{\partial p}{\partial y}=T_{yx}+\frac{\partial}{\partial z}\left[\left(\frac{\mu}{\rho_b}+\nu_T\right)\frac{\partial v}{\partial z}+\frac{\tau_{yz}}{\rho_b}\right] \tag{11.42}$$

where ρ_b is the bulk density of the water-sediment mixture (see below), $\partial p/\partial x$ and $\partial p/\partial y$ are the horizontal pressure gradients, μ is the molecular (dynamic) viscosity of water and ν_T is the eddy viscosity. Diffusion processes in the horizontal plane are represented by T_{xy} (see Eq. (11.25)). The shear stresses τ_{xz} and τ_{yz} include effects of high sediment concentrations on the (non-Newtonian) stress-strain relations of the mixture. For instance τ_{xz} becomes:

$$\tau_{xz}=\mu_a\left(\frac{\partial u}{\partial z}+\frac{\partial w}{\partial z}\right)\approx\mu_a\frac{\partial u}{\partial z} \tag{11.43}$$

in which we assume $\partial w/\partial x \ll \partial u/\partial z$, following from an analysis of the relevant scales (shallow water approximation). Further, the apparent viscosity μ_a is used (see Sec. 7.3). This is also the case for Bingham and similar models with a yield strength:

$$\tau_{xz}=\tau_B+\mu\frac{\partial u}{\partial z}=\mu_a\frac{\partial u}{\partial z}\quad\text{thus}\quad\mu_a=\frac{\tau_y+\mu\partial u/\partial z}{\partial u/\partial z} \tag{11.44}$$

Note that shear rates are not defined in the Bingham model for $\tau < \tau_B$. Therefore, Papanastasiou and Alexandrou (1987) proposed a modification to the Bingham model:

$$\tau = (1 - \exp\{-m\partial u/\partial z\})\,\tau_B + \mu\partial u/\partial z \tag{11.45}$$

where m is a user-defined empirical parameter quantifying shear rates at very small shear stresses (thus yielding a very high apparent viscosity).

Transport equation for SPM

The physical processes in soft muddy beds and in the water column close to the bed largely depend on the volume and properties of mud flocs. The transport equation is therefore written as a function of the volumetric concentration of fine sediment $\phi_s(= c/\rho_s)$ reads (see also Sec. 5.1):

$$\begin{aligned}&\frac{\partial \phi_s}{\partial t} + u\frac{\partial \phi_s}{\partial x} + v\frac{\partial \phi_s}{\partial y} + w\frac{\partial \phi_s}{\partial z} - \frac{\partial}{\partial z}\left(\Xi_s \phi_s\right)\\ &\quad - T_x - T_y - \frac{\partial}{\partial z}\left[\left(D_s + \Gamma_T + \Gamma_c\right)\frac{\partial \phi_s}{\partial z}\right] = 0\end{aligned} \tag{11.46}$$

in which the molecular diffusion D_s, eddy diffusivity Γ_T and the consolidation parameter Γ_c have been linearly added, with the understanding that Γ_T will tend to zero under high-concentration conditions, and T_x and T_y represent diffusion/dispersion in the horizontal plane (Eq. (11.31)). This equation may be expanded for multiple fractions (Eq. (11.7)), generally assuming that the diffusion coefficients are identical for all fractions.

The advection term Ξ_s consists of a hindered settling term $f_{h,s}$ and a hydraulic conductivity term f_c (the latter representing the first phase of consolidation — see also Sec. 5.1):

$$\begin{aligned}&\Xi_s = f_{hs} + \frac{f_c}{1 + \eta f_c}, \quad \text{with}\\ &f_{hs} = w_{s,0}\frac{\left(1-\phi_f\right)^2\left(1-\phi_s\right)}{1 + 2.5\phi_f}, \quad \text{and}\\ &f_c = \frac{\rho_s - \rho_w}{\rho_w}k\phi_s\end{aligned} \tag{11.47}$$

in which η is an empirical parameter and where we have implicitly assumed self-similarity (Sec. 5.1):

$$k = K_k(\phi_s)^{-\frac{2}{3-n_f}}, \quad \sigma_{zz}^{sk} = K_p(\phi_s)^{\frac{2}{3-n_f}} \quad \text{and} \quad \Gamma_c = \frac{2}{3-n_f}\frac{K_k K_p}{g\rho_w} \tag{11.48}$$

This continuous approach is completed with the equation of state (Eq. (11.10)), and a turbulence model, such as the k–ε model (Eq. (11.14)) and (11.15) and its low-Reynolds variant (Eq. (7.33), Sec. 7.5.1). Soil mechanical processes, other than consolidation, are not accounted for in this approach, such as over-consolidation, crust formation, soil deformation, etc.

Boundary conditions

The kinematic boundary conditions for the momentum equations of the continuous model are identical to those for the 3D shallow water equations (Eq. (11.29)). The authors are not aware of generalized dynamic boundary conditions, i.e. valid in both a turbulent and non-Newtonian flow regime. Therefore, we propose a simple superposition of these two regimes (as in Eq. (11.3)):

$$\left[\left(\frac{\mu}{\rho_b} + \nu_T + \frac{\mu_a}{\rho_b}\right)\frac{\partial u}{\partial z}\right]_{z=Z_b} = \tau_{b,x,turb} + \tau_{b,x,lam} \tag{11.49}$$

with the understanding that ν_T will be zero in high-density non-Newtonian conditions, whereas μ_a will approach the kinematic viscosity in Newtonian, turbulent flows. The turbulent bed shear stress $\tau_{b,x,turb}$ follows from Sec. 3.3 as in the 3D shallow water model.

The laminar part, $\tau_{b,x,lam}$, follows from Skelland (1967) who introduced a generalized (apparent) Reynolds number Re_a:

$$Re_a \equiv \frac{\rho_b U \ell}{\mu_a} \tag{11.50}$$

where U is a characteristic (local) velocity, ℓ a characteristic length scale, ρ_b is the bulk density of the sediment-laden fluid, and μ_a the apparent viscosity. Note that these formulations have been derived for pipe flow, whence ℓ refers to the pipe diameter. For open water conditions, the thickness of the non-Newtonian layer seems to be an appropriate length scale. As this thickness depends on the flow properties, ℓ becomes an implicit parameter in the continuous model.

The laminar bed shear stress becomes:

$$\tau_{b,x,lam} = \frac{1}{8} f \rho_b u^2 \quad \text{with} \quad f = \frac{R_a}{16} \tag{11.51}$$

where f is the Darcy–Weisbach friction factor (see Sec. 3.3.1). For Newtonian flows, $\mu_a = \mu$, and Eq. (11.51) represents common laminar friction.

The dynamic boundary conditions in y-direction are similar to the above.

Further, one assumes no normal transport of momentum (i.e. $w = 0$) and matter through the rigid bed:

$$\left[(w - w_s)c + (D_s + \Gamma_T + \Gamma_c)\frac{\partial c}{\partial z} = 0\right]_{z=Z_b} \tag{11.52}$$

Also, boundary conditions for SPM at the rigid bed now yield zero transport through that bed; in other words, water-bed exchange processes are switched off.

The boundary conditions for the (low-Reynolds number) k–ε turbulence model are identical to those of the standard model, given by Eq. (11.33), in which u_* refers to the turbulent part of the bed shear stress only.

Chapter 12

Model Schematization

12.1 Sub-grid effects

In every system description, aggregated parameters are used. For instance, whence referring to temperature or turbidity of an aquatic system, this implies a characteristic value, realizing that these parameters may vary over time and space. In numerical models, aggregation is formalized by the computational grid and the time step applied. The finer the grid and the smaller the time step, the higher the resolution and the smaller the degree of aggregation — spatial (and temporal) variations are explicitly described. Of course grid resolution and time step are closely interlinked, both from the physical and the numerical points of view (see also Sec. 13.3.2).

3D vs. 2DH vertical SPM fluxes

We start our discussion on aggregation by comparing SPM fluxes computed in a depth-averaged (2DH) and three-dimensional (3D) framework (see Secs. 11.2 and 11.3). As an example, we consider the settling of fine sediment in still water (Fig. 12.1, as in a settling column). In case of a mono-disperse suspension, settling implies a lowering of the interface, below which the SPM-concentration remains constant over time, equal to its initial value c_0. The deposition flux D thus reads (Secs. 4.2 and 4.4):

$$D_{3D} \equiv w_s c_{b,3D} = w_s c_0 \tag{12.1}$$

where w_s is the constant settling velocity and $c_{b,3D}$ the near-bed SPM-concentration in a three-dimensional representation, which, in this example, remains identical to the initial concentration c_0. However, in a depth-averaged framework, the near-bed SPM-concentration equals its depth-averaged value, no interface is formed and the (near-bed) SPM-concentration $c_{b,2Dh}$ decreases exponentially over time:

$$c_{b,2Dh} \equiv \bar{c}(t) = c_0 \exp\left\{-\frac{w_s t}{h}\right\} = c_{b,3D} \exp\left\{-\frac{w_s t}{h}\right\} \tag{12.2}$$

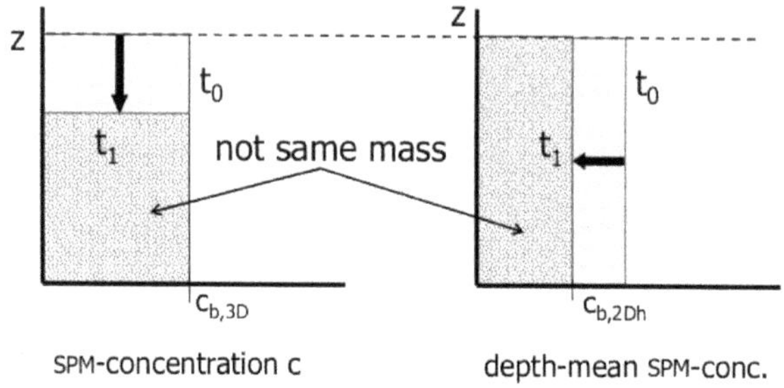

Fig. 12.1. Diagram showing the settling of SPM in still water (left panel) and its depth-averaged representation (right panel).

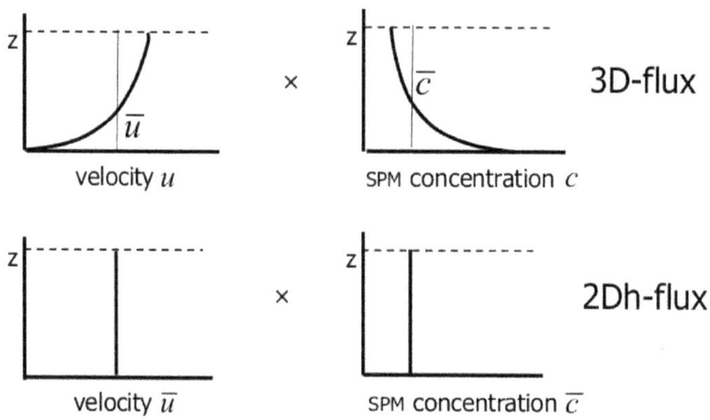

Fig. 12.2. Diagram depicting the flux of fine sediment in a fully three-dimensional and depth-averaged representation.

in which h is water depth. Thus the sedimentation flux in a depth-averaged framework reduces over time, contrary to the 3D case, and becomes:

$$D_{2Dh} = w_s c_{b,2Dh} = D_{3D} \exp\left\{-\frac{w_s t}{h}\right\} \tag{12.3}$$

In a three-dimensional framework, deposition is complete after a time $T_D\,(= h/w_s)$, whereas in a two-dimensional framework, only 36% of the original suspension has settled within that time.

3D vs. 2DH horizontal SPM fluxes

The horizontal flux in a depth-averaged framework also differs from its three-dimensional representation. This is illustrated in Fig. 12.2, where the flux is defined as the product of the velocity u and SPM-concentration c (integrated over the water depth h). To quantify these 2DH- and 3D-fluxes, we apply a decomposition procedure similar to the one used in analyzing turbulent transport (i.e. Eq. (3.1) and (3.54)):

$$u(z,t) = \bar{u}(t) + u'_v(z,t) \quad \text{with} \quad \int_h u'_v \, \mathrm{d}z = 0 \tag{12.4}$$

where $\bar{u}$ is the depth-averaged velocity, which may vary over time, and u'_v is the vertical deviation from that mean. Applying a similar decomposition to the SPM-concentration c, the SPM-flux becomes:

$$\overline{uc} = \bar{u}\bar{c} + \overline{u'_v c'_v} \tag{12.5}$$

Thus the three-dimensional and depth-averaged flux differ by a term $\overline{u'_v c'_v}$.

Though these two examples are rather trivial, they show clearly that resolution and aggregation are important aspects in setting up a numerical model and in the interpretation of the model results. This is further elaborated in Chapter 13.

Longitudinal dispersion by vertical shearing

Let us elaborate further on the previous example, as this is important for numerical modeling. The term $\overline{u'_v c'_v}$ is known as dispersion, or dispersive flux, and represents a diffusion-like spreading of matter with respect to the mean advective velocity. It can been shown (see below) that this term can be modeled as a diffusion term, relating dispersion to the gradient of mean SPM-concentration. Thus, Eq. (12.5) becomes:

$$\overline{uc} = \bar{u}\bar{c} - D_x \frac{\mathrm{d}\bar{c}}{\mathrm{d}z} \tag{12.6}$$

in which D_x is the so-called dispersion coefficient, with the same dimensions as a diffusion coefficient (i.e. [m^2/s]). The negative sign in Eq. (12.6) derives from the fact that $\overline{u'_v c'_v}$ is generally negative.

Dispersion was studied extensively in the 1950s – 1970s (e.g. Elder (1959); Taylor (1954) and many others), parameterizing spatial effects in the low-resolution analytical/numerical models operational at that time. Currently, with ever-increasing computational power, dispersive effects are often explicitly resolved in numerical models, but not always, in particular in engineering studies. Today, we refer to sub-grid effects, i.e. effects that cannot explicitly be resolved within the computational grid. Many such sub-grid effects can be accounted for through a diffusion concept, but not all (see also Sec. 8.3). Therefore, an understanding of such sub-grid effects is necessary for good modeling practice. This can be obtained from e.g.

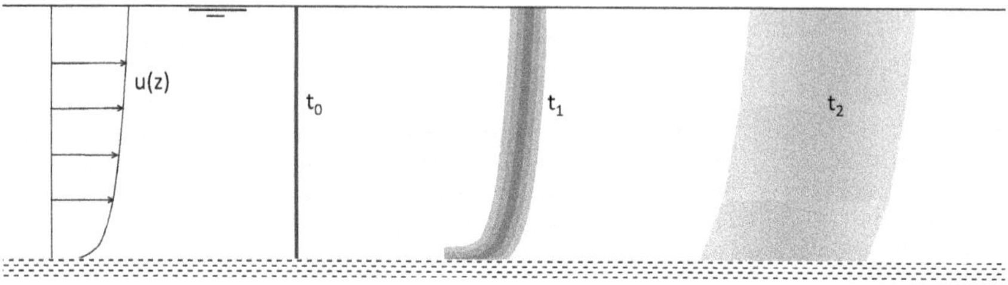

Fig. 12.3. Longitudinal dispersion of dissolved matter in open channel shear flow at two times after release at t_0 — stretching and vertical mixing lead to dispersion.

elaborating on the net transport of matter in open channel shear flow, a configuration resolved by Elder in his famous 1959 paper. Fig. 12.3 sketches this configuration, explaining how advective stretching and vertical mixing induce diffusion around an initially homogeneous (passive) distribution of matter. The relevant "grid" size is the water depth, and the dispersion coefficient derived below has been used extensively in one-dimensional and depth-averaged numerical modeling.

In his derivation, Elder started from the one-dimensional (horizontal) momentum and mass balance equation for uniform, stationary flow with a passive substance (hence not sediment with its gravitational settling):

$$\frac{\partial u}{\partial x} = 0 \tag{12.7}$$

$$\frac{\partial c}{\partial t} + u\frac{\partial c}{\partial x} - \frac{\partial}{\partial x}\varepsilon_x\frac{\partial c}{\partial x} - \frac{\partial}{\partial z}\varepsilon_z\frac{\partial c}{\partial z} = 0 \tag{12.8}$$

in which ε_x and ε_z are the horizontal and vertical eddy diffusivities, respectively. Similar to the derivations for the turbulent contributions, the velocity u and concentration c are decomposed to depth-mean values $\bar{u}$ and $\bar{c}$, and their variation u' and c' over depth h, e.g. Eq. (12.4), in which we omitted subscript $\cdot_v$. The continuity equation then becomes:

$$\frac{\partial u}{\partial x} = \frac{\partial \bar{u}}{\partial x} = \frac{\partial u'}{\partial x} = 0 \tag{12.9}$$

Substitution into the mass balance equation (12.8) and integration over the water depth, yields the depth-averaged mass balance for dissolved matter:

$$\frac{\partial \bar{c}}{\partial t} + \bar{u}\frac{\partial \bar{c}}{\partial x} + \frac{\partial \overline{u'c'}}{\partial x} - \frac{\partial}{\partial x}\varepsilon_x\frac{\partial \bar{c}}{\partial x} = 0 \tag{12.10}$$

Using the Boussinesq approach again, introducing Fick's law (i.e. assuming diffusive, gradient-type transport), we obtain:

$$\frac{\partial \bar{c}}{\partial t} + \bar{u}\frac{\partial \bar{c}}{\partial x} - \frac{\partial}{\partial x}\left(D_x + \varepsilon_x\right)\frac{\partial \bar{c}}{\partial x} = 0 \tag{12.11}$$

in which D_x is the longitudinal dispersion coefficient, defined as $\overline{u'c'} = -D_x \partial\bar{c}/\partial x$, as before.

It is illustrative to quantify D_x, as elaborated by Elder, assuming a logarithmic velocity profile (and thus a parabolic eddy viscosity profile). First, the depth-mean mass balance equation (12.10) is subtracted from the full mass balance (i.e. with $\bar{c}$ and c' terms), which gives the balance equation for c':

$$\frac{\partial c'}{\partial t} + (u' - \bar{u})\frac{\partial \bar{c}}{\partial x} - \frac{\partial}{\partial z}\varepsilon_z \frac{\partial c'}{\partial z} = 0 \tag{12.12}$$

Next, a Lagrangian coordinate system is introduced ($\tau = t$ and $\xi = x - \bar{u}t$), and assuming that longitudinal eddy diffusion is small compared to advection — this implies that dispersion is analyzed with respect to advection with the mean flow. Substitution into Eq. (12.12) gives:

$$\frac{\partial c'}{\partial \tau} + u'\frac{\partial \bar{c}}{\partial \xi} - \frac{\partial}{\partial z}\varepsilon_z \frac{\partial c'}{\partial z} = 0 \tag{12.13}$$

We assume steady conditions (within the Lagrangian framework), i.e. the vertical concentration distribution does not vary in longitudinal direction, nor in time. Then c' can be determined by integrating Eq. (12.13) over the water depth:

$$c' = \int_z \frac{1}{\varepsilon_z}\left(\int_z u'\frac{\partial \bar{c}}{\partial \xi}\mathrm{d}z\right)\mathrm{d}z, \quad \text{hence:} \tag{12.14a}$$

$$\overline{u'c'} = \frac{1}{h}\int_h u'\left[\int_z \frac{1}{\varepsilon_z}\left(\int_z u'\frac{\partial \bar{c}}{\partial \xi}\mathrm{d}z\right)\mathrm{d}z\right]\mathrm{d}z \tag{12.14b}$$

Thus the Fickian assumption follows directly from Elder's derivation. From the logarithmic velocity profile (3.20) u' is found:

$$\frac{u}{u_*} = \frac{1}{\kappa}\ln\frac{z}{z_o} \quad \text{thus } u'(z) = \frac{u_*}{\kappa}\ln\left(\frac{z}{h} + 1\right) \tag{12.15}$$

Substitution of the parabolic eddy viscosity (3.22) into Eq. (12.14) and integration over depth gives:

$$\overline{u'c'} = \frac{0.404\sigma_T}{\kappa^3} h u_* \frac{\partial \bar{c}}{\partial \xi} \tag{12.16}$$

With the Von Kármàn constant $\kappa = 0.41$ and the turbulent Prandtl-Schmidt number $\sigma_T = 0.7$, we find

$$D_x = 5.86\sigma_T h u_* = 4.1 h u_* \tag{12.17}$$

which differs by a factor 0.7 ($= \sigma_T$) from Elder's original result. For typical values $h = 5\,\mathrm{m}$ and $u_* = 0.05\,\mathrm{m/s}$ in open channel flow, D_x would amount to about $1\,\mathrm{m^2/s}$, thus three orders of magnitude larger than the horizontal eddy diffusivity.

Longitudinal dispersion by horizontal shearing

In the horizontal plane, cross-shore gradients in bathymetry, bank irregularities, such as groins, harbor basins, etc., can largely contribute to the longitudinal dispersion of matter in a river. A conceptual example is sketched in Fig. 12.4. Because of (horizontal) turbulent mixing, matter enters a harbor basin and remains there for some time, before returning to the main stream. The initially confined patch of matter becomes elongated, which can be described with an augmented longitudinal dispersion coefficient. The reader is referred to literature (e.g. Fischer, 1976) for quantification of these sub-grid (in this case extra-grid) effects.

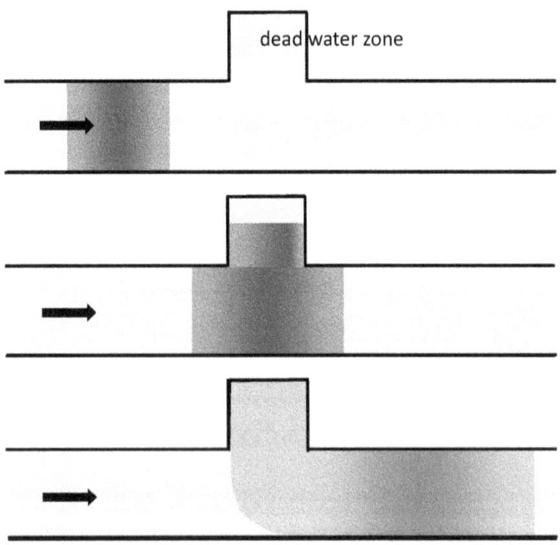

Fig. 12.4. Schematic of longitudinal dispersion of matter in a river with harbor basin.

Table 12.1. Typical longitudinal dispersion coefficients in one-dimensional horizontal (1DH), depth-averaged (2DH) and three-dimensional (3D) models, for homogeneous and fresh/saline systems.

	Homogeneous	Saline
1DH	$100 < D_x < 1000\,\mathrm{m}^2/\mathrm{s}$	$500 < D_S, x < 3000\,\mathrm{m}^2/\mathrm{s}$
2DH	$5 < D_x \& D_y < 20\,\mathrm{m}^2/\mathrm{s}$	$500 < D_S, x < 3000\,\mathrm{m}^2/\mathrm{s}$
3D	$D_x \& D_y \approx 1\,\mathrm{m}^2/\mathrm{s}$	–

A zero-order estimate of the magnitude of the dispersion coefficients for rivers or estuaries can be obtained from the mixing length concept introduced by Prandtl, e.g. Eq. (3.15): $D \propto u'\ell$. A typical velocity in open water systems measures 1 m/s, thus $u' \approx 1$ m/s. In a one-dimensional approach, i.e. averaging over the rivers cross section, the larger length scale yields the rivers width b, i.e. $\ell \approx b$, thus a few 100 m for smaller rivers in general. In a depth-averaged approach, the characteristic length scale is the water depth h, hence of the order of 10 m. These considerations explain the typical values of the coefficients for the dispersion of passive matter in open waters in engineering models as given in Table 12.1. The coefficients for 3D simulations follow from the size of the computational grid.

One has to bear in mind that higher model resolution yields smaller length scales and smaller gradients in velocity and matter concentration. Then, by explicitly resolving sub-grid effects in more and more detail, the required dispersion coefficients is reduced considerably. Further note that it is not common to account for cross terms, such as D_{xy}.

Finally, it is noted that simulation of salinity distributions in estuaries and bays with a 1DH or 2DH approach requires very large dispersion coefficients, as summarized in Table 12.1 (salinity intrusion is assumed in x-direction), see also Sec. 9.1.1 and Kuijper and Van Rijn (2011).

12.2 Schematization of bed processes

This book treats the transport and fate of SPM in shallow waters, and the exchange of fine sediment between the water column and the bed plays a major role, as discussed in Chapter 1 (see Fig. 1.2). In particular, if mixtures of sediment and/or long-term (seasonal) variations are studied, variations in composition and properties of sediment within the bed have to be accounted for. Such bookkeeping is done with multi-layer bed models, the first of which were probably published by Ariathurai (1974) and Hayter (1986). In the

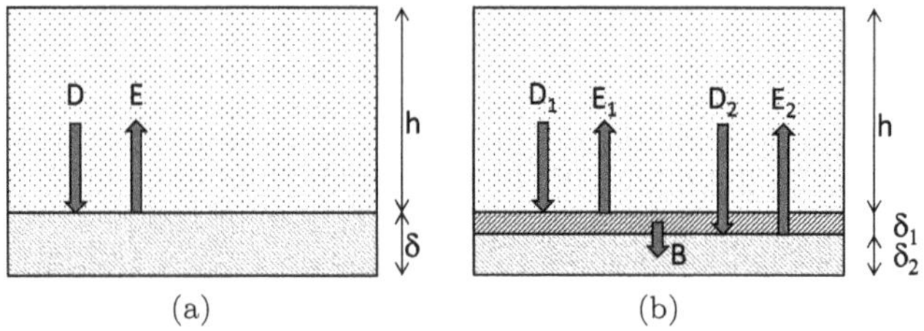

Fig. 12.5. The two-layer buffer model, with active layer $\cdot_1$ and buffer layer $\cdot_2$ (b) compared with the Partheniades–Krone approach (a).

following, we presume morphostatic conditions, i.e. the morphology remains constant — see also the last paragraph of this section.

The most simple multi-layer bed model is a two-layer approach, referred to as the buffer model (Van Kessel *et al.*, 2011a), as sketched in Fig. 12.5. In this figure, h is water depth, δ is the thickness of the bed, δ_1 is the thickness of the active layer (see below) and δ_2 is the thickness of the buffer layer. E, D and B are erosion, deposition and burial terms, respectively (see below).

The deposition and erosion fluxes D and E that were discussed in Secs. 4.2 and 6.3.2, i.e. Eqs. (4.20) and (6.15), are repeated for convenience:

$$D = p_d w_{s,b} c_b \tag{12.18}$$

$$E = M_E(\tau_b - \tau_e) \tag{12.19}$$

where p_d represents the coefficient for reduced deposition (Sec. 4.2). The SPM mass balance for Fig. 12.5(a) in a depth-averaged representation reads:

$$h\frac{\mathrm{d}\bar{c}}{\mathrm{d}t} = E - D \tag{12.20}$$

omitting horizontal advection and diffusion. This is basically the traditional Partheniades–Krone approach discussed in previous chapters. A problem with this approach is that the mass balance equation (12.20) yields equilibrium only when $E = D$. Otherwise c becomes 0 or ∞, apart from the effects of advection and diffusion. Thus bookkeeping of the amount of sediment in the bed becomes meaningless, except over smaller time scales.

Therefore, the initial (pre-equilibrium) SPM fields computed with numerical models depend largely on the initial amount of fine sediment on/in the bed specified by the modeler. This is one reason why we advocate the "empty bed" concept, as elaborated upon in Sec. 12.3. The empty-bed approach is in particular useful when modeling non-capacity conditions in large domains, which generally exhibit large memory effects.

The buffer model

The two-layer buffer model yields the simplest bookkeeping of the amount, composition and properties of sediment in a bed. This model concept was first developed to simulate bed level changes in rivers (thus morphodynamic simulations, see Hayter, 1986; Sloff *et al.*, 2006). For fine sediment, the buffer model is for instance applicable in case of:

(1) Simulating long-term transport and fate of mixed sediment, such as sand–mud mixtures (e.g. Secs. 6.6 and 8.3).
(2) Simulating transport and fate of SPM over spring-neap cycles, where daily responses to tidal velocity variations are governed by the properties of the fluff layer, and spring-neap variations are affected by the properties of the buffer layer (see Sec. 6.5).

The concept of this buffer model is sketched in Fig. 12.5(b). In morphodynamic simulations, the active layer thickness is a key parameter to steer either bed level or bed composition changes. A thin active layer suppresses bed level changes and favors bed composition changes. A thick active layer suppresses bed composition changes and favors bed level changes. Unfortunately, there is currently no experience with models which include both features.

For illustrative reasons, we elaborate on the first application, as in Van Kessel *et al.* (2011a). We are only concerned with the transport and fate of SPM — the dynamics of the sand is not of interest in this example. Hence, we presume a sand matrix in the buffer layer, containing some fine (mud fraction ξ_m), whereas the active layer only contains mud.

Fine sediment settle mainly in the active (fluffy) layer with thickness δ_1. However, a fraction α also settles in/on the buffer layer, as a result of which the fine sediment fraction in the buffer layer ξ_m may vary over time (the burial term in Fig. 12.1). Because of erosion, the fluff layer may be become depleted of fines, which is retarded by a linear/zero-order erosion approach (see below). When stresses are large enough, fines may also be eroded from the buffer layer (in fact washed out when the sandy bed is mobilized by e.g. waves). The mass balance for depth-averaged SPM in the water column then becomes, again omitting horizontal advection and diffusion:

$$h\frac{\mathrm{d}\bar{c}}{\mathrm{d}t} = E_1 + E_2 - D_1 - D_2 \tag{12.21}$$

Deposition D_1 of SPM on the active layer is given by

$$D_1 = p_d\,(1-\alpha)\,w_{s,b}c_b \tag{12.22}$$

where p_d is the reduced deposition coefficient (see Sec. 4.2) and α represents the SPM fraction that settles on/in the buffer layer. As discussed in Sec. 4.2, one should bear in mind that because of bed irregularities etc., the active layer may be non-existent at some locations of the bed, which justifies this "α-approach". Further, also a burial term B is shown in Fig. 12.5b. Here, we do not elaborate further on B; the reader is for instance referred to Winterwerp *et al.* (2018). However, from a modeling point of view in the present application, α and B fulfill the same role.

The deposition flux on the buffer layer D_2 thus becomes:

$$D_2 = p_d \alpha w_{s,b} c_b \tag{12.23}$$

The coefficient of reduced deposition p_d in the reduced deposition approach was introduced in Sec. 4.2.

The active layer is generally thin, and depleted frequently by erosion. As explained in Secs. 6.5 and 9.2.2, it is merely the time phasing within a tidal cycle at which the fluff layer is eroded, more than its rate. The erosion rate of the active layer is therefore modeled with the formulation for surface erosion (Eq. (6.15)), i.e. $E_1 = M_{E,1}\left(\tau_b - \tau_{e,1}\right)$. However, as the active layer may become very thin, the mass balance in the numerical model might be violated when more sediment would be picked up in a computational time step than available in the active layer. This numerical artifact can be prevented by introducing a first-order erosion formulation, reducing erosion rates at small layer thickness (in fact its mass):

$$E_1 = m_1 M_1 \left(\tau_b - \tau_{e,1}\right) \quad \text{for } \tau_b > \tau_{e,1} \tag{12.24}$$

where M_1 is the first-order resuspension flux parameter. However, from a physical point of view, this first-order erosion approach also reflects the uneven spatial distribution of fines on the bed. In that sense it can be regarded as the counterpart of the reduced deposition approach — see the discussion in Sec. 4.2 and Fig. 4.11.

This approach can be applied if the fluff layer is depleted rapidly every tidal cycle, hence remains thin. If not, the erosion rate E_1 would continue to increase with m_1, eventually exceeding the zero-order erosion rate of Eq. (6.15). This can be prevented through maximizing the first-order erosion rate by introducing a threshold value m_{cr} for m_1, with a linear transition from first- to zeroth-order erosion:

$$E_1 = \min\left(1; \frac{m_1}{m_{cr}}\right) M_{E,1}\left(\tau_b - \tau_{e,1}\right) \quad \text{for } \tau_b > \tau_{e,1} \tag{12.25}$$

where $M_{E,1}$ is the zero-order erosion flux, as in Eq. (12.19) (thus $M_1 = M_{E,1}/m_{cr}$), and $\tau_{e,1}$ is the critical shear stress for erosion of the active layer. The user-defined (to be calibrated) parameter m_{cr} is the threshold mass in the active layer beyond which the zero-order erosion formula is applied.

The buffer layer generally contains abundant sediment and will not be depleted easily (in fact should not be). A linear/zero-approach is therefore not necessary. In case the buffer layer is cohesive/muddy, its erosion rate can be described with the surface erosion model (6.15):

$$E_2 = \xi_2^m M_{E,2}(\tau_b - \tau_{e,1}) \quad \text{for } \tau_b > \tau_{e,2} \quad \text{and} \quad \tau_{e,2} > \tau_{e,1} \tag{12.26}$$

where ξ_2^m is the mud content in the buffer layer.

However, it the buffer layer represents a sandy substrate, it is assumed that fines are eroded from the buffer layer, when that sand is mobilized (see also Sec. 6.6). This is described with the Van Rijn (1993) pick-up formula, defining M_2 [kg/m^2/s] and $\tau_{e,2}$:

$$E_2 = \xi_2^m M_2 \left(\frac{\tau_b}{\tau_{e,2}} - 1 \right)^{1.5} \quad \text{for } \tau_b > \tau_{e,2} \quad \text{and} \quad \tau_{e,2} > \tau_{e,1} \tag{12.27}$$

This formulation was applied in a 3D hydrodynamic and sediment transport model by Van Kessel *et al.* (2011a) to simulate SPM-values measured in the sandy Dutch coastal zone of the North Sea over a period of eight months in 2001. This period includes numerous spring-neap cycles and storm periods.

The buffer model was calibrated with a 1DV point model using two fractions ("fine" and "coarse"), yielding the parameters given in Table 12.2 — transport and fate of the coarser sand fraction is not explicitly modeled.

Table 12.2. Parameters for 1DV simulations.

	Fine	Coarse	Unit
$\tau_{e,1}$	0.1	0.1	Pa
$\tau_{e,2}$	1.5	1.5	Pa
α	0.05	0.0125	–
w_s	0.125	1.0	mm/s
M_1	$5.8 \cdot 10^{-6}$	$2.3 \cdot 10^{-5}$	s/m
M_2	$3.5 \cdot 10^{-7}$	$3.5 \cdot 10^{-7}$	kg/m/s^2
δ_2	0.1	0.1	m
h	18	18	m

Note that for this particular application, the fluff layer remains thin, and Eq. (12.24) can be used. As the data were collected near (1 m below) the water surface, the 1DV model was run with 10 vertical layers, assuming a constant vertical eddy diffusivity of $\varepsilon_z = 0.01\,\mathrm{m}^2/\mathrm{s}$. The reader is referred to Van Kessel *et al.* (2011a) for details on the calibration procedure.

Next, the buffer model and calibrated settings were deployed in a 3D schematization of the Dutch coastal zone, see also Chapter 14. The effects of tide, waves and wind are all accounted for. Computational results at Noordwijk station are presented in Fig. 12.6, and compared with data. The spring-neap variations, and incidental storm-induced SPM peaks are well reproduced with the buffer model. SPM-concentrations near the water surface are quite low, but their variation over time is important for water quality evaluation (primary production — see Sec. 3.5).

It appeared possible to reproduce the observed dynamics with the Partheniades–Krone approach as well, i.e. the without buffer model. However, these results appeared very sensitive to the model settings, and small

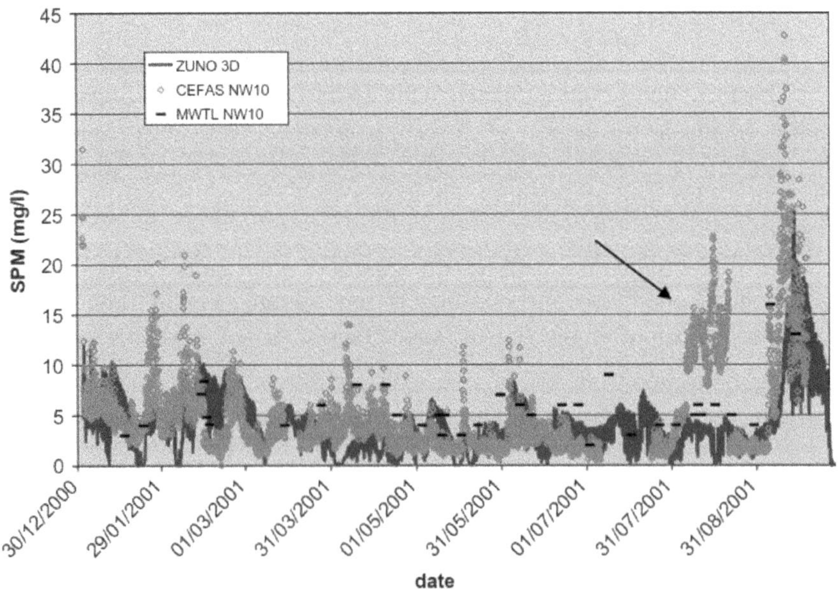

Fig. 12.6. Simulation of high-frequency near-surface SPM measurements at Noordwijk, The Netherlands with a 3D model using the buffer model described above (Van Kessel *et al.*, 2011a) — data early August 2001 (see arrow) are unreliable.

changes yielded very different results (see Van Kessel *et al.*, 2011a). The buffer model, on the other hand, is far more robust with respect to its model settings.

Multi-layer bed models

Greater vertical resolution within the bed to account for e.g. multiple time scales, variations in erosion properties and/or morphodynamic simulations requires multi-layer models, as developed by for instance Ariathurai (1974); Le Hir *et al.* (2011); Sanford (2008); Zhou *et al.* (2016) and Winterwerp *et al.* (2018).

We will not go into into the details of these multi-layer models — the reader is referred to one of the references above. However, there is one aspect in these models which deserves further attention, and that is the use of fixed or moving frames of reference, relevant in particular if bed level changes have to be predicted.

Accretion and erosion of a bed can be treated in a Lagrangian or Eulerian framework, as shown in Fig. 12.7. In a Lagrangian framework, the thickness of the layers remains constant and the set of layers moves up and down with accretion/erosion. This migration induces an artificial vertical velocity. The advantage of this method is that the computational grid itself does not change with respect to the accrediting/eroding bed. This means that the thickness of the layers can be kept constant, which allows higher resolution near the bed surface at all times, while lower resolution deeper within the bed may be deployed. The drawback is that, depending on the numerical scheme, the artificial vertical velocity may induce artificial diffusion between the layers smoothing stratification within the bed.

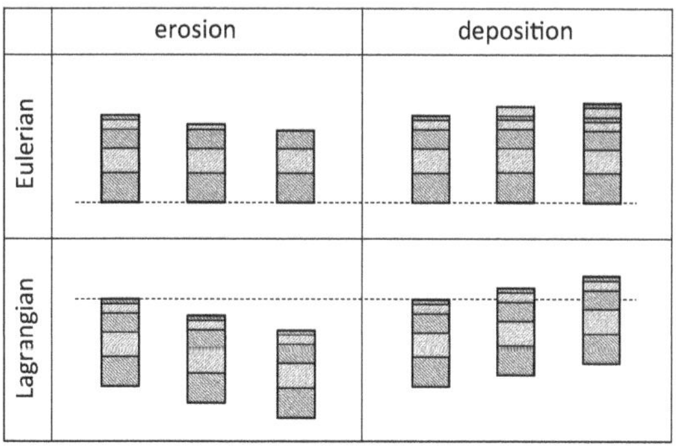

Fig. 12.7. Multi-layer bed model in Eulerian and Lagrangian frames of reference.

In an Eulerian framework, the position of the layers is kept constant with respect to a bed-independent frame of reference. Deposition/erosion is accounted for by changing the thickness of the top layer(s). In case the top layer becomes too thick due to deposition, it must be split. In case the thickness tends to zero due to erosion, this layer is merged with the layer below. The drawback of this approach is that the top layer has no fixed thickness, and vertical resolution varies over a simulation period. This limitation does not only influence the numerical accuracy of a model, but also affects the (model) time scales of the system.

A combination of Lagrangian and Eulerian approaches would resolve these limitations largely. This was done for instance by Winterwerp *et al.* (2018) in a semi-equilibrium consolidation model for simulating the morphodynamic developments of cohesive sediment-rich systems and by Braat *et al.* (2017) in a schematized simulation of the long-term morphodynamic evolution of the Scheldt estuary.

As stated at the beginning of this chapter, we presume morphostatic conditions. In morphodynamic simulations, variations in the amount of sediment (mud) in the bed imply bed level changes. As a result, the transport and fate of SPM become linked to these bed level changes, and equilibrium becomes possible, as shear stresses increase with decreasing water depths, and vice versa. So in morphodynamic simulations with sand-only or mud-only sediment, the only degree of freedom is the bed level, hence bed composition is fixed. In morphostatic computations the only degree of freedom is bed composition, as the bed level is fixed. In reality both vary, and the system response becomes complex, in particular in case of sediment mixtures. The reader is referred to e.g. Roelvink and Reniers (2012) for further reading on this subject.

12.3 Schematization long-term SPM transport

As natural systems are dynamic with considerable variations in SPM-concentrations, the major time scales for marine systems are:

(1) the quarto-daily tidal signal with four slack waters per day (for diurnal tides, of course a semi-diurnal signal) of about 6 hrs,
(2) the spring-neap cycle of about 14 days,
(3) seasonal variations with a period of about 6 months,
(4) extreme events with a return period of maybe a few years.

Depending on the engineering question to be addressed, it may be necessary to study the transport and fate of SPM at all these time scales. As

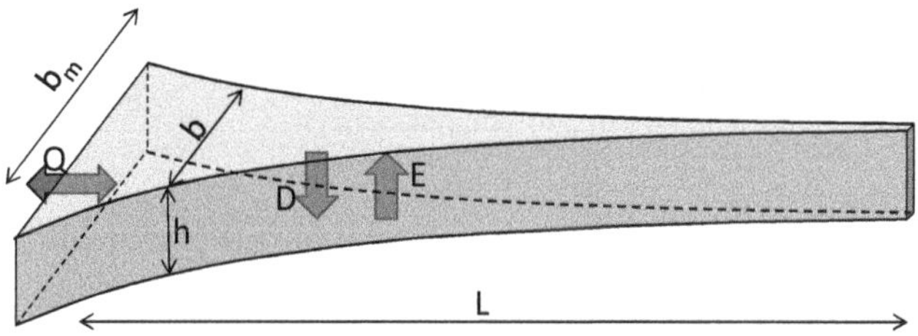

Fig. 12.8. Zeroth-order mass balance of semi-enclosed aquatic system.

time scales are closely linked to spatial scales (De Vriend, 1991b), large temporal scales imply large spatial scales, and the other way around.

We have argued (see Chapter 1 and Sec. 12.2) that SPM-concentrations in shallow water are largely governed by water-bed exchange processes. However, at longer times scales, sediment exchange with the environment (thus over the open boundaries of a computational domain) by advective processes becomes important, as these external sources ultimately determine the amount of sediment in an aquatic system.

Zeroth-order mass balance

To sustain this argument, we analyze the zeroth-order mass balance of the schematized estuary in Fig. 12.8, which can be regarded as a zeroth-order representation of the Sea Scheldt, without any intertidal areas though.

The variation in estuary-averaged SPM-concentration C reads:

$$\begin{aligned} V\frac{\mathrm{d}C}{\mathrm{d}t} &= Q\left(C_{sea} - C\right) - A(w_s C - E) \\ &= \left(QC_{sea} + AE\right) - \left(Q + Aw_s\right)C \end{aligned} \tag{12.28}$$

where V and A are the volume and projected bed surface area of the basin, h is a characteristic (mean) water depth ($h = V/A$), Q is the flux of water exchanging with the ambient waters, thus over the open boundary of the model domain, b_m is the width of the open boundary, C_{sea} is the SPM-concentration in the ambient water, and E and D $(= w_s C)$ are erosion and deposition fluxes, representing the exchange of SPM between bed and water column.

The solution to Eq. (12.28) reads:

$$C = C_e\left(1 - \exp\left\{-\left(\frac{Q}{V} + \frac{w_s}{h}\right)t\right\}\right) \tag{12.29}$$

with the equilibrium SPM-concentration C_e:

$$C_e = \frac{QC_{sea} + AE}{Q + Aw_s} \tag{12.30}$$

Let us quantify these equations for a large model domain, such as the schematic Sea Scheldt (in Belgium) with a part of the Western Scheldt (in The Netherlands — see Fig. 12.8, Winterwerp and Wang, 2013). This funnel-shaped estuary has a length $L = 104\,\text{km}$ (Antwerp would be at 30 km from the mouth of this schematic channel). The cross-section of the estuary is prismatic, with vertical walls, a horizontal bottom and 6 m water depth. At the mouth, the width of the estuary measures $b_m = 3$ km, exponentially decaying as $b(x) = b_m \exp(x/L_b)$ with $L_b = 30\,\text{km}$, so that at its head $b_r = 90\,\text{m}$. The surface area of this estuary measures $A = \int_0^L b\,\mathrm{d}x = 1.97 \cdot 10^8\,\text{m}^2$, with a volume $V = 1.18 \cdot 10^9\,\text{m}^3$. We assume a typical (tidal-)mean velocity U in the estuary mouth of $0.7\,\text{m/s}$, thus $Q(= b_m hU) = 1.26 \cdot 10^4\,\text{m}^3/\text{s}$. Further, we assume $E = 10^{-4}\,\text{kg/m}^2/\text{s}$, thus $AE = 1.97 \cdot 10^4\,\text{kg/s}$, $w_s = 0.5\,\text{mm/s}$ and $C_{sea} = 100\,\text{mg/L}$. Substitution into Eq. (12.30) gives the equilibrium SPM-concentration of this configuration:

$$C_e = \frac{QC_{sea} + AE}{Q + Aw_s} = \frac{1.3 \cdot 10^3 + 2.0 \cdot 10^4}{1.3 \cdot 10^4 + 9.9 \cdot 10^4} \approx 190\,\text{mg/L} \tag{12.31}$$

The time scale to reach this equilibrium is $T_e = V/(Q + Aw_s)$, which is in fact the e-folding time of Eq. (12.31), computed at 3 hrs only. However, in Eq. (12.31 we assume a homogeneously mixed Scheldt. Though this is far from the actual situation, C_e gives a reasonable first order estimate of typical SPM-concentrations in the Scheldt River. However, this assumption fails in assessing the accompanying time scale for equilibrium.

Mixing within the Scheldt estuary is governed by tidal dispersion, i.e. mixing by tidal eddies with a maximum length of the tidal excursion L_T; a typical value of which in the Scheldt yields about 14 km. The mean size of these eddies thus measures about 7 km, while the relevant time scale equals half the tidal period T, i.e. from slack to slack water. Further to Sec. 12.1, we may thus compute a dispersion coefficient for tidal mixing as $D_{tm} = (L_T/2)^2/(T/2) \approx 2.3 \cdot 10^3\,\text{m}^2/\text{s}$. The time scale T_m for mixing the ambient SPM with concentration C_{sea} over the Scheldt thus becomes $T_m = L^2/D_{tm} = 4.7 \cdot 10^6\,\text{s}$, which is almost 2 months, thus two orders of magnitude larger than T_e.

The actual time for attaining equilibrium for dissolved matter will amount to a few times T_m, and for suspended matter even a more, owing to the fact that most of the dynamic stock of fines sediment resides within the bed. A realistic 3D model of the Scheldt estuary, for instance, requires 5 years of simulation time to attain equilibrium (Chapter 15 and Van Kessel *et al.* 2011a).

The empty bed approach

Let us analyze these numbers a bit further — bearing in mind that we applied a zero-dimensional analysis, the numbers can only be approximate. Equation (12.31) shows that C_e is determined mainly by the ratio of E and $w_s c$. Indeed, if Q and/or C_{sea} become zero, C_e changes a few percent only. This result is consistent with our presumption that in shallow waters, the water-bed exchange determines the SPM-concentrations in the water column. Hence, horizontal mixing of C_{sea} appears not necessary to determine C_e and T_e.

However, E depends on the availability of fine sediment in the bed, which is unknown, but should depend somehow on C_{sea}, certainly over longer time scales. Moreover, even if we would know the actual fine sediment distribution in the Scheldt bed, there is no guarantee that this distribution matches the solution of our numerical model (we need a spin-up time).

In other words, the initial conditions of the fine sediment distribution (and composition and properties) within the Scheldt estuary bed for numerical model simulations is unknown. Such a distribution, consistent with the solutions of that model, can only be established by the numerical model itself. Of course, this requires bookkeeping of the fine sediment in the bed, as discussed in Sec. 12.2.

However, a numerical model needs initial conditions to get it running, for which two options exist:

(1) Prescribe a finite or infinite distribution of fines in the bed, homogeneous or not, based on measurements or expert judgment (heuristic distribution) — we refer to the "full-bed" concept.
(2) Start without any fines on/in the bed — we refer to the "empty-bed" concept.

It is emphasized that the empty- and full-bed approach refer to the fine fraction only. In case of a sandy bed, sand is supposed to be present, and is not imported into a model domain over its open boundaries. When possible, we advocate the use of the empty-bed concept for a number of reasons:

(1) The empty-bed approach allows for below-capacity conditions, typical for many muddy systems.
(2) The full-bed approach generally requires longer computational times to obtain (dynamic) equilibrium (see below).
(3) The full-bed approach, in combination with the zeroth-order Partheniades–Krone approach may yield highly unrealistic results, as initially a model predicts SPM-concentrations as a function of the initial (user-defined) conditions. When the bed becomes depleted, SPM-levels drop dramatically.
(4) In a large computational domain, as the Scheldt estuary, part of the bed may remain unaffected (stable) for a majority of a simulation period, becoming mobilized during extreme conditions only. Sediment, eroded from dynamic locations within a model may accumulate at unrealistic rates at low-dynamic locations, further increasing computational times for reaching equilibrium.
(5) Using observations to initiate bed composition suffers from the same problems as application of a uniform initial bed layer, though likely to a lesser extend.
(6) Simulations with the empty-bed concept are more easy to interpret, as we may assume that the model does not contain "hidden" sources of sediment which may affect model results — hidden means locations which have not yet been inspected through model output.

Of course, both options should yield identical results at equilibrium — our engineering models should not depend on user-defined initial conditions.

Our arguments above are further sustained with two numerical simulations with a 2DV Delft3D model (i.e. with 23 vertical layers, but no variations over the width of the model) for the schematic estuary described above (Fig. 12.8). At the mouth we prescribe an S_2 and S_4 tide (1.07 m and 0.12 m) with a relative phase difference $2\phi_2 - \phi_4 = 3°$, an SPM-concentration $C_{sea} = 0.2$ g/L, and a salinity $S_{sea} = 20$ ppt, using a Thatcher–Harleman time lag $t_{TH} = 0.5$ hr (see Sec. 11.2). At the upstream model boundary, we prescribe a constant river flow of $Q_r = 30\,\mathrm{m^3/s}$. A constant Chézy coefficient of $60\,\mathrm{m^{1/2}/s}$ is applied, while D_x is set at $20\,\mathrm{m^2/s}$. We use the traditional Partheniades–Krone boundary conditions with the following settings: $w_s = 0.5$ mm/s, $c_{gel} = 100$ g/L, $\tau_e = 0.5$ Pa, and $M_E = 10^{-4}\,\mathrm{kg/m^2/s}$.

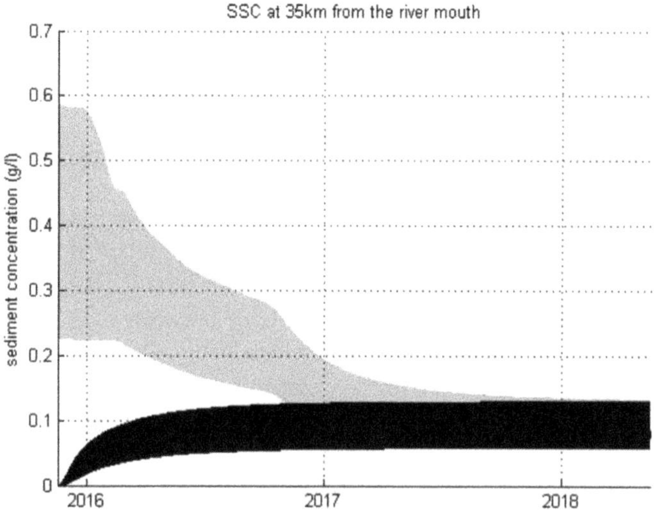

Fig. 12.9. Time evolution at a station 35 km from the river mouth of SPM-concentration starting from "full" (gray curves) and "empty" (black curves) bed configuration — courtesy M. de Lucas Pardo.

Finally, both simulations are carried out including the effects of hindered settling and sediment- and salinity-induced stratification.

This configuration was also used to demonstrate the role of water-bed exchange processes in the net transport of fine sediment by tidal asymmetry in Sec. 9.2.

Figure 12.9 shows the time variation over 2.5 years of the computed SPM-concentration at 35 km from the mouth, starting with a full (initial mud thickness of 0.12 m) and empty bed. The wide bands of the computational results reflect tidal SPM variations, indistinguishable at the time scale of the figure. From this figure, we can make the following observations:

(1) Both simulations converge to the same dynamic equilibrium SPM-value of about 100 mg/L.
(2) For the empty-bed simulation, equilibrium is approached gradually (exponentially), though in the first month SPM-values build up rapidly. The value for T_m derived above gives a reasonable estimate of the time to equilibrium. Complete equilibrium, though, is only attained after about 1–1.5 years.
(3) For the full-bed simulation, the time to full equilibrium measures about 1.5–2 years. Likely, T_m can be reduced through better estimates of the

initial amount of sediment in the full-bed approach. Obviously, one will not gain computational time, while requiring extra simulations.

(4) More important than T_m, however, is the route towards equilibrium. Equilibrium is attained gradually and smoothly for the empty-bed simulation, and after a few months, computed SPM values do not differ too much from their equilibrium values. However, for the full-bed simulation, SPM values differ largely from equilibrium during the first year of the simulation. Moreover, the results show a number of steps/kinks in the results. These steps are to be attributed to the nonlinear approach to equilibrium, induced by the interplay between SPM transports by gravitational circulation and tidal asymmetry.

Thus, we may conclude that this simple simulation sustains our arguments above, that from a practical point of view, starting with an empty bed is generally advisable for fine sediment simulations. From a fundamental point of view, the empty-bed approach is consistent with below-capacity conditions. Full-bed simulations may yield large errors during the first phase of a simulation, and sometimes during later phases as well, when episodic events are elaborated upon. This is of course no longer true if the real bed contains of unlimited amounts of cohesive sediment.

Further, these simple examples show that long simulation times are required to attain results with a numerical model which can be interpreted with confidence.

Modeling low dynamic systems with small external SPM input form an exception. Sediment buildup in such systems may have taken centuries. Computational times then become prohibitive, and a pragmatic approach for setting initial conditions is required (Chapter 17). Finally, in case of multiple simulations, one would use the computed equilibrium bed composition from previous runs as initial conditions.

Reducing spin-up times with 2DH approach

From a computational time point of view, 2DH models are (much) faster than 3D models. It may therefore seem attractive to calibrate model parameters and the initial sediment distribution in the bed with a numerical model in 2DH mode, and then switch to 3D mode, using these 2DH results. However, in Sec. 12.1 we described some fundamental differences between a 2DH and 3D schematization of flow and SPM distributions. SPM distributions computed with a 2DH model are therefore never identical to those computed with a 3D model. To attain full equilibrium, simulation times for a 3D simulation can therefore not be shortened largely by establishing the initial conditions from a 2DH SPM field. However, when such full equilibrium is not required in an engineering study, the calibration process for a 3D model study can be shortened considerably with some initial 2DH-simulations. Also a first understanding the model response to variations in parameter settings can be done in 2DH mode, though with care as results may be misleading (see Chapter 13).

Chapter 13

Good Mud-Modeling Practice

13.1 Introduction

A model is a simplified/schematic representation of the real world. Models may exist in various complexities, but all aim to schematize/simplify, better understand and quantify interactions within a complex system or between phenomena. Models may be purely descriptive (explaining in words or graphically how variables are interrelated) or mathematical (where relationships are formalized by a set of equations). The focus in this book is on the setup and use of complex, process-based mathematical models, describing the transport and fate of fine sediment in open waters at the level of the individual transport processes as described in the previous chapters. As the relevant equations are generally nonlinear, they have to be solved numerically — we refer therefore to numerical models from hereon. However, we argue in the present and following chapters, that descriptive models (we refer to conceptual models, see Sec. 13.2) are an essential element of the setup and use of any numerical model. In fact, a numerical model is a mathematical formalization of the conceptual model.

A numerical model consists of the numerical code solving the set of shallow water equations in Chapter 11, with input parameters such as bathymetry, roughness coefficients, sediment settling velocities, etc., and boundary conditions such as the time-varying water level at an open model boundary. Examples of numerical codes are Delft3D, Telemac, CMS, SedCom, etc. Though the arguments in this and the next chapters are code-independent, the examples given are based mainly on Delft3D simulations.

Fine sediment transport models can be set up in many ways, differing in size, dimensions, formulations, approaches and time scales. This chapter aims at providing a framework for setting up numerical models for simulating the transport and fate of fine (cohesive) sediment. In The Netherlands, Stowa (1999) introduced a framework for good modeling practice for hydrodynamic models (Fig. 13.1). Seven steps were defined to be

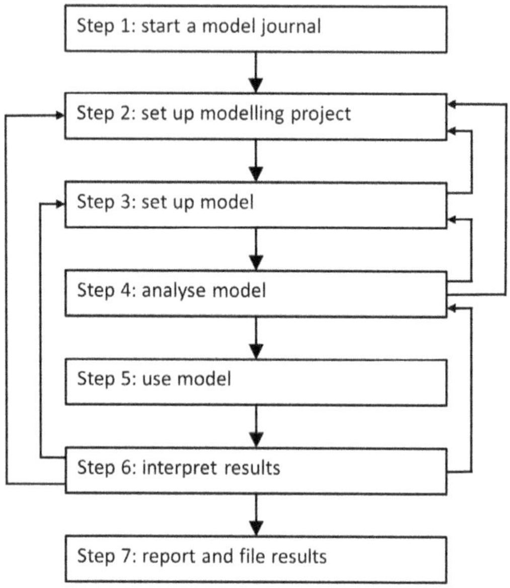

Fig. 13.1. Seven steps in good modeling practice, as defined by the Good Modeling Practice Handbook Stowa (1999).

followed throughout model setup, application, and reporting. In this book, we address the sections relevant for the actual model setup and calibration (Steps 3 and 4). Step 1 and Step 2 relate to user- or organization-specific logistic procedures (amongst which the model journal/logbook, archiving all steps and the variety of modeling runs), whereas the actual model application (Steps 5–7) is beyond the scope of this chapter — these are addressed in the next chapters. Step 3 involves the setup of the conceptual model (Sec. 13.3) and of the actual numerical model (Sec. 13.4), whereas model calibration and validation (Sec. 13.5) are part of Step 4. We conclude this chapter with practical guidelines for setting up a numerical model and assessing parameter settings for simulating the transport and fate of cohesive sediment in open waters, including some thoughts on how to use and interpret the modeling results (Sec. 13.6), i.e. Good Modeling Practice (GMP). These guidelines will be illustrated with five case studies described in the chapters hereafter.

Before elaborating on the description of Good Modeling Practice, it is important to point out that predicting transport and fate of fine sediment with a numerical model is often challenging. As discussed in previous chapters , sediment properties are generally poorly known, and may vary largely in time and over space, influenced by hydrodynamics (turbulence, salinity), biological and soil mechanical processes, etc. Net transport rates are small compared to gross sediment fluxes, which in turn are affected by subtleties in hydrodynamics and driving forces (see Secs. 9.4, 10.4 and 13.6.2). One of the larger uncertainties stem from the water-bed exchange processes, which are difficult to quantify, and generally vary over time and space as well, affected by the state of consolidation (over-consolidation!) and pore water under-pressures (pseudo-cohesion) (Chapter 6). The number of potential variables is therefore large, and many parameters will remain poorly known. Fine sediment transport models can therefore not provide a perfect mathematical representation of all governing physical processes, i.e. of the real world.

Thus a modeler has to make choices on what to model, and how, and what not ("fit for use"). This choice shall be based on a proper analysis of the questions to be addressed, a proper system understanding, a proper understanding of the relevant physical (and biochemical) processes, and on the data available. In other words, "modeling is an art", and will remain so.

These choices are to be conceptualized in a so-called conceptual model, informing modeler and his stakeholders (users, clients, scientists, ...) on the consequences of the choices made. Diagrams, cartoons, etc. may help to visualize this, as shown below.

Later in this chapter we describe the importance of formulating model objectives. This does not imply that the model itself is an objective. A (numerical) model is always a tool to help answering engineering or management questions, or to help understanding natural processes, i.e. to advance science.

13.2 The conceptual model

The conceptual model is a key concept in numerical modeling. This section discusses how research questions and system understanding are at the heart of a conceptual model, which is schematically displayed in Fig. 13.2, and elaborated below.

The most important step in any numerical modeling study is setting proper and realistic numerical modeling objectives. These objectives follow from a detailed formulation, formalization and analyses of the question(s) to be addressed. These questions may follow from engineering

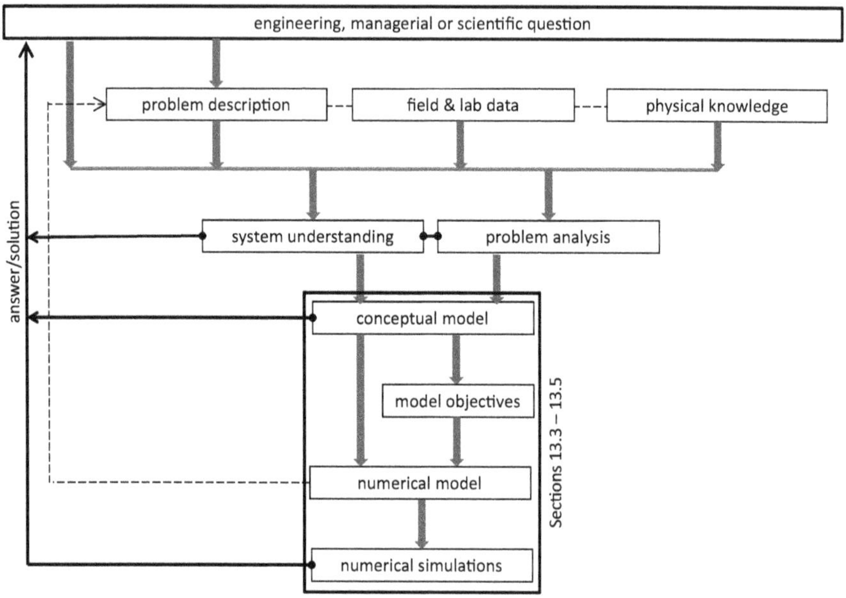

Fig. 13.2. Steps towards modeling objectives and conceptual model.

problems, management issues, scientific research, or otherwise. This phase is known as problem definition and problem description. An important issue concerns the different "languages" and jargon used by various stakeholders. Many modeling studies have failed because the problem was not properly defined and/or not properly communicated with the client/stakeholder — in extreme cases, the wrong question may have been addressed in the end.

Also, distinction has to be made between input- and output-controlled questions. Engineering and managerial problems are often output-controlled: for instance a client needs numbers for the siltation rates in a new fairway to be constructed. Scientific research is generally input-controlled, formulating research questions: for instance how does pH affect consolidation rates? Output-controlled questions are generally set "in concrete", whereas input-controlled questions may evolve during a research project in conjunction with the development of the science itself. A final crucial distinction is on how a numerical model is to be used, i.e. in predictive or exploratory mode (see below).

An essential element of a conceptual model is understanding the aquatic/sedimentary system. We refer to system understanding, which should give a first qualitative idea of the relevant parameters and variables, and their mutual interactions. System understanding is based on three pieces of information:

(1) the problem definition, problem description and problem analysis themselves,
(2) data available, from earlier studies and reports/publications, from literature, satellite images, etc.,
(3) in-depth knowledge/know-how on the transport and fate of SPM in open water, as discussed in the previous 12 chapters.

Problem definition, description and analysis, and system understanding are formalized in a so-called conceptual model. A conceptual model describes how an aquatic system functions, thus which natural processes are to be addressed, which processes can be omitted, and visualizes the consequences of these choices. A conceptual model defines the (dominant) conditions to be studied, such as mean hydrodynamic forcing, extreme conditions, or seasonal variations, etc. It also defines which (additional) measurements have to be taken at which locations, which model simulations have to be carried out for which period of time, which output from the model is to be obtained, etc. In other words, a conceptual model is a framework for thinking and interpretation, and for communication with the stakeholders.

A conceptual model may be purely descriptive (a piece of text) or visualized in the form of diagrams, cartoons or drawings. As an example, Fig. 13.3 presents an artist impression of the most relevant processes governing the dynamics of a muddy coastal system. This figure was used for communicating the complicated physical system with a client, supporting arguments on modeling choices and the use of data in an engineering study. Such a conceptual model also helps to maintain focus in the numerical modeling study, in particular for complex systems. A more common way of visualizing a conceptual model is a flow chart relating processes and/or variables, including input–output relations and positive/negative feedback mechanisms.

Next, proper and realistic goals and sub-goals for the modeling work have to be set. Often, a conceptual model is made implicitly, in particular by inexperienced, but also by very experienced engineers/scientists. Our experience is that formalizing a conceptual model explicitly helps improving modeling results, its reporting and communication. The present and next chapters aim to aid modelers in this.

Often, one's system understanding is not static, but evolves during the execution of the study. The results of a numerical model study themselves

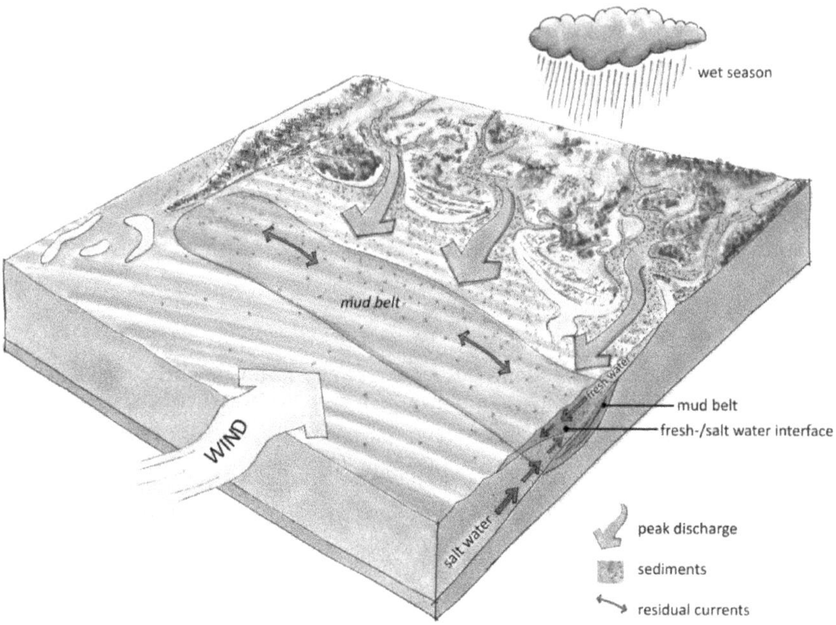

Fig. 13.3. Artist's impression of a conceptual model visualizing the sediment dynamics along a muddy coast.

may also enhance system understanding. Whether or not there is time to include such new insights into the conceptual model depends on the time available for the study. If not, the consequences of such new insights should be addressed in a discussion on the modeling results. In scientific studies, the evolution in understanding is generally part of the research itself, and may even modify the objectives of the input-controlled work.

13.3 From a conceptual to a numerical model

On the basis of our system understanding and conceptual model, one can start designing the lay-out of a numerical model to address the formulated questions/problems. This section discusses the decisions to be taken and the underlying argumentation.

13.3.1 *Model type*

The numerical models addressed are known as process-based models. They are generally characterized by the numerical solution of the governing equations of Chapter 11 on a detailed bathymetry with a detailed formulation of the underlying physical processes in the form of algorithms. These models may be more or less complex, depending on the processes accounted for, such as waves, flocculation, consolidation, etc. They generally require considerable computational times for simulating large-scale SPM dynamics (e.g. Sec. 12.3). Also calibration of these models is often an issue. A calibrated model is fit for its calibrated conditions, but may fail for other conditions or time scales — we refer to over-calibration, which is further discussed in Secs. 13.5.4 and 13.5.6.

The most common use of a numerical, process-based model is to simulate real world cases for solving e.g. engineering problems. However, process-based models can also be used as research tools themselves when run in exploratory mode, enhancing one understanding of a sedimentary system. Applications may include simplified geometries, providing insight into the morphodynamic behavior of estuarine (Hibma *et al.*, 2004; Van der Wegen and Roelvink, 2008), riverine (Van Maren, 2007; Van Maren *et al.*, 2009a) or deltaic (Edmonds and Slingerland, 2007) systems. Also more complex topographic model configurations can be used in exploratory mode, by systematically investigating the role of individual processes (Van Maren *et al.*, 2015a,b).

Water-bed-sediment interactions

One of the first decisions to be made in setting up a numerical sediment transport model is to include or not:

(1) bed level updates, i.e. morphodynamic modeling,
(2) sediment–fluid interactions (SPM-induced buoyancy and rheology, see Chapter 7).

Including bed level updates (morphodynamic modeling) is essential when bed level variability influences the water motion. This is the case for most sand-dominated environments covering time scales exceeding (typically) one year. Bed level variability is of little importance in low-concentration fine-sediment environments, where the time scale of interest (months-years) is often shorter than the time scale where morphological developments become significant (years-decades). For turbid environments, however, bed levels changes may be relevant on tidal Van Maren *et al.* (2015b) or river flood Van Maren *et al.* (2009a) time scales. In this book, we do not address the ins and outs of morphodynamic modeling; the reader is referred to e.g. Roelvink and Reniers (2012).

Next, formulations for the water-bed exchange processes themselves have to be selected, in conjunction with a bed model (buffer model, multi-layer bed model, etc.), accounting for the soil mechanics time scales, which generally differ from those of the hydrodynamic and sedimentological processes.

Coupled approach

Sediment-fluid interactions involve the effect of (suspended) sediment on density, viscosity and yield strength, and dynamics of the sediment-water mixture. From a process-point-of-view, these interactions are discussed in detail in Chapter 7. It was demonstrated that even at moderate concentrations of several 0.1 kg/m^3, vertical density gradients in a water-sediment mixture increases so much that turbulent mixing can be significantly damped. Sediment-induced damping of turbulent mixing may lead to rapid settling and hence the formation of (mobile) fluid mud layers. Such highly concentrated near-bed layers may be essential to establish siltation rates in ports and waterways (Winterwerp and Van Kessel, 2003), and modeling the morphodynamic behavior in high-turbid rivers (Van Maren *et al.*, 2009b): "mud attracts mud". Suppression of turbulence also leads to a reduction in near-bed velocity gradients, hence reduced effective hydraulic resistance (Sec. 3.3.3), which may enhance flow rates and/or depth-averaged flow velocities. Resolving these processes in a numerical model requires inclusion of SPM in the equation of state (Eq. (11.10)). This implies that hydrodynamics and sediment transport are to be resolved simultaneously (Sec. 13.3.1) at high vertical resolution (Sec. 13.4.2) — we refer to a coupled modeling approach.

Uncoupled approach

One major advantage of excluding bed level updates and sediment-fluid interactions is that sediment transport computations may be decoupled from hydrodynamic computations — we refer to an uncoupled modeling approach. In this case, the flow field needs to be computed only once, the results of which are stored on file. Sediment transport computations can then be repeated infinitely with these hydrodynamics, even on aggregated grids and/or larger time steps. Of course, relevant velocities, water levels, bed shear stresses, etc. need to be stored at sufficient resolution (Van Kessel *et al.*, 2011a; Van Maren and Winterwerp, 2013; Van Maren *et al.*, 2015b). The uncoupled approach allows much longer simulations than feasible with a fully coupled model. This is in particular important in case simulations for dynamic equilibrium conditions are required (see Sec. 13.4.1). In systems where morphodynamics and sediment–fluid interactions have a moderate impact, it is the conceptual model which determines the choice between long simulation times at the cost of physical accuracy (uncoupled mode) and a more complete physical description, at the cost of simulation time (coupled mode).

Non-Newtonian effects

At large sediment concentrations, generally around c_{gel} and beyond, stress–strain relations become modified through an increase in the fluid apparent viscosity and the development of yield strength (Sec. 7.3). Such non-Newtonian effects have been implemented into a variety of research models (Sec. 11.4), but are currently too costly and complicated to be used in large-scale engineering studies. Sometimes, a full Newtonian approach can be parameterized, even at concentrations of several 100's kg/m^3, such as occurring in the Yellow River (Van Maren *et al.*, 2009b). We do not further elaborate on the setup and calibration of such non-Newtonian models.

Model dimensions

Models can be setup in one-dimensional (1D), two-dimensional (2D), or three-dimensional (3D) mode. For 1D models, distinction is made between 1DH and 1DV models, the first solving cross-sectional averaged hydrodynamic and sediment transport equations in longitudinal direction, whereas the second solves the equations in vertical direction at one point only (see Sec. 11.4). 2D models may resolve processes in longitudinal and lateral direction (2DH, Sec. 11.3), or in longitudinal and vertical direction (2DV). Best known examples of 1DH models are relatively simple river models (not addressed in this book), but also include fairly sophisticated models for tidal basins (e.g. Van Prooijen and Wang, 2013). 1DV models are fast and flexible tools to investigate vertical mixing and settling processes at high vertical resolution in the water column (e.g. Winterwerp, 2001), and are convenient in assessing model parameters (Sec. 13.6.3), as elaborated in Secs. 13.5 and 17.4.3.

Conditions at which sediment–fluid interactions (both SPM-induced buoyancy and rheology) and salinity-induced stratification are important always require resolution of the processes over the water column, i.e. a 1DV, 2DV or 3D modeling approach.

Systems where the hydrodynamics and sediment dynamics follow well-known equilibrium profiles (such as the logarithmic velocity and Rouse concentration profile) may be approximated with 2DH models. 2DH models are commonly applied to simulate the morphodynamic behavior of sand-dominated coastal or riverine systems, requiring long simulation times.

13.3.2 *Modeling domain*

Sediment transport models may cover a wide range of spatial and temporal scales and may differ greatly in complexity. This variability reflects the various objectives for which such models are setup. "Universal" sediment transport models, setup and calibrated prior to formulating the questions to be addressed, often lead to disappointments. Models need to be tailor-made in space (model domain, resolution), time (with a time step sufficiently short to resolve relevant physical phenomena but with a simulation period long enough to answer the relevant question) and complexity. Models should be sufficiently complex to simulate the physics relevant to answer the questions, but not too complex — adding complexity makes a model slower to run, more difficult to interpret its results, and sometimes even less accurate.

The relevant space and time scales are largely, and often uniquely correlated (Fig. 13.4). However, apparent secondary (small scale) processes may have large scale impact because of nonlinear interactions/feedbacks

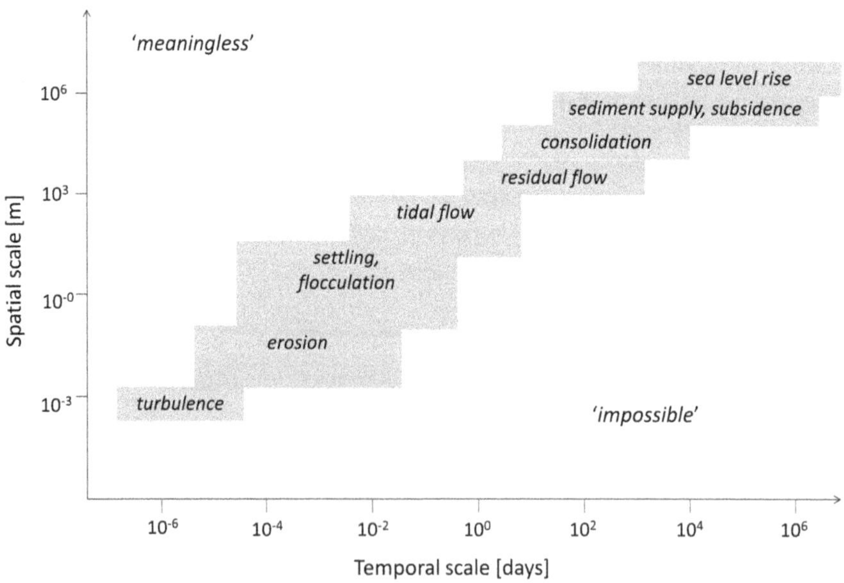

Fig. 13.4. Coupling of model complexity, time and spatial scales involved in fine sediment transport.

between processes at different scales. Within a realistic space–time scale continuum, the modeler has to look for an optimum in every model application, because processes operating at time/spatial scales smaller than the scale of interest may reflect noise only, whereas larger time/spatial scales should be treated as extrinsic boundary conditions (De Vriend, 1991b; see also Sec. 13.6.2).

13.3.3 *Below-capacity conditions*

Sand transport is generally determined fully by the turbulent water motion and sediment grain size (distribution). This yields a fundamental difference with the transport of SPM, as this is usually limited by the availability of sediments (see Chapter 1). For capacity conditions, sediment transport is determined by the transport capacity of the flow, and not by the amount of sediment on/in the bed. Spatial and temporal peaks in SPM then coincide with peaks in bed shear stress, and SPM and velocity are more or less in phase (Sec. 13.5.2). Transport of SPM may be supply-limited (not enough mud available, for instance in sandy environments with only a limited amount of fines) or erosion-rate limited (beds with increasing strengths, such as within (over-)consolidated beds), as discussed in Chapters 1 and 5. Under such conditions, SPM-concentrations depend only weakly on the local flow velocities. For below-capacity conditions, sediment availability is highest in areas where fine sediments converge through hydrodynamic processes (such as the ETM, Sec. 9.1.1). Only in such areas, where the availability of fine sediment in/on the bed is large, SPM-values synchronize with local hydrodynamics.

Models can be setup for capacity or below-capacity conditions, depending on modeling objectives (see Table 13.1). For capacity conditions, an

Table 13.1. Model aspects for equilibrium and non-equilibrium bed conditions.

Modeling approach	Capacity conditions	Below-capacity conditions
Model setup	Fast	Time consuming
Sediment on bed	Full-bed concept	Empty-bed concept
Dominant physics	Local erosion/ deposition	Gravitational circ., tidal pumping and residual flow
Time scales	Days–weeks	Weeks–years
Applications	Local scour and rapid assessments	Transport of disposed sediment, ETM formation, effect of channel deepening, water quality and turbidity, coastal erosion

abundant initial amount of sediment is put in/on the bed, while the erosion parameters are calibrated to match computed SPM-concentrations with observed values. We refer to the "full-bed" concept (Sec. 12.3). Setting up full-bed models is fast, and sufficiently accurate when addressing questions related to local resuspension. As a result, they are common practice in many engineering studies, in particular when relevant time scales are short.

While easy to setup, the full-bed concept often becomes problematic for below-capacity conditions. This is well-illustrated by a virtual simulation of ETM formation. In the full-bed approach, an ETM is attained by prescribing a certain amount of easily erodible sediment on the bed at the location where the ETM is expected to reside. Erosion of this bed leads to high local SPM-concentrations, and over short periods this may yield realistic modeling results, even in a 2DH-approach.

However, location and SPM-concentrations in an ETM vary with river flow and tidal range (spring-neap cycle), which would require other full-bed settings. Obviously, the full-bed approach cannot predict the long-term dynamics of an ETM. Interventions in channel geometry cannot be evaluated either with a full-bed approach. For instance, deepening would lead to smaller bed shear stresses, therefore less resuspension, and consequently lower SPM-concentrations. In reality however, deepening increases gravitational circulation effects, potentially enhancing SPM-concentrations in an ETM.

There is a second, more fundamental reason why the full-bed approach is not advocated, even if data on the actual sediment composition are available. Each model has its own "reality", i.e. the solution of the set of equations resolved. This implies that in general the equilibrium bed conditions of a model are not identical to those in the real world, even if a model would be well-calibrated.

Models for below-capacity conditions are preferably initialized with the "empty-bed" approach (Sec. 12.3), where sediment enters the model domain only through its open boundaries. When sediment properties and hydrodynamic processes are properly represented, sediment transport processes will lead to a gradual convergence in SPM-concentration and bed composition, as explained in Sec. 12.3. Driving processes are residual transport mechanisms by baroclinic and barotropic processes (Chapters 9 and 10). Water-bed exchange cannot be modeled morphostatically with the standard Partheniades–Krone approach: attaining dynamic equilibrium is difficult due to the binary long-term prevalence of this approach for either erosion or deposition (Sec. 4.2). This requires the use of a multiple-layer bed model (or a morphodynamic approach). In general, at least a two-layer buffer model is necessary to account for temporal variations in driving forces (Sec. 12.2). Empty-bed models with multiple-layer bed modules are time-consuming to setup and execute, but essential for many applications.

An important implication of the empty-bed approach is that residual fluxes are largely determined by the sources and sinks within the model domain, i.e. locations and rates of sediment supply and accumulation (Sec. 9.4). This is illustrated with a simple analysis on SPM dynamics in a low-concentration system. Barotropic effects (Sec. 9.2) initially transport SPM into a preferential direction, for instance up-estuary. If there are no SPM sinks, sediment cannot settle, and SPM-concentrations rise. This generates a horizontal concentration gradient and therefore a diffusive SPM flux in opposite direction (in this case down-estuary). When diffusion balances the barotropic flux, the residual SPM flux becomes zero. Important sinks for fine sediment are tidal flats and marshes, and harbor basins. Over long time scales, tidal flats typically maintain their shape and hence are not net sinks of fine sediments. Only when the boundary conditions change, such as with sea level rise or development of vegetation on the intertidal, new net sinks/sources are formed.

Though the empty-bed approach is conceptually superior to the full bed approach for below-capacity conditions, its application may not be practical. For very large open domains, e.g. the entire North Sea, or closed systems with negligible sediment input (Chapter 17), attaining equilibrium may take many years to decades, yielding prohibitive computational times. Another approach is then required. Section 13.6.3 presents some thoughts on how the bed sediment distribution can then be obtained efficiently.

13.3.4 *Data requirements*

The setup of a sediment transport model traditionally consists of two steps (see Sec. 13.4):

(1) a hydrodynamic model simulating water levels, flow velocities, waves, bed shear stresses, salinity, etc., and
(2) a sediment transport model, which may or may not be coupled to the hydrodynamic model (Sec. 13.3.1) — the coupled/uncoupled approach.

Both steps need data for three purposes: (1) for conceptual model development and the subsequent setup of the numerical model configuration;

(2) providing values for model input parameters; and (3) for calibrating and validating the model. Data collection for the hydrodynamic model is generally fairly straightforward and often standardized, and today many data can be found on the internet. Data collection for the sediment transport model, however, is less straightforward, be it only because some of the parameters required cannot be measured directly. Appendix A presents some instructions and considerations on how to measure a number of sediment parameters in the laboratory.

In the following, we refer to a few field measuring techniques, without discussing details of the pros and cons of these techniques — this would require a textbook by itself, while the developments today are fast.

The hydrodynamic model

The hydrodynamics in shallow open water in general, and velocities in particular, are very sensitive to the bathymetry of the aquatic system, i.e. its topography with the land boundaries and water depths. The (initial) bathymetry of an aquatic system is therefore possibly the most important data for any hydrodynamic model setup. Bathymetry is commonly measured with a multi-beam echo-sounder, whereas today, topography follows from remote sensing.

(Tidal) Water levels are generally easy to measure from land-based stations or fixed poles, except in remote areas such as the mouth of the Amazon River; today, also satellites can provide water level data as well, though not yet at the resolution required. Generally, data bases exist with long-term tide recordings, and/or tidal constituents can be found in literature. A cheap, but not always sufficiently accurate source of tidal data, in particular for model boundaries, is found in global tidal models (e.g. the Goddard Ocean Tide Model).

In shallow waters, such as the North Sea (Sec. 10.2.2 and Chapter 14), overtides are important drivers of residual sediment transport (Sec. 9.2). If these overtides are not (only) generated within the modeling domain, they have to be prescribed at the model boundary as well. Note that tidal constituents published in literature generally do not contain information on these overtides.

However, sediment dynamics are driven by velocity and bed shear stresses — velocities have therefore to be measured at various locations, preferably their variation over the vertical and over (a) tidal cycle(s). Velocity measurements are carried out in the form of point measurements with ADCP or ADV. Drogues can be used for obtaining a more holistic picture of the (residual) flow field — the drogues are followed with GPS or drones. Velocity measurements are generally expensive, and can never cover an entire model domain. System understanding and the conceptual model are necessary to set up an efficient and cost-effective measuring program, and interpretation of the data.

The phasing of velocity and SPM gives important information on the relevant transport processes (Sec. 13.5.2). Velocity and SPM measurements (see below) should therefore preferably be carried out simultaneously and at the same locations — in tidal environments at least over a full tidal cycle, whereas seasonal variations may have to be measured as well.

Large-scale (ocean) currents may also affect SPM transport in coastal waters (see Chapter 10). These are notoriously difficult to measure, and literature and atlases are often the only data sources available.

As discussed in Sec. 3.2.4, waves are important in particular for stirring up sediment from the bed (bed shear stresses). The wave field is therefore important data, and have to be measured, again at strategic locations. Sometimes, use can be made of global wave models (such as WaveWatch III) to drive a global SWAN-model or otherwise. The weather, and therefore also the wave climate generally exhibit strong seasonal variations, in particular in more temperate climate zones. Long-term wave data are therefore often a necessity for proper system understanding and modeling.

In Chapters 9 and 10, the role of salinity (and temperature) gradients on the residual transport and fate of SPM is discussed. Salinity data are therefore often crucial. However, salinity fields in shallow waters are difficult to quantify, be it only because of the large inertia of the water bodies: the horizontal and vertical extension and stratification of a salinity field are largely determined by historic hydrodynamics (including wind fields) and fresh water input. However, salinity measurements (in fact conductivity measurements, generally with CTD) have to be carried out in conjunction with velocity and SPM measurements. Interpretation of these measurements and the construction of a large-scale salinity field is done on the basis of numerical model output, system understanding and the conceptual model. Today, remote sensing techniques may help to obtain a holistic picture of the salinity field, for instance on the basis of temperature differences between saline and fresh water, though these do not give information on the vertical salinity structure.

The argumentation above also implies that long-term data on fresh water input are often crucial information to both system understanding and model calibration/simulations.

Finally, data on wind speed and direction are often crucial as well, in particular in shallow areas with intertidal flats and in the case of stratified flow conditions.

The sediment transport model — field data

The main parameters to calibrate fine sediment transport models are SPM-concentrations, SPM-fluxes, bed composition, the mud balance, and bed level variations. A wide range of instruments exist to measure SPM-concentrations, primarily based on acoustic and optical techniques (OBS). These sensors have to be calibrated always against local water samples. Point measurements are needed for quantitative model calibration, in conjunction with velocity and salinity measurements, as discussed above. Vertical SPM profiles (measured through casting or remote, using acoustics) maybe less suitable for quantitative calibration, but provides important insight into the governing sediment dynamics (Sec. 13.5.2).

SPM values follow stochastic patterns, driven by the local and more remote turbulent water movement. Moreover, the availability of erodible sediments, thus SPM values at a particular time also depend on the history of the driving forces, such as the time elapsed after a storm (e.g. Sec. 12.3) and the soil mechanics state of the bed. Long-term SPM data are therefore in general necessary to obtain meaningful information on characteristic SPM values, and for the interpretation of SPM point measurements, see also Sec. 17.4.3.3.

SPM fluxes are often measured with ADCP, using the ADCP velocity data together with its back-scatter signal. Calibration of the latter to SPM is site-specific and requires extensive calibration for every deployment. Though difficult to reproduce in detail and/or at high accuracy, assessment of SPM fluxes provides important information for evaluating model performance. Note that these measurements generally provide gross sediment fluxes, and are usually not sufficiently accurate and/or representative to determine net SPM fluxes: these are often much smaller than the gross fluxes, hence within measurement accuracy (Secs. 9.4, 10.4 and 13.6.4).

Bed composition may be used as input parameter in the full-bed approach. However, since most models are setup to answer questions requiring dynamic equilibrium (thus an empty-bed approach), we advocate data on bed composition to be used to calibrate a model. The bed composition can be obtained through soil sampling, which is costly and time-consuming though. Bed sampling can be combined with back-scatter information from a multi-beam signal. More accurate than the multi-beam back-scatter is the use of side-scan sonar, which also requires ground-truthing with (grab) samples.

Determination of the particle size distribution of these samples is addressed in Appendix A.3, and is not straightforward.

In case considerable amounts of fluid mud are to be expected, dual frequency echo-sounders are an essential instrument in quantifying extend and thickness of these layers (see Chapter 7 and in particular Sec. 7.6). These echo sounders measure at two frequencies, with the lowest frequency typically reflecting on the solid bed (consolidated mud or sand), whereas the high frequency reflects on the interface between water and fluid mud. An example is given in Fig. 13.5. The large scatter in the reflection of the 33 kHz signal is caused by density gradients within the fluid mud and commonly encountered. Therefore, we recommend inspecting the raw signal always before using the mean values generated by the instrument software (see also Sec. 7.6).

A mud balance cannot be measured directly, but follows from the interpretation of all information available (*in-situ* data, numerical model results

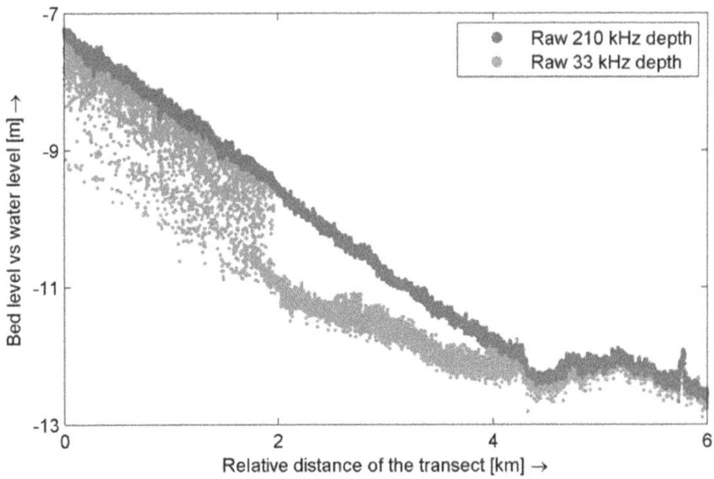

Fig. 13.5. Transect with low- (33 kHz) and high-frequency (210 kHz) bed registration using a Knudsen 33–210 kHz dual-frequency echo sounder, revealing mud deposits between 2 km and 4 km. The scattered 33 kHz reflection between 0 km and 2 km is indicative for mud deposits with internal density gradient; the collapse of the 33 and 210 kHz signal between 4 km and 6 km for strongly consolidated or sandy sediment.

and the conceptual model). A mud balance is one of the final pieces of the system-understanding puzzle. Patterns in clay mineralogy can contribute to understand the long-term (decades — millenniums) mud balance, e.g. marine or riverine sources as in Verlaan (2000); or locations at which clay minerals are eroded, e.g. Adriaens *et al.* (2018). Over short (tidal) time scales, SPM fluxes can be inferred from through-tide ADCP transects. However, while such estimates provide realistic gross transport rates, they poorly represent residual transport as (1) an ADCP misses near-bed SPM dynamics, and (2) sub-tidal variations requires (unrealistically) long measuring periods (see also Secs. 9.4 and 10.4).

Maintenance dredging volumes form an important data source to establish a mud balance at intermediate time scales (month, years), see Van Maren *et al.* (2015a) and Chapter 14.

The sediment transport model — sediment properties

Three main model input parameters for the Partheniades–Krone model are M and τ_e (for erosion of sediment, Chapter 6) and w_s (for settling of sediments, Chapter 4). All three parameters can be measured in the field or in the laboratory, while all methodologies yield specific advantages and disadvantages. However, measurements are difficult, and the parameters are subject to substantial uncertainty and variations.

The settling velocity of fine sediment is largely affected by flocculation and may vary widely (Sec. 4.1). Flocs are fragile, and easily destroyed during sampling and subsequent transport to the laboratory. Settling velocities measured in the laboratory can therefore be very misleading. Moreover, many of the laboratory techniques require pre-treatment of the samples, further affecting the sediment's floc structure (Appendix A). We believe that the best method to measure settling velocities of flocs in the laboratory is by using a settling column, the procedure of which is described in Appendix A.8.

Four techniques exist to measure settling velocities *in-situ*, as described in Sec. 4.1, using an *in-situ* camera or the LISST settling column, or indirectly from ADV-data or vertical SPM profiles.

Section 6.1 describes a number of techniques to measure the erodability of cohesive sediment beds. These can be grouped as follows:

(1) Onset of erosion on intertidal areas, e.g. the CSM (Cohesive Strength Meter, Paterson, 1989; Tolhurst *et al.*, 1999) and the ISIS (Instrument

for Measuring Erosion Shear Stress *In-situ*, Williamson and Ockenden, 1996).
(2) *In-situ* flumes. On the intertidal, the ISEF (*In-situ* Erosion Flume, Houwing and Van Rijn, 1998). Under water, the Sea Carousel (Maa *et al.*, 1993) and the Benthic Annular Flume (Widdows *et al.*, 1998).
(3) Erodability of cores. The Sedflume (Jepsen *et al.*, 1997; McNeil *et al.*, 1996), the Erodimètre (Le Hir *et al.*, 2006, 2008) and the U-GMES, which is a modification of the Gust Microcosm (Gust and Müller, 1997).
(4) Laboratory erosion tests on placed or deposited beds, as in a rotating annular flume.

The majority of these techniques are cumbersome, do not provide in-situ information, or only on the intertidal, characterized by profound over-consolidation (Sec. 5.2). We believe that measuring erodability on cores is doable from a cost-benefit point of view, while providing reasonable information on soil behavior, though at isolated points only. In general, we recommend the procedure described in Secs. 6.3 and 6.4, which relates *in-situ* erosion parameters to bulk soil mechanical parameters which can be determined easily.

In Sec. 4.2, the reduced deposition coefficient p_d was introduced. There is no way to measure this parameter *in-situ*, or in the laboratory, as it parameterizes a number of processes not explicitly modeled. p_d has therefore to be determined from sensitivity analyses with the numerical model (calibration procedure, Sec. 13.5.5), and one's system understanding (Sec. 15.4). If the buffer model (Sec. 12.3), or any other multiple bed layer model is used, parameters for the transport of sediment between the various layers are required, which again cannot be measured directly, but follow from sensitivity analyses with and calibration of the numerical model (1DV-approach, Chapters 14 and 17). In both cases, a sound system understanding is necessary for determining the relevant parameters.

At higher SPM-concentrations, hindered settling becomes important (Sec. 4.4), requiring values for the gelling concentration c_{gel}, which can be determined from settling column tests (Appendix A.8).

Depending on the sediment modules activated, many more values for input parameters my be required, such as burial coefficients in case of using the buffer model, permeability and effective strength parameters, in case of using a consolidation model, etc. The reader is for instance referred to Winterwerp *et al.* (2018) for a discussion on such parameters.

13.4 Setup of a numerical model

13.4.1 *Spatial and temporal scales*

The spatial domain covered by a numerical model is determined by the engineering/managerial/research question(s) to be addressed (see Sec. 13.3.2) and by the scales of the governing processes. In Sec. 13.3.2, we also argued that simulation period and domain size are closely related, because of the correlation between spatial and temporal scales of physical processes.

However, there is a third, numerical requirement in defining the model domain, related to the model boundary conditions. The boundary conditions of a model generally do not (exactly) match the solution of a numerical model. As a result spurious circulations, unrealistic salinity patterns, etc. are generally induced in the vicinity of these boundaries. Obviously, such circulations and salinity patterns may not affect the domain of interest. Therefore, adaptation space is required. In tidal environments, this adaptation space covers at least one tidal excursion. For steady flow, the bathymetry is more leading, for instance the presence of tributaries, bank irregularities, etc.

For large-scale applications (e.g. a shelf sea) these boundary condition effects generally lead to a relatively small increase in model size, whereas for small-scale applications (e.g. scour around a construction in a tidal environment) the adaption space may constitute the majority of a model domain.

A numerical model has to be initialized, for instance with a uniform water level and zero velocity, and SPM-concentrations. Also these initial conditions generally do not match the solution of a numerical model. Therefore, a numerical model has to be run for some time before its results become meaningful. This period is referred to as the "spin-up time". Often, numerical model studies require attaining a dynamic equilibrium, i.e. repetitive results (SPM-values, mud balance, ...) over a specific period, which may comprise a tidal cycle, a spring-neap cycle, seasonal cycles or otherwise. In that case, spin-up times are much longer, and may even cover many months to years, depending on the size of the model domain — this is particularly the case for the empty-bed approach.

In the latter case, the spin-up time is largely determined by the inertia and history of the relevant processes. Spin-up times for the hydrodynamic themselves are relatively short, but generally much longer for salinity and SPM, as these have to be internally redistributed, depending on the hydrodynamic exchange processes and, for sediment, the water-bed exchange processes.

13.4.2 *Horizontal and vertical grid resolution*

In a numerical model, the relevant equations are discretized and resolved on a computational grid, its size determining the accuracy of the numerical solution (see also Sec. 12.1). The horizontal grid resolution (i.e. grid size) is determined by the questions to be addressed and the hydrodynamic processes driving sediment transport in the area of interest. For large-scale model applications, this may lead to unpractical computational times if a small grid size would have to be applied throughout that domain. Then numerical techniques, such as nesting, domain decomposition and/or an irregular computational grid (see Chapter 18) become mandatory.

A proper resolution to resolve horizontal circulations over a river cross-section or tidal channel or of stable channel morphodynamics requires at least six grid cells, though 10–20 cells is advisable.

The vertical resolution of a numerical model should be optimized to simulate (the steepest) vertical gradients in hydrodynamics, salinity and SPM. Sometimes, these gradients coincide, sometimes not. The larger vertical gradients in SPM are generally found near the bed. In that case, a high near-bed resolution is advisable. However, at high SPM-concentrations (HCMS-conditions), large SPM and hydrodynamic gradients coincide (Chapter 7) and SPM-induced stratification may occur throughout the water column. Also salinity-induced stratification interacts with the local hydrodynamics, reducing for instance vertical mixing, in particular around the larger salinity gradients (Sec. 3.2.3).

Large gradients in bathymetry (such as along the continental shelf, but also in estuaries and coastal seas with intertidal flats) may further complicate the grid design. For non-stratified flows over complicated bathymetries, the so-called σ-transformation was developed, keeping the number of vertical grid cells (layers) constant over the computational domain. The thickness of these layers thus varies with water depth, including tide-induced water level variations. Such a σ-grid is sketched in Fig. 13.6. Though efficient from a computational point of view, the σ-approach has two major disadvantages:

(1) The near-bed resolution, often crucial for sediment transport models, varies over the model domain — note that such variations may be very large, and
(2) The thickness of a salinity-induced stratification is often more uniform in space than the bathymetry. The resulting spatial variation in σ-cell

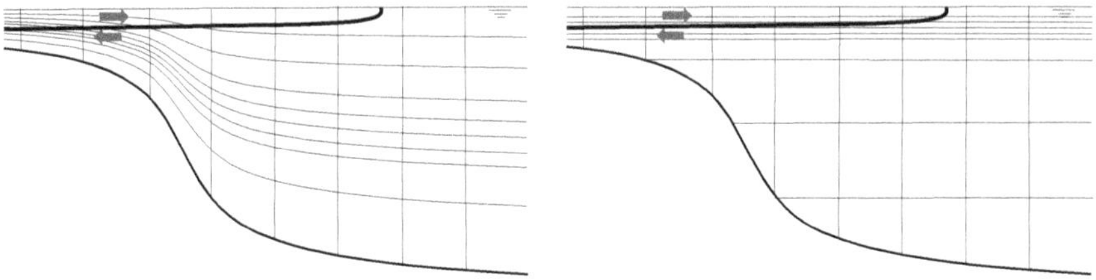

Fig. 13.6. Vertical grid coordinates with σ-layers (left) and z-layers (right) to capture, e.g. a river plume.

thicknesses leads to spatial variations in the interface depth (Fig. 13.6), and numerical mixing as a consequence.

The second disadvantage can be resolved using horizontal grid layers, the so-called z-approach, or z-model, as shown also in Fig. 13.6. A disadvantage of the z-model is that the near-bed computational cell thickness varies even more over space than in the σ-model. This may induce strong spatial variations in the computed bed shear stress, leading to inaccuracies in sediment transport and morphologic developments.

Obviously, a more flexible approach is advisable, in which the near-bed resolution is kept constant, for instance five layers of constant thickness, following the bed topography in detail, whereas higher in the water column, grid layers are kept horizontal, allowing an accurate representation of salinity-induced stratification.

Proper modeling of high-concentrated mud suspensions (HCMS — Sec. 7.1) forms another challenge, as significant stratification may occur throughout the water column. Simulating such conditions requires high vertical resolution throughout the water column, with 40–100 vertical layers. Fortunately, HCMS-occurrences are generally well restricted in space, for instance in fairways or small estuaries (Winterwerp and Wang, 2013), thus also limiting the domain where such high resolution would be required.

From the above we have to conclude that no general recipes can be given for setting the vertical resolution in numerical sediment transport models. For engineering applications, a ten-layer vertical resolution may often suffice, but again, there are exceptions, and the user has to include the consequences of limited vertical resolution in the conceptual model.

13.4.3 *The bed*

As discussed in Chapter 1, on shorter time scales, water-bed exchange processes are generally dominant in shallow waters. These exchange processes do not only largely govern short-term variations in SPM-concentrations, but also affect long-term net sediment fluxes (see Sec. 9.2). Further, we have discussed the role of fluff layer dynamics in the net fluxes of fine sediment during moderate hydrodynamic conditions (see Sec. 6.5), and the role of over-consolidation on intertidal flats (Sec. 5.2) and pseudo-cohesion in case of large erosion rates (Sec. 6.3).

However, many natural systems are characterized by mixtures of sediment. Here we address sand–mud mixtures only (Chapter 1). Even if the engineering question to be addressed relates to the mud fraction only, it may be necessary to account for the sand behavior as well (Sec. 6.6). Sedimentary beds exhibit either cohesive or non-cohesive properties, as discussed in Chapter 1 and Sec. 2.2. Fortunately, this transition is generally very distinct because of sorting and feedback processes (Sec. 6.6), or in other words, the actual clay content ξ^{cl} in the bed is either well below or well above its critical value ξ_0. Of course this implies that ξ^{cl} should be tracked during a simulation.

We also argued that the silt–clay ratio in a specific aquatic system is generally constant (Sec. 2.2). This implies that for sand–mud mixtures we may use the mud content (or its complement, the sand content) to distinguish between cohesive and non-cohesive behavior, which is favorable from a modeling point of view.

The arguments above, in conjunction with our preference for the empty-bed approach (Sec. 12.3) implies that the traditional approach of a uniform one-layer bed schematization is only suitable for a limited number of practical applications. In general, a two-layer or multi-layer bed model is required, which can account for vertical and small-scale horizontal (see Sec. 4.2) gradients in sediment composition and properties. The two-layer buffer model is a simple bed model which can account for at least two different time scales. Multi-layer and two-layer models can more realistically account for longer time scales, including episodic events, and sediment properties. They may even account for consolidation processes in a parameterized way (Sec. 12.2). The choice between these various approaches is largely driven by the questions to be solved, as one can argue that, from a physical point of view, multi-layer bed models are always preferable.

Apart from large computational times, another disadvantage of using a two- or more-layer bed model involves the model parameters required. Whereas the uniform bed model approach requires values for M and τ_e only, which may vary over the modeling domain, multi-layer bed models require substantially more physical input parameters, some of which can be measured in the laboratory, others have to be set during the calibration phase, as argued by Winterwerp *et al.* (2018). Adding more parameters also complicates the calibration procedure of a model (see Sec. 13.5), as more parameters have to be tuned. Obviously, the selection of a bed model is important, and may largely affect the amount of work and time required to complete a modeling study. This issue is addressed in Sec. 17.4.3.

13.4.4 *SPM fractions*

Even when flocculated, SPM is characterized by a distribution in particle/floc size (psd — particle size distribution, Sec. 4.3), ranging from very fine primary particles of only a few μm to large flocs of a few mm diameter, and the flocculi (floc building stones) in between. The finer particles govern turbidity and water quality issues (Sec. 3.5), whereas the coarser fraction is more relevant for assessing harbor and fairway siltation (see also Sec. 4.1). Hence, the question(s) to be addressed determine the SPM fraction for which transport and fate has to be simulated.

We have discussed (Sec. 4.3) that the psd of SPM may vary over time and space, because of flocculation and sorting. However, flocculation dynamics are still difficult to quantify, and expensive to model (Sec. 4.3), even if a limited number of floc classes are accounted for. A more pragmatic approach is accounting for 2–4 floc size classes, characterized by equivalent settling velocities. This approach works quite well if SPM-concentrations are relatively small, as then flocculation times are long enough to presume a static psd (Sec. 4.3). Changes in psd are then caused by sorting only. Typical values for the settling velocity w_s are presented in Sec. 4.1.

Application of a more-fraction approach requires bookkeeping of the fractions depositing on/in the bed. While residing in/on the bed, it is (has to be) assumed that the properties of these fractions do not change, nor that the various fractions interact mutually. Upon erosion, sediment fractions are then re-mobilized in proportion to their availability in the bed (Sec. 6.6). Note that currently even the most sophisticated flocculation models do not contain modules to account for in-bed aggregation/breakup processes, no doubt because the underlying physical processes are still unknown. From a modeling point of view, this implies that all sediment fractions follow the same erosion law with the same model parameters, though these may differ over the depth of a (multi-layer) bed model.

Because of the large volume of mud flocs, hindered settling becomes relevant at a few 100 mg/L, as near-bed concentrations then may attain values of a few g/L (Sec. 4.4). Hindered settling may largely reduce near-bed settling velocities and deposition rates (see also the discussion on reduced deposition, Sec. 4.2). Then, the effects of hindered settling have to be included in numerical models for fine sediment transport. As the inclusion of a hindered settling formula in a numerical model is simple and computationally not expensive, we recommend to always account for hindered settling.

13.5 Model calibration

In numerical modeling studies, distinction is made between two classes of variables:

(1) Independent variables which steer the numerical model and/or model processes.
(2) Dependent variables which are the outcomes of the model, and are the parameters used to address the questions to be solved.

Examples of independent variables are tidal constituents at the model boundaries, riverine fresh water and sediment input, soil mechanical parameters used for assessing erodability, etc. Flow velocities and SPM-concentrations are examples of dependent variables. However, in practice, this distinction is less sharp. For instance, flow velocities are dependent variables in the hydrodynamic simulations, but independent variables in determining erosion rates. Experienced modelers often play unconsciously with this ambiguity, but it is good modeling practice to formulate this distinction explicitly in every modeling step.

Calibration is the adaptation/modification of independent variables to obtain a match between observed and computed distributions of dependent variables Oreskes *et al.* (1994). Traditionally, model validation compares the calibrated model for other conditions, i.e. against an independent second dataset, but with the model settings derived from the calibration phase.

This traditional calibration/validation procedure is only possible if accuracy in model description and input parameters and in (field) data is

Table 13.2. Steps in calibrating a numerical sediment transport model.

Phenomenological calibration	Compare phenomena in real world and model
Quantitative calibration	Compare model output with all data available
Uncertainty analysis	Analyze possible errors inherent in data and model
Sensitivity analyses	Assess sensitivity of model output to model parameters

large, and uncertainties are small, while data sets themselves are large (Sec. 17.4.3). This is for instance the case with tidal prediction models. However, this is generally not the case for sediment transport models, and we need to adopt another strategy. This strategy makes use of all data available, of our system understanding and understanding of the relevant physical processes, thereby using the conceptual model described above as a key tool. Table 13.2 summarizes the four steps in this strategy. These steps are not necessarily consequential, and often iterative.

We have stressed a number of times that (conceptual and numerical) model setup and applications depend strongly on the questions to be addressed. The same holds for assessing the calibration of a numerical model. This implies that models which reproduce e.g. SPM-concentrations best, are not necessarily the best models to quantify the impact of (engineering) interventions in an aquatic system. We return to model evaluation in Sec. 13.5.6.

13.5.1 *Calibration of hydrodynamics*

The setup of any sediment transport model is preceded by the setup and calibration of a hydrodynamic model. For details on hydrodynamic modeling, reference is made to the large number of textbooks covering this discipline. However, it is crucial to calibrate, or at least evaluate, the hydrodynamic model against parameters relevant for fine-grained sediment transport. Important dependent parameters for calibration are:

(1) *Water levels*: Fine-grained sediment transport is driven by flow velocities; residual transport arises from time and spatial asymmetries in the flow field generated by tides and residual flows (Sec. 9.2). Emphasis in water level calibration should therefore be on phase difference between the main tidal constituents (typically M_2 and M_4 on mid-latitudes; additionally the combination O_1, K_1, and M_2 at lower latitudes) and its overtides, when relevant.

(2) *Current velocity*: Calibration of flow velocities is often hampered by limited availability of data. Further, flow velocities are often strongly influenced by small-scale bathymetrical gradients, not resolved through model topography. Determination of higher harmonics from velocity data is often cumbersome, because of the rotational character of the velocity vector, in particular in coastal waters. In that case, the phasing of the velocity magnitude with respect to the water level variations becomes an important parameter for assessing the quality of an hydrodynamic model.
(3) One of the most important parameters in a fine sediment transport model is the *bed shear stress*. Therefore, maps with mean and maximum and minimum bed shear stresses as function of the tidal phase are important tools for evaluating hydrodynamic models — unfortunately, these cannot be measured in the field, and these maps have to analyzed on the basis of system understanding.
(4) *Residual flows*: Residual flows are important for the long-term transport and fate of fine-grained sediment. However, residual currents are difficult to measure, in particular over longer periods of time, and good quality residual flow data is therefore rarely available. Computing residual flows requires long time-series, involving at least several tidal cycles, but generally much longer (seasons to years). A large problem in assessing residual flows is their inherent inaccuracy — residual flows and fluxes are generally the small difference between two large counter-flowing water masses, e.g. ebb and flood currents (see Secs. 9.4 and 10.4).
(5) *Waves*: Wave measurements involve converting high-frequency water level variations into standard wave quantities such as the significant wave height H_s and peak wave period T_p. These may then be compared to computed wave quantities in the time or frequency domain. Note that this may not be straightforward in very shallow water because of the proximity of wave and turbulence frequencies. The wave fields are used to compute wave-induced bed shear stresses, and sometimes wave-induced mixing (Sec. 3.2.4) and streaming (Sec. 7.5.2) — again maps with mean and maximum and minimum wave-induced bed shear stresses have to be inspected.
(6) *Salinity*: Salinity gradients determine net SPM fluxes by gravitational circulation (Secs. 9.1 and 10.2). Residual and sub-tidal flows strongly influence salinity. Though salinity point measurements are relatively easy, measuring entire salinity fields is cumbersome, be it only because

of its large inertia. Assessment of computed salinity fields generally requires proper system understanding, next to sufficient data.

(7) Aggregated parameters such as mixing and residence time, the water balance (including origin of water masses). Such parameters can be measured, however sophisticated techniques are required (for instance tracers). Hence, evaluation of these parameters from model results is often done qualitatively, based on system understanding.

13.5.2 *Phenomenological calibration of sediment transport*

The next step in the calibration of a sediment transport model is a qualitative assessment on how physical phenomena, identified in the conceptual model, are reproduced by the numerical model. We refer to phenomenological calibration, which is generally subject to expert judgment. In general, the following phenomena should be reproduced:

(1) Typical levels of SPM-concentrations, including vertical, seasonal, spring-neap and tidal variations, and peaks during e.g. storms. Such comparisons can be aided by surface sediment concentrations obtained from satellite images (often from the visible light spectrum), providing a qualitative measure of the accuracy of the spatial distribution of fine sediments (e.g. De Boer, 2009; Van Maren *et al.*, 2014). Frequency analysis of long-term SPM time series (e.g. years) allows determination of the contribution of overtides, principal tides, spring-neap variations and the hydrological regime on turbidity (Jalón-Rojas *et al.*, 2017), which can be used to calibrate numerical models as well, see also Sec. 17.4.3.3.

(2) The phasing of SPM-concentrations in relation to variations in tidal velocity. When in phase, advective SPM transport and bed erosion are likely dominant factors, whereas a 90°-phase difference indicates local mixing and resuspension. Further, if the SPM peak is attained well before maximum velocity, supply limited conditions are to be expected.

(3) Ebb-flood asymmetries in both SPM-concentrations and SPM fluxes give information on preferential transport directions. Such comparisons can be carried out in the time domain, and in the frequency domain. Converting observed and computed SPM time series into average concentrations relative to the tidal phase (Vanlede *et al.*, 2019) provides valuable information on the accuracy of computed tidal variability of SPM, including tidal asymmetry (Fig. 13.7).

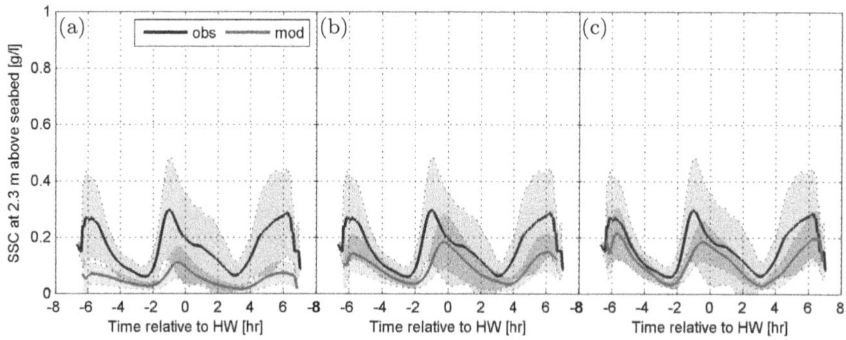

Fig. 13.7. Computed (mod) and observed (obs) SPM-concentration in the North Sea (station MOW1, near the Port of Zeebrugge, Belgium), ensemble-averaged over the summer season and relative to High Water (HW). The three panels show three different model realizations, compared to the same observation (Vroom *et al.*, 2017).

(4) Sources and sinks. Although absolute transport fluxes are generally unknown, areas where SPM transport pathways converge or diverge can often be identified. These areas represent sources and sinks for SPM within the modeling domain (think of e.g. a silting-up fairway, or an intertidal mudflat). In estuarine environments, sediment convergence often gives rise to an ETM (Sec. 9.1.1), which is an essential features to be reproduced by a numerical model.

(5) Water-bed exchange, such as the residence time of fines within the bed, the buffer capacity of the bed (the sediment stock), or overall variability in bed levels give qualitative information on the performance of a model.

13.5.3 *Quantitative calibration*

The phenomenological calibration phase can be further formalized using metrics quantifying the agreement between observed and computed variables. Common metrics to quantitatively compare computed (M — modeled) and observed (D — data) SPM-concentrations are the model bias B, the Root Mean Square Error (RMSE), the correlation coefficient R^2 and the skill score S. The normalized unbiased RMSE* and B* are defined as follows:

$$B^* = \frac{\overline{M} - \overline{D}}{\sigma_D} \tag{13.1}$$

$$\text{RMSE}^* = \frac{\text{sgn}(\sigma_M - \sigma_D)}{\sigma_D}\sqrt{\overline{\left[(M_n - \overline{M}) - (D_n - \overline{D})\right]^2}} \tag{13.2}$$

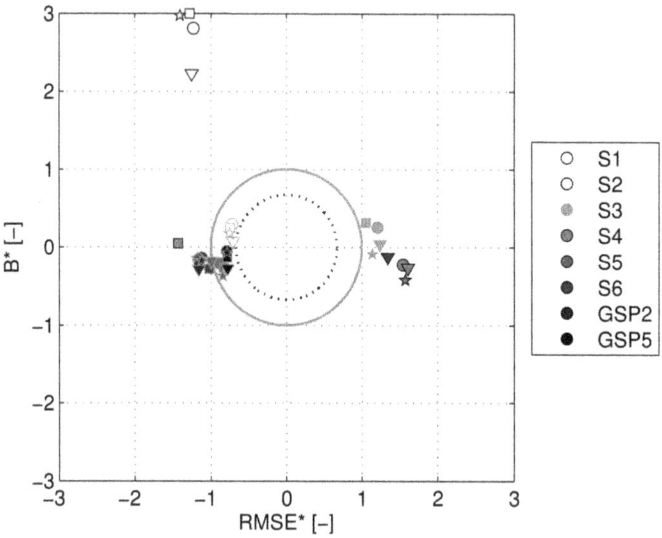

Fig. 13.8. Example of target diagram illustrating B* and RMSE* for various model realizations (triangles, squares, circles and stars) at eight different locations (various symbols). See Van Maren and Cronin (2016) for details.

where an overbar indicates mean values, σ_D and σ_M are the standard deviations of the observed and computed values (respectively) and $\text{sgn}(\sigma_M - \sigma_D)$ reflects the sign of the standard deviation difference.

B* and RMSE* can be conveniently visualized in so-called target diagrams (Jolliff *et al.*, 2009; Los and Blaas, 2010). Target diagrams relate the model bias B* on the ordinate to the unbiased root-mean-square-error RMSE* on the abscissa. Such plots concisely present the accuracy of a great number of model realizations or output locations. Figure 13.8 shows an example of an arbitrary model application for various model runs. This diagram shows that average values are well reproduced (low B*) but that the variability in the data is poorly reproduced (high RMSE*).

The goodness of fit R^2 and the skill score S (Willmott, 1981) represent metrics for the mean square error of model results. Computed values are in perfect agreement with data when $R^2 = S = 1$, while a value of 0 indicates complete disagreement:

$$R^2 = \left(\frac{\overline{\left((M_n - \overline{M}) \, (D_n - \overline{D}) \right)}}{\sigma_M \sigma_D} \right)^2 \tag{13.3}$$

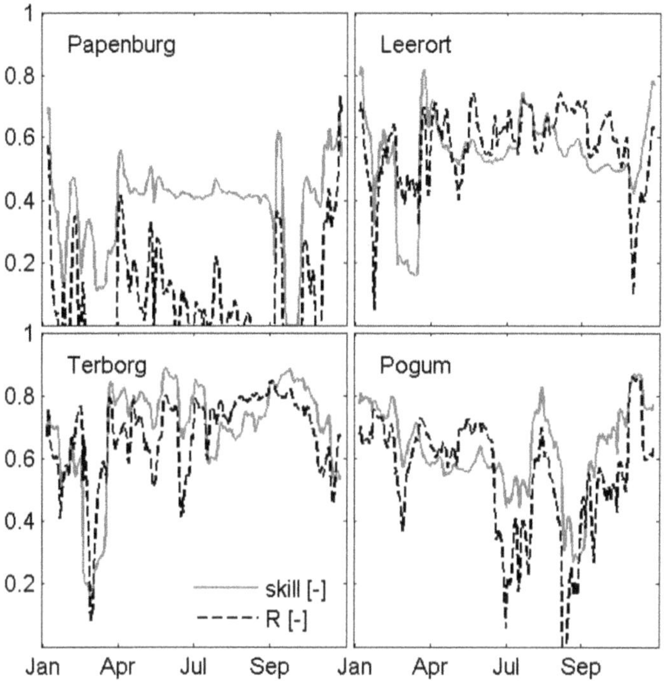

Fig. 13.9. Example of S- and R^2-scores for a model application simulating SPM-values in the Ems-River (Van Maren *et al.*, 2015b).

$$S = 1 - \frac{\overline{|M_n - D_n|^2}}{\overline{\left(|M_n - \overline{M}| + |D_n - \overline{D}|\right)^2}} \tag{13.4}$$

An example of such S- and R^2-scores is presented in Fig. 13.9 for a model application for a one year SPM simulation in the River Ems, showing S- and R^2-values at four different stations along the river; see Sec. 16.4.3 and Van Maren *et al.* (2015b) for further details.

These metrics can be used to quantify model accuracy for a time series at a specific measuring station, and to assess time variations in model accuracy itself (as in Fig. 13.9). Such visualizations conveniently illustrate when a model performs properly, but does not provide an overall criterium on whether a model is "good", "reasonable", or "bad" — see Sec. 13.5.6. Moreover, assessing model performance should be parameter-dependent, as for instance water levels, flow velocities, and SPM-concentrations are progressively more erratic and thus more difficult

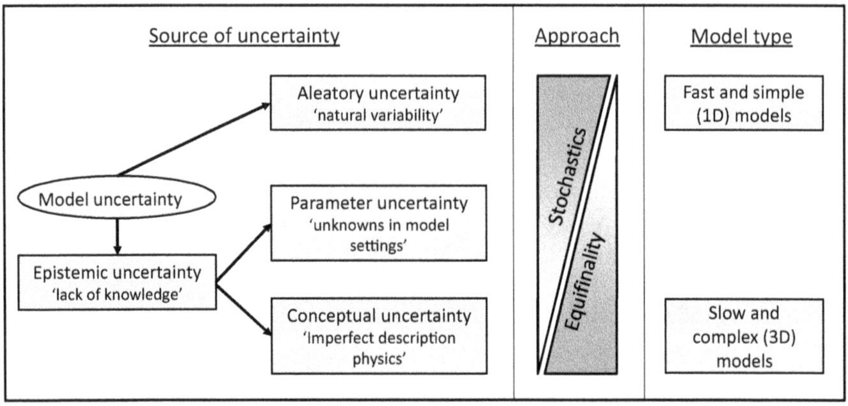

Fig. 13.10. Relation between sources of uncertainty, modeling approach, and model type.

to reproduce accurately. For SPM-concentrations, a reliable accuracy score is currently not available.

13.5.4 *Model uncertainty*

Numerical sediment transport models are subject to several sources of uncertainty (see Fig. 13.10). Aleatory uncertainty relates to the stochastic nature of physical processes, varying from turbulent fluctuations to the variability in wind and wave forcing. Epistemic uncertainty accounts for limitations in understanding relevant physical processes, and uncertainty in model input parameters (Walker *et al.*, 2003). The influence of aleatory uncertainty and input parameter uncertainty on model performance can be quantified fairly straightforward in disciplines where fast computational models are available, such as in hydrology, where 1D approaches are common.

Such analyses require large amounts of simulations and are therefore not possible for numerical SPM transport models, often requiring three spatial dimensions and long simulation periods to achieve dynamic equilibrium. Stochastic analyses on model input uncertainty for sediment transport studies has been limited to 2DH model approaches (Van der Wegen and Jaffe, 2013; Van Vuren, 2005).

A second disadvantage of stochastic methods to quantify uncertainty in input parameters is that these methods assume that all input parameters to be varied are mutually independent. In reality these are not, of course. For instance, combining a very low critical bed shear stress with a high-erosion parameter yields physically meaningless high-erosion rates and SPM-concentrations which provide a false sense of model uncertainty.

A computationally more efficient method to quantify uncertainty is non-uniqueness or equifinality, which also overcomes the problem of parameter interdependency. In case data is limited, a model can be calibrated to a set of available data using more than one set of model input parameters. As an example, increasing the erosion flux parameter jointly with the critical bed shear stress, and/or reducing the settling velocity may lead to comparable SPM levels. The fact that different combinations of model input parameters may lead to identical results (equifinality) may be interpreted as a weakness in the predictive capacity of a model. However, equifinality also provides a measure of model accuracy. A model may be calibrated to a certain set of measured variables using an infinite number of parameter settings, and this becomes more true if the number of parameters increases. This range of model parameters influences the importance of individual sediment transport processes, e.g. by strengthening burial in the seabed, settling from suspension or erosion, etc. When a model is used to address questions outside the model calibration realm, a substantial uncertainty is introduced in the model reliability. However, when the results of such simulations are comparable for all equifinal model settings, the model is apparently not sensitive to input parameter uncertainty, at least, not with respect to the question resolved with the model (see Van Maren and Cronin, 2016).

Note that such an equifinality analysis is useful only when data for calibration are scarce. If for instance data on bed composition are available, next to SPM data, the parameter space becomes more confined, reducing the degree of variability in model parameters. Then it will be difficult to establish sets of model parameters which can give identical/similar model results for all parameters (see for instance the analysis on residence times in Chapter 14). Note that for instance additional calibration for variations in the fine fraction of bed sediment may substantially reduce the number of parameter sets.

More problematic are limitations in our understanding of the governing physical processes, and/or ignoring physical processes in the modeling setup, such as erosion, flocculation, consolidation, and biological processes — see Orvain and Sauriau (2002). This conceptual uncertainty may have a large, but unknown impact on the model applicability, and can, by definition, not be quantified. Conceptual uncertainty has been acknowledged as a main source of uncertainty, yet is rarely considered in environmental modeling (Refsgaard *et al.*, 2006). A formal uncertainty analysis on input parameters and boundary conditions may then be misleading. As the physical effects of many of these processes are well understood (as described in the first part of this book), the modeler may make an educated guess on the effect of the ultimate model results. In other words, here also the conceptual model is an essential tool in addressing uncertainty by expert judgment.

13.5.5 *Sensitivity analyses*

Given all uncertainties in the setup and calibration of a numerical fine sediment transport model, it is essential to get a feeling on the sensitivity of model output (the dependent variables) to variations in input variable (the independent variables). This is done with a so-called sensitivity analysis. Note that an equifinality analysis can be executed fairly easy with the results of a proper sensitivity analysis.

A sensitivity analysis implies that model input parameters are varied around a set of reference parameters, assessing model response on SPM-concentrations and fluxes, siltation rates, bed composition, etc. Such sensitivity analyses should always be part of the calibration procedure, and their results should always be included in the interpretation of model results.

In case model output variability as function of model parameter variations is small compared to model output variability as function of variations in driving forces, sensitivity analyses give a good indication of the spread in model results. If not, this range is difficult to assess.

13.5.6 *Model evaluation*

As explained in the previous sections, model performance can be evaluated in many ways. Quite often, the most quantitative calibration method is considered to be the most valuable means of evaluating model performance. This is not necessarily true. In fact, as illustrated in Sec. 13.5.4, this may even be misleading. For instance, a model may be a factor of two to three off in reproducing observed SPM-concentrations (thus a poor quantitative calibration), but still be very useful in predicting the effect of interventions. This is the case for instance if a model accounts for the relevant transport mechanisms. On the other hand, a model which perfectly reproduces calibration data may be completely off in predicting impacts when not accounting for the relevant transport processes (as exemplified with the ETM dynamics in relation to capacity and below-capacity transport, see Sec. 13.3.3).

Therefore, there are three other, more qualitative criteria for evaluating the quality and applicability of a numerical sediment transport model:

(1) *Representativeness*: Does the model represent the proper physical processes, and do the boundary conditions represent the dominant conditions, which are representative for the questions to be addressed? Representativeness is discussed in Secs. 13.5.2 and 13.5.5.
(2) *Reliability*: The questions to be addressed generally imply extrapolation of a model beyond its domain of calibration (for instance, changes in bathymetry, the effect of episodic events, ...). Reliability implies that one can have confidence that the model will indeed provide reliable predictions for these extrapolations. Reliability is at the heart of system understanding and the conceptual model, and has been discussed partly in Sec. 13.5.2.
(3) *Credibility*: Sometimes there are just too few data to perform a proper calibration of a sediment transport model. The quality of a model has then to be evaluated on the basis of expert judgment. Do the modeling results make sense, are the results qualitatively in agreement with expectations? In other words, are the model results credible? Credibility largely depends on expert judgment.

The model criteria above should be carefully evaluated and discussed with relevant stakeholders. Stakeholders are often not clearly defined in case of academic studies. In case of commercial modeling projects, the main stakeholder is the paying client: is the model in accordance with the client/user expectations? This is not trivial, as the main stakeholder may not always be able to interpret and understand the technical details but rather focus on the implications of the model results.

13.6 Good Modeling Practice guidelines

Based on the considerations above, we define five steps in Good Mud Modeling Practice. However, to put a modeling exercises in perspective, one has to realize that:

(1) Models may not become more accurate with increasing complexity, e.g. by accounting for more processes. A model is only useful as long as the user understands its input, the model itself, and its output. When setting up a model, start simple and add complexity step by step.
(2) "All models are wrong, but some are useful" (Box, 1976). Models are in particular useful if they can predict changes in relevant dependent

parameters relative to a reference case. For instance, what is the relative increase in siltation rates if a fairway is deepened? Absolute model results, such as computed SPM-concentrations or bed level changes may deviate from observations (or observations may even be non-existent or not consistent), while the response to changes in model input, parameter settings or processes may be realistic — see also Sec. 13.6.5.

(3) Model setup takes time. Calibrating and understanding sediment transport models requires many model runs. Most 3D model applications are computationally demanding, and therefore the calibration period takes considerable time. Models should therefore be setup at the lowest resolution possible resolution, accounting for the relevant physics, though. Note that reducing dimensions (2DH) may be misleading (Sec. 12.1).

(4) Also the setup of a conceptual model takes time. However, this is almost always a good investment, as it reduces the changes on erratic paths largely.

The five steps for good mud modeling practice are summarized in Fig. 13.11. We use Roman numerals to distinguish from the more general good modeling practice in Fig. 13.1. All steps are elaborated upon in this chapter, albeit not exactly in the order of this framework.

13.6.1 *GMP — Step I: Conceptual model*

Step I involves the conceptual model development (Sec. 13.2). Every study starts with a clear definition of the underlying objectives and research questions, i.e. the aims and objectives of a study. Then data are collected, starting the conceptualizing of the aquatic system in a quantitative way, identifying the relevant processes. If necessary, new data are collected. This is followed by a formalization of purpose and system understanding into a conceptual model, describing (1) which processes are relevant for answering the engineering/research questions, (2) which are the relevant spatial and temporal scales, defining model domain and simulation times and (3) the sediment transport modeling approach and formulations. The system understanding itself, and thus the conceptual model can evolve throughout the course of model development, especially during Steps II, III and V.

The conceptual model largely determines the setup of a numerical model, and may therefore evolve only slightly during the course of a study. To address other study goals, a model may need to be modified. However, the choice between an full-bed or empty-bed approach, for instance, has

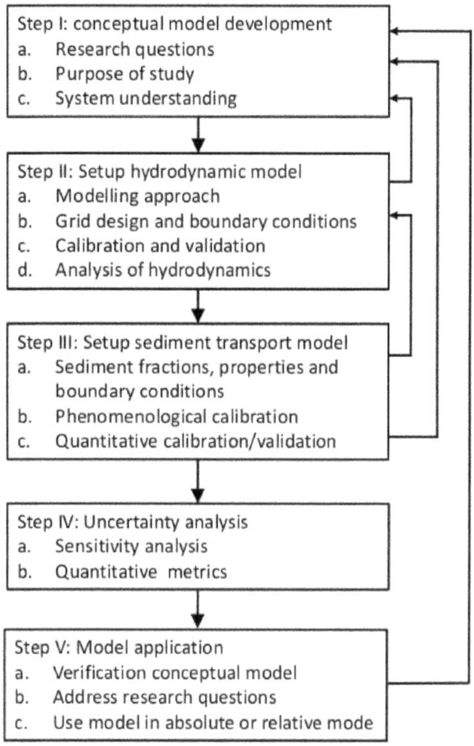

Fig. 13.11. Five steps in Good Mud Modeling Practice (GMP).

a large impact on the simulation period and therefore feasible grid size and resolution. Then, it is difficult to switch from one approach to another towards the end of a study.

13.6.2 *GMP — Step II: Hydrodynamic model*

The conceptual model defines model domain and simulation times, and the hydrodynamic processes to be included (tides, waves, salinity, turbulence model, ...), see also Sec. 13.3. This assessment includes considerations on the sediment transport processes, because sediment may have to be integrated to various degrees of complexity within the hydrodynamics, e.g. with/without water-sediment interaction (and/or morphodynamic updates). These considerations provide requirements in terms of scales and processes, which need to be converted into a numerical grid. Grid design (see Sec. 13.4.2) yields a balance between accuracy and computational time, as many complex numerical models require days to weeks to run. These conflicting demands may require local grid refinement, for instance through nesting small, high-resolution sub-models into a large low-resolution model. Next, appropriate bathymetry and boundary conditions need to be collected. Note that navigational charts generally provide depths of shallows, which may not be representative for flow models.

A hydrodynamic model has to be calibrated (Sec. 13.5.1). When the bathymetry and model geometry are reasonably accurate, the only independent parameters that can be tuned are:

(1) Bed roughness; see Sec. 3.3 for an overview of roughness parameters that can be used. For smaller models, bed roughness is not too important, as the flow is mainly governed by the model boundary conditions. The hydrodynamics over intertidal areas may be important for the fine sediment dynamics, but are often difficult to model and to calibrate, owing to a lack of field data. Note that under these shallow water conditions, the functional relations between bed shear stress and water depth differ for the various roughness parameterizations.

(2) Boundary conditions are always important, but in particular for smaller models. Boundary conditions for tides and waves are relatively easy to obtain, from direct measurements, permanent tide and wave stations, or from global tide and wave models (e.g. Topex Poseidon and Wave Watch III, respectively). Boundary conditions for residual currents can be crucial for fine sediment transport, but are more difficult to acquire. The accuracy of Global Ocean Models such as NEMO and HYCOM is steadily increasing, providing a means to prescribe the residual flow resulting from large-scale ocean currents. Especially in larger models, wind stresses may induce the relevant residual currents and circulations locally.

    In small models, acquiring proper boundary conditions for the salinity field may be highly problematic, in particular when large simulation times are required.

(3) Initial conditions for the hydrodynamics are, in general, not too important. If a model domain contains many intertidal areas, starting simulations at high water is recommended to prevent spurious waves during model spin-up.

    The initial conditions for salinity fields are important, owing to the large memory of salinity fields. As data are difficult to acquire, long simulation times are often necessary to attain a proper salinity field.

(4) Horizontal dispersion coefficients may be important for resolving large-scale circulations, and the horizontal distribution of salinity. These coefficients should be chosen on the basis of the grid size selected (Sec. 12.1).

Generally, calibration of the hydrodynamic model starts with water levels, which is usually fairly straightforward and can be accurately predicted. Instantaneous velocities and water fluxes are more complex, being strongly influenced by local bathymetrical gradients, often at sub-grid scale. Of great importance for fine sediment transport are the salinity fields and residual flow, though field data are often difficult to acquire. Step II concludes with an analysis of the model results, such as tidal asymmetry, residual flow, and bed shear stress distributions, which are mandatory for an efficient setup of the sediment transport model.

13.6.3 *GMP — Step III: Mud model*

Sediment parameters

The first step in the setup of a sediment transport model is the definition of the number of sediment fractions and their properties, which includes the relevant process formulations, in conjunction with relevant sediment boundary conditions. The number of fractions follows from the conceptual model. Input parameters are based on available data, but also on hydrodynamic model results, further to the analyses of the hydrodynamic model in Step II. If no further information is available, a good starting point for the sediment properties is given in Table 13.3.

In shallow open water systems, subject of this book, local SPM-concentrations are generally determined largely by local water-bed exchange processes. A 1DV approach is therefore very useful for a first assessment of model parameters to reproduce SPM-concentrations — this

Table 13.3. Characteristic/starting values for mud properties for Partheniades–Krone bed boundary conditions.

	Coarse SPM	Fine SPM	Unit
w_s	1.0	0.1	mm/s
c_{gel}	100	150	kg/m^3
τ_e	0.5	0.1	Pa
M	10^{-4}	10^{-3}	kg/m^{2}s
τ_d	∞	∞	Pa

procedure is elaborated upon in Chapters 14 and 18. Note that more or less identical results can be obtained with many different combinations of parameter settings (equifinality, Sec. 13.5.4). The problem of non-uniqueness of model settings can be reduced by including rates of changes in SPM-concentrations in the calibration phase, next to their absolute values, as discussed below.

Initial bed composition

In Sec. 12.3, we have advocated to use the "empty-bed" approach for the sediment transport model, as this approach is consistent with below-capacity conditions and the control of model performance is much easier than with the "full-bed" approach. However, for very large models, or when sediment supply is very small, the empty-bed approach may be prohibitive for practical application, as spin-up times to reach dynamic equilibrium by natural processes become too large. However, spin-up times may be reduced through one of the following approaches:

(1) Observations and the conceptual model may give a general pattern on the presence of mud in the bed. Such patterns can be used as initial conditions, bearing in mind that these patterns most likely do not coincide with the equilibrium bed composition of a model.
(2) From an inspection of computed bed shear stress distributions, sedimentation and erosion areas can be identified to first order. Some mud can be allocated to these areas as initial conditions, anticipating rapid dispersion to equilibrium patterns within a model.
(3) Most problematic, yet important for model performance, are temporary deposits, where fine sediment can be stored during calm hydrodynamic conditions, being remobilized during dynamic events. Such areas can be identified with synthetic model simulations, under calm and dynamic conditions. Computed temporary deposits can be thickened artificially to realistic values.
(4) Spin-up times to reach equilibrium can be reduced by designing coarser models for spin-up (in 2DH-mode and/or reduced model resolution) or gradually increase the thickness of the active bed layer in multiple model runs.

In case of (almost) closed systems, such as lakes, redistribution times are generally much smaller than time scales for sediment input from ambient sources. Then an initially spatially homogeneous mud distribution on the bed may be a good approach to initiate a model.

Calibration
If data are available for prolonged periods, a formal division between calibration and validation is feasible. The data set is split, and the model is calibrated against the first part of the data, and validated against the second part, using the same parameter settings, as outlined in Sec. 13.5. If data are scarce, as is generally the case, such a division is contra-productive, as all data are required for system understanding and phenomenological calibration.

The phenomenological calibration, varying model input parameters to match observations (Sec. 13.5.2), can be done at incremental levels of complexity or immediately at full model complexity. An example of incremental complexity is to start with modeling sediment transport with the 1DV-approach advocated, neglecting advection and diffusion, and then run the full 3D model, followed by including sediment sources and sinks.

An important parameter for model calibration is the phasing of SPM-concentrations with the tide, as discussed in Sec. 13.5.2, Fig. 13.7. This comparison may enhance the credibility of fine sediment transport models largely.

When sufficient data are available, the phenomenological calibration is followed by a quantitative calibration (Sec. 13.5.3), and possibly validation. In case bed–sediment–fluid interactions (buoyancy destruction) are included, the hydrodynamic calibration should be verified again, as both bed level variations and turbulence damping influence flow velocities and vertical mixing, thus SPM distributions.

13.6.4 *GMP — Step IV: Uncertainty*

Quantifying model uncertainty (Sec. 13.5.4) is an important step, even if not formally possible. Its simplest form is a systematic sensitivity analysis, which is generally carried out during model calibration. More quantitative measures on model uncertainty can be carried out using equifinal or stochastic techniques. Quantifying uncertainty is especially important if a model is applied to scenarios significantly different from the reference conditions.

In general, gross SPM fluxes are many orders of magnitude larger than net fluxes, as illustrated in Secs. 9.4 and 10.4. This has profound consequences for the accuracy of process-based numerical models, which compute the gross fluxes only. The accuracy of the net fluxes (and therefore of the model) can be obtained from a simple error propagation analysis of the sediment dynamics in a semi-enclosed basin, as in Fig. 9.4. Let us denote the importing flux (over an arbitrary period) with F_I, and the exporting flux with F_E, so that the net flux becomes $F_N = F_I - F_E$. It is further assumed that the error in F_I and F_E scales with its magnitude, i.e. $F_I = \mu_I \pm \sigma_I = \mu_I \pm \varepsilon_I F_I$, where μ_I is the mean of F_I, σ_I its standard deviation, and ε_I a proportionality coefficient. Assuming that F_I and F_E are uncorrelated and $F_I \wedge F_E \gg F_N$, while their error is Gaussian, the error σ_N in F_N becomes:

$$\sigma_N = \sqrt{(\varepsilon_I \mu_I)^2 + (\varepsilon_E \mu_E)^2} \tag{13.5}$$

If $\varepsilon_I \approx \varepsilon_E \wedge \mu_I \approx \mu_E$, Eq. (13.5) reduces to

$$\sigma_N \approx \varepsilon_I \mu_I \sqrt{2} \approx \varepsilon_E \mu_E \sqrt{2} \tag{13.6}$$

implying that $\sigma_N \gg \mu_N \wedge F_N$, i.e. the error is likely much larger than the mean value of net flux, and the model results are not very accurate, to make an understatement.

Only when one of the gross fluxes is small, as for instance F_E in case of harbor siltation, calibration of F_I implies calibration of F_N, thus siltation rates in the basin. This is the reason that calibrated, process-based numerical models can do a reasonable job in assessing absolute values of harbor siltation. It also further supports the importance of the phenomenological calibration, which typically addresses model aspects driven by net fluxes.

The final step in the calibration procedure is to verify the conceptual model and to answer the research questions formulated in Step I.

13.6.5 *GMP — Step V: Application*

Absolute vs. relative effects

Model applications and interpretation of the model results are hampered by the accuracy issues above. However, the error in model applications can be reduced considerably, if models are used in comparative mode, i.e. results for scenarios are compared to a reference case.

Let us assume that the impact of a relatively small intervention in an open water system has to be assessed. The focus is again on the fluxes, and the reference situation is identified with subscript $\cdot_0$, and the new situation, after intervention, with subscript $\cdot_1$. It is reasonable to assume that, given the smallness of the interventions, that F_{I_0} and F_{I_1} are correlated, i.e. $F_{I_1} \approx \alpha F_{I_0}$, and similar $F_{E_1} \approx \beta F_{E_0}$ (with α the modification in import and β the modification in export due to the interventions — other parameters are defined above). This implies:

$$\begin{aligned} F_{N_1} - F_{N_0} = \Delta F_N &= (F_{I_1} - F_{E_1}) - (F_{I_0} - F_{E_0}) \\ &\approx (\alpha - 1)\, F_{I_0} - (\beta - 1)\, F_{E_0} \end{aligned} \tag{13.7}$$

and further to Eq. (13.6):

$$\sigma_{\Delta F_N} \approx \sqrt{((\alpha - 1)\, \varepsilon_I \mu_{I,0})^2 + ((\beta - 1)\, \varepsilon_E \mu_{E,0})^2} \tag{13.8}$$

Again assuming $\varepsilon \approx \varepsilon_I \approx \varepsilon_E \wedge \mu_I \approx \mu_E$, gives:

$$\sigma_{\Delta N} \approx \varepsilon \mu_{I,0} \sqrt{(\alpha - 1)^2 + (\beta - 1)^2} \approx \varepsilon \mu_{E,0} \sqrt{(\alpha - 1)^2 + (\beta - 1)^2} \tag{13.9}$$

For small interventions, we may assume $(\alpha - 1) \wedge (\beta - 1) \ll 1$, implying that the relative error is much smaller than the absolute error: $\sigma_{\Delta N} \ll \sigma_{F_N}$. This is the reason that process-based numerical models are very useful in establishing the relative impact of interventions in open water systems.

If the absolute impact needs to be assessed as well, it is often good strategy to add the computed relative impact to observational data on the reference situation, accounting for natural variability as well.

Examples

The practical guidelines itemized in Fig. 13.11 are applied in a series of model applications, summarized in Table 13.4, illustrating the various issues discussed above. Except for the Ems-application, all examples deal with below-capacity conditions.

Each of these chapters exemplifies a physical environment dominated by fine sediments (varying from lakes to coastal seas), illustrating the importance of good modeling practice, explaining how the complex systems can

Table 13.4. Examples of model applications.

Chapter 14	Shallow sea	North Sea
Chapter 15	Estuary	Western Scheldt
Chapter 16	Tidal River	Ems River
Chapter 17	Lake	Lake Markermeer
Chapter 18	Harbor	Deurganckdock

be simplified and which processes have to be accounted for to address the engineering questions at hand. All these examples are situated (at least partly) in The Netherlands, mainly because of copyright and confidentiality issues. Although all modeling studies presented hereafter are carried out with the Delft3D software system, the modeling platform itself is irrelevant for the Good Modeling Practice advocated.

Chapter 14

Modeling SPM in Shallow Seas: The North Sea

14.1 Introduction

This chapter describes the setup for a fine sediment transport model of the North Sea, and two applications. Not all aspects are discussed in detail, as this would require a book in itself. The key elements of this chapter are the formulations for water-bed exchange of fines in a predominantly sandy seabed and their consequence for SPM dynamics and transport. Parts of this chapter have been published before in abbreviated form by Van Kessel *et al.* (2011c).

The North Sea model has been setup to address a number of managerial questions, which are discussed in the following section.

14.2 Conceptual model

14.2.1 *Model purpose*

The purpose of the North Sea SPM model is to increase the understanding of SPM dynamics in the Dutch coastal zone and provide quantitative information on the spatial distribution and temporal variation of SPM as a function of hydro-meteo forcings (tide, wind, waves, freshwater discharge). Further, changes in SPM due to human interventions are to be established, among which:

(1) coastline realignment such as the Maasvlakte-2 construction (i.e. the seaward extension of Rotterdam Port),
(2) large-scale sand mining, and
(3) maintenance dredging in the various ports along the coastal zone, and subsequent dumping of the dredged sediment.

The fine sediment dynamics model is an essential part within the ecological effect chain model of the North Sea, assessing primary production and other water quality parameters. Although the applications are limited to the Dutch coastal zone, the model domain covers the entire southern North Sea to account for meteorological effects. However, a higher resolution of the computational grid is set in the coastal zone.

14.2.2 *System understanding*

The Dutch coastal zone is shallow, with depths up to about 40 m, see Fig. 14.1. The mean tidal range varies from 1.9 m near Vlissingen (Flushing) to 0.7 m near Den Helder. The tide is semi-diurnal with a profound spring-neap variation. Along the Dutch coast, it propagates from south to north, induced by the Kelvin wave in the North Sea (see also Sec. 10.2.2). Also a small northward residual current exists, driven by the Atlantic Ocean, surrounding the North Sea. Meteorological effects are important, with substantial seasonal variations in wind speed and direction, and in wave height, period and direction.

The Dutch coastal zone seabed is generally sandy, containing a few percent of fines only (Bockelmann *et al.*, 2018; Van Heteren *et al.*, 2007). Fines are buffered into and washed out of the seabed by physical and biological processes (Sec. 6.6). The fines originate from the English Channel (both from the Atlantic Ocean and coastal erosion), and from erosion of the seabed off the Belgian coast (Flemish Banks, e.g. Eisma (1981); Fettweis and Van den Eynde (2003); Van Alphen (1990). Input from the River Rhine is small. SPM values in the North Sea generally do not exceed a few 100 mg/L, and are on average of the order of about 50 mg/L, with near-surface values almost an order of magnitude smaller (Sec. 12.2), while temporal and spatial variations are large, though. Note that this range of SPM values implies that the fine sediment dynamics in the North Sea are characterized by below-capacity conditions.

Long-term observations suggest that the SPM dynamics are governed by at least two fractions, a "fine" fraction with a slow response to changes in hydrodynamic forcing, and a "coarse" fraction with a much faster response. An overview of characteristics of fine sediment transport in the North Sea can be found in Van der Kaaij *et al.* (2003).

The coastal zone is strongly influenced by the Rhine ROFI (region of freshwater influence — see also Sec. 10.2.2). The discharge of the River Rhine induces substantial cross-shore salinity gradients along the Dutch coast, resulting in near-coast trapping of SPM by gravitational circulation and thermal wind effects. The ROFI is influenced by tide and wind.

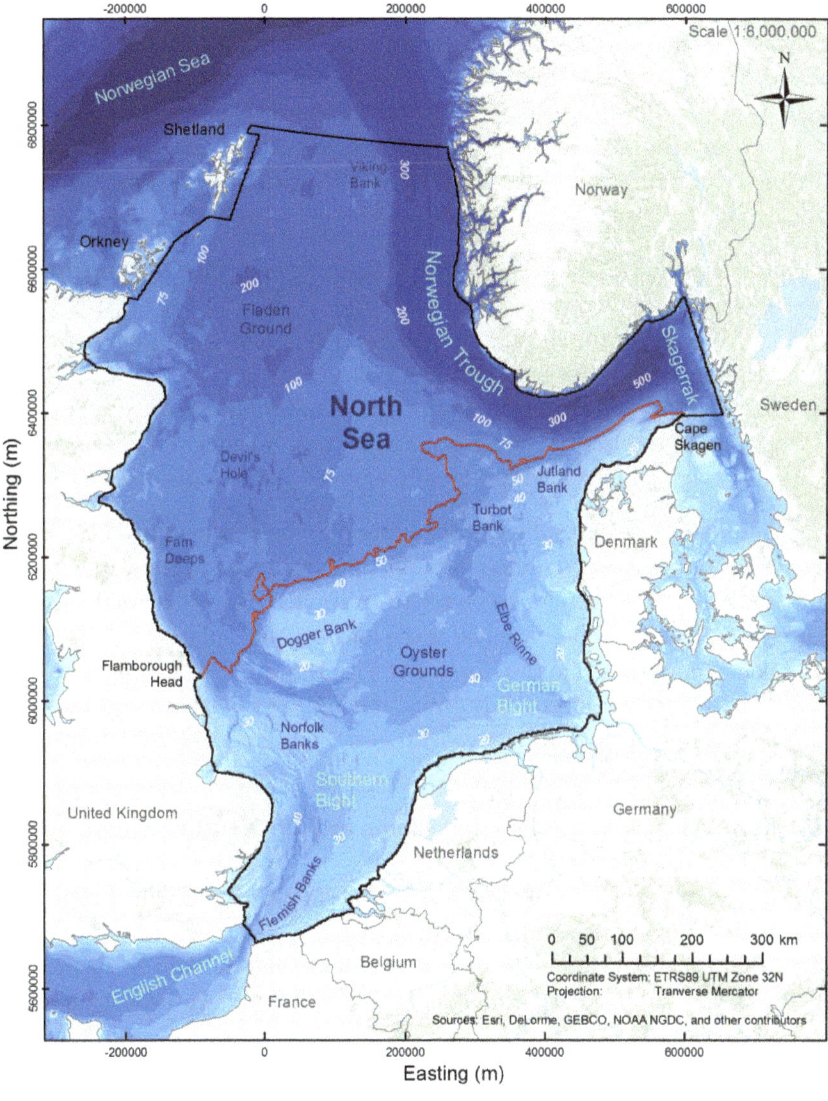

Fig. 14.1. Overview of North Sea with bathymetry (Bockelmann *et al.*, 2018).

During neap tide the coastal zone is generally fairly stratified, whereas during spring tides more mixing occurs. Offshore and onshore winds strain the stratified flow in different ways: offshore winds enhance stratification whereas onshore winds reduce stratification. The Rhine ROFI is also known as the coastal river, and plays an important role in the net northward transport of fines. Field observation on cross-shore SPM gradients are discussed by Suijlen and Duin (2002). Waves are the main driver for sediment resuspension from the seabed: during storms sediment resuspension is important and SPM levels peak (see also Sec. 12.2). Although the tidal forcing is deterministic, SPM-concentrations are highly variable because (Blaas and Van den Boogaard, 2006):

(1) the meteorological forcing is highly variable over the seasons with significant yearly variations,
(2) the fresh water discharge of the River Rhine has a profound seasonal variation, and varies over years as well,
(3) the supply of fines varies over irregular time scales, and
(4) because of its size, the North Sea has a large "memory", i.e. actual SPM values and salinity field are largely affected by previous events.

Meaningful model results can therefore only be obtained from the interaction between hydrodynamics and SPM on the long term (multiple years). However, the time-average response to diurnal dynamics should be well-represented in the model to make these long-term simulations viable.

The memory effects in the North Sea with respect to SPM-concentrations can be understood conceptually by analyzing the transport and fate of a certain amount of fines in the coastal zone, e.g. fines released from sand mining activities:

(1) the fines may stay in the water column and affect turbidity until flushed out of the system,
(2) they may deposit to the bed quickly, permanently buried, thus having only a brief impact on turbidity, or
(3) an intermediary behavior may occur: fines may remain in suspension for some time, while available for wave-induced resuspension upon deposition.

From data analyses and our physical understanding outlined in previous chapters, it is anticipated that the intermediate behavior (3) is to be expected. This behavior therefore forms the basis for the model setup, and has to be reproduced with the model. As the natural variability in SPM-concentrations is large, the impact of human interventions can therefore only be assessed when the model is used in comparative mode (Sec. 13.6.5). Of course, it is almost impossible to assess such impacts from field measurements alone.

14.2.3 *Model requirements*

With respect to the requirements for SPM modeling, discussed in Chapter 13, the following features are relevant:

(1) reproduction of characteristic SPM values and spatial SPM gradients, and
(2) reproduction of tidal, neap-spring, seasonal and storm-driven variations of SPM.

In addition, data on SPM fluxes and bed composition should be reproduced as well:

(3) residual SPM flux along the Dutch coast,
(4) distribution of fines on/in the seabed,
(5) net siltation in harbors or naturally sheltered areas, and
(6) amount and residence time of fines in the seabed available for resuspension — the "mud buffer".

In the majority of the North Sea and Dutch coastal zone, SPM-concentrations are so low that they do not affect the hydrodynamics (Sec. 7.1), nor seabed levels (morphostatic conditions). For addressing the practical questions above, possible sand-related morphodynamic evolutions of the seabed are neither important. Therefore hydrodynamics and SPM dynamics can be modeled separately, i.e. an uncoupled approach is feasible in which the results of hydrodynamic simulations are stored, and used repeatedly for SPM simulations. In this way, multiple long-term simulations become feasible, and model sensitivity to process parameters can be studied efficiently.

Note that locally sediment-fluid interactions may become important, for instance in quantifying the sediment flux into the Port of Rotterdam (see Winterwerp and Van Kessel, 2003). Hence, such processes cannot be studied with the current model.

A model which predicts SPM-concentrations satisfactory does not necessarily perform well with regard to residual SPM fluxes, the sediment balance and/or residence times, as outlined in Chapter 13. Typically, in literature most focus is on the first aspect, whereas the other parameters are often not addressed. This limits the model applicability for impact assessments, though remains unnoticed, while models are often applied beyond their range of validity. Therefore, SPM fluxes and residence times are explicitly addressed, as part of our system understanding.

14.3 Hydrodynamic model

The North Sea sediment transport model is based on the ZUNO-DD hydrodynamic model, an acronym for Southern North Sea model using Domain Decomposition on the Delft3D software platform (Deltares, 2018; Lesser *et al.*, 2004). In what follows, we refer to the ZUNO-model. This is a 3D model with 12 sigma layers over the water depth. Domain decomposition (DD) is used to refine the computational grid in two sub-models, yielding a horizontal resolution ranging between 250 and 500 m along the Dutch coast. Domain decomposition is a form of on-line nesting of small, finer-grid models in larger, coarse-grid models. Figure 14.2 shows the model grid and bathymetry.

For a discussion on the setup, calibration and validation of this model; see Cronin (2013); Van der Kaaij *et al.* (2017). For the SPM simulations, the most important criteria for the calibration of the hydrodynamic model are (see also Fig. 14.1):

(1) residual water flux through the English Channel,
(2) longshore and cross-shore salinity gradients in the Dutch coastal zone,

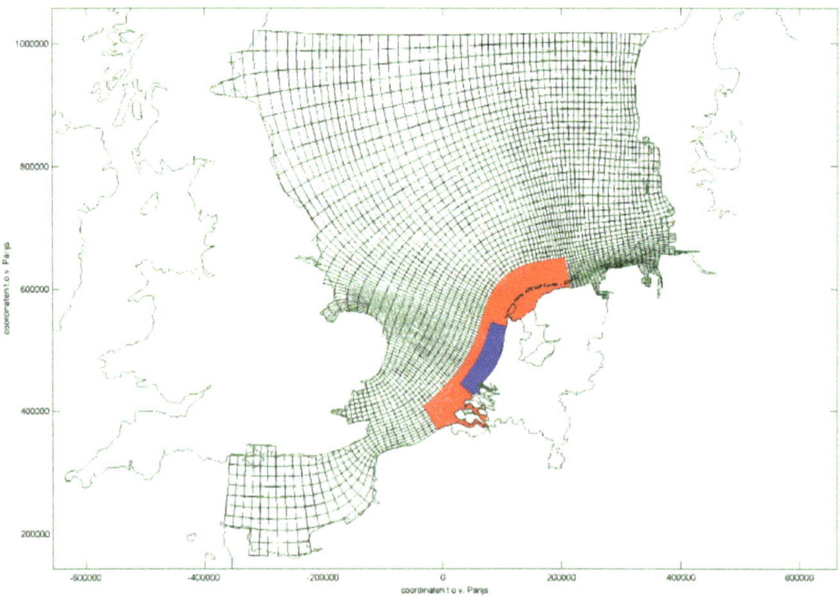

Fig. 14.2. ZUNO-DD model grid. The intermediate and fine grid sub-models are indicated in red and blue, respectively.

(3) residual currents along the Dutch coast,
(4) large-scale salinity field, and
(5) residence time of different water masses (e.g. from Rhine discharge and from the southern boundary).

Tidal asymmetry is not relevant. These parameters may vary from year to year. To account for this, a hydrodynamic database was built, originally containing eight years of model results (1996–2003) driven by tidal constituents, and hourly wind, wave and river discharge data. Later, this database was extended with the hydrodynamic forcing of the period 2004–2011 and subsequently 2012–2015. All results presented in this chapter are based on multiple-year subsets of this 20-year data base, or on single years (2001 or 2007) judged representative with regard to wind forcing and river discharge.

In Van Kessel *et al.* (2006) and Keetels *et al.* (2012) further details are given on the hydro-meteo forcing in this period and their effects on residual currents and the salinity distribution. The impact of these variations on SPM transport is discussed in Brière and Van Kessel (2006).

Wave-induced bed shear stresses were computed using an assimilation method combining hourly wave observations at North Sea stations with year-average values computed with the wave simulation model SWAN, without wave–current interaction (off-line wave modeling). This was done as running the wave model on hourly basis would be too computationally demanding, while the period of longer waves (swell) would be underestimated, resulting in too little SPM resuspension in deeper water. This assimilation technique is described in detail by Bruens *et al.* (2007).

14.4 Sediment transport model

14.4.1 *Model setup*

As discussed above, the fine sediment transport model is run decoupled from the hydrodynamics, using velocities, bed shear stresses, water levels and salinity values, computed with the hydrodynamic model, as input.

To account for long-term temporal variations in SPM-concentrations, bookkeeping of the amount of fines in the bed is required. We refer to the buffer capacity of the seabed. In the present application, the two-layer buffer model is used to account for these memory effects, prescribing the bed-boundary conditions (Sec. 12.2, Fig. 12.5). This is an efficient tool to model buffering into and remobilization of fines from the sandy seabed as function of the hydrodynamic forcing, accounting for at least two time scales.

For convenience, we repeat the relevant deposition and erosion formulations (see Fig. 12.5), and Eqs. (12.21)–(12.23):

$$\begin{aligned} D_1 &= p_d\,(1-\alpha)\,w_{s,b}c_b \\ D_2 &= p_d\alpha w_{s,b}c_b \end{aligned} \tag{14.1}$$

where D_1 is the deposition flux towards the fluff layer 1 and D_2 is the deposition flux towards the buffer layer 2. At dynamic equilibrium,[1] the time-average deposition flux is balanced by the time-average erosion fluxes E_1 and E_2 from the fluff and buffer layer, respectively (see Sec. 12.2)[2]:

$$\begin{aligned} E_1 &= m_1 M_1\left(\frac{\tau_b}{\tau_{e,1}}-1\right) \quad \text{for } \tau_b > \tau_{e,1} \\ E_2 &= \xi_2^m M_2\left(\frac{\tau_b}{\tau_{e,2}}-1\right)^{1.5} \quad \text{for } \tau_b > \tau_{e,2} \text{ and } \tau_{e,2} > \tau_{e,1} \end{aligned} \tag{14.2}$$

Concurrent erosion is possible from the fluff layer and buffer layer. The buffer layer is more difficult to erode, parameterizing reworking within the bed (see also Sec. 8.3). Note that for North Sea hydro-sedimentological conditions, fines do not accumulate in the fluff layer. A transition from first to zeroth-order erosion formulation (Eq. (12.25)) is not necessary, as it is the timing of erosion of the fluff layer, more than its rate (see also Sec. 9.2.2).

The seabed is sandy, and therefore cannot represent a permanent sink or source; thus it acts as either a transient sink or source (time-dependent buffer). Only in model areas where the critical shear stress for resuspension is never exceeded, fines may accumulate indefinitely as no morphological update is made. Everywhere else, a long-term dynamic equilibrium between erosion and deposition will be established. In this approach, permanent sources, such as from the erosion of geological deposits (e.g. Flemish Banks) are prescribed as user-defined input time series.

As the seabed composition is an integral part of the model, it is computed rather than user-prescribed. The seabed level (bathymetry), however, is user-defined and fixed, i.e. the model is run in morphostatic mode. Of course, initial conditions for the seabed composition have to be specified.

---

[1]Note that this equilibrium does not relate to capacity conditions.

[2]Note that the excess shear stress differs from the one in Eq. (12.24), hence M_1 has dimension [1/s].

However, if the model is run sufficiently long the results become independent from these initial conditions. Therefore initially no sediment is placed on/in the bed ("empty-bed" approach, Sec. 12.2), and sediment gradually enters the model domain via its boundaries until a dynamic equilibrium has been reached. The time scale to reach equilibrium depends on the residence time of mud in the domain and can be in the order of years, and locally even much longer for large models such as the North Sea model. Here, the model is spun-up for a period of eight years to reach dynamic equilibrium conditions for SPM and bed composition.

As discussed in Sec. 12.3, this "empty-bed" concept is important, as user-prescribed initial conditions may substantially deviate from the computed bed composition at dynamic equilibrium. Such initial conditions may therefore yield misleading results (see also Chapter 17).

14.4.2 *Calibration*

As discussed in Sec. 14.2.2, modeling SPM dynamics in the Dutch coastal zone requires at least two fractions, which is assumed in the following. Model calibration was carried out against the following data:

(1) the high-frequency frame and buoy data from monitoring station Noordwijk-10 with temporal SPM fluctuations, notably the neap-spring tidal cycle and the response to calm and rough meteo conditions,
(2) characteristic cross-shore gradients of SPM based on long-term, low-frequency ship-based observations,
(3) the mean longshore SPM flux along the Dutch coast,
(4) the residence time of fines within the seabed,
(5) the buffer capacity of the seabed, and
(6) the spatial distribution of mud content in the seabed.

More information on these data is given by Van Kessel *et al.* (2012) and Cronin (2013). It is noted that most of the calibration and validation effort was focused on the Dutch coastal zone (up to 50 km offshore). Little attention was paid to SPM dynamics farther offshore. In this region, the model should be considered as under-calibrated (or even not calibrated).

Temporal variations

As a first step, a 1DV model was set up to calibrate the model parameters (see also Sec. 12.2). This is a very effective approach, as water-bed exchange processes dominate short-term SPM variations in shallow water (Chapter 1),

Table 14.1. Calibration parameters with values after calibration with the 3D model.

w_s	Settling velocity	0.125 & 1	mm/s
p_d	Reduced deposition coefficient	1	–
$\tau_{e,1}$	Critical shear stress for erosion fluff layer 1	0.1	Pa
M_1	Resuspension parameter for fluff layer 1	(5.8 & 23) $\times 10^{-6}$	1/s
$\tau_{e,2}$	Critical shear stress for erosion buffer layer 2	1.5	Pa
M_2	Erosion parameter for buffer layer 2	1×10^{-4}	kg/m^2/s
α	SPM fraction that settles on buffer layer 2	0.05 & 0.0125	–
δ_2	Thickness of buffer layer 2	0.1	m

while the number of calibration parameters is large. This approach is superior to an initial calibration step in 2DH-mode, as discussed in Sec. 12.1. High frequency data from an anchoring station at Noordwijk, 10 km offshore (Noordwijk-10) was used in this 1DV-approach. For the two sediment fractions, there are eight adjustable parameters, given in Table 14.1.

The settling velocity w_s of the finer fraction is obtained from the observed SPM decay after storms. The settling velocity of the coarser fraction is obtained from observed vertical SPM-concentration gradients during storms. The critical shear stress for erosion of the fluff layer $\tau_{e,1}$ is derived from the observed onset of SPM after slack water. The first-order resuspension parameter M_1 should be set such that the fluff layer becomes thick enough to reproduce the observed tidal SPM fluctuations properly (M_1 can be assessed from δ_1, Fig. 12.5). To first order, $\delta_1 \approx \Delta c/h\,\mathrm{kg/m^2}$, where Δc is a characteristic tidal SPM fluctuation and h is water depth. If M_1 is too small, δ_1 becomes thick and the first-order erosion rate assumption cannot be met. Also, for large δ_1 the buffering function of the buffer layer underneath is affected, as the fluff layer then accumulates too much fines. Otherwise, the model is quite insensitive to the value of M_1.

Regarding the parameter settings for the buffer layer 2, $\tau_{e,2}$ follows from the properties of the sandy seabed, i.e. $D_{50,sand}$ (Shields curve). The resuspension parameter M_2 is set to match the observed peaks in SPM levels during storms. The parameter α steers the computed equilibrium mud fraction in the seabed to match observations for a typical year-averaged SPM-concentration. Finally, the layer thickness δ_2 is chosen to match the measured (or assumed) residence time of fines in the seabed, thus establishing the buffer capacity of mud in the seabed. In this setup, it was not necessary to apply the hindered deposition approach, thus p_d was set to unity.

It appears that only one set of parameter values for buffer layer 2 can be found with which storm-induced peak SPM-concentrations, equilibrium mud fraction and mean residence time in the seabed can be properly predicted. Equations (14.1) and (14.2) therefore do not contain redundant parameters, and equifinality (see Sec. 13.5.4) does not apply here. However, if data on bed composition and residence time would be missing, the model would be under-defined and equifinality may occur. Such data are therefore essential for a unique calibration of the model settings.

Next, the model parameters obtained with the 1DV point model are used in a first 3D simulation, which is then further tuned to reproduce the observations. As expected, the optimal settings for the 3D model are close to the settings derived with the 1DV model, owing to the shallowness of the coastal zone, the final parameters thus obtained are given in Table 14.1. As an example, Fig. 14.3 shows the observed and computed SPM-concentrations at station Noordwijk-10 for the calibrated parameter settings. The model predicts both storm peak as neap-spring variations.

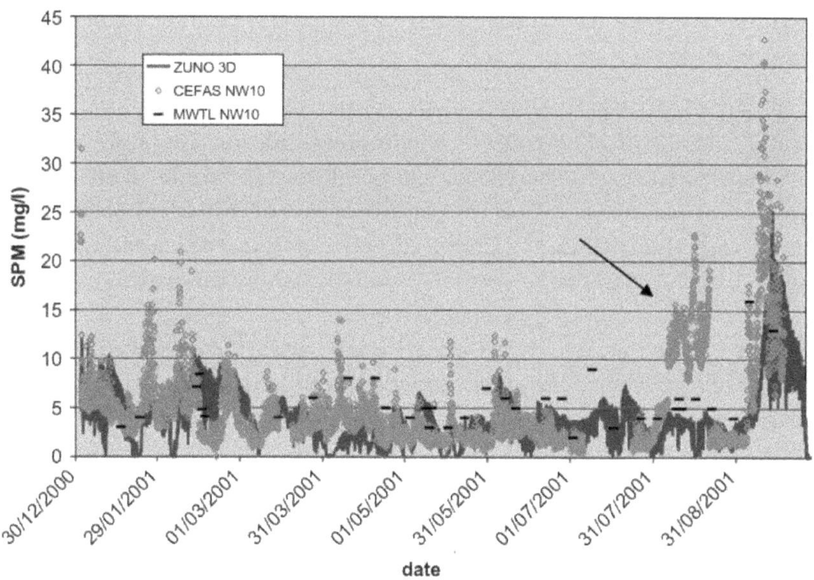

Fig. 14.3. Observed and computed SPM concentration at station Noordwijk-10 between 1/1 and 1/10/2001. The black arrow indicates a period with unreliable observations (probably because of OBS sensor fouling).

For a more detailed and quantitative discussion of the calibration procedure of a 3D model, using a 1DV point model, the reader is referred to Chapter 17.

Spatial distribution

The spatial distribution of SPM follows from the dynamics in hydrodynamic forcing and is tuned on the basis of a phenomenological calibration (Sec. 13.5.2). In the more traditional "full-bed" approach (Sec. 12.3), SPM levels can be increased by adding more sediment to the bed in case of supply-limited conditions (Chapter 1) or by reducing the critical shear stress for erosion in case of erosion-limited conditions. This approach may be effective for short-term simulations, but becomes troublesome for long-term simulations. For the "empty-bed" approach , spatial gradients in SPM are steered by the settling velocity, affecting local trapping:

(1) in case the vertical concentration gradients become more pronounced (larger w_s), water column processes such as gravitational circulation and tidal straining become more effective (Secs. 9.1.1, 10.2.2), and
(2) in case the deposition rate increases (larger w_s), water-bed exchange processes that may cause residual transport become more effective (Sec. 9.2).

A larger settling velocity will therefore result in stronger spatial gradients. This would, of course, require recalibration (possibly with the 1DV-model) to regain the proper temporal SPM variations. Note that the effects of pronounced vertical SPM gradients can also (partly) be accounted for through the coefficient of reduced deposition p_d, which is however set to unity in this chapter.

Figures 14.4 and 14.5 present the computed cross-shore SPM distributions in 2001 along four transects compared with the observed long-term average distributions. The data are well reproduced, though the model calibration was based on the data at Noordwijk-10 station only. This supports our recommended approach using a 1DV point model as a first step in model calibration.

Sediment fluxes

Table 14.2 presents the computed residual SPM fluxes through a 30 and 70 km wide stretch along the Dutch coast in the period 2003–2011, yielding about 10–20 Mton/year, which is consistent with estimates based on field observations. The majority of transport takes place within the 30 km stretch, consistent with the width of the coastal river (Sec. 10.2.2).

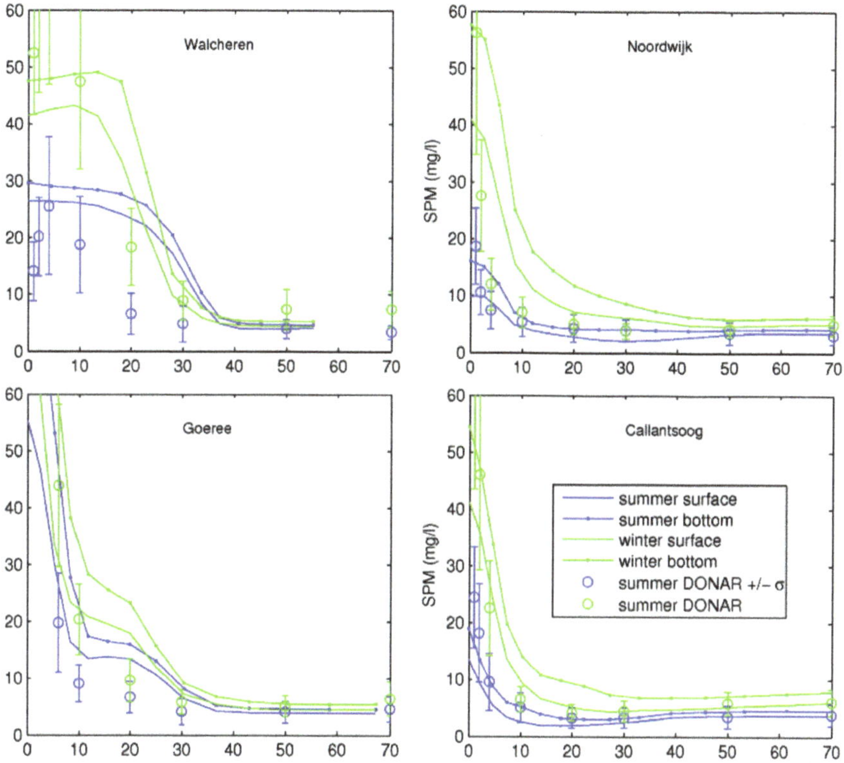

Fig. 14.4. Observed (o) and computed (–) winter and summer mean SPM-concentrations along four transects in the Dutch coastal zone.

However, the computed numbers show large variations from year to year, depending on the hydro-meteo forcings, which is likely one reason that it is difficult to measure the actual SPM flux accurately.

Maintenance dredging in the Port of Rotterdam amounts to about 3 Mton/year, which implies that about 20% of the fine sediment flux is captured. The influence of the port on the SPM dynamics is therefore substantial, see Sec. 14.5.2 for further discussion.

The model was tuned to reproduce this siltation rate of about 3 Mton/year. The grid resolution in the harbor area, however, is insufficient to resolve the local sediment dynamics in detail. Also, the model does not include sediment-induced density flows, which largely influence the residual sediment transport towards Rotterdam harbor (Winterwerp and Van Kessel, 2003). Therefore, a parameterized approach is required

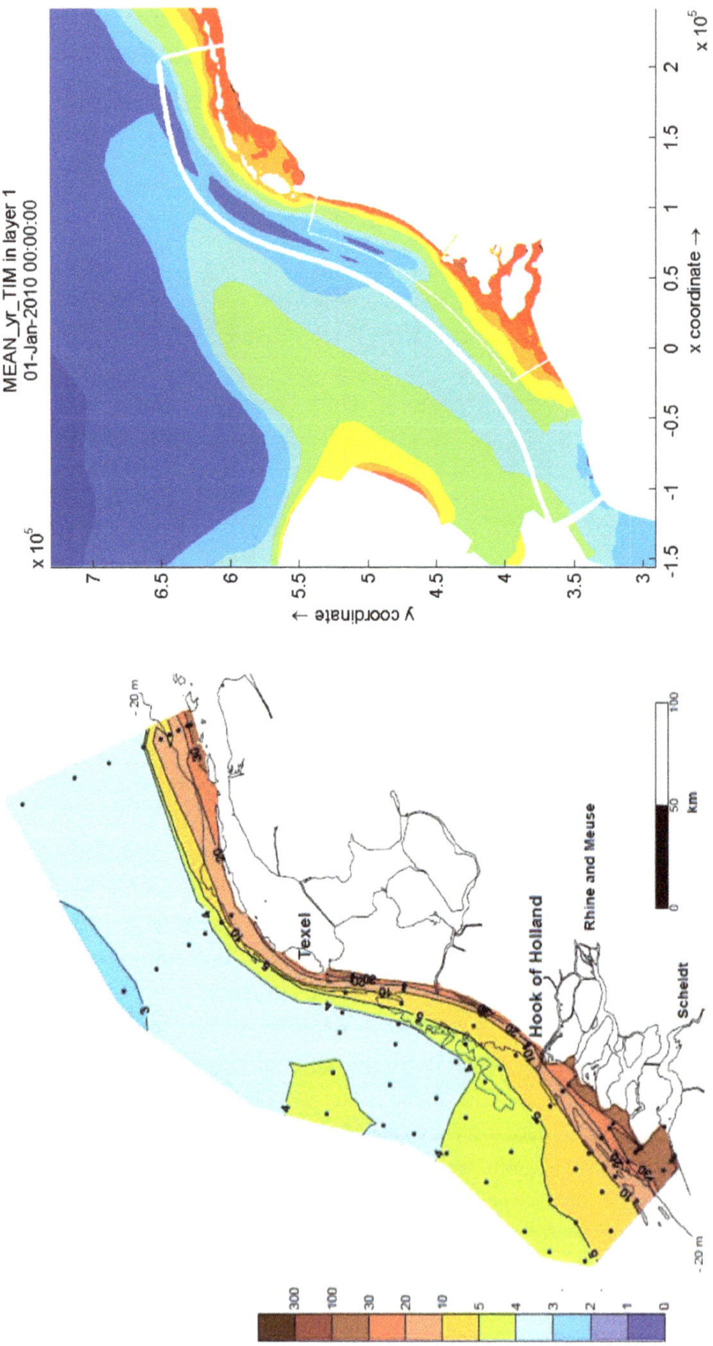

Fig. 14.5. Observed (left) and computed (right) year-average SPM-concentration [mg/L] along the Dutch coast.

Table 14.2. Computed year-averaged residual SPM flux (Mton/year) along the Dutch coast in the period 2003–2011.

	Walcheren		Schouwen		Ter Heijde		Noordwijk		Egmond		Callantsoog	
Year	30 km	70 km	30 km	70 km	30 km	70 km	30 km	70 km	30 km	70 km	30 km	70 km
2003	9.7	12.4	10.0	12.2	7.0	7.5	7.3	8.0	9.5	10.3	13.0	14.2
2004	11.3	13.5	11.4	12.9	7.8	8.4	7.7	8.6	10.5	11.4	14.9	15.9
2005	9.5	11.1	9.4	10.6	6.1	6.6	6.2	6.7	8.1	8.7	11.4	12.2
2006	14.9	16.7	14.6	15.6	10.2	11.1	10.0	10.9	11.8	12.7	15.3	16.1
2007	13.5	15.6	13.5	14.7	9.2	10.3	9.4	10.5	12.0	13.2	16.7	17.7
2008	16.9	19.2	16.9	17.9	12.2	13.5	12	13.3	14.3	15.6	19.4	20.3
2009	16.6	18.3	16.2	17.0	10.9	11.7	10.2	11.1	11.3	12.2	14.6	15.4
2010	9.1	10.6	9.2	10.1	6.3	6.8	6.4	7.3	8.0	8.8	10.4	11.3
2011	19.0	21.5	18.6	19.8	12.8	14.1	12.2	13.6	13.8	15.1	18.1	19.0
Mean	13.4	15.4	13.3	14.5	9.2	10.0	9.0	10.0	11.0	12.0	15.0	15.7

and to this order, the critical shear stress for erosion was increased locally in the harbor basins.

Seabed composition and exchange with water column

The computed equilibrium long-term bed composition is shown in Fig. 14.6. This map is a reflection of the bed shear stress climate (governing resuspension) and SPM levels in the water column (governing deposition). Areas with high bed shear stress and low SPM have low mud contents, as in the Dutch coastal zone, whereas areas with low bed shear stress and high SPM have high mud contents in the bed. The latter occurs only in a still water zone north of the entrance jetties to the Port of Rotterdam, and beyond the Dutch coastal zone. To illustrate this, the results for the entire model are shown. Areas with both high bed shear stress and high SPM or both low bed shear stress and low SPM show intermediate contents, depending on the balance between deposition and resuspension.

It is noted that the average value of the computed bed composition is a function of the bed model parameters. For instance, a twofold higher resuspension parameter would yield a 50% lower equilibrium mud content. However, the spatial distribution of the bed mud content is steered in a qualitative sense by the hydrodynamic processes represented in the model.

The same bed composition can be obtained for different water-bed exchange fluxes. If deposition and resuspension rates would both be halved or doubled, the equilibrium composition will not change. However, the computed residence time of fines in the seabed will change. This has little impact on short-term simulations, but has a major impact on long-term simulations. The residence time in the seabed can be adjusted by the thickness of the buffer layer, which will also affect its buffer capacity.

Residence time analysis

In this section, we discuss the role of the residence time and buffer capacity in more detail. Distinction should be made between the residence time of fines in the seabed (i.e. the time between deposition and resuspension), and the residence time of fines in a specific area of the North Sea (i.e. the time between the entrance of a mud particle to and its exit from this area). The residence time is an important parameter:

(1) to enhance one's understanding of the physical system, such as the contribution of various mud sources to the bed sediment and SPM in a system, or the response time of the system to events, and

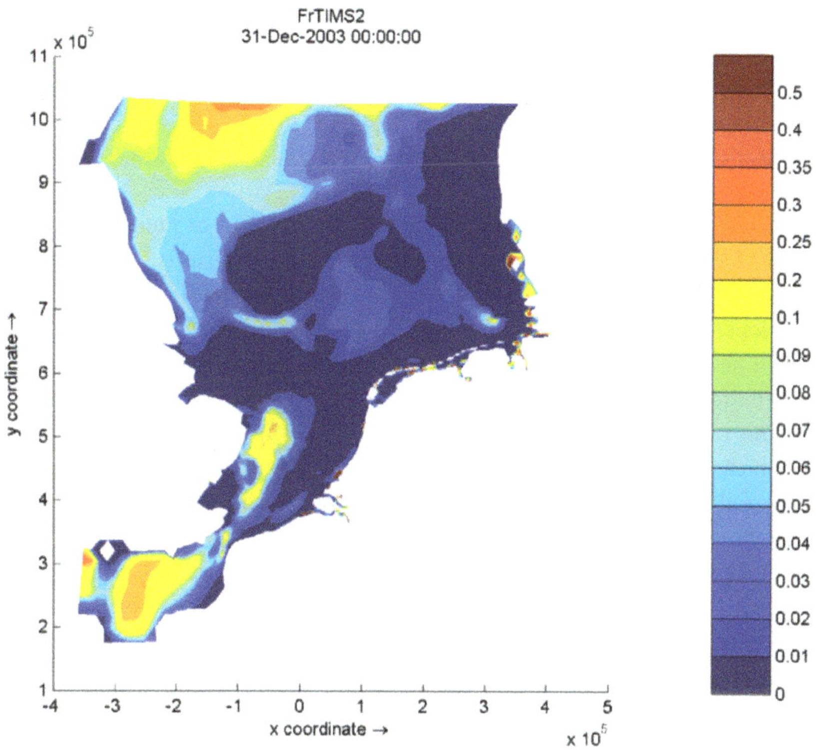

Fig. 14.6. Equilibrium mud fraction in the seabed computed with the ZUNO model.

(2) to assess the large-scale and long-term impact of human interventions in the system, such as the effects of dredging plumes on turbidity.

Intuitively, the longer the residence time the larger the impact on turbidity, as fines released remain longer in the system. However, larger buffer capacity yields smaller turbidity, as storage in the seabed reduces the amount of fines in the water column. Also, a large buffer capacity implies that the amount of fines released during dredging operations contribute less to the natural SPM levels. As residence time and buffer capacity are interrelated, the combined effect on excess turbidity is not obvious. An analysis using the buffer model can clarify this.

The most important parameters to consider in the analysis are:

(1) F_e, the sediment exchange flux [kg/m^2/s] between the seabed and the water column,

(2) m_2, the buffer capacity of the buffer layer [kg/m^2],
(3) R_{Tb}, the local residence time of fines in the bed [s],
(4) R_{Ts}, the overall residence time of fines in a specified area/domain [s], either in the water column, or in the bed, and
(5) R_{Tw}, the overall residence time of water in a specified area/domain [s].

The buffer capacity m_2 is a measure of the amount of mud stored in the sand skeleton of the buffer layer under average conditions, and is defined as $m_2 = \xi_2^{m,eq}\delta_2\rho_{dry}$; $\rho_{dry} = (1-n)\rho_s = \phi_s\rho_s$ is the dry density of bed sediment [kg/m^3]; $n = 1 - \phi_s$ is the bed porosity; ρ_s is the specific density of bed sediment [kg/m^3]; $\xi_2^{m,eq}$ is the equilibrium fraction of mud in the bed (buffer layer); and δ_2 is buffer layer thickness [m].

The residence time R_{Tb} is defined as the time that a particle resides within the bed until resuspended. It can be determined locally (for example measured at monitoring stations), or can be determined for an entire area (e.g. the entire Dutch coastal zone). In the latter case, the residence time accounts for multiple deposition–resuspension cycles of fines during their journey through the domain. The local residence time R_{Tb} is defined as the ratio between the buffer capacity and the time-averaged sediment exchange rate:

$$R_{Tb} = \frac{m_2}{F_e} = \frac{m_2}{\alpha w_s C} \tag{14.3}$$

with C a characteristic SPM-concentration. The overall residence time of sediment R_{Ts} partly determines the system response to changes in boundary conditions, among which (temporary) extra sources of fines due to e.g. sand mining. For short residence times, response times of the system to changes become short as well.

The overall residence time of a sediment particle in the water column is determined by the residence time of water R_{Tw} divided by the probability that a particle is in suspension: only sediment in suspension moves through the system. Sediment in the bed does contribute to the residence time R_{Ts}, but not to its transport, nor to turbidity in the water column.

We illustrate this with a simple analytical example. Consider a coastal water body with length L, width W, depth h, residual velocity parallel to the coast U_{res}, and average SPM-concentration C. The average residence time of water R_{Tw} thus reads:

$$R_{Tw} = \text{Volume/Flux} = LWh/WhU_{res} = L/U_{res} \tag{14.4}$$

The average overall residence time of fines R_{Ts} in the coastal zone reads:

$$R_{Ts} = p_2 L / U_{res} \tag{14.5}$$

where p_2 is the probability that fines are in suspension, given by:

$$p_2 = \frac{Ch}{Ch + m_2} = \frac{1}{1 + k_B} \tag{14.6}$$

in which $k_B = m_2/Ch$ is a dimensionless buffer capacity factor specifying the ratio of time-averaged sediment mass in the seabed and in the water column. Hence, p_2 reflects the equilibrium between fines in the water column and the bed, and k_B steers the probability distribution of a large number of particles residing in the water column.

The intensity of water-bed exchange over a time period Δt is given by the parameter p_1, defined as:

$$p_1 = \frac{F_e \Delta t}{m_2} = \frac{\Delta t}{R_{Tb}} \tag{14.7}$$

Typical numbers for the Dutch coastal zone read: $L = 100\,\mathrm{km}$; $U_{res} = 0.04\,\mathrm{m/s}$; $C = 100\,\mathrm{mg/L}$; $h = 20\,\mathrm{m}$; $p_1 = 0.01$; $\delta_2 = 0.3\,\mathrm{m}$ (see also Sec. 8.3); $\rho_s = 2600\,\mathrm{kg/m^3}$; and $n = 0.4$. This gives $m_2 = 4.7\,\mathrm{kg/m^2}$; $p_2 = 0.04$; and $R_{Tw} \approx 29$ days and $R_{Ts} \approx 2$ years. Within a distance of $W = 20\,\mathrm{km}$ from the coast, the total active mass of fines amounts to $WL(Ch + m_2) = 9.8$ Mton — this is the mass that can be mobilized by waves and currents. The coastal flux of fines is estimated at $U_{res}CWh = 160\,\mathrm{kg/s} = 5\,\mathrm{Mton/year}$. In reality, this flux is higher Eisma (1981); Van Alphen (1990), as both C and U increase during storms and spring tide. This can be achieved with either a smaller R_{Ts} or a larger m_2.

It is remarkable that, according to Eq. (14.5), R_{Ts} does not depend on the local residence time of fines in the seabed, but only on its buffer capacity. This can be understood as follows. If the exchange rate between the water column and the buffer layer is large, fines will exchange many times between water column and bed during their journey along the Dutch coast. This is reflected by the probability of water-bed exchange p_1. Thus for large p_1, the distribution of the residence time R_{Ts} of SPM in the water column becomes narrow, and approaches R_{Tw}. The number of exchanges is irrelevant, only the time that fines are in suspension or remain in the bed matters.

On the other hand, if the number of water-bed exchanges of a particle traveling a distance L becomes smaller (small p_1), the distribution of

R_{Ts} becomes wider, and will exceed largely R_{Tw} and more skewed with a long tail.

This is further illustrated by a simulation with a schematic stochastic model of the Dutch coastal zone, with its characteristic hydrodynamic settings. The model is run with 1000 individual particles traveling along the Dutch coast while jumping between the water column and the bed, with a probability of water-bed exchange p_1 and probability of being in the water column p_2. The model is initiated with a uniform dimensionless SPM-concentration $C = 1$, representing these 1000 particles. Then at $t = 0$, the SPM-concentration at the model boundary ($x = 0$) is decreased in one time step from $C = 1$ to $C = 0$. Such step-forcing illustrates the model response to e.g. a sudden change in sediment supply, as would occur when stopping a prolonged period of sand mining.

The results are presented in Fig. 14.7, showing the concentration response at $x = 50\,\text{km}$ (thus halfway the domain of analysis) for various settings of p_2 and p_1. The response curves reflect the fraction of particles released at $t = 0$ at $x = 0$ which have not yet reached position $x = 50\,\text{km}$ as a function of dimensionless time t/R_{Ts}. With more particles, the curves would become slightly smoother, maintaining the same trends, though. As the time axis is normalized with R_{Ts}, the response is approximately independent of p_2. However, the residence time distribution (the steepness at which the concentration curve falls) depends strongly on p_1, the probability of water-bed exchange. The smaller the exchange, the wider the residence time distribution, as argued before.

The local residence time of fines in the seabed R_{Tb} differs substantially from the overall residence time R_{Ts} of fines in a coastal zone, and can be much shorter. In case of a dynamic equilibrium between erosion from and deposition to the buffer layer, it is determined by $R_{Tb} = m_2/(\alpha w_s C)$ with w_s the settling velocity of fines in the water column and α the fraction of the deposition flux $w_s C$ towards the buffer layer. With $\alpha = 0.05$, $w_s = 0.5\,\text{mm/s}$ and the above-used ratios, $R_{Tb} = 0.6$ years, whereas we found before $R_{Ts} = 2$ years. This would mean that a mud particle traveling along the Dutch coast exchanges on average only three to four times between the water column and the seabed (buffer layer). This is equivalent to $p_1 = 0.1$ in Fig. 14.7, yielding a wide residence time distribution: shortly after the concentration step, the concentration starts to decrease at $x = 50\,\text{km}$, but after $4R_{Ts}$ there are still particles in the water column. Note that exchange with the fluff layer is far more frequent (up to four times per day).

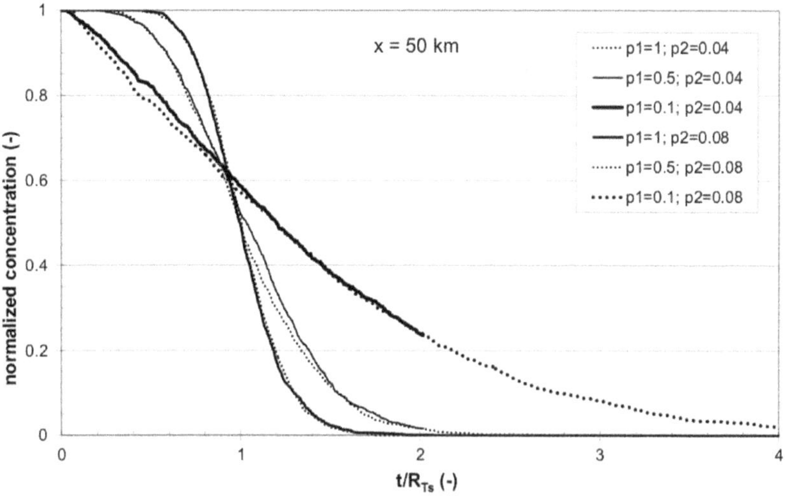

Fig. 14.7. Normalized concentration decrease at 50 km from the model boundary as function of p_2 and p_1 for $\Delta t = 0.7$ day (distance traveled at $U_{res} = 2.5$ km).

The residence time of fines in the water column amounts to $R_{Tw} = \alpha h w_s$ or about 9 days in this example. During this time, particles travel about 32 km with the residual flow.

Summarizing, two parameters are important to define the water-bed exchange of fines, namely:

(1) the buffer capacity m_2, or in dimensionless form, the buffer capacity factor $k_b = m_2/Ch$, and
(2) the local residence time in the bed $R_{Tb} = k_b h/(\alpha w_s)$.

For horizontal transport, also the global residence time in the area of interest $R_{Ts} = L/(1 + k_B)U_{res}$ is important, for instance to assess the impact of a release of fines on turbidity. We have argued that the buffer factor k_b determines how much of the released mass of fines remain in suspension, affecting turbidity. The global residence time R_{Ts} determines the response to a fine sediment pulse in the system, and R_{Tb} determines the width of the residence time distribution.

Typical values for the Dutch coastal zone are $k_b = 20$, $R_{Tb} = 0.5$ years and $R_{Ts} = 2$ years. The exchange rate follows from the ratio between the buffer capacity and the residence time (m_2/R_{Tb}), thus is not an independent parameter.

Values for the parameters k_b and R_{Tb} can be derived from dispersion experiments with a tracer, provided its amount is accurately known, and released in a period brief compared to the response time of the system.

In the next section, this theoretical and schematized analysis is applied to an engineering study on the effects of sand mining, while contributing significantly to the understanding of the North Sea SPM system.

14.5 Model application

We discuss two applications of the fine sediment transport model of the North Sea:

(1) the effect of sand mining on SPM levels, and
(2) the effect of harbor maintenance on SPM levels and harbor siltation.

14.5.1 *Effects of sand mining*

The Dutch coastal zone is rich in high quality sand, a valuable resource, which is therefore mined frequently. This sand contains about 3% fines. During mining, these fines are released at the dredgers suction head, through overflow from the hopper (generally the larger contribution) and during transport and dumping. These fines unfavorably affect turbidity levels, thus water quality and primary production (Sec. 3.5). As sand mining is an ongoing activity in the Dutch coastal zone, the main interest is in the long-term and far-field impact on water quality, while near-field effects are deemed less critical. Therefore, turbidity levels within individual plumes are not studied, but instead the cumulative effects of thousands of plumes on the overall SPM-concentration. Note that studying near-field effects would require another model setup.

As discussed above, the far-field impact of these mining operations is determined largely by water-bed exchange processes. Plumes from which no sediment deposits behave as a solute, thus are diluted and washed out of the coastal zone with the mean water movement. Plumes from which sediment is permanently deposited/buried will have a local impact only, notably the sedimentation footprint.

When the mobilized fines are deposited and resuspended alternately, they will gradually disperse, but at a much slower pace than non-depositing plumes. Deposited sediment does not contribute to turbidity nor transport. However, when remobilized during e.g. storms, it will.

To quantify this buffering effect, the model is used to simulate the turbidity induced by a 2 Mton/year release of fines, a typical value generated during the sand mining activities for the extension of the Port of Rotterdam (Maasvlakte 2). This is a substantial load compared to the natural alongshore flux of about 10 Mton/year, hence its impact on the North Sea water quality may be significant. The model was run for a period of 8 years, during the first 7 years of which the additional release of 2 Mton/year due to sand mining activities has been added. The natural/background SPM-concentration is also included in the model, which required another period of 8 spin-up years prior to the start of sand mining to reach dynamic equilibrium (bed and water column).

Two simulations are carried out, one without sediment buffering and a second with moderate buffering, i.e. with $m_2 = 0$ and $5\,\mathrm{kg/m^2}$, respectively. The latter value was obtained from the calibration of the ZUNO-model above. Figure 14.8 shows the results of these simulations averaged over the sixth year after the start of sand mining. The large difference between both scenarios illustrates the importance of buffering on the environmental impact of sand mining, reflected in the settings for water-bed exchange. Erroneous settings may result in a large over- or under-estimation of the far-field turbidity impact of sand mining. As emphasized in the previous sections, model calibration on observed SPM time series is insufficient to calibrate the buffer capacity of the seabed properly. Data on bed composition are thus required as well.

Another illustration of the large impact of buffering in the seabed comprises the behavior of a short-term release of a fine sediment tracer brought by the River Rhine into the Dutch coastal zone. Figure 14.9 shows the computed dispersion of these fines within the southern North Sea after six months for a simulation with and without seabed buffering, again using $m_2 = 0$ and $5\,\mathrm{kg/m^2}$. Without buffering, all sediment travels rapidly towards the German Bight, and virtually no fines are found in the Dutch coastal zone. This pattern would be characteristic for a dissolved tracer, but is highly unrealistic for fine sediment. When buffering into the seabed is accounted for, a more realistic pattern is obtained with fairly high SPM-concentrations in the Dutch coastal zone (see Fig. 14.5), comparable with observations. The practical relevance of this scenario concerns the dispersion of fines contaminated with river-borne industrial waste, such as heavy metals or PCB's, either in solution (left panel of Fig. 14.5), or bound to fines (right panel of Fig. 14.5).

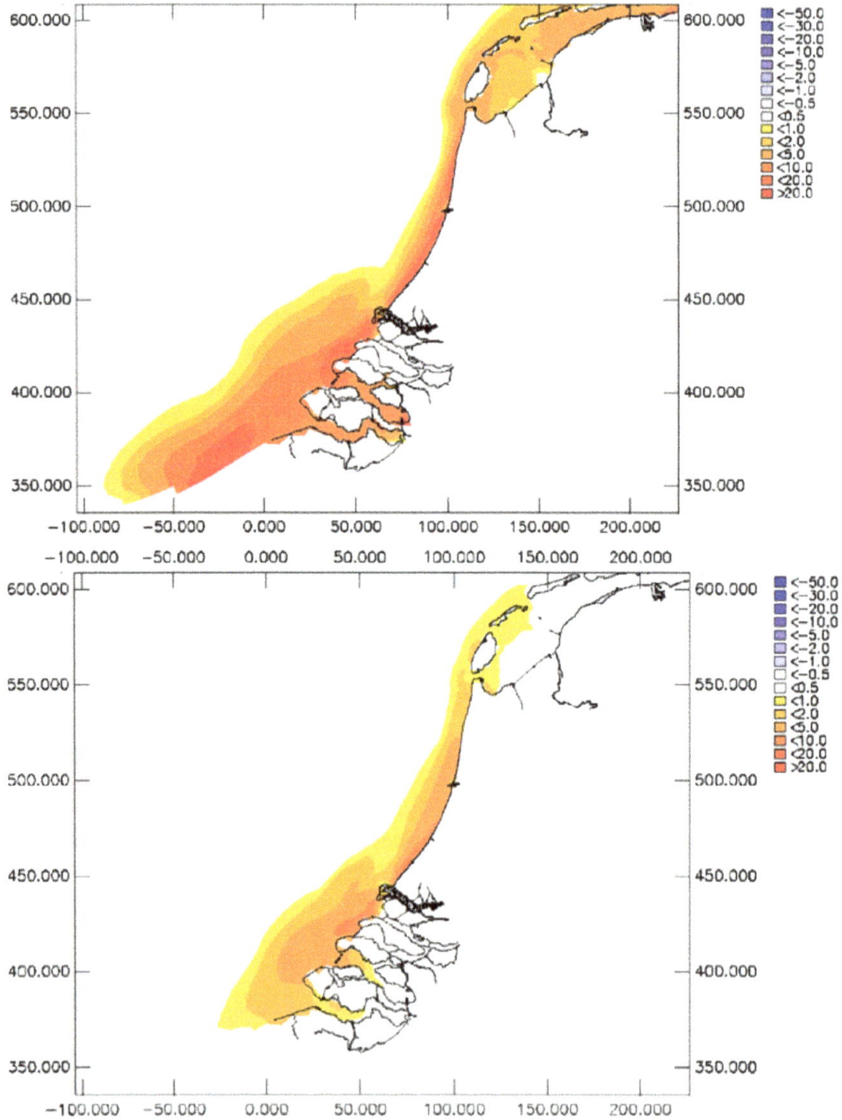

Fig. 14.8. Computed increase in near-surface SPM levels averaged over the 6th year after the start of sand mining. Upper panel: model computations without buffer capacity; lower panel: the more realistic setting with buffer capacity $m_2 = 5\,\mathrm{kg/m^2}$.

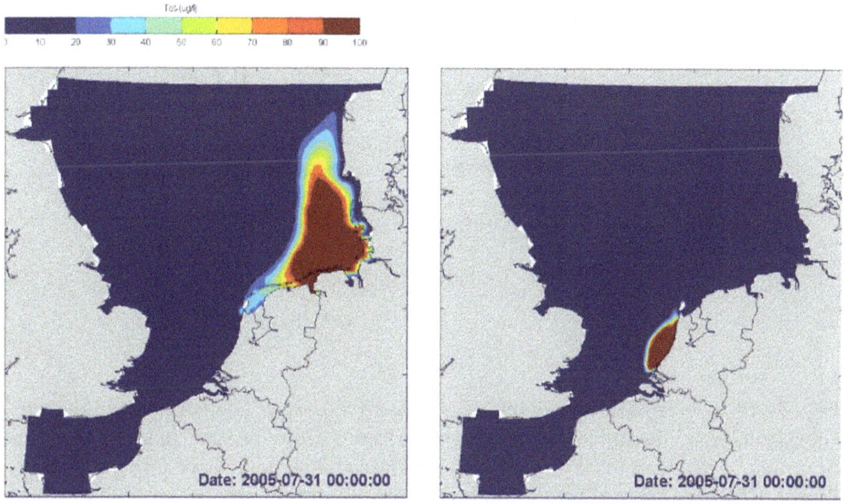

Fig. 14.9. Dispersion of River Rhine fines (near-surface concentrations) six months after release. Left panel: without buffering (representative for dissolved matter), right panel: with buffering (representative for particulate matter).

14.5.2 *Effect of harbor maintenance*

The second model application addresses the influence of harbor maintenance, i.e. the release of dredged material on the SPM dynamics along the Dutch coast, and its recirculation towards the harbor. We focus on the Port of Rotterdam, with its basins in the mouth of the Rhine-Meuse River (to be more precisely, its northern branch, the Rotterdam Waterway).

For this study, a dredging and dumping module was implemented in the model, in which mud deposited in harbor basins is removed at user-prescribed intervals and put back into the lowest computational water layer at user-defined dumping locations. In this way, mass conservation is guaranteed. The dumped mud may subsequently be transported, deposited and/or resuspended. By labeling the dumped mud, its dispersion can be tracked and recirculation towards the harbor basins can be quantified.

The resolution of the computational grid in the harbor area/basins is insufficient to resolve the details of local sediment dynamics (see also Sec. 7.1 and Winterwerp and Van Kessel, 2003). Siltation rates within the harbor basins can therefore not be resolved properly. However, the far-field effects of dumping and recirculation can be resolved, which is the focus of this application. The siltation within the harbor basins is

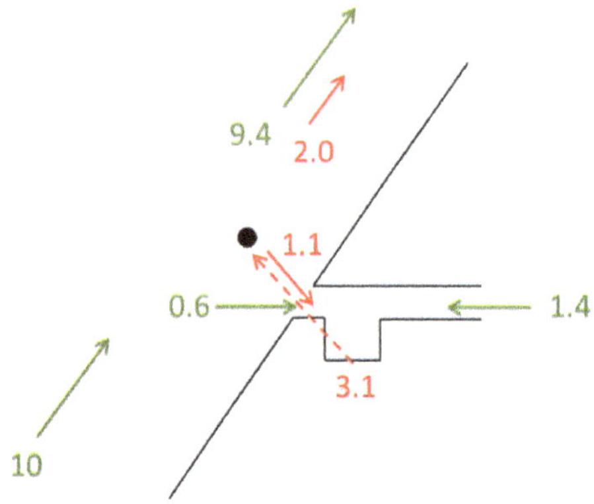

Fig. 14.10. Schematic of computed sediment fluxes in the Rhine-Meuse mouth (in Mton/year) including anthropogenic fluxes from harbor maintenance and dumping of dredged material at a disposal site (black dot). Green arrows: natural fluxes. Red arrows: anthropogenic fluxes. Recirculation rate = 1.1/3.1 = 0.36.

simulated through a local increase of the critical shear stress for erosion (see Hendriks and Schuurman, 2017) to attain the actual dredging rates of about 3 Mton/year. Figure 14.10 shows a schematic of the computed SPM fluxes along the Dutch coast near Rotterdam harbor entrance at the river mouth, as affected by dredging and dumping. The lower value of the natural SPM flux is used, reflecting the relative vicinity to the coastal waters from which sediment enters the harbor area.

The impact of dumping on turbidity levels and recirculation is simulated with the calibrated ZUNO-model, which is run for the years 2007 (repeated ten times to reach dynamic equilibrium) and for the years 2012–2015. The results for the hydrodynamic forcing of 2007 (10th cycle) are presented. Several simulations have been carried out:

(1) scenario 1, which represents the current situation for which the model is calibrated, with the present dumping location ("Verdiepte Loswal" = "Verdiepte LW");
(2) scenarios 1a–f, which are identical to scenario 1 apart from alternative dumping locations (LW1–LW4, LW–NW & LW–West);
(3) scenario 2, a hypothetical condition representing the coastal zone without any sedimentation and dredging in the Port of Rotterdam, nor dumping in the North Sea, i.e. a more or less "pristine" situation.

In scenario 1a–f simulations, the influence of the dumping location on the recirculation is studied by placing it closer or further from the harbor

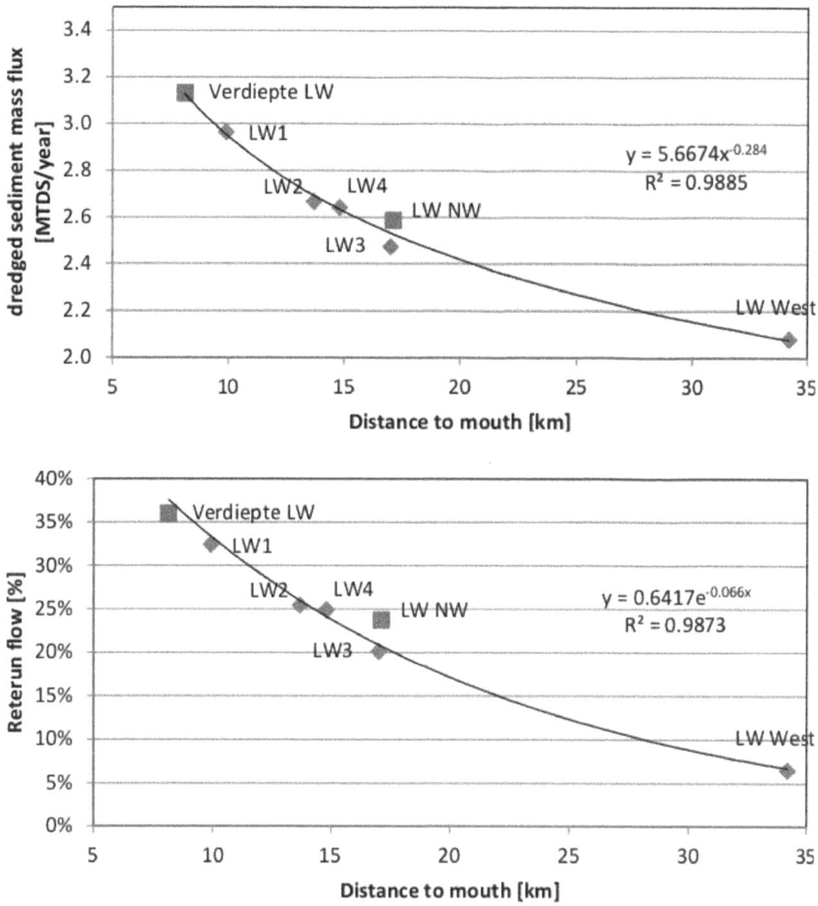

Fig. 14.11. Dredging volume (top panel) and return percentage (bottom panel) as a function of distance from harbor mouth.

basins. The simulations are continued till a new dynamic equilibrium is attained. Figure 14.11 presents the computed dredging volumes and recirculation rates as a function of the distance of the dumping site from the mouth of the River Rhine. Dumping locations close to the harbor basins induce larger recirculation, hence larger maintenance volumes. Obviously, recirculation rates and dredging volumes decrease when the dumping location is positioned further offshore. The model predicts that currently about 1/3 of the maintenance efforts is to be attributed to recirculation of the dumped sediment, back towards the harbor basins.

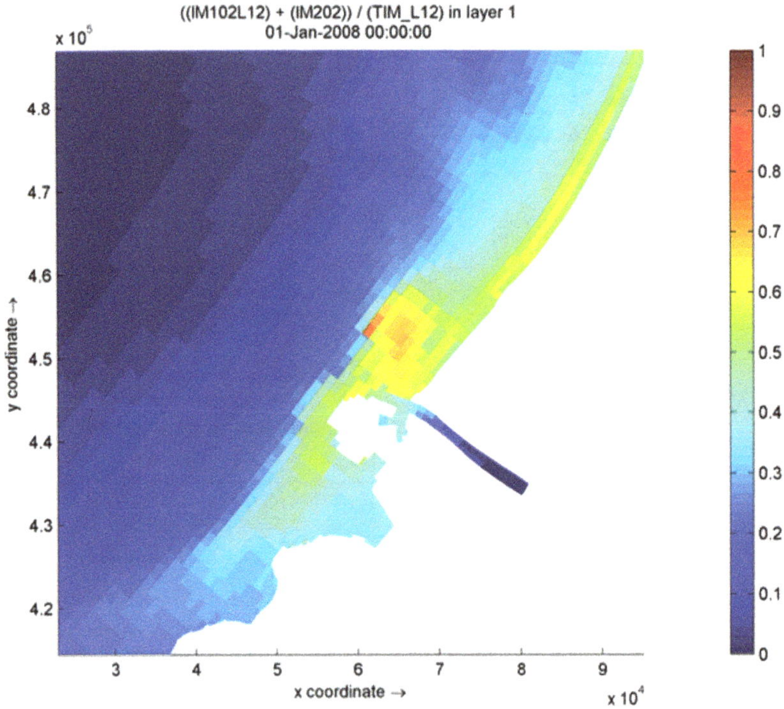

Fig. 14.12. Mud fraction in the seabed originating from present dumping location at equilibrium (after 10 years).

Figure 14.12 shows the fraction of mud originating from the present dumping location after 10 years of ongoing dredging and dumping (scenario 1). This period is required to reach a new equilibrium, as the replacement of natural mud by dumped mud is a slow process in accordance with the long residence time of mud in the Dutch coastal zone. It is concluded that the mud content has increased largely over a considerable area, consistent with observations, and that a large part of the mud around Rotterdam harbor is recirculated by the dredging-dumping cycle. So the effect of human intervention on the mud fluxes is large. This demonstrates that optimization of the release location is feasible. The larger the distance from the harbor mouth, the smaller the return flow hence the lower the sedimentation rate. As the return percentage is about 1/3, harbor siltation may be reduced by this number according to the model results. However, this would require longer sailing distances for the dredging vessel, or removal of sediment from the system. Both are costly options.

A most economically beneficial release location can be established if financial parameters such as sailing cost per unit volume per unit distance are included within the analysis. However, this is beyond the scope of this book. In general terms, a release location very close to the mouth is not a good idea because of a large return flow (so increasing maintenance volume) and a release location very far from the mouth is not a good idea because of prohibitive sailing costs. The sweet spot lies between both extremes.

Also, a most ecologically beneficial release location can be established. This would be defined as the location yielding smallest impact on SPM levels compared to the no-harbor scenario 2. The ecological and economical optima are generally not identical. However, from an ecological point of view neither a location very far away nor one very close is desirable. A distant location would induce a concentration decrease near the harbor and an increase near the dumping location and/or a large sedimentation footprint, compared to the no-harbor scenario. A nearby location would induce a strong concentration increase near the harbor and little changes further away.

This type of modeling allows analyzing scenarios to reduce dredging costs and the impact of dredging. However, there are a few caveats. The return flow is sensitive to hydro-meteo conditions (river discharge, tide, wind) and to the settling and resuspension properties of the dumped sediment. So theoretically, it is different for each trip. In practice a single location shall have to be designated which is (close to) optimal for average conditions and properties. Also, it may take a number of years before the benefits of a new release location can be demonstrated from the actual dredging records, because:

(1) sedimentation volumes exhibit large natural variations because of seasonal and yearly variations in driving forces and sediment availability, and
(2) the long response time of the system, related to the residence time scales elaborated above — a new equilibrium may take many years to establish, and become visible in dredging numbers/costs.

Finally, we discuss the effect of the siltation, dredging and release cycle on SPM levels by comparing simulation 1 with simulation 2. Figure 14.12 suggests that this effect is very large, as more than 50% of the mud around Port of Rotterdam originates from the dumping location and has thus been affected by this cycle. However, for an important part this excess

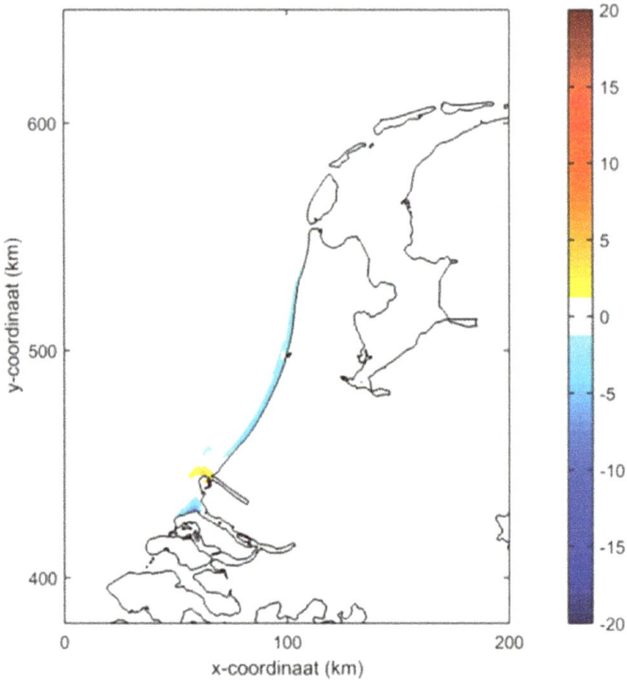

Fig. 14.13. Concentration difference (mg/l) scenario 1–scenario 2. Effect of siltation, dredging and release on year-averaged near-surface SPM levels. A negative value signifies a concentration increase by the siltation, dredging and release cycle.

SPM-concentration is balanced by the effect of harbor siltation, which captures an important part of the natural mud flux in the coastal system. The net effect is therefore a concentration increase by only a few mg/L in a narrow band along the Dutch coast (Fig. 14.13). This effect may be further reduced when the dumping would be shifted to a location about 5 km further offshore.

This text is a short and incomplete overview of fine sediment transport modeling studies along the Dutch coast during the past two decades. The reader interested in more detail is referred to Cronin and Blaas (2015); Hendriks and Schuurman (2017); Van Kessel *et al.* (2011c).

14.5.3 *Crucial system understanding aspects*

This chapter describes the setup, calibration and application of an engineering model to address management questions related to the transport and fate of fine sediment in the Dutch coastal zone (North Sea) at larger spatial and temporal scales. The most crucial elements in our understanding of the North Sea system, accounted for in the model, comprise:

(1) The majority of the seabed consists of sand, which buffers fines during shorter or longer periods of time as a function of the hydrodynamic driving forces, thus varying over the seasons. This variation is resolved with the use of the two-layer buffer model.
(2) As accumulation of fines do not accumulate in the fluff (active) layer on the seabed, transition from first- to zeroth-order erosion does not have to be accounted for in the buffer model.
(3) The amount of fines buffered in the seabed is unknown. The empty-bed approach is used to obtain a dynamic equilibrium bed mud contents consistent with the model formulations, allowing long-term simulations.
(4) The transport and fate of SPM is not only governed by water-bed exchange processes, but also by tide, stratification effects (ROFI) and waves, which are therefore accounted for through 3D modeling.
(5) Because of its size, temporal scales in the sediment dynamics in the North Sea are large, yielding substantial memory effects on the SPM-concentrations and bed mud content distribution. A thorough analysis of the residence time of fines in the water column and the seabed give credibility to the model results. The spin-up times of the model thus becomes eight years.
(6) Entrance and harbor basins of the Port of Rotterdam, their maintenance and the subsequent dumping of the dredged sediment have a significant effect on the net fine sediment fluxes along the Dutch coast. Accurate modeling of the siltation rates in the port is beyond the resolution of the model, and is therefore parameterized by modifying water-bed exchange rates. Also a dredging-dumping module was developed to study the recirculation of dumped sediment back to the port area.
(7) The sediment transport model is calibrated on the basis of variations in SPM-concentrations, bed mud content and residence times, thus not (only) on the SPM-concentrations themselves. This gives favorable results for long-term simulations. Because the Dutch coastal zone is shallow, a first calibration step has been carried out with a 1DV-model, enabling a rapid convergence in the large parameter space.

Chapter 15

Modeling SPM in Estuaries: The Scheldt Estuary

15.1 Introduction

Estuaries are natural systems where fresh and seawater meet and mix. Across the world, they have experienced (strong) anthropogenic interventions, as they are often densely urbanized and host substantial port and waterway infrastructure. Embankments, land reclamation, channel deepening, and the loss of (inter)tidal flats have enhanced tidal effects, salinity intrusion and SPM-concentrations (Chapter 9). Recurring managerial questions are to which extent hydrodynamics and sediment transport have been affected, and what will happen if these interventions continue, or discontinue in response to estuarine restoration works.

Hydrodynamic and sediment transport models are important tools to address these questions, in combination with field observations. The present chapter discusses the approach for a model-based analysis, with the Scheldt estuary as example.

The first step in this approach is to setup a conceptual model, describing the main driving forces in a qualitative way. As elaborated in Chapter 13, such a conceptual model is a prerequisite to make all implicit and explicit assumptions visible, and to communicate and discuss results with stakeholders, all within the context of the management/engineering questions to be addressed.

15.2 Conceptual model

15.2.1 *Purpose of the study*

The purpose of the model study is to provide a quantitative framework for evaluating the present state and future developments of the estuary with respect to its hydro-sedimentological dynamics, with focus on fine sediment, to be acceptable to both its many stakeholders and the scientific community. Two major managerial questions to be addressed are:

(1) How do SPM levels vary in response to the evolution of the estuary? — This is relevant with respect to its ecosystem (turbidity, light climate).
(2) How do dredging volumes develop, and how can maintenance costs be minimized in conjunction with water quality criteria?

The relevant time scale for the management of the estuary is about one to a few years. This implies that not only SPM-levels themselves are important, but also the amount of fines within/on the bed (the mud content of the riverbed), as SPM-levels and mud content are "communicating vessels".

Note that ongoing fairway maintenance involves dredging and relocation of sandy sediment as well, which has considerable also effects on the ecosystem of the Scheldt estuary. This is however not addressed in this chapter.

15.2.2 *Research questions*

The most important research questions to be addressed are:

(1) What are the dominant fine sediment transport processes in the Scheldt estuary, in particular in relation to the formation of its estuarine turbidity maximum (ETM, Sec. 9.1.1)?
(2) What is the mud balance in the Scheldt estuary, and how does this interact with turbidity levels (SPM levels)?
(3) What is the effect of interventions, such as channel deepening, harbor extension and land reclamation on the mud balance, SPM levels and maintenance dredging volumes?

Further to Chapter 13, other research questions would lead to a different setup of the model, as elaborated in Chapter 18.

15.2.3 *System understanding*

The Scheldt River, with a catchment of about 22,000 km^2, flows through the north of France and Belgium before discharging into the North Sea through the Western Scheldt Estuary in The Netherlands (see Fig. 15.1). This chapter deals with the tidal part of the river, downstream of Ghent, referred to as the Scheldt estuary, with the Sea Scheldt in Belgium and the Western Scheldt in The Netherlands. The Sea Scheldt is generally sub-divided into

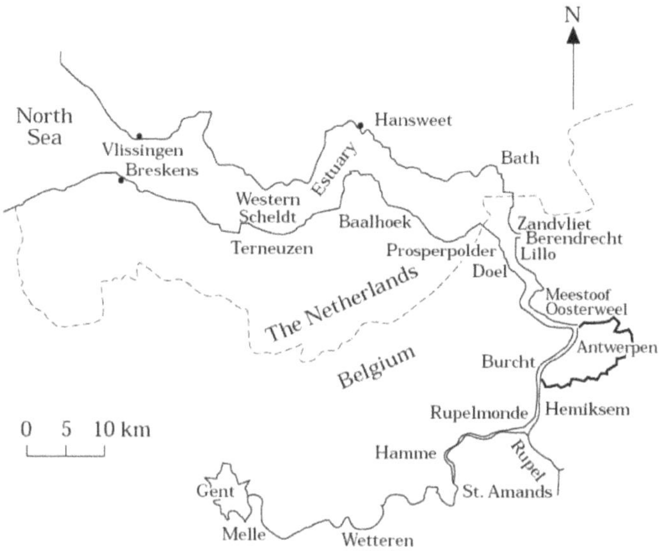

Fig. 15.1. The Scheldt estuary (from Fettweis *et al.*, 1998).

the Upper Sea Scheldt, which contains mainly fresh water, and an estuarine part, the Lower Sea Scheldt, down-estuary of Rupelmonde, where fresh and saline water mix. The average freshwater discharge is around 100 m^3/s, with minimum and maximum values between 20 and 600 m^3/s (Fettweis *et al.*, 1998). The tides in the Western Scheldt are meso-tidal and semi-diurnal with spring/neap ranges of about 3–4.5 m in its mouth (Flushing), amplifying in the estuary to a mean tidal range exceeding 5 m near Antwerp. Fine sediment near Antwerp is both of fluvial and marine origin (Verlaan, 2000). About 0.75–2.2 Mton is annually supplied by the Scheldt River (Fettweis *et al.*, 1998). An additional amount of sediment originates from the North Sea and is transported upstream, mostly through tidal asymmetry (Sec. 9.2). A large fraction of the fines accumulates in the estuarine turbidity maximum (ETM, Sec. 9.1.1), located somewhere between the Deurganckdok (a large harbor basin near "Doel" — Fig. 15.1) and the city of Antwerp. SPM-concentrations in the Western Scheldt characteristically amount to several 10 mg/L up to about 100 mg/L, but are higher in the Sea Scheldt, up to a few 100 mg/L, in particular in the ETM. Note that this range of SPM values implies below-capacity conditions throughout the estuary, with possible exceptions SPM dynamics in the ETM and some entrances to the various locks along the river.

The bed of the Western Scheldt estuary channel is primarily composed of sand. In the Sea Scheldt, a more solid substrate is found (strongly consolidated tertiary clay, shell fragments and gravel) reflecting large flow velocities IMDC (2013). Here, muddy sand and sand deposits are found on the inner bends and on sills, and (sandy) mud deposits are encountered in the entrance to the locks and harbor basins. In the main channel, no significant deposits of soft mud are found. The ETM of the Scheldt therefore only interacts with the bed in a limited way: most SPM is in suspension or stored in a thin fluff layer on the bed around slack water (Sec. 6.5). Harbor basins, abandoned channels and intertidal areas provide significant storage space for mud.

At the larger spatial and temporal scales addressed in this chapter, SPM-concentrations and riverbed mud content are controlled mainly by residual transports. These do not necessarily synchronize with gradients in transport capacity, as is the case for capacity-conditions governing sand transport and related morphodynamic developments. SPM-concentration levels and residual transport in the Scheldt are driven by a multiple asymmetries steering transport and concentration, induced by the tide (barotropic effects, Sec. 9.2), freshwater discharge (baroclinic effects and mixing, Sec. 9.1), storm surge, spatial and temporal variations in sediment availability and properties, etc., or in other words, a complex interplay between horizontal advection, vertical settling-mixing and deposition-resuspension, i.e. the water-bed exchange processes.

High SPM-concentrations can for instance be maintained by residual circulations, which may occur in the absence of any gradients in transport capacity. Examples are the ETM and the dredging-dumping cycle:

(1) An ETM is formed and maintained at the location where up-estuary, near-bed gravitational circulation balances down-estuarine near-surface gravitational circulation and riverine flushing. Exchange with the riverbed is not necessary to maintain an ETM, as shown in Sec. 9.2, Fig. 9.11. Thus SPM levels may be high at small bed mud content, and the governing circulation takes place in the vertical plane.
(2) SPM accumulation may also be induced by horizontal circulations, either natural, or anthropogenic driven. The first occurs behind obstacles in a river (jetties, breakwaters, land heads, ...), the second as a result of recirculation of dumped material towards the location of dredging, i.e. the dredging-dumping cycle (see also below).

Riverbed mud content may be high or low, depending on the local flow velocities.

On the other hand, high bed mud contents may be formed over time, even at low SPM-concentrations, in net deposition zones. These can be found where transport capacities become small, e.g. in harbor basins and on inter-tidal flats.

It is of utmost importance to recognize which of these processes controls observed SPM dynamics and bed mud content, and consequently tune model formulations and parameters to reproduce these effects correctly (system understanding). As discussed in Chapter 13, on the short term, locations of and SPM-concentrations in an ETM may be simulated correctly by tuning river bed mud content locally. If the ETM dynamics are governed by vertical circulations only, this approach will fail on longer time scales.

Of course, bed levels in harbor basins are generally maintained by dredging. Therefore, sediment accumulation may continue indefinitely, disrupting the morphological feedback that would ultimately end this accumulation. In the Scheldt estuary, the dredged fine sediment is dumped at some distance from the basins, resulting in a strong local horizontal circulation, which is not caused by peculiarities of the local hydrodynamic forcing, but is supply-driven as long as the dumping rate remains below the local transport capacity of the river. However, the local hydrodynamics determine the enhanced SPM-concentrations and their gradients.

An important management question concerns the optimizing of dumping locations for the dredged sediment with respect to their distance from the harbor basin (see also Sec. 14.5). Short distances induce strong return fluxes (recirculation) to the harbor basin, increasing dredging volumes. Large distances yield long operating cycles, thus increasing dredging costs. The optimal distance can be determined with an SPM transport model, as illustrated later in this chapter.

In the conceptual model, three SPM transport processes are identified, operating at three different time scales, governing the SPM-concentrations in the water column, as well as the mud content in the riverbed. These factors are (see Fig. 15.2):

(1) The quantity of fine sediment, some of which is in suspension and some is (temporarily) stored in the riverbed, whether easily erodible or not.

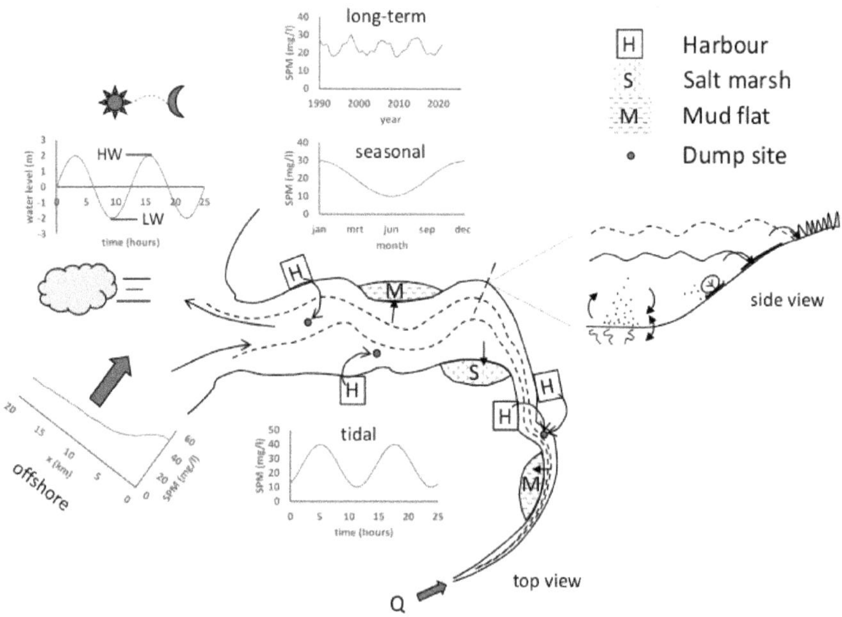

Fig. 15.2. Schematic diagram of the most important processes affecting the fine sediment dynamics in the Scheldt estuary.

(2) The sediment properties determine both horizontal and vertical transport rates (i.e. sedimentation, erosion and transport), mainly characterized by settling velocity, critical shear stress for erosion and erosion rate.
(3) The hydrodynamic forces driving the sediment, steered by tide, wind, waves and freshwater discharge.

With regard to the first two points, fine sediment dynamics differ from sand dynamics, as elaborated in Chapters 1 and 2. Sand transport is generally transport/capacity-limited, i.e. the carrying capacity of the flow determines the transport rate, rather than the supply of sand. The transport of fine sediment is generally supply-limited, i.e. the supply rather than the transport capacity determines its transport rate. Also, the properties of sand are closely linked to the grain size distribution solely, whereas the properties of fine sediment vary largely depending on stress history, degree of aggregation and consolidation, etc.

The three most important time scales are:

(1) short: tidal period to spring-neap cycle (0.5–14 days);
(2) intermediate: seasonal to multiple year;
(3) long: >5 years, i.e. the morphodynamic time scale, at which bed level changes affect the hydro-sedimentological system considerably.

These short, intermediate and long time scales are closely related to the spatial micro-, meso- and macro-scales of the estuary (De Vriend, 1991b). At the short time scale, water-bed exchange processes, such as deposition and resuspension, and exchange between the Scheldt and North Sea of SPM take place. At the very long time scale (decades), interactions between major hydrological changes, riverbed mud content and bed levels are observed — these may be driven by e.g. changes in tidal volume as a consequence of land reclamation or channel deepening. Observations reveal a third time scale, which ranges from seasonal to several year, where the "memory" of the system is important. This memory is rather long for the Scheldt estuary. There are year-to-year variations in mud content of the bed and SPM-concentration, which may exhibit auto-correlation time scales of more than one year.

As a consequence, long periods of above-average and below-average SPM-concentration levels occur. This time scale is often neglected in conceptualizations of mud dynamics, yet plays an important role in variations within observational time series. This time scale is implicitly resolved for in the "empty-bed" approach, advocated here, and applied in all present SPM models of the Dutch coastal waters (Western Scheldt, North Sea, Wadden Sea and Ems-Dollard, see Sec. 12.3).

The processes operating at the shortest time scales are generally well-known and incorporated within most SPM models — model calibration is then more or less straightforward, basically against short time series of SPM observations. In the Scheldt estuary, this part of the calibration process is based on the short-term effects of tide, wind and waves on the vertical profile of SPM, erosion and deposition processes, and deviations from average transport pathways.

The incorporation of intermediate-scale phenomena, i.e. processes responsible for seasonal variations and multi-year trends in time series of SPM and mud content, is fraught with more difficulties. Conceptually, the relative importance of physical and biological processes operating at this scale is not well-understood. Examples are the feedback between mud content and erodability of the bed, bed roughness, and biological processes, such as the development of microphytobenthos mats, activity of filter feeders and bioturbators, etc. Many of these interactions are steered by seasonally varying factors such as light and temperature, though may be affected also over longer time scales by weather/climate patterns. If, for instance, summer storms erode microphytobenthos mats (and subsequently resuspend large amounts of fines), redevelopment of the microphytobenthos may take a long time (Chapter 8), during which large amounts of sediment might be lost from the system. Then changes at the scale of the estuary may be induced.

In this way, short time scale events may induce large-scale effects, thus the long-term development of an estuary. Though this nonlinear behavior is well-known, elaborated upon for morphodynamics by De Vriend (1991b), its consequences are still difficult to grasp, especially as the relevant time scale is at or beyond the simulation times for which process-driven models can be run in practice (several years). Currently, only 3D SPM models are available to study estuarine response at these intermediate time scales, and their consequences for longer time scales have to be assessed by extrapolation, based on a thorough system understanding. Important uncertainties in these are:

(1) seasonal changes or long-term trends in sediment properties induced by physical, chemical and/or biological factors (properties),
(2) uncertainty in the amount of "active" mud (i.e. easily erodible) and its residence time in the system (quantity),
(3) limitations in the accuracy and predictability of hydrodynamic forcing, which have cumulative effects on residual transport and the evolution of the mud balance (forcing).

These uncertainties limit the predictability of process-based SPM transport models, while model performance is likely to reduce further over time. Obviously, more advanced/complete process formulations cannot improve performance with respect to the uncertainties in sediment properties and driving forces, but may be a prerequisite to enhance the credibility of model results (Sec. 13.5). As these uncertainties are inherent to any natural system, successful modeling is only possible if based on well-formulated engineering/research questions, as elaborated upon in Chapter 13.

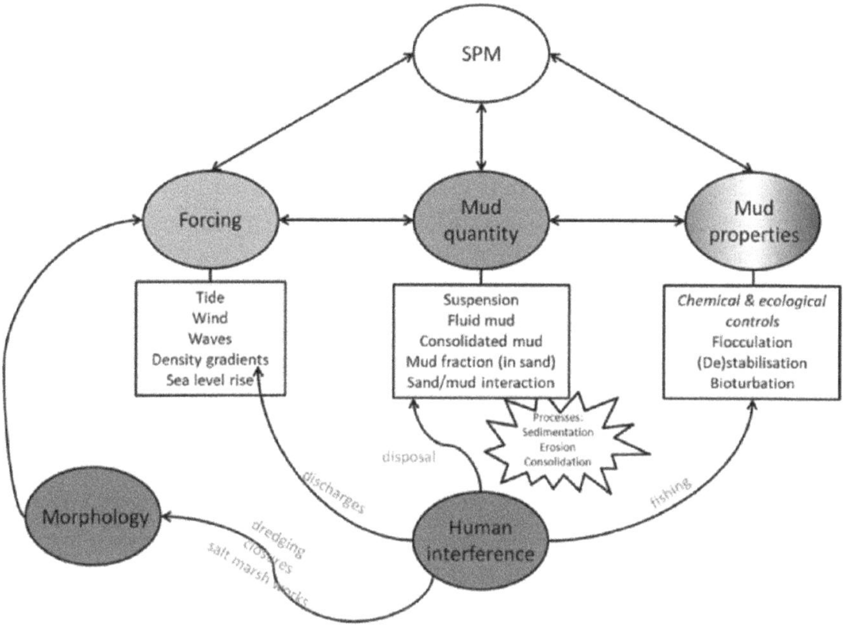

Fig. 15.3. Schematic and generic diagram of the most important processes affecting SPM levels.

The diagrams of Figs. 15.2 and 15.3 sketch the major relevant processes and their interaction, determining SPM-concentrations in the Scheldt estuary. At short time scales, the amount of fine sediment and its properties are more or less constant, and SPM dynamics are therefore mainly driven by hydrodynamic forcing inducing horizontal transport and sedimentation/ erosion. At long(er) time scales also variations in sediment quantity and properties become important, thus increasing the level of system complexity and uncertainty significantly.

At intermediate and long time scales, SPM dynamics in the Scheldt estuary are largely determined by SPM processes in the North Sea, in particular with respect to the availability of fine sediment and its timing. Further upstream, fluvial mud supply also becomes important. The following elements are important:

(1) The residual flux of SPM along the Dutch coast (about 10–20 Mton/year — see also Chapter 14) determines the availability of fines

in the mouth of the estuary — note that this flux varies largely over the years (Cronin and Blaas, 2015).

(2) The influence of tidal currents and waves on sedimentation and resuspension in the North Sea coastal zone, also affecting SPM levels in the Scheldt mouth.
(3) Gradients in salinity and SPM are major factors affecting SPM dynamics and concentration distributions in the North Sea; in particular the offshore turbidity maximum near the mouth of the Western Scheldt (off Zeebrugge) plays an important role.
(4) Long-term, as well as seasonal variations in SPM in the coastal waters.
(5) Short-term interactions between tide and weather (waves and wind) are also important elements determining SPM time series in the North Sea.

Fresh water input has a profound effect on SPM levels at short time scales within the upper reaches of the Sea Scheldt (Upper Sea Scheldt). Further down-estuary, the effects of river flow input becomes smaller, and important only at longer time scales (e.g. during a dry year).

Also biological processes (microphytobenthos, filter feeders, bioturbation, see Chapter 8), in particular on salt marshes, are indicated in conjunction with tidal phenomena in the diagram of Fig. 15.3. These induce strong seasonal effects, but may also steer long-term developments, as exemplified above. A major anthropogenic disturbance of the meso-scale fine sediment dynamics in an estuary are dredging of fairways and harbor basins, and the subsequent dumping of the dredged sediment. Dredging volumes in the Scheldt estuary have gone up significantly since the opening of the Deurganckdok (a large tidal dock on the left bank of the estuary, close to its ETM), totally amounting to about $10 \cdot 10^6$ m^3 for all maintenance works in the Scheldt estuary in 2017.

On the long term, embankments, land reclamation and channel deepening become steering parameters. The morphodynamic response of the estuary to these interventions can be large, but are beyond the scope of this book.

The objective of the modeling study is to reproduce the mud balance of the estuary as a convoluted result of the sediment dynamics over time. This implies that apart from SPM levels, also the mud content of the bed (with different sub-levels of exposure to hydrodynamic forces and erodability), the exchange of fines between water column and seabed and the residence times in the different compartments are key elements for the long-term mud dynamics and its response to interventions or natural changes. In this context, SPM variations at smaller (tidal) time scales are less important, and are not resolved if that would be at the cost of model performance at larger spatial and temporal scales.

15.2.4 *Model requirements*

The setup of the numerical model is determined by the research questions to be addressed and our system understanding of the processes relevant for answering these questions. This results in the following requirements:

(1) A sufficient vertical resolution is needed to resolve the effects of vertical stratification and residual flows (gravitational circulation). Mixing needs to be modeled with a turbulence closure scheme resolving the effects of buoyancy destruction, such as the k–ϵ turbulence closure model (Sec. 3.2.3).
(2) The horizontal resolution needs to be sufficiently detailed to reproduce (i) observed spatial gradients in SPM-concentration and (ii) the transient and long-term residual sediment balance of the system.
(3) The turbidity maximum offshore near the mouth of the Western Scheldt and the ETM-dynamics in the Lower Sea Scheldt near Antwerp should be captured. Note that these turbidity maxima are not not in equilibrium with the bed sediment (Chapter 1), but are governed by the convergence of sediment pathways, which is an essential feature of supply-limited transport regimes, as in the Scheldt. Hence, these dynamics should not depend on water-bed exchange processes. Such a model setup requires simulation periods of months to years to attain dynamic equilibrium.

Owing to the relative low SPM-concentrations, in conjunction with large flow velocities in the estuary, interaction between SPM and turbulence is fairly small in the majority of the river (Sec. 7.1). An uncoupled approach is therefore feasible (Sec. 13.3.1), i.e. the hydrodynamic and SMP transport model can be calibrated and run independently, i.e. an uncoupled approach. This also implies that morphodynamic developments are not accounted for, i.e. the model is run in "morphostatic mode".

Attaining dynamic equilibrium in SPM conditions in the Scheldt (requirement 3) demands long simulation times (typically several months to a year), which matches the managerial time scales described in Sec. 15.2.1. However, this conflicts with requirements on vertical (requirement 1) and horizontal (requirement 2) resolution, owing to the large computational times involved.

The choice of resolution is therefore a compromise, sacrificing horizontal resolution (see below). In principle, for large-scale applications it is not problematic to loose details at a small scale, unless small-scale processes affect the large-scale response of the system. In that case, either the model resolution has to be increased, or sub-grid formulations should be included to account for effects of unresolved small-scale process on the large scale (see Sec. 12.1).

To investigate the effect of horizontal resolution, the Scheldt model is run at two different grid sizes to check convergence (with different time steps, obviously). The computational results of the high-resolution model are aggregated on the lower resolution grid. For a fully convergent model, these results should be identical. If not, tests with an even higher resolution model would be required, which may be prohibitive because of computational times. Here, we accept some non-convergence, i.e. the model calibration is grid dependent, a pragmatic, engineering approach — these numerical inaccuracies are likely much smaller than uncertainties in mud-related model parameters.

Van Kessel *et al.* (2011a) argue that the SPM model should qualitatively reproduce the following features observed in the Scheldt estuary:

(1) spatial SPM distributions, notably near the ETM near Antwerp and offshore near Zeebrugge;
(2) vertical SPM-concentration gradients;
(3) temporal SPM-concentration fluctuations in the estuary, including seasonal dynamics;
(4) bed composition, i.e. the spatial distribution of sandy and muddy areas;
(5) the rates of siltation on intertidal flats and in harbor basins;
(6) the ratio between fluvial and marine mud;
(7) the total amount of "active" mud moving around in the estuary;
(8) the long-term mud balance of the estuary.

15.3 Hydrodynamic model

Though a decoupled approach is applied, setup and calibration of the hydrodynamic model are steered by the requirements for the SPM model. The hydrodynamic model covers the complete estuary ranging from the weir at

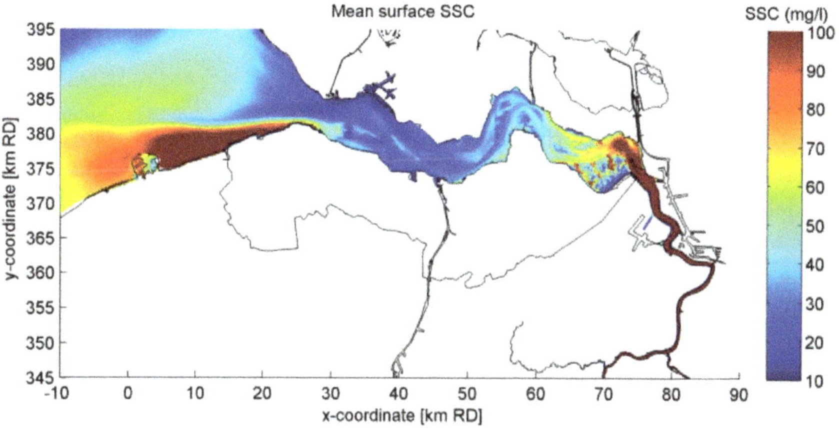

Fig. 15.4. Computed annual-mean near surface SPM concentration distribution for 2006.

Ghent to and beyond its mouth near Zeebrugge (Fig. 15.4). We use the 3D hydrodynamic NeVla model on the basis of Delft3D. Setup, calibration and validation are described in Vanlede *et al.* (2015). Water levels and currents in the main channels are well resolved by this model. However, the model performance in some shallow areas is less accurate, which is explained by (i) limitations in resolution, and by (ii) outdated and less accurate data on bathymetry and bed roughness. Longitudinal salinity gradients and their variation with freshwater discharge are well reproduced; over the water depth, the computed salinity field is well-mixed, as it should be.

This model is used as a basis for fine sediment transport computations. It is driven by water level time series at its open sea boundary, obtained from the larger Continental Shelf Model (Vanlede *et al.*, 2015), and hourly variations in the forcing by wind, waves and freshwater discharge. The hydrodynamics for two years are simulated, using bathymetry/geometry and boundary conditions of 2006 and 2014, which are both typical years, i.e. without extreme events, and more or less average river flows. Computational results of water level, velocity, salinity and mixing are stored on disk (ca 1 TB) and used to steer the sediment transport model. Although the bathymetry for both years is somewhat different, the hydrodynamics in the area of interest between the estuary mouth and Antwerp are similar on a year-average scale. Within shorter periods, differences may be more pronounced because of different phasing of river discharge or wave conditions.

15.4 Sediment transport model

15.4.1 *Sediment fractions, properties and boundary conditions*

The SPM model covers the same domain as the hydrodynamic model, is steered by the hydrodynamic parameters computed with the hydrodynamic model, and sediment properties obtained from calibration for the year 2006. The sediment model is setup according to the same philosophy and using the same process formulations described in Chapter 14, including a 1DV-phase, which is not presented again, though. The step-by-step approach presented in Chapter 13 is followed. Specific modeling details for the Scheldt estuary are discussed in Van Kessel *et al.* (2011a). To attain (dynamic) equilibrium with the empty-bed approach, simulation times of six years are required, using the same one-year hydrodynamic input six times, i.e. the SPM model requires a spin-up time of five years.

The model consists of two sediment fractions (micro- and macro-flocs) which do not interact, though some exploratory simulations with floc interactions have been carried out. Both fractions can be of marine or fluvial origin, allowing validation of the fluvial-marine mixing curve found by Verlaan (1998); Verlaan *et al.* (1998). These various fractions are labeled to track their pathways and fate, which is not further elaborated here. In this way mixing between fluvial and marine sediment can be investigated, in conjunction with the pathways of dredged and dumped sediment within the estuary, assessing its contribution to overall SPM levels.

Seasonal-varying sediment boundary concentrations were derived from observations. Because of model size and its memory, short-term concentration variations at the sea boundaries do not affect model response and have not been included. Initially, spatial uniform parameter settings and SPM-concentration field were prescribed. Thus temporal and spatial variations and gradients in SPM-concentrations are internally generated by the model.

A dredging-dumping sub-routine was developed to guarantee mass conservation for simulating the dredging-dumping cycle. Sediment deposited in harbor basins (e.g. in Zeebrugge, Vlissingen, Terneuzen and Antwerp) is "dredged" and "dumped" at user-defined times and locations within the model domain (see also Chapter 14). This is an important model feature, as the net deposition rate in the various harbor basins (>5 Mton/year) is large compared to the net rates elsewhere in the estuary (100s kton/year).

The SPM model was calibrated against observations in the year 2006. Next, these settings were used in simulations for the year 2014. A summary

Table 15.1. Parameter settings of the 3D Scheldt model.

Parameter	Value	Unit
w_s	0.5 % 2	mm/s
p_d	0.1	–
α	0.1	–
$\tau_{e,1}$	0.2	Pa
$\tau_{e,2}$	1	Pa
M_1	2	1/day
M_2	10^{-3}	kg/m^2/s
$M_{E,1}$	10^{-3}	kg/m^2/s

of calibrated parameter settings reflecting the sediment properties is given in Table 15.1.

15.4.2 *Phenomenological calibration*

The model calibration was aimed at meeting the objectives in Sec. 15.2.4. The most challenging aspect of the calibration process turned out to be a proper simulation of the ETM near Antwerp in combination with the proper overall mud balance (Cleveringa, 2013; Van Maldegem, 2004). In the first phases of calibration, the model was initiated with the prescribed mud distribution in the bed. For these conditions, either the ETM was well represented but the mud balance showed a strong export, or the mud balance was realistic at the expense of the proper reproduction of the ETM. However, the first approach did not work either, as it could not be maintained until dynamic equilibrium was reached, as the bed would become depleted. Also, an ETM does not develop if the empty-bed approach is followed as advocated earlier.

The solution to this controversy appeared to enhance the gross vertical fluxes in the water column without changing net deposition-erosion rates. This was done by decoupling settling and deposition through the application of the reduced deposition concept, which allows for larger near-bed SPM-concentrations — see Sec. 4.2 and Chapter 17 for further details. Next to setting $p_d = 0.1$, the erosion rate parameters were reduced. This approach parameterizes the dynamics of a fluff layer in the ETM and allowed for a proper representation of mean SPM-concentrations in the ETM near Antwerp in combination with a realistic overall mud balance, i.e. requirement (1) and (8) of Sec. 15.2.4.

Figure 15.4 shows the computed annual-mean spatial distribution of near-surface SPM-concentrations along the estuary. The two observed turbidity maxima are predicted: one near the mouth of the estuary, around Zeebrugge, and an ETM near Antwerp at the head of salinity intrusion. Occasionally, larger SPM-concentrations are also observed in the upper reaches of the Sea Scheldt after high river flow events, following prolonged periods of low river flow (Chen *et al.*, 2005; Cox *et al.*, 2019). However, the model resolution in these reaches is too low to resolve these SPM dynamics properly. The applicability of the model is therefore limited to the Western Scheldt and the Lower Sea Scheldt up to Rupelmonde.

Figure 15.5 shows the computed and observed tidal variations in SPM-concentrations at 0.8 m above the bed (m.a.b.) at Boei 84 (an anchor station near Antwerp) for the first three months of 2006. The black and blue dots indicate computed and observed concentration levels which have been time-averaged with respect to HW. The + and – symbols represent the standard deviations. Although mean SPM levels are comparable to observed values, the computed phasing and amplitude differ significantly. This is to be attributed mainly to a lack in resolution: using a model with higher resolution locally improves the model-data comparison greatly, as elaborated upon in the siltation study for Deurganckdok, shown in Fig. 18.7,

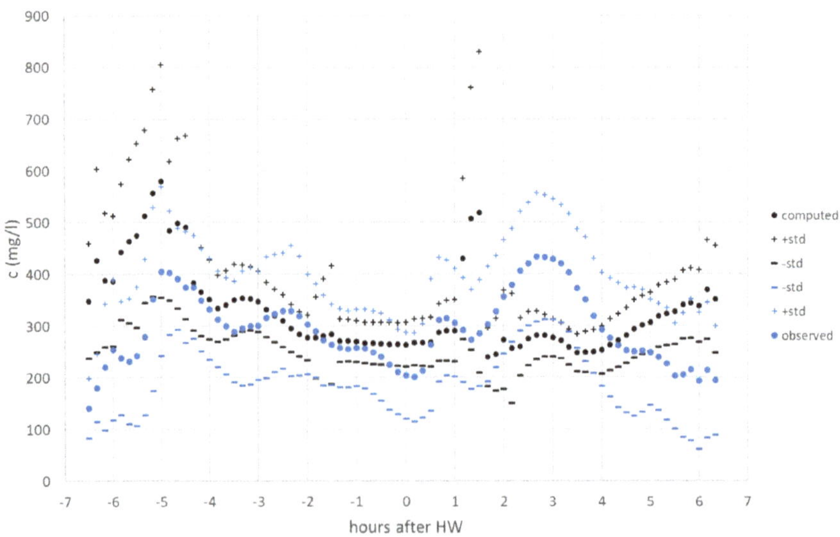

Fig. 15.5. Computed and observed near-bed SPM-concentrations in first three months of 2006 relative to HW at station Boei 84.

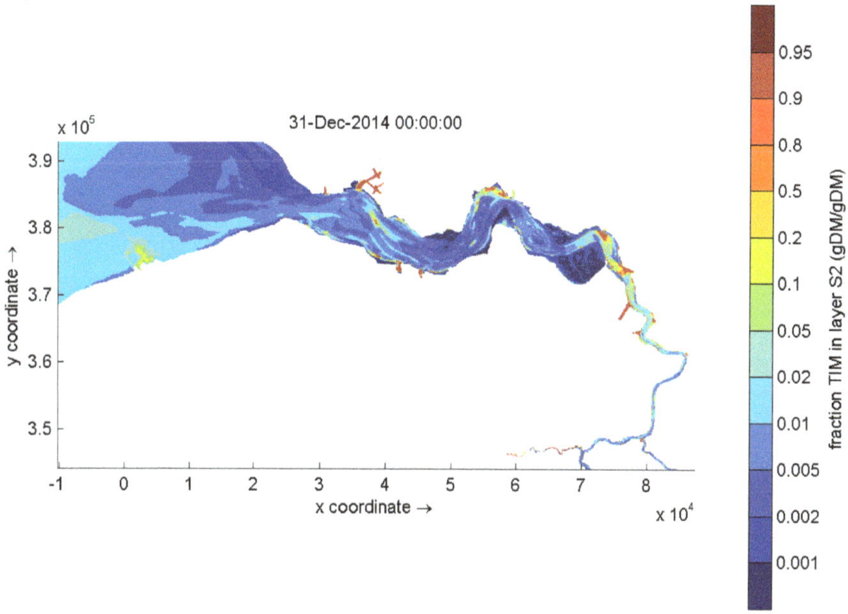

Fig. 15.6. Computed equilibrium bed composition at the end of 2006 simulation ("fraction TIM in layer S2" represents the river bed mud content fraction).

and discussed in Sec. 18.4.2. As the focus of the current model application is on long-term large-scale behavior of the SPM dynamics in the Scheldt estuary, the resolution in the model was not refined, preventing large computational times. Hence, in interpreting the results, the inaccuracies of the computed ETM dynamics have to accounted for.

Figure 15.6 presents the sediment composition in the bed along the estuary, as computed with the coarse-grid model. Characteristic patterns are well represented, with low mud content in the main channels, a higher mud content in sheltered shallow areas and the highest mud content in harbor basins. Also, the larger mud content in and around the turbidity maxima is predicted — note that the model was initiated with the empty-bed approach, hence the larger mud content here reflects SPM exchange with the ETM. From observations (McLaren, 1994) similar patterns of bed sediment composition are found. At smaller scale, the model predictions are less accurate, which is to be attributed to the spatial scale at which the bed composition varies in conjunction with a lack of resolution in computational grid size, local bathymetry, hydrodynamic and wave forcing.

Table 15.2. Observed and computed yearly siltation rates (kton/yr) on mudflats in the Western Scheldt and Sea Scheldt, and in various harbor basins — simulations for hydrodynamic conditions of 2006 and 2014.

	Observations	2006 model	2014 model
W-Scheldt mudflats	200	100	20
Sea Scheldt mudflats	100	1000	200
Zeebrugge	3100	1000	1100
Sloe	600	200	100
Braakman	800	100	800
Terneuzen	250	200	90
Zandvliet sluice	n.a.	800	440
Deurganckdok	2000	1800	680
Kallo sluice	n.a.	700	320

Table 15.2 presents the observed and computed siltation rates in harbor basins and on tidal flats. The observations represent mean data from long-term observations and the computed values follow from the 2006 and 2014 simulations. The model predicts the siltation rates in most harbors to a reasonable degree, appreciating the limited resolution of the model. The computed siltation rates on the tidal flats in the Western Scheldt are underestimated, while overestimated in the Sea Scheldt. The 2006 and 2014 results reveal a substantial difference, which is to be attributed to:

(1) yearly variations in freshwater discharge and in wave wind and wave conditions on the North Sea, both resulting in significant variations in mud supply, and
(2) yearly variations in local wind and wave dynamics, resulting in variation in local mud trapping.

Note that the hydrodynamics are not much affected by these variations, as largely driven by the deterministic tide. This is an important observation, explaining the large yearly variations in SPM dynamics and hence in e.g. siltation rates, as experienced by yearly variations in maintenance requirements. This also reflects the large memory effects in large systems such as the Scheldt estuary.

The last part of the phenomenological model calibration concerns the fine sediment mass balance. The computed mud balance for 2006 and 2014 is presented in Fig. 15.7, showing that (1) the gross SPM fluxes largely exceed the net fluxes, as explained in Secs. 9.4 and 10.4, and (2) that year-average net fluxes differ significantly between 2006 and 2014, consistent with the

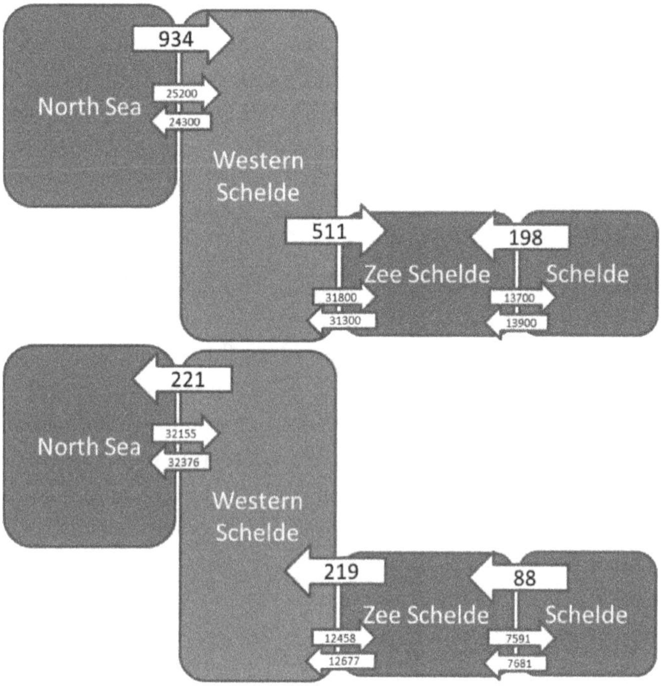

Fig. 15.7. Computed gross and net sediment fluxes in the Scheldt estuary (Schelde) in kton/yr for 2006 (upper panel) and 2014 (lower panel).

discussion above. Thus, the assessment of a residual mud balance for the Scheldt (and basically all estuaries) is only meaningful over long time scales, i.e. at least multiple years. At short time scales, covering a few seasons or a single year only, residual fluxes are strongly influenced by meteorological conditions and the order of events, and may be either importing or exporting. A second reason for these large differences is the large memory effect of the estuary, which is steered for instance by sediments temporarily deposited on mud flats, etc. One year as averaging period is therefore insufficient, as the differences between the 2006 and 2014 results illustrate.

Although not all calibration steps of Chapter 13 have been discussed, the model meets the targets of Sec. 15.2.4. As the model is to be applied at estuarine scale, most attention was paid to large-scale model behavior. Detailed reproduction of SPM values at individual points is therefore not a necessity. The Scheldt SPM model is thus suited to elaborate on the managing/engineering questions of Secs. 15.2.1 and 15.2.2, provided that computational results are evaluated at an aggregated level only, and accounting for the large variability over different years as well. Further details on model setup, calibration and validation can be found in a series of technical reports (e.g. Van Kessel and Vanlede (2010); Van Kessel *et al.* (2011b), available on www.scheldemonitor.be).

15.5 Model application

As mentioned (Fig. 15.7 and Table 15.2), the siltation rates in the Deurganckdok, which is situated near the border between the Western and Sea Scheldt, are large compared to the net SPM fluxes in the river, but small compared to the gross fluxes. It is therefore not obvious whether or not opening of the Deurganckdok has affected the overall mud balance in the estuary. As managing authorities were interested in an order of magnitude estimate only, the model was run for a hydrodynamic forcing of three months, i.e. the last quarter of the year 2006, for two scenarios, i.e. with (the current, thus reference situation) and without Deurganckdok. Of course, the results have to be interpreted in comparative mode (Sec. 13.6.5). Focus is on the large-scale impact of the dock on the mud balance, SPM levels and dredging volumes. The local exchange flow between the estuary and the dock are discussed in Chapter 18.

The construction of a large dock has three major consequences for the sediment dynamics in the estuary:

(1) A dock traps a large amount of sediment and therefore reduces SPM levels in its vicinity.
(2) Trapped sediment is removed by maintenance dredging. This material is relocated elsewhere within the estuary. SPM levels in the vicinity of the dumping location(s) will increase.
(3) Dumped sediment may recirculate back to the harbor basin, thereby increasing the maintenance volumes.

These consequences are quantified with the SPM model for the Scheldt estuary. Figure 15.8 shows the computed relative difference in time-averaged, depth-averaged SPM-concentrations between a simulation without and with tidal dock at location "D", and without and with dumping at location "R" (i.e. (without – with)/with). As expected, SPM-concentrations are lower without dock, with reductions between 5% and 15% in a substantial part in the Lower Sea Scheldt between the tidal dock and Rupelmonde,

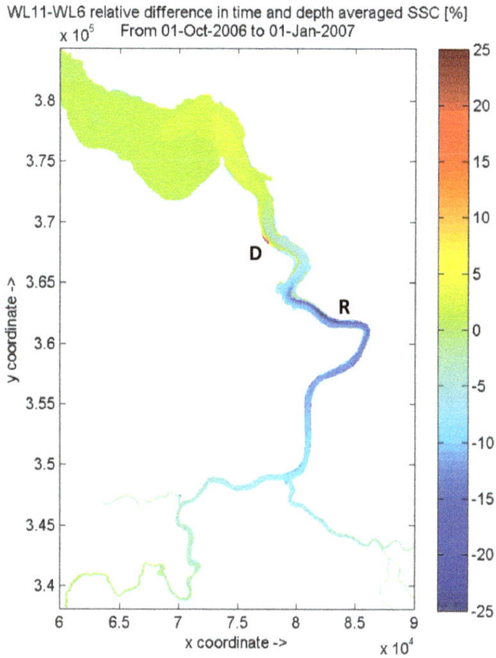

Fig. 15.8. Impact of Deurganckdok on time- (three months) and depth-averaged SPM-concentration [%]. "D" represents dock location and "R" dumping location. Negative values reflect smaller SPM-concentrations for the reference case (no dock).

with the largest difference near location "R". In this part of the estuary, the increase in SPM-concentration by dumping dominates the concentration-decreasing effect of sediment trapping by the dock. On the other hand, down-estuary of the dock, the opposite is observed. In the scenario without dock, the SPM-concentration is up to 5% higher.

These results suggest that it might be beneficial to relocate dumping down-estuary of the Deurganckdok. However, this would imply a location close to or even across the Belgium–Dutch border, thus subject to substantial legislative/political complications.

Comparison of the computed net mud balance (Fig. 15.7) and harbor siltation rates (Table 15.2) shows that anthropogenic-induced sediment fluxes are locally much larger than the "natural" net sediment fluxes, and form a substantial part of the gross fluxes. Note that these "natural fluxes" are

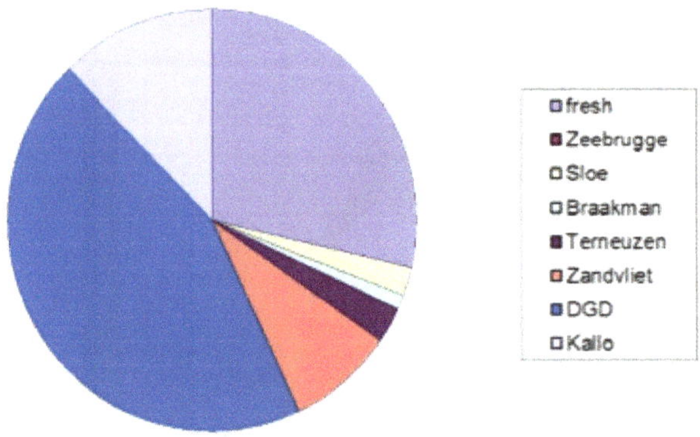

Fig. 15.9. Computed average mud composition at Boei84, close to Antwerp harbor. "Fresh" reflects natural mud not originating from dumping of dredged sediment from one of the harbors.

affected by previous human interventions in the river as well. This is further sustained by a simulation in which the various mud sources (natural or originating from maintenance dredging) have been labeled to identify these individually (Fig. 15.9). This simulation was restarted multiple times forced by the last three months of 2006 hydrodynamics to reach a dynamic steady state for the labeled sediment fractions. Only 30% of the mud is "fresh", i.e. not related to the dumping and dredging cycle in the estuary. About 40% (i.e. nearly 60% of the "anthropogenic" mud) originates from sediment mobilized from the dumping location for Deurganckdok mud. This demonstrates the large potential for optimization the local dredging and dumping strategy with respect to its environmental impact, i.e. on SPM levels. It is noted that this does not imply that the construction of the Deurganckdok did increase SPM levels with 40%, as the dock also is a sediment sink — see Fig. 15.8.

Ultimately, the fate of dredged and subsequently dumped sediment is:

(1) recirculation towards the harbor basins and access channels — these are maintained, so part of the sediment remains within a dredging-dumping cycle;
(2) deposition on mudflats, shallows or abandoned side channels, which are net sinks;
(3) export towards sea. Note that this is possible only if SPM-concentrations would increase over the larger part of the estuary.

The analysis above can be quantified further with a resilience parameter R, defined as:

$$R = \frac{B}{B + N} \tag{15.1}$$

where B is the total volume of maintenance dredging and N is the volume of "natural" sediment in the estuary (i.e. in the absence of anthropogenic effects). If R is close to zero, the sediment balance is governed by natural fluxes and an additional sediment source by e.g. dumping of dredged material has little impact on SPM levels, inducing a slight increase in B only. Likely, recirculation effects are small as well. However, if R is close to unity, the sediment balance is mostly steered by anthropogenic effects and an additional sediment source is likely to increase SPM levels further. Also recirculation effects may become more important, being highly sensitive to the release location.

One could say that estuaries with $R \to 1$ lack a safety valve (lack resilience), i.e. small interventions have an immediately impact on turbidity levels, thus water quality. In case $R \to 1$, dumped mud can only flow back to harbors or towards sea, but the latter can only occur after a substantial SPM-concentration gradient has been established to force that seaward flux. This may, however, result in high SPM levels inducing substantial ecological damage and prohibitive dredging costs. For the Scheldt estuary near Antwerp, $B \approx 3.3$ Mton and $N = 0.3 - 1$ Mton, thus R is estimated at around 0.8 to 0.9 (see Table 15.2). Hence, anthropogenic effects are dominant, and the river is not very resilient anymore to further interventions.

15.5.1 *Crucial system understanding aspects*

This chapter describes the setup, calibration and application of an engineering model to address management questions related to the transport and fate of fine sediment in the Scheldt estuary at larger spatial and temporal scales. The most crucial elements in our understanding of the Scheldt estuary, accounted for in the model, comprise:

(1) The Scheldt estuary is large with a number of rather different subdomains, i.e. part of the North Sea, the wide, multiple channel Western Scheldt, and the Sea Scheldt, with the saline/brackish Lower Sea Scheldt and the fresh water Upper Sea Scheldt. From a practical point of view, it is not possible to resolve the hydrodynamics and sediment transport at full detail over the entire model domain. A compromise

in resolution is therefore based on the engineering/management questions to be addressed, omitting dynamics at small temporal and spatial scales where these hamper model performance at the larger scales.

(2) At the larger scales addressed in this chapter, sediment-turbulence interactions can be ignored, and an uncoupled-approach is feasible. Note the difference with the small-scale application for the Scheldt estuary addressed in Chapter 18.

(3) Because of its size, the Scheldt model requires about five years spin-up time before a dynamic equilibrium is attained. These large simulations times, containing multiple seasons (wind, waves, river flow) require the use of the two-layer buffer model.

(4) The sediment transport model is calibrated on the basis of variations in SPM-concentrations, bed mud content and the overall mud balance, thus not only on the SPM-concentrations themselves. This gives favorable results for long-term simulations. Because the Scheldt estuary is shallow, a first calibration step has been carried out with a 1DV-model, enabling a rapid convergence in the large parameter space.

(5) The estuarine turbidity maximum (ETM) near Antwerp is governed by convergence of fine sediment transport, and not by water-bed exchange processes.

(6) Because of limited resolution, these convergence processes cannot be resolved accurately with a classical approach. However, parameterizing the vertical SPM gradients in the water column with the concept of hindered deposition allows for a proper simulation vertical SPM gradients and with that of the location of the ETM, thus transport convergence, and reasonable mean SPM-concentrations.

(7) The overall mud balance of the estuary is largely governed by the difference between gross and net sediment transport fluxes. Temporal or permanent sources within the system, comprising of intertidal mudflats and harbor basins play therefore a prominent role. Because of the large memory of the Scheldt estuary, small variations in hydrodynamic forcing and sediment availability, and their synchronization induce large variations in year-averaged siltation rates.

Chapter 16

Modeling SPM in Tidal Rivers: The Lower Ems River

16.1 Introduction

Tidal rivers are defined as natural waterways influenced by tides but not or negligibly affected by salinity gradients (Hoitink and Jay, 2016). They form the transition between estuaries and rivers. Tidal rivers may extend many 100s of km inland, with the Yangtze River and Amazon Rivers as extreme examples. Over such large distances, tides experience friction and convergence, resulting in tidal amplification and/or damping, while slopes in bed level also affect their propagation. Though tides induce oscillations both in water level and velocity, velocity reversal may take place in only part of a tidal river, owing to the unidirectional river discharge. Often, compound tides (with lower frequencies, such as fortnightly constituents) and overtides (with higher frequencies, such as the M4-component) are induced by non-linear interactions, as discussed in Sec. 9.2. Sediment transport is strongly driven by such tidal asymmetries (in space and/or time) in combination with river-induced residual flows.

An example is the Lower Ems River, a 60-km long tidal river discharging into the Ems estuary close to Emden (Fig. 16.1). The Ems is not a typical tidal river, as it is narrow with little intertidal, fairly short because of a weir near Herbrum, and with low river discharge. Particularly its hyper-turbid state complicates detailed modeling, and its dynamics and morphodynamics are not yet fully understood (Dijkstra *et al.*, 2019a–c, 2018; Talke *et al.*, 2008; Wang, 2010; Wehr and Malcherek, 2012). Yet, the engineering and managerial questions can be addressed with state-of-the-art modeling, based on system understanding and a proper conceptual model. Though saline water enters the river, its effects are not important in the

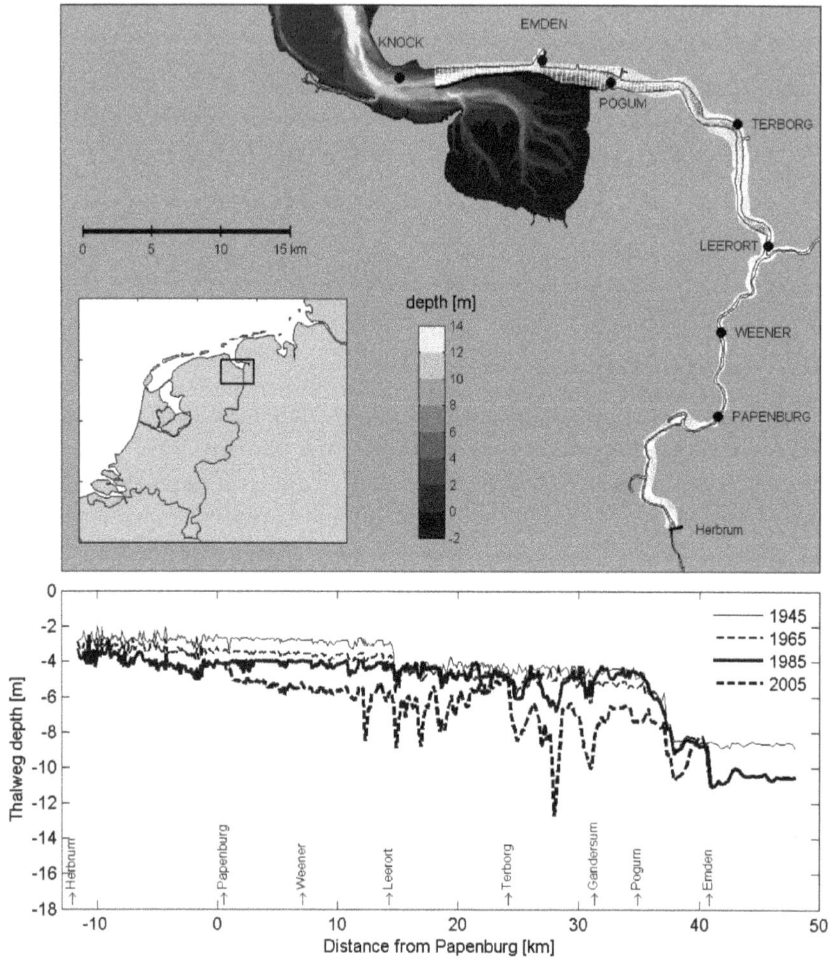

Fig. 16.1. Map and bathymetry of the Lower Ems River (beyond Emden), from the weir at Herbrum to Emden (depth in meter below MSL). The model grid (in white) is also plotted on this map. Observation points are denoted with dots. The lower panel provides the observed bathymetry along the thalweg of the Lower Ems River (2005) and reconstructed depth for 1945, 1965, and 1985.

larger part of the Lower Ems River (see below), which can therefore be treated solely as a tidal river. The Lower Ems River therefore provides an excellent case study to illustrate the importance of the steps outlined in Chapter 13.

16.2 Conceptual model

16.2.1 *Purpose of the study*

The Ems estuary suffers from very high SPM-concentrations, and requires regular maintenance dredging. Water quality conditions do not fulfill requirements set by the European Water Framework Directives; in particular SPM values are too high, affecting the light climate, thus primary production (Sec. 3.5). The high concentrations in the Ems Estuary are probably related to the hyper-turbid state of the more up-estuary Lower Ems River (De Jonge *et al.*, 2014). However, to what extend the sediment dynamics in the Lower Ems River influence those in the outer estuary is insufficiently known. Sediment may be flushed from the tidal river into the estuary, but alternatively, the tidal river is also a net sink of sediment as material dredged from the tidal river is brought on land. These interactions may have changed as a result of channel deepening, which was probably the main reason for the present-day extremely high sediment concentrations. The purpose of this study is to better understand the transport mechanisms in the Lower Ems River in response to deepening, and whether or not the river affects down-estuary conditions.

16.2.2 *Aims and research questions*

The main aim of this study is to develop a model to quantify changes in SPM in the Lower Ems River in response to deepening. Deepening directly influences tidal propagation and thereby turbidity, but also sets in motion a more complex feedback mechanism through reduced effective bed roughness and vertical mixing (Winterwerp and Wang, 2013). The relevant research questions to address therefore are:

(1) What are up-river transport mechanisms and rates?
(2) What is the role of deepening and of hydraulic roughness on up-river sediment transport and accumulation?
(3) What triggered the transition from a system with fairly moderate sediment concentration towards a hyper-turbid state?

These management questions are addressed with the numerical model described in this chapter, which is primarily based on Van Maren *et al.* (2015b).

16.2.3 *System understanding*

The Lower Ems River is funnel-shaped, decreasing in width from 700 m at the mouth to 70 m at the tidal limit (the weir at Herbrum). The depth decreases from ~7 m below MSL at Pogum to ~3 m below MSL at the tidal limit (Fig. 16.1). The present-day tidal range increases up-river, peaking at Papenburg at 3.6 m. As this tidal range is comparable to the water depth, the tidal wave is strongly distorted with a short period of high flood flow velocities and a long period of weaker ebb flow velocities, inducing flood-dominant conditions (see Fig. 9.9 and Wang, 2010).

Today, the Lower Ems River is characterized by thick, soft mud deposits with concentrations up to 500 kg/m^3, which probably remain mobile up to concentrations of about 200 kg/m^3 (Papenmeier *et al.*, 2013). Turbid water (concentrations up to several 10 kg/m^3) migrates up- and down-river with the tide over several 10s of km. Owing to the large mud availability, bed conditions can be considered alluvial, and (near-)capacity conditions prevail. This implies that the peak flow asymmetry resulting from the pronounced tidal distortion is an efficient mechanism to generate an up-river residual transport of fine sediment (Sec. 9.2.1). Close to the tidal limit, flood-dominance decreases rapidly owing to fresh water input. With a cross-sectional area of several 100 m^2, river-flow-induced velocities are around 1 m/s during peak discharges. The river-flow-induced velocity decreases down-river to ~5 cm/s during peak discharges, owing to the wider and deeper cross-section.

The river discharge ranges from several 10s m^3/s during summer conditions to several 100s m^3/s during winter. Saline water intrudes halfway into the river during summer conditions, but salt water is completely flushed out during winter high discharge events. Salinity gradients are important in the deeper part of the river (~10 m) up to Emden (Fig. 16.1), e.g. Dijkstra *et al.* (2019b); Talke *et al.* (2008). Further up-river, where the river becomes progressively more shallow, the effects of gravitational circulation (Sec. 9.1.1) are much smaller than those by tidal asymmetry, and can be ignored.

The Lower Ems River was deepened from a water depth of ~4 m below HW to ~8 m below HW between the 1930s and 1994 — see Fig. 16.1 for an estimated change in bed levels. This deepening has led to strong tidal amplification: the tidal range at Papenburg has increased from 1.6 m in 1950 to 3.6 m in 2010 (Krebs and Weilbeer, 2008). Until 1990, the tidal range peaked at Emden, and the tide was damped in up-river direction. Since then, the tide beyond Emden is amplified, and within 20 years the tidal range at Papenburghas became 50% larger than at Emden. Deepening also led to a more pronounced flood-dominant tidal asymmetry (Chernetsky *et al.*, 2010; Herrling and Niemeyer, 2008) . The increasing flood-dominated asymmetry enlarged various sediment importing mechanisms such as resuspension, vertical mixing, and flocculation (Winterwerp, 2011) and lag effects (Chernetsky *et al.*, 2010).

The Lower Ems River is not unique. Many tidal rivers have been deepened in the past decades, allowing ship access to inland ports. Both deepening and reclamation of intertidal areas have led to increases in tidal range (Winterwerp and Wang, 2013; Winterwerp *et al.*, 2013). In some rivers, the turbidity also increased significantly, in line with the relation between tidal range and turbidity levels documented by Uncles *et al.* (2002). Winterwerp *et al.* (2013) identified a positive feedback mechanism between tidal deformation (notably resulting from channel deepening and reclamation of intertidal areas), sediment import (leading to increasing SPM-concentrations) and hydraulic drag (e.g. Sec. 3.3.3): see Fig. 16.2. This feedback may initiate a regime shift in which the tide becomes progressively more asymmetric, while more and more sediment is imported, increasingly reducing the effective hydraulic drag.

The Lower Ems River experienced such a regime shift. Since the 1980s, it has evolved from a system with moderate SPM-concentrations (several 100 mg/L) at its ETM to a hyper-turbid system with thick layers of mobile fluid mud and SPM-concentrations up to several 10s kg/m^3 (Talke *et al.*, 2008). This transition was accompanied by a sharp increase in tidal range.

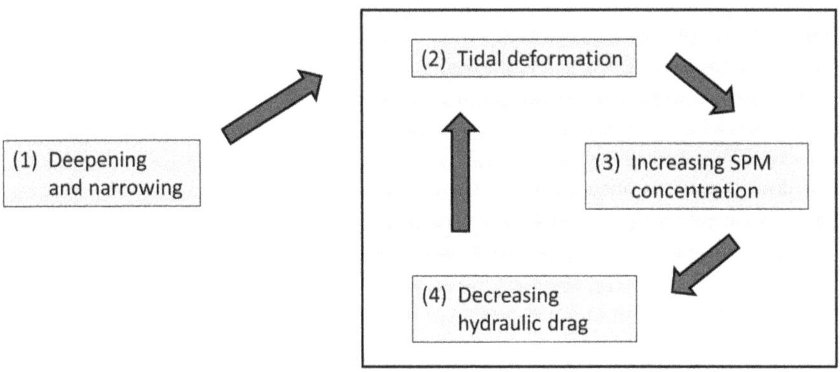

Fig. 16.2. Positive feedback loop of estuarine response to channel narrowing and deepening, with increasing sediment import and reducing hydraulic drag, leading to progressively larger tidal deformation.

For further details on the evolution towards hyper-turbidity, the reader is referred to e.g. (Dijkstra *et al.*, 2019a–c, 2018).

16.2.4 *Model requirements*

To investigate the sediment dynamics in the Lower Ems River (and the impact of deepening thereon — see Sec. 16.1, it is not necessary to model all subtleties of the sediment dynamics in the hyper-turbid river. However, large-scale evolution in tidal dynamics, SPM-concentrations and gross fluxes should be captured.

Winterwerp *et al.* (2017) argue that at average river flows, the fluid mud layer accelerates after the turn of the tide, becoming turbulent, thus being mixed over the water column. This is characterized by a strong asymmetry in SPM over the tidal cycle (Fig. 9.9), owing to the large asymmetry in peak flood and ebb velocity. Entrainment erodes the fluid mud layer rapidly, faster during flood than ebb. At larger river flows, peak flood velocities decrease, reducing entrainment rates, and thus SPM values. During elongated periods of high river flow, as in winter, the fluid mud layers are no longer eroded and can therefore consolidate. Modeling the effects of variations in river flow and seasonal variations therefore would require the modeling of consolidation.

Above we have reasoned that the interaction between hydraulic drag and SPM plays an important role in the Lower Ems River. This interaction is largely driven by sediment-induced stratification of the water column, a mechanism well-described in the k–ε turbulence closure model (Sec. 11.1). As SPM-concentrations are high, hindered settling (Sec. 4.4) affects the sediment dynamics largely, and has to be accounted for as well.

The conceptual model sketched above provides the framework for the model setup. The essential elements for the engineering model are as follows:

(1) Several historic scenarios need to be evaluated, at least covering the conditions before and after deepening.
(2) The intra-tidal variability in SPM is very high (several 10s kg/m^3) due to entrainment of the fluid mud layer. This needs to be parameterized, as not properly accounted for in state-of-the-art numerical models within acceptable computational times.
(3) At concentrations of several kg/m^3, there is a feedback between hydrodynamics, turbulence and SPM-concentration (Winterwerp, 2001). Simulating this interaction requires a 3D modeling approach in which SPM influences the density of the suspension and thereby turbulent

mixing through buoyancy destruction (Sec. 7.1), i.e. a coupled approach is required (Sec. 13.3.1).

(4) The effects of hindered settling have to be included (Sec. 4.4).

(5) The effect of hydraulic roughness needs to be accounted for, either as a concentration-related apparent roughness (as in Winterwerp *et al.*, 2009) or parameterized.

(6) Bed and suspended sediment in the model should be in dynamic equilibrium. This requires a simulation period of at least one to two years, almost independent of the initial conditions (Secs. 12.3 and 13.3.3). This dynamic equilibrium needs to be attained for all historic scenarios. Computational time is therefore a key element in designing model setup.

(7) As tidal asymmetry is a key driver of SPM transport in the Lower Ems River, it should be reproduced properly, both asymmetries in peak velocity (Sec. 9.2.1) as in slack water duration (Sec. 9.2.2). This implies also that higher harmonics should be prescribed at the model boundary, and properly reproduced within the model domain.

16.3 Hydrodynamic model

16.3.1 *Grid resolution*

The computational grid has been designed to resolve the relevant processes within acceptable computational times. The main considerations are as follows:

(1) At least ten vertical layers are needed to represent vertical gradients in flow and sediment transport with smooth near-bed behavior, which is best resolved using σ-coordinates (Sec. 13.4.2), bearing in mind the large tidal amplitude. At the high SPM-concentrations typical for the Lower Ems River, sediment mixing and settling processes take place throughout the water column. Therefore, equidistant vertical spacing is used, rather than an exponential distribution of the vertical grid, as is more common in sediment transport models.

(2) The main channel of the Lower Ems River is to be represented by at least six cells over its width to resolve channel morphodynamics (Sec. 13.4.2). To include the rivers floodplain, and accounting for channel meandering, the entire channel cross-section is schematized with about 20 grid cells.

(3) Along-channel and cross-channel grid cells are of similar size to accurately model channel curvature.

(4) A one-year simulation with the numerical model should not take more than a few days on a fast PC.

The designed model grid is visualized in Fig. 16.1. Historical model bathymetries were constructed by modifying the measured 2005-bathymetry with anecdotal information on channel deepening and straightening, thus obtaining synthetic bathymetries for 1945, 1965, 1985 and 2005 (see Van Maren *et al.* (2015a) for details).

In a coupled approach, the hydrodynamics and sediment dynamics are mutually dependent, and can therefore not be calibrated independently. However, in the following we describe the calibration of hydrodynamics and sediment dynamics sequentially for didactic reasons.

16.3.2 *Calibration and validation*

The model is forced with observed water level time series at the mouth of the Lower Ems River (Knock station) and river discharges at the landward boundary, i.e. the weir at Herbrum using data for the year 2005. The hydrodynamic model was calibrated to water level observations in two steps. The first step involves a detailed calibration of water levels for 2005, for which a large amount of data is available, including bathymetry. In the second step, the model is compared with historical water levels, using the synthetic bed levels; see also Fig. 16.1.

Water levels are primarily calibrated by varying the Manning roughness parameter n, comparing computed and observed tidal constituents at the stations indicated in Fig. 16.1. Such a comparison in the frequency domain reveals to what extent (1) tidal amplification/damping and (2) tidal asymmetry are reproduced by the model. The largest asymmetry component in the Lower Ems River is the M_4 overtide. Therefore the asymmetry of the tide is quantified with the amplitude ratio a_{M4}/a_{M2} and the phase relation $2\phi_{M2} - \phi_{M4}$. Figure 16.3 provides an example of the model-data comparison for Papenburg station with Manning's $n = 0.01$, which represents observations best. The computed water levels and phases of the main tidal constituents are in good agreement, implying that the overall tidal flow is sufficiently represented. The higher harmonics (of which M_4 and MS_4 are the larger) are mostly generated within the model domain. The computed water level amplitudes deviate less than 5% from observations and computed phases less than 10°. Note that a Manning's n of 0.01 is very low, in line with drag reduction by the high SPM-concentrations and fluid mud occurrences (see Sec. 3.3.3).

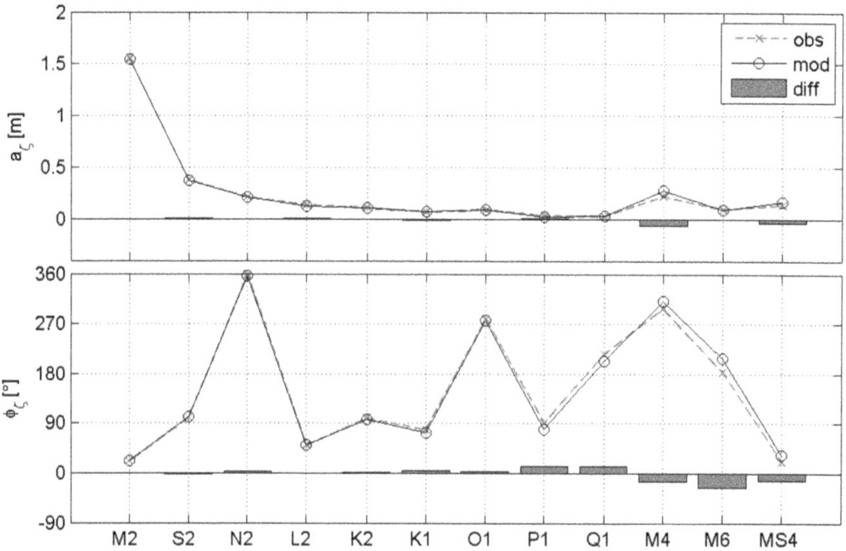

Fig. 16.3. Harmonic water level amplitude (top panel) and phases (lower panel): computed (solid lines with circles), observed (dashed lines with crosses), and difference between observations and model (gray bars). Papenburg station for the year 2005, using a Manning's $n = 0.01$.

Next, the model was modified implementing the historic synthetic bed levels at 20 year intervals. The same boundary conditions are used, but n-values are varied to match observations. For these periods, only time series of high and low water are available, which does not allow model-data comparison in the frequency domain. In the past, the tidal amplitude was smaller, especially near and beyond Papenburg (Fig .16.4). The observed historical increase in tidal amplitude is partly predicted by the model by changing the bathymetry only, implying that deepening partly induced the observed amplification. However, the model approximates historical observations much better if also the bed roughness is modified. For the 1945 bathymetry, observed high and low waters are best reproduced with $n = 0.02$, whereas best results for 2005 are obtained with $n = 0.01$. Closest agreement for intermediate years was obtained using an intermediate value of $n = 0.015$. The higher n-value is typical for sandy systems, whereas the lower value is typical for muddy, stratified systems. This decrease in bed roughness is supported by anecdotal evidence by Krebs and Weilbeer (2008) that the Lower Ems River has transformed from a low-concentration river with a sandy bed and bed forms into its current hyper-turbid state.

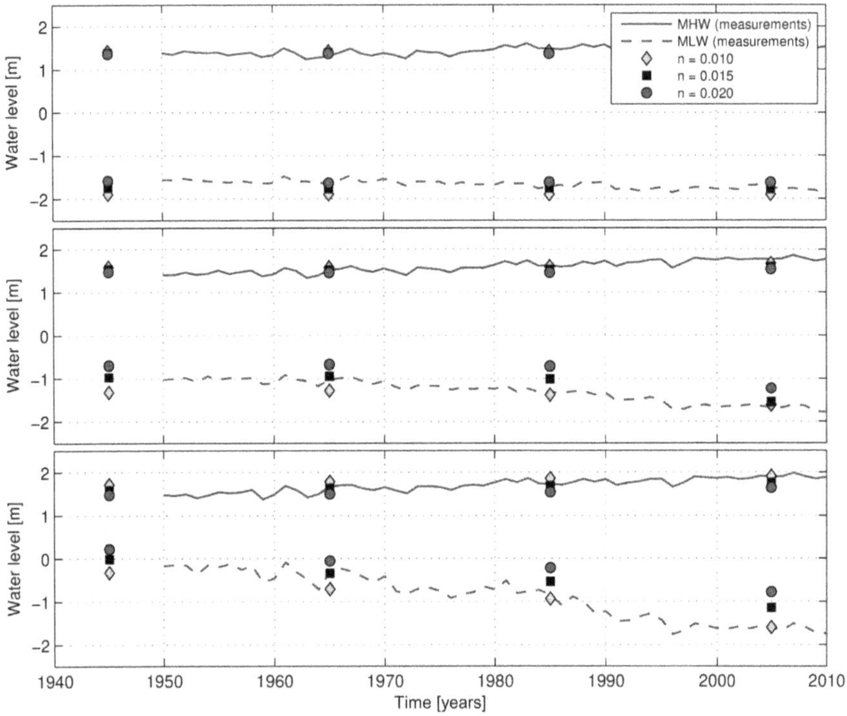

Fig. 16.4. Computed (using Manning's $n = 0.01$ (light grey triangle), 0.015 (black square), and 0.02 (dark grey circle)) and observed high water (HW) and low water (LW) at Pogum (top panel), Leerort (middle panel), and Papenburg (lower panel). HW (LW) is defined as the yearly average of every high water (low water) per tidal cycle.

16.3.3 *Analysis hydrodynamics*

The results of the hydrodynamic model are further elaborated with respect to input values for the sediment transport model, and the dominant sediment transport processes. First, the computed bed shear stress τ_b is analyzed, in which the direction of the flow is accounted for, i.e. $\tau_b > 0$ implies up-river (flood) flow, while $\tau_b < 0$ implies down-river (ebb) flow. Figure 16.5 shows these stresses in the river's thalweg at Weener for the four historic bathymetries as a function of the river discharge at Herbrum.

The bed shear stress is composed of a tidal and a residual component. The first exhibits asymmetry in peak flow velocity, whereas the second is always in ebb-direction. In 2005, maximum values for τ_b occur during flood for most of the discharge conditions, reflecting the flood dominance of the

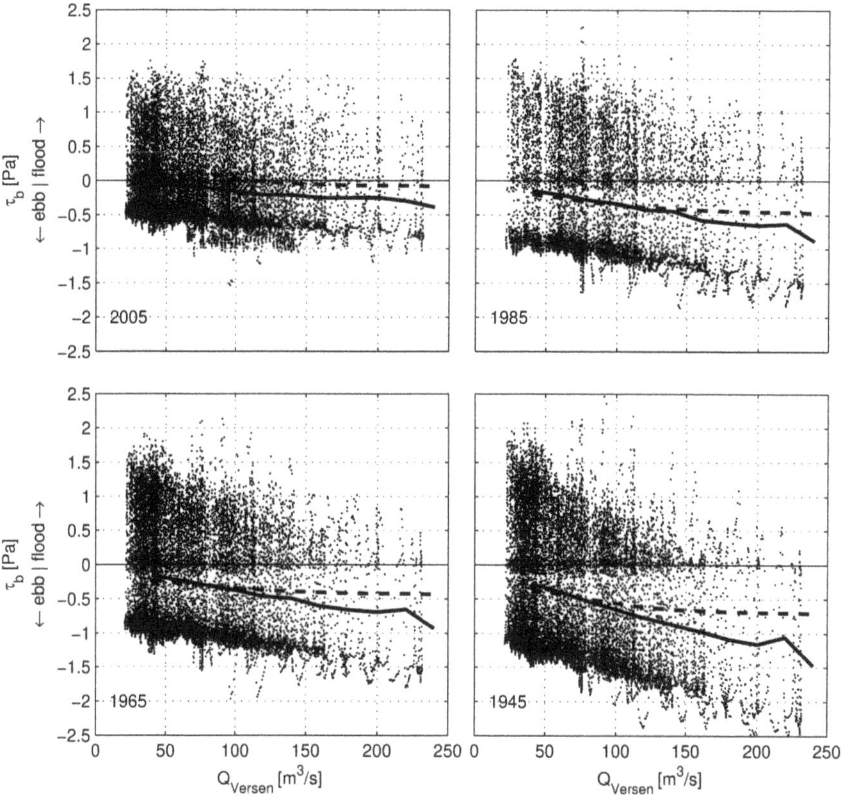

Fig. 16.5. Computed bed shear stress τ_b (positive τ_b is flood/up-river, negative is ebb/down-river) in the channel thalweg at Weener as a function of discharge Q at Versen, computed for 2005 ($n = 0.01$), 1985 ($n = 0.015$), 1965 ($n = 0.015$), and 1945 ($n = 0.02$). The solid line reflects the average bed shear stress $\langle\tau_b\rangle$ and the dashed line the average excess bed shear stress $\langle\tau_b - \tau_e\rangle$.

tidal currents. In 1945 the bed shear stress was highest during flood only at very low Q, and strongly ebb-dominant for high Q. Patterns for 1965 and 1985 are in-between. The average bed shear stress (solid line in Fig. 16.5) is, for all years, in the ebb-direction. In a tidal river, the time-averaged bed shear stress is typically in the ebb-direction, induced by the river flow.

However, the transport of sediment is driven by the excess bed shear stress, defined as $(\tau_b - \tau_e)$. For larger value of the critical bed shear stress τ_e, the effect of tidal asymmetries is more pronounced, with respect to both peak flow and slack tide asymmetry. The value for τ_e can be estimated from SPM-concentration observations, revealing that sediment is eroded (or entrained) from the bed during both ebb and flood conditions. Therefore τ_e should not exceed 0.5 Pa (top left in Fig. 16.5). Computing $(\tau_b - \tau_e)$ as a function of the river discharge Q (dashed line in Fig. 16.5) suggests that (1) between 1985 and 2005 the tidal river switched from a net exporting to a net importing system, as suggested by Krebs and Weilbeer (2008); and (2) the present-day tidal river becomes exporting beyond a critical Q, as suggested by Spingat and Oumeraci (2000). Note that this latter may not be true for consolidated mud beds, as stronger river flows lead to a reduction in maximum τ_b (+1.5 Pa at low Q, −1 Pa at very high Q; top left panel in Fig. 16.5). This is elaborated in detail by Winterwerp *et al.* (2017).

Variations in tidal asymmetry over time and space are examined in more detail in Fig. 16.6. Near the river mouth, the phase difference $2\phi_{M2} - \phi_{M4}$ is close to 180°. For standing wave conditions this corresponds to high water slack asymmetry — see Sec. 9.2.2). Up-river, the amplitude ratio increases

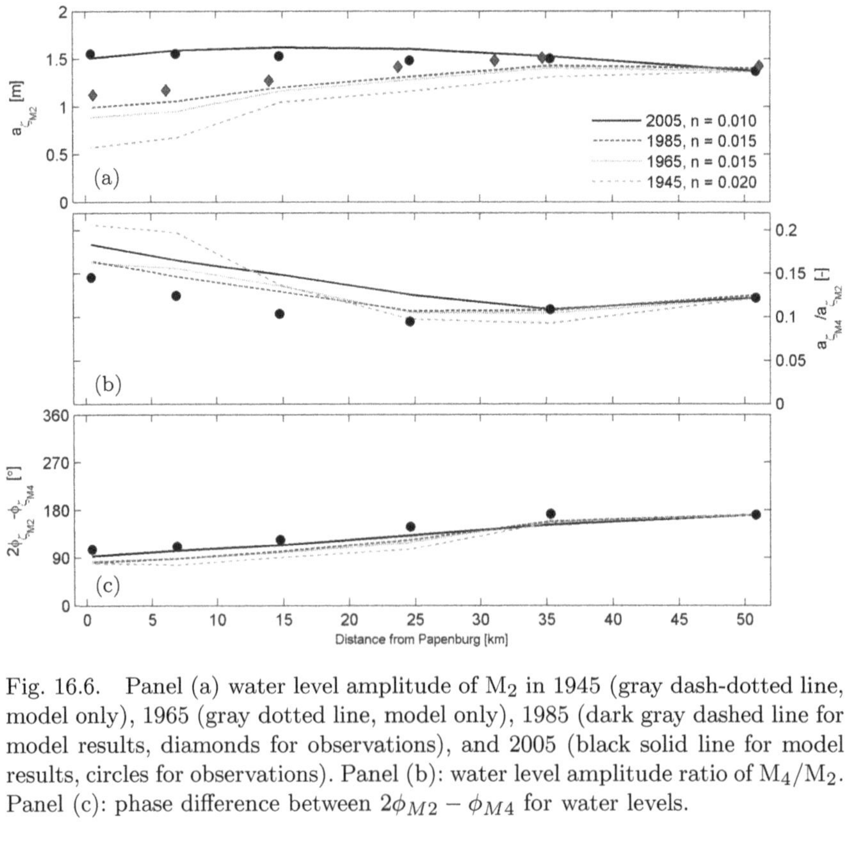

Fig. 16.6. Panel (a) water level amplitude of M_2 in 1945 (gray dash-dotted line, model only), 1965 (gray dotted line, model only), 1985 (dark gray dashed line for model results, diamonds for observations), and 2005 (black solid line for model results, circles for observations). Panel (b): water level amplitude ratio of M_4/M_2. Panel (c): phase difference between $2\phi_{M2} - \phi_{M4}$ for water levels.

while $2\phi_{M2} - \phi_{M4}$ decreases to 90° (corresponding to peak flood flow asymmetry, see Sec. 9.2.1). High water slack asymmetry is an efficient up-river transport mechanism for fine-grained sediment under below-capacity conditions. With a large amount of sediment present in the form of fluid mud, the pronounced peak flood flow asymmetry is an efficient mechanism transporting sediment up-river. Note that the tidal asymmetry did not change significantly in response to the deepening (panel (b) ad (c) in Fig. 16.6). However, the tidal amplitude changed significantly (panel (a)), and thereby the relative contribution of the tide to the total transport.

16.4 Sediment transport model

16.4.1 *Model setup*

The hydrodynamic model results (Fig. 16.5) suggest that tide and river flow determine the residual SPM transport in the Lower Ems River. Riverine effects become progressively smaller down-river of Weener (see also Sec. 9.2.1). The hydrodynamics in the river are energetic enough to mobilize sediment in the fluid mud layer (Winterwerp *et al.*, 2017), resulting in high SPM-concentrations and mobile fluid mud layers with concentrations up to 200 kg/m^3 (Papenmeier *et al.*, 2013). The dynamics of this fluid mud have been elaborated by Wang (2010); Winterwerp *et al.* (2017). Modeling the dynamics of these fluid muds involves solving consolidation processes (Sec. 5.1) and fluid mud dynamics (Chapter 7), including entrainment processes and non-Newtonian stress-strain relations. Our physical understanding of these topics has advanced sufficiently to account for these processes, but they require high spatial and temporal model resolution. This has been done in research models, simulating part of the river over short time scales, for instance by Knoch and Malcherek (2011); Wehr and Malcherek (2012). However, for the current application, the resulting computational times are in conflict with the required simulation periods (Sec. 16.2.4) and grid requirements (Sec. 16.3.1). Therefore the behavior of fluid mud needs to be resolved in a parameterized way.

Detailed observations of the fluid mud in the Lower Ems River by Wang (2010) and Winterwerp *et al.* (2017) suggest that:

(1) longitudinal sediment transport within the fluid mud is small with respect to transport in the water column,
(2) exchange of sediment between the fluid mud layer and the water column occurs beyond a critical flow velocity, and
(3) the fluid mud is internally well mixed.

Therefore, a first-order modeling approach is to assume that the fluid mud can be represented by an easily erodible soft muddy bed. Exchange of sediment between the fluid mud and overlying water column is therefore described with the classical Partheniades–Krone model (Sec. 4.2 and 6.1), using the following settings:

(1) The influence of changes in fluid mud thickness on the overall hydrodynamics is accounted for by morphodynamic feed-back (updating bed levels), using a low dry density of the muddy bed (100 kg/m^3, close to the gelling point). This low density allows for a dynamic equilibrium in up- and down-river SPM transport, as affected by tide, river flow and water depth.
(2) The critical bed shear stress τ_e is set relatively high for such soft sediment (Table 16.1), mimicking the reduction in bed shear stresses by sediment-induced stratification (Sec. 3.3.3). The value of τ_e is set to 0.5 Pa, based on the hydrodynamic model results. This high τ_e generates large flood-dominant transport (Sec. 9.2, Fig. 9.9).
(3) The erosion parameter M is large (Table 16.1), implying that sediment can be eroded rapidly once the critical bed shear stress is exceeded. This results in high SPM-concentrations after the turn of the tide, as observed.
(4) A large erosion flux requires a large settling flux, for which a moderately high settling velocity w_s (1 mm/s) is used (Table 16.1). This further strengthens the dependency of net transport on hydrodynamic asymmetries (Van Maren and Winterwerp, 2013). A second sediment fraction (w_s = 0.1 mm/s) is especially needed to reproduce the baseline sediment concentrations around slack tide (the fraction with $w_s = 1$ mm/s completely settles during the slack water).
(5) Hindered settling is described with the Richardson and Zaki model (Sec. 4.4; Eq. (4.34)) using $n = 5$ and $c_{gel} = 100$ kg/m^3. This results in a smooth transition of SPM from the bed to the water column.

Table 16.1. Settings Lower Ems River model.

Parameter	Unit	Fraction 1	Fraction 2
w_s	mm/s	1	0.1
t_e	Pa	0.5	0.5
M	kg/m^{2}s	0.01	0.01
c_{gel}	kg/m^3	100	100
r_{dry}	kg/m^3	100	100

(6) Suspended sediment influences the water density through the equation of state (Sec. 3.2.3; Eq. (3.26); also Chapter 11), leading to damping of turbulent mixing by vertical SPM gradients. However, because of the parameterization of the fluid mud dynamics, turbulent mixing is not damped sufficiently (in fact too small vertical SPM gradients), which is compensated by a large τ_e, as explained above.
(7) The model is initialized with an empty bed (Sec. 12.3), i.e. all sediment enters the model domain through its boundaries. At the down-river model boundary, continuous SPM readings at Knock station are prescribed. No sediment enters with the river flow at Herbrum, the landward limit of the tidal river.

16.4.2 *Qualitative calibration*

Because of the strong feedback between hydrodynamics and sediment transport, the 1DV calibration step, advocated in Chapters 14 and 17 is not efficient. Calibration of the sediment parameters is therefore done with the entire 3D model. Also bed composition is not relevant, as the entire Ems River bed consists of soft/fluid mud.

The model is calibrated using SPM-concentration time series observed at the stations depicted in Fig. 16.1. These observations are compared with the computed SPM-concentrations averaged over the lower three layers in Figs. 16.7 and 16.8. The major calibration parameter appeared to be the erosion parameter M.

SPM-observations are well-predicted in the lower part of the river (Pogum and Terborg): both the peak in SPM and the spring-neap variability are resolved (Fig. 16.7). Figure 16.8 gives an indication of the vertical gradients in computed SPM, showing further that the double peak in observed SPM at Pogum is represented by the model, but not at Terborg.

Further up-river the model agrees less with available data (Fig. 16.7). In particular the observed large SPM-concentrations during the first week of December are not predicted by the model. It is likely that the model cannot reproduce the large entrainment rates when the flow becomes internally super-critical (e.g. Winterwerp *et al.*, 2017). However, such conditions would require detailed modeling of the fluid mud layer itself, including the transition between turbulent and laminar flow conditions, which is in conflict with the time and spatial scales of this modeling study (see above).

Over longer time scales, the model reveals a rapid up-river transport of sediment during summer conditions, and seaward flushing (up to

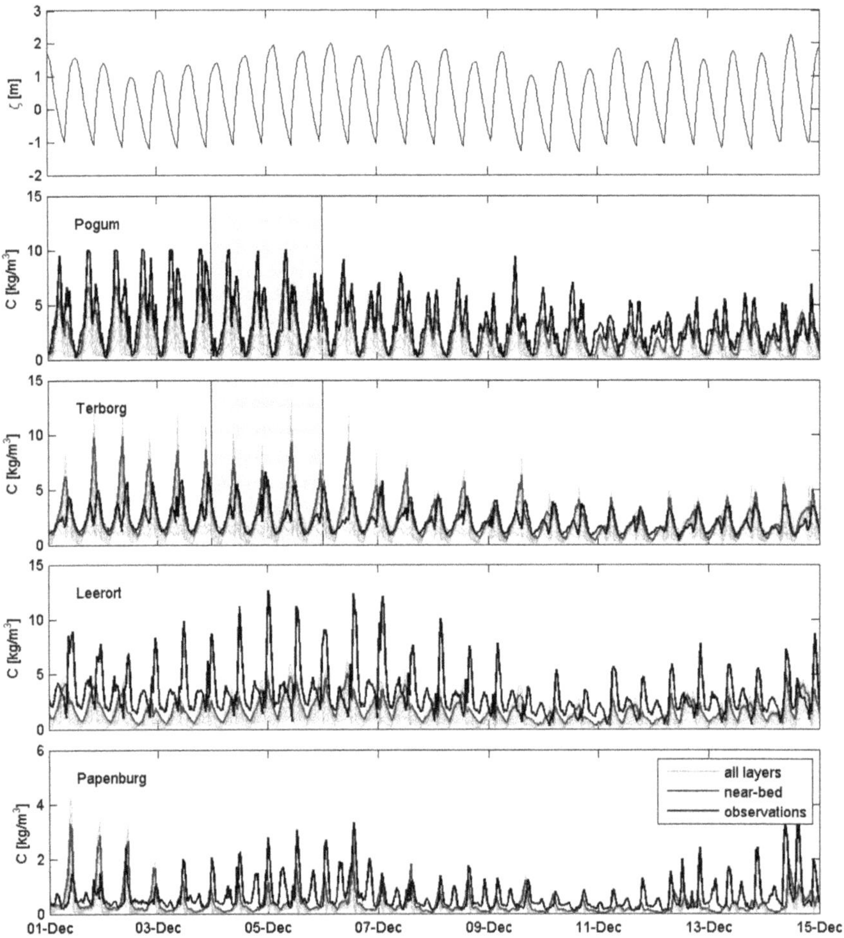

Fig. 16.7. Computed water level at Pogum (top panel) and SPM-concentration in December 2005 at Pogum (2nd panel), Terborg (3rd panel), Leerort (4th panel) and Papenburg (bottom panel). The observed SPM-concentration are in black, computed SPM-values at all vertical grid cells are in light gray, whereas the dark gray line represents the average computed SPM-concentration in the lower 10–30% of the water column, which best approximates the location of the sensor. A zoom-in of SPM-concentrations highlighted by the gray boxes is presented in Fig. 16.8.

Terborg/Pogum) during high discharge winter conditions (see Van Maren *et al.*, 2015b). This behavior depends strongly on the choice of τ_e. With the current setting of $\tau_e = 0.5\,\text{Pa}$, sediment is always eroded (during ebb and flood, throughout the year). As such, sediment is remobilized in the

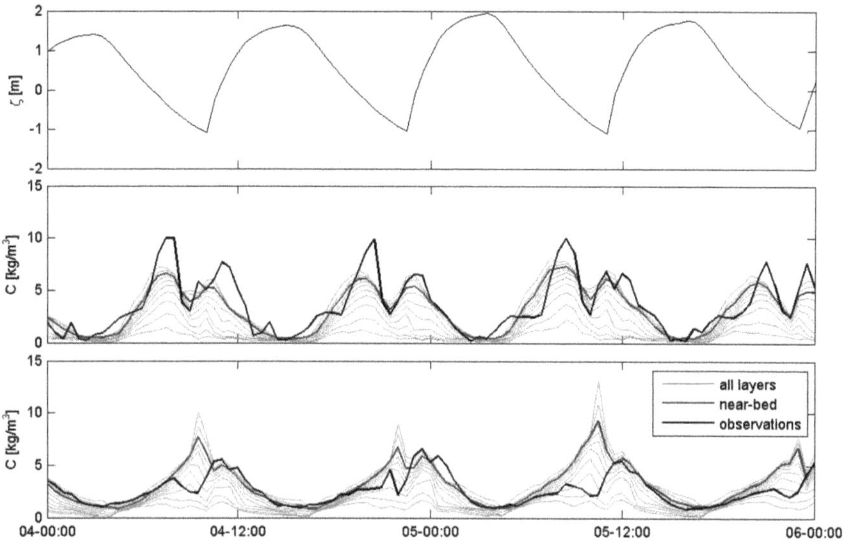

Fig. 16.8. Same as Fig. 16.7, for 4 and 5 December, and stations Pogum (middle panel) and Terborg (bottom panel).

upper tidal river and transported seaward by the residual flow. This behavior is supported by a reduction in SPM in the upper tidal river during winter, and an increase in SPM further down-river. However, analyses by Winterwerp *et al.* (2017) suggest that sediment is not flushed seaward. An alternative explanation for SPM reduction in the upper tidal river is that the reduction in maximum bed shear stress (see Fig. 16.5) during high discharge actually stabilizes the mobile fluid mud. Both options are still under investigations, and without such *a priori* physical understanding of the system, this model may only be used carefully to investigate the role of river discharge on sediment flushing.

16.4.3 *Quantitative calibration*

The quality of the model calibration has been quantified in Fig. 16.9 through model skill S and goodness of fit R analyses (Sec. 13.5.3). This analysis confirms that the model is most accurate near the river mouth, while exhibits a significant seasonal variability in model accuracy. Near the river mouth (Pogum) the model agrees better with data in winter, whereas further up-river (Leerort and Terborg) summer conditions are more accurately resolved. The absolute values of the quantitative metrics are difficult to

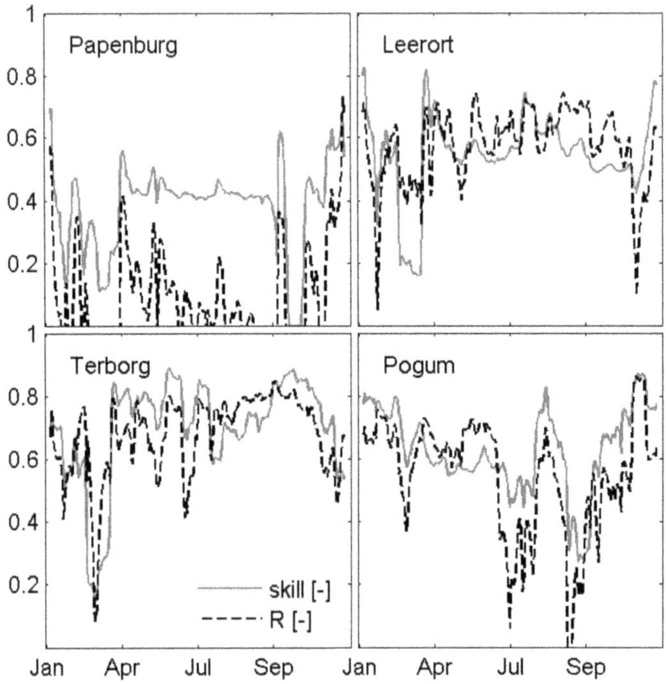

Fig. 16.9. Time variation of the goodness of fit R and model skill.

interpret, as no references or standards exist to evaluate the accuracy of computed SPM-concentrations. Note that both R and S for SPM cannot be compared to water level benchmarks as SPM-concentration are more erratic than water levels.

The quantitative analysis supports the qualitative analysis in the sense that:

(1) The intra-tidal variability is well reproduced, in which the high concentration peaks directly after low water slack are important, since this drives residual transport. This is essential for the management question related to sediment important.
(2) The seasonal variation is reproduced. This is a key observation for management questions on sediment flushing. However, as indicated in the previous section, uncertainties remain in the role of discharge on sediment flushing.

In spite of these shortcomings, it was concluded that this numerical model can be applied to address the management questions, as discussed in the next section.

16.5 Model application

The hydrodynamic model has been calibrated for a range of years from 1945 till 2005, while the sediment transport model is calibrated for the year 2005. Though comparison with data is not perfect, the model is good enough to carry out a series of exploratory simulations to investigate the role of deepening and hydraulic roughness on SPM-concentrations. From the hydrodynamic part of the modeling study it was concluded that the hydraulic roughness must have decreased over time, the effect of which on tidal amplification is comparable to that of deepening.

Next, the effect of depth and roughness on SPM-concentrations are investigated, to determine the effect of deepening in more detail. With a low roughness ($n = 0.01$), computed SPM-concentrations are very high in 2005, as well as in 1965 and 1985 (top panel of Fig. 16.10). However, for $n = 0.015$, SPM-concentrations are only high in 2005 (lower panel in Fig. 16.10), and not in previous years. These simulations suggest that:

(1) the Lower Ems River may have transformed from a low to a high turbid state, while its hydraulic roughness remained relatively high (i.e. the river bed was still sandy), and
(2) stopping maintenance dredging will not lead to lower SPM-concentrations: even when the bed level would return to 1965- or 1985-conditions, SPM-concentrations remain high when the system remains muddy with the accompanied low hydraulic roughness.

These conclusions confirm the conceptual model, based on Winterwerp and Wang (2013) analyses on the effects of deepening and sediment-induced stratification on tidal amplification and deformation and SPM dynamics. With respect to the original management questions (Sec. 16.2.2), the model provides important insights into the functioning of the system.

First, sediment is transported into the tidal river by tidal asymmetry (high water slack asymmetry in the lower tidal river, and peak flow asymmetry further up-river). With the large amount of bed sediment presently available, sediment is efficiently transported up-river by peak flow asymmetry. This sediment may be flushed from the upper to the lower tidal river

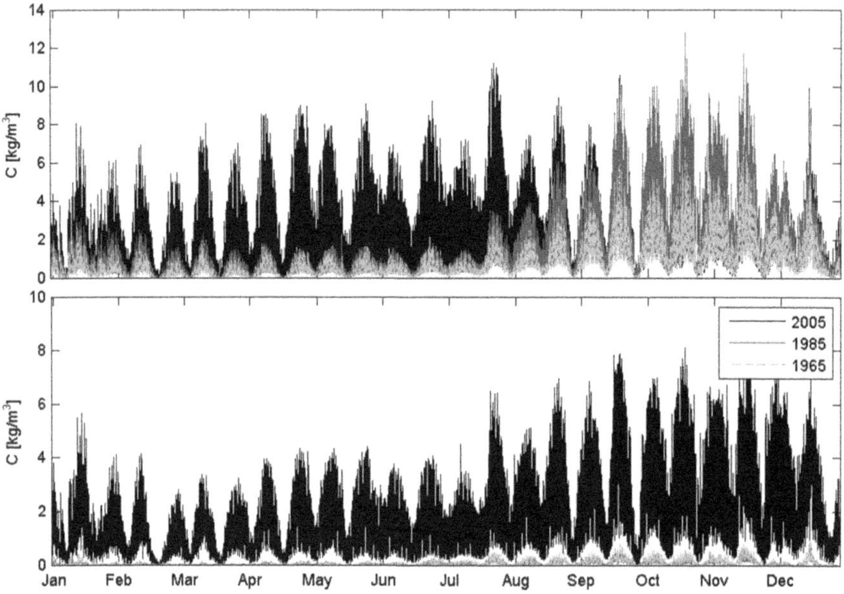

Fig. 16.10. Computed SPM concentration at Leerort in 1965, 1985, and 2005 for $n = 0.01$ (top panel) and 0.015 (bottom panel). Results for 1965 are poorly visible, large overlapped by the 1985 results.

during high discharge events. The present model suggests this is indeed the case. However, alternative hypotheses on sediment transport dynamics, suggesting that erosion rates are actually reduced during higher discharges, do not result in seaward flushing. Such behavior may also be achieved with the current model with higher critical bed shear stresses (requiring a new model calibration, though). The model can therefore not be used to evaluate the importance of sediment flushing during high discharge.

Secondly, deepening has led to an increase in sediment transport by (1) an increase in residual tidal transport and (2) reduced flushing by river flow. The increase in residual tidal transport is the result of larger gross transports (and not so much a change in asymmetry), strengthened by a positive feedback with SPM-concentrations: tidal amplification leads to more sediment import, hence higher sediment concentrations and reduced hydraulic roughness, which in turn further amplifies the tides.

16.5.1 *Crucial system understanding aspects*

This chapter describes the setup, calibration and application of an engineering model to address management questions related to the transport and fate of fine sediment in the Lower Ems River at larger spatial and temporal scales. The most crucial elements in our understanding of the Scheldt estuary, accounted for in the model, comprise:

(1) The Lower Ems River is in a hyper-turbid state, with a profound interaction between the tidal water movement, SPM, stratification and the fluid mud bed, including the effects of non-Newtonian stress–strain relations and consolidation and entrainment of the fluid mud layer, and a coupled approach is mandatory. Though high-resolution research models start to become sufficiently accurate to simulate some of these interactions, these are unsuitable for addressing engineering/management questions because of prohibitive computational times. Parameterization of various processes is therefore necessary.
(2) Because of the coupling between hydrodynamics, turbulent mixing and SPM, an in-depth analysis of the hydrodynamics is required, up to much larger detail than in the Chapters 14, 15 and 17.
(3) Net sediment transports in the Lower Ems River are governed more by tidal asymmetry, than by gravitational circulation. The boundary conditions to the model therefore contain measured overtides, and the model is calibrated against main and overtide amplitude and phase.
(4) Where possible, the hydrodynamics are calibrated and analyzed in the frequency domain.
(5) The SPM induced stratification cause large damping of vertical mixing, yielding:
   - significant drag reduction, which is modeled through a small Manning coefficient,
   - reduced erosion/entrainment of the fluid mud layer, which is modeled by a large critical shear stress for erosion.
(6) However, when the threshold shear stress is exceeded, the fluid mud layer erodes/entrains rapidly, which is modeled with a large erosion flux parameter.
(7) The long-term ebb-flood dominance induced by tidal asymmetry can be analyzed from the mean values of the excess bed shear stress $(\tau_b - \tau_e)$.
(8) A smooth transition between sediment concentrations in the water column and the fluid mud layer is modeled with hindered settling.

Chapter 17

Modeling SPM in Shallow Lakes: Lake Markermeer

17.1 Introduction

This chapter discusses the setup and calibration of a fine sediment transport model for a large, shallow lake, i.e. Lake Markermeer in The Netherlands. Lakes are generally low dynamic aquatic systems, containing a more or less constant amount of sediment, particularly over shorter time scales. Wind-induced processes, such as waves and large-scale circulations redistribute that sediment over the lake. This has major consequences for the setup of a sediment transport model. Often, lakes contain also large amounts of organic material, which generally requires modeling of more than one SPM fraction. As explained in the introduction, the chemical conditions in lakes differ considerably from those in marine waters. Being a former sea-arm, Lake Markermeer is an exception, and we do not elaborate on the chemistry of lakes.

Lakes may also freeze in winter, but this is not accounted for in the current approach, as neither temperature-induced density currents (also under ice!) are elaborated upon.

17.2 Conceptual model

Lake Markermeer is a shallow lake in the center of The Netherlands, close to Amsterdam. It measures about 720 km^2, with an average depth of 4 m (Fig. 17.1). It is part of the Lake IJssel freshwater system, formed in 1932 by a 30 km closure dike ("Afsuitdijk") isolating a former lagoon, known as the "Zuiderzee" from the North Sea. In anticipation of the construction of the polder "Markerwaard", a compartment dike was constructed in 1972 between Lelystad and Enkhuizen, creating Lake Markermeer. However, in the 1980s it was decided not to reclaim the lake.

Lake Markermeer is characterized by average SPM-concentrations of 50 mg/L, varying between less than 10 mg/L during prolonged periods with

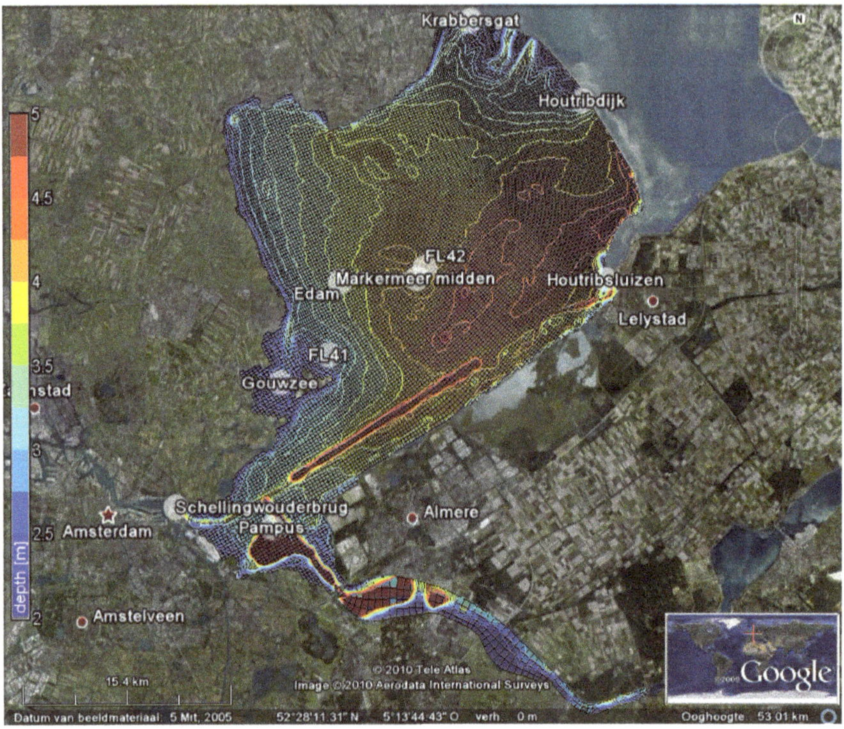

Fig. 17.1. Lake Markermeer, area of interest, bathymetry and computational grid. The position of the two measuring sites FL41 and FL42 and water level stations are indicated. The narrow arm in the south of the lake is known as the "Eemmeer". The length of the color bar represents a distance of about 35 km.

little wind, and more than 150 mg/L during storms — all these values reflect below-capacity conditions. Though these concentrations are not very large in an absolute sense, turbidity levels are high. Under extreme conditions, near-bed SPM values of a few g/L have been measured (Vijverberg *et al.*, 2011), which implies that locally capacity conditions may have been met.

17.2.1 *Purpose of the study*

The interest in Lake Markermeer sediment dynamics increased since it was registered as a Natura 2000 nature conservation area (http://ec.europa.eu/environment/nature/natura2000/index_en.htm). Nature conservation goals had therefore to be set. However, presently the ecosystem is degrading, with a declining number of birds, fish, mussels, etc. (Noordhuis and Houwing, 2003). One of the factors responsible for this decline is sought in the high turbidity level of Lake Markermeer (Secchi depth a few decimeter, at most). Apart from its ecological value, Lake Markermeer is also important for recreation, and is a potential target area for urbanization and infrastructure, alleviating pressure on Amsterdam.

The objective of the present study is twofold:

(1) Improve our understanding of the fine sediment dynamics in the lake, based on a combination of field observations and process-based numerical modeling.
(2) Develop that process-based numerical model for evaluating measures and interventions in the lake, aimed at improving its water quality.

17.2.2 *Aims and research questions*

A fine sediment transport model was setup to quantify the general understanding of the fine sediment dynamics in the lake and to evaluate measures to improve its water quality. At the time this model was developed, two managerial questions had to be addressed:

(1) How large the expected dredging maintenance is after deepening and widening the $>$ 6 m deep navigation channel along the south-east border of the lake, between Amsterdam and Lelystad, see Fig. 17.1.
(2) A wetland, known as the "Markerwadden" of about 5000 ha is being built in Lake Markermeer with the objective of enriching its ecological value. One question to be addressed is how this wetland affects turbidity levels and fine sediment dynamics in the lake.

From a scientific point of view, a sediment transport model is required to quantify and formalize the understanding of the fine sediment dynamics in Lake Markermeer over time scales of one to a few years.

17.2.3 *System understanding*

For long-term water quality studies, water fluxes (and accompanying nutrients, etc.) related to ground water processes, drainage from neighboring polders, precipitation and evaporation have to be accounted for. However, from a hydrodynamic point of view, Lake Markermeer is to be considered as a closed water system, with some water exchange at Krabbersgat and Houtrib (Fig. 17.1). Although the lake bottom is very flat, there are some man-made exceptions in the form of navigation channels and deep pits resulting from sand mining. These pits can reach depths of 20 m. As the sediment demand (both for sand and clay) has increased because of the need to strengthen the dikes around Markermeer and the decision to construct wetlands in the lake ("Markerwadden"), the area with depths larger than 4 m has increased as well.

Our system understanding has been obtained mainly from analyzing data of total suspended matter (SPM) and bed sediment data in Lake Markermeer. The following data are available:

(1) low-frequency water sample data on SPM and turbidity (MWTL, ref: www.waterbase.nl),
(2) high-frequency data from measurement rigs FL41 and FL42 (Blok, 2008),
(3) remote-sensing data (Kamps *et al.*, 2007), and
(4) bed sampling data (RWS-IJG).

Low-frequency SPM-data

The so-called "MWTL"-data is acquired within the framework of the standard monitoring program of Rijkswaterstaat (Dutch Ministry of Public Works). These data cover a long history; measurement series are available from 1974 onward at a number of locations. A limitation is their low frequency of once in 14 days, or less. This is too low to evaluate the short-term/quick response of SPM to changes in wind conditions. Also, in 1982 the number of measurement locations was reduced to four, and in 1992 further reduced to one. This remaining station is referred to as "Markermeer Midden" (MKM), in the middle of the lake near station "FL42" (Fig. 17.1).

The long-term average SPM-concentration in Lake Markermeer is about 50 mg/L. The 10-percentile value is just under 10 mg/L, whereas the 90-percentile values are well above 100 mg/L. The long-term spatial distribution of SPM is quite uniform, but values are much lower in the IJmeer (about 20 mg/L), the south-western extension of Lake Markermeer (Fig. 17.1). However, instantaneous SPM-concentrations often show strong spatial gradient and show strong variations in time driven by prevailing wind speed and direction. Note that these concentrations are not very large in an absolute sense. However, the sediment is very fine (see also Table 17.1), and hence affects turbidity levels largely (Sec. 3.5).

Figure 17.2 shows all measurements for station MKM, the only station with data from 1/1/1982 to 1/1/2008. The SPM values exhibit a large scatter

Table 17.1. Parameter settings based on (1) analytical expressions from 1DV case (using the τ_b-values of FL41, and T_{r2} = 1 year); $m_1 = 0.5\,\text{kg/m}^2$; $m_{tot} = 23.4\,\text{kg/m}^2$ (equivalent to $\xi_2^m = 0.3$); $C = 60\,\text{mg/L}$; $\Delta C = 50\,\text{mg/L}$; $h = 4\,\text{m}$; and (2) after calibration of the 3D model.

Parameter	Value 1DV	Value 3D	Unit
w_{s01}	$2.31 \cdot 10^{-5}$	$2.31 \cdot 10^{-5}$	m/s
w_{s02}	$4.63 \cdot 10^{-6}$	$4.63 \cdot 10^{-6}$	m/s
C_0	0.025	0.025	kg/m^3
$M_{E,1}$	$1.0 \cdot 10^{-6}$	$1.0 \cdot 10^{-6}$	$\text{kg/m}^2\text{s}$
m_{cr}	1.37	4.02	kg/m^2
M_2	$7.5 \cdot 10^{-6}$	$1.4 \cdot 10^{-6}$	$\text{kg/m}^2\text{s}$
τ_{e1}	0.05	0.05	Pa
τ_{e2}	0.1	0.1	Pa
n_2 (porosity)	0.4	0.4	—
δ_2	0.05	0.05	m
V_b	$1.44 \cdot 10^{-6}$	$9.06 \cdot 10^{-7}$	1/s
P	—	$3.17 \cdot 10^{-8}$	$\text{kg/m}^2\text{s}$
n (manning)	0.024	0.024	$\text{s/m}^{1/3}$

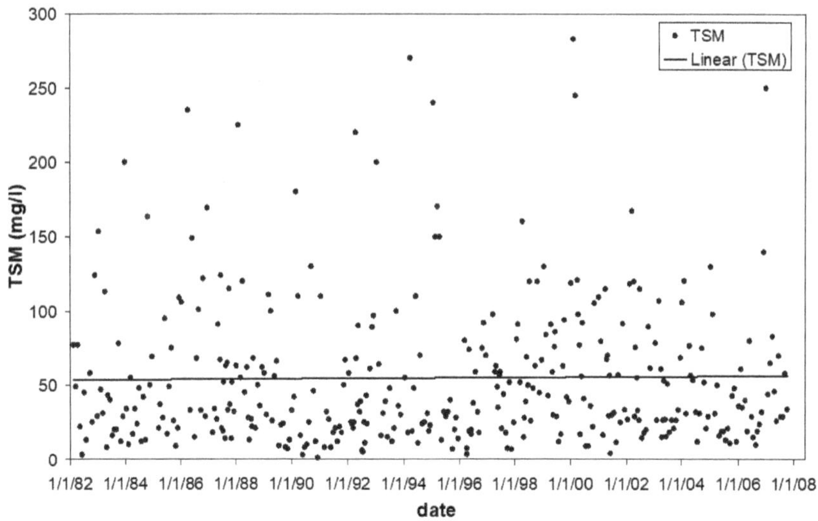

Fig. 17.2. Overview of all SPM (mg/L) observations at station MKM in the period 1/1/1982–1/1/2008 (MWTL-data).

in the range 0–300 mg/L in response to the different wind conditions during sampling. No significant long-term trend in SPM levels is observed unless normalized with wind speed during sampling (not shown).

High-frequency SPM-data

To complete the long-term MWTL-data with regard to sampling frequency, two semi-permanent observation stations have been installed in Lake Markermeer. Figure 17.1 shows their locations. Station FL41 is positioned close to the island of Marken, whereas station FL42 is positioned in the middle of the Lake Markermeer close to MWTL station Markermeer Midden. The data can be found at www.opendap.deltares.nl.

Details of the setup, instrumentation, calibration and servicing of the rigs is reported by Blok (2008). Here, only the issues most relevant for the calibration of the hydrodynamic and fine sediment model are discussed, i.e.:

- wind speed and direction,
- wave height and period,
- current velocity, and
- turbidity.

The first parameter is used to force the model, the next two parameters to calibrate the hydrodynamic model and the last to calibrate the sediment transport model. Turbidity sensors (Optical Back Scatter, OBS) were mounted at two levels, one about 1 m from the surface, the second 0.2–0.4 m above the bed. The OBS sensors were cleaned at an interval of 14 days, but in summer this turned out to be insufficient, at times. Therefore, gaps in the data exist in the period from 1/8/2007 until 1/9/2008, addressed later in this chapter.

The measurement frequency was set at 10 min to enable monitoring the short-term SPM response to short-term changes in wind. Figure 17.3 shows the observed SPM concentrations from 1/8/2007 until 1/9/2008 at location FL42. These observations are compared with the MWTL-observations at Markermeer Midden, close to location FL42. Figure 17.3 demonstrates how much more information is retrieved on the suspended sediment dynamics of Lake Markermeer from the high-frequency observations compared with the MWTL-observations.

The vertical concentration gradient (i.e. the difference in SPM level between top and bottom sensors) in combination with the local current conditions can provide information on the settling velocity of the suspended sediment. The vertical concentration gradients for both FL41 (not shown)

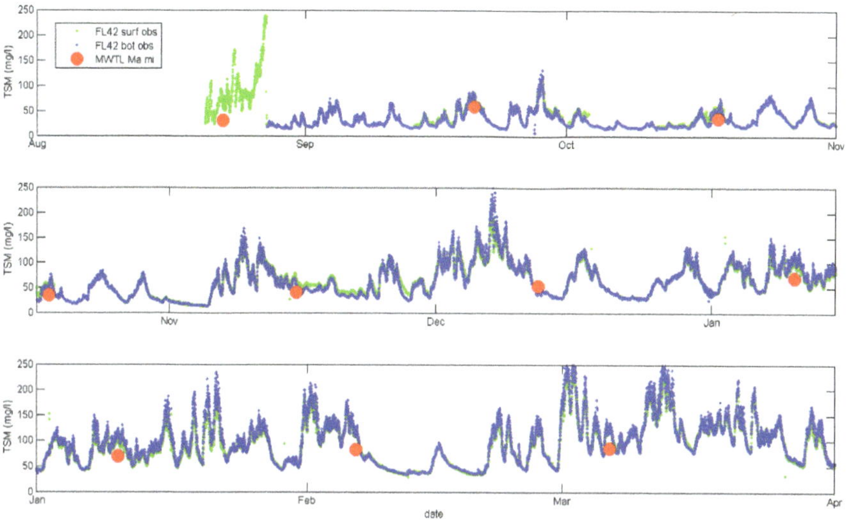

Fig. 17.3. Measured near-surface and near-bottom SPM concentration (mg/L) at location FL42 in the period 1/8/2007–1/4/2008. MWTL data points at MKM are indicated by red dots.

and FL42 turns out to be small. However, large concentration gradients may exist in the lower meter of the water column during extreme storms (Vijverberg *et al.*, 2011).

Wind-induced waves are the dominant agent for sediment resuspension, whereas wind-induced currents are dominant for horizontal sediment transport. Both wave height and direction, and current speed and direction are very sensitive to wind speed and direction, and its duration. The surface and bottom current direction may differ significantly, thus inducing complex three-dimensional flow patterns in the lake. During storms water level setup and setdown effects are observed, with large setup at the Houtribdijk for SW gales and large setup in the “IJmeer” for NE gales.

Remote-sensing data

Remote-sensing data have been used to explore spatial variability of SPM-concentrations in Lake Markermeer, additional to the data from the two high frequency observation points. Thirty images covering various wind conditions are available for 2006 (Kamps *et al.*, 2007). Figure 17.4 presents two examples, one on 16 July 2006 (NE-wind, 4 m/s) and another at 17 April 2006 (W-wind, 8 m/s). These demonstrate distinct spatial variability in SPM-concentrations, depending on wind direction.

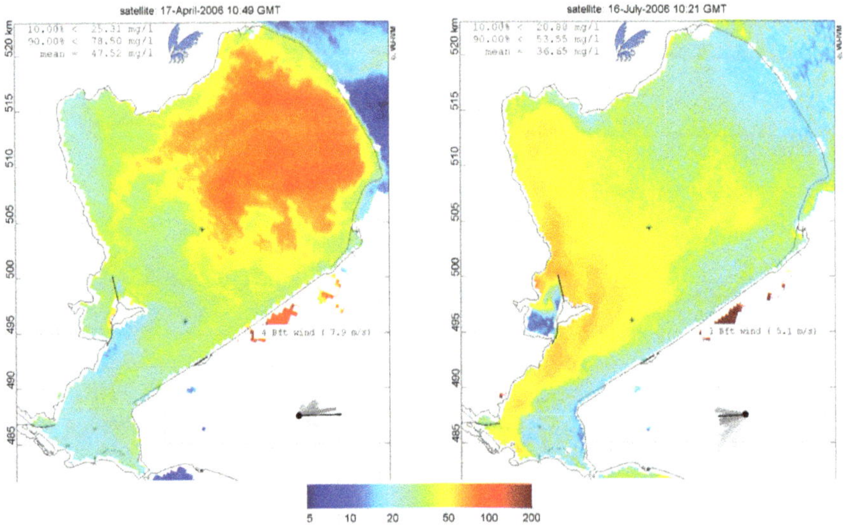

Fig. 17.4. Near-surface SPM distribution (mg/L) from remote sensing data. Left: at 17/4/2006 (wind W, 8 m/s); right: at 16/7/2006 (wind NE, 5 m/s).

Bed data

Additional to the water column data, bed data are relevant. Field observations show that the bed of Lake Markermeer consists of a layer of soft mud (fluff layer) atop a solid Holocene bed (see Sec. 8.3). The thickness of the soft layer varies from less than a few cm in most parts to about 0.4 m in the south-eastern part of the lake (Vijverberg, 2008), with an average of about 0.2 m. The upper few mm of the otherwise anoxic fluff layer is oxygenated, exhibiting a distinct different color (Fig. 8.6). The following data have been used for setting up and calibrating the SPM model:

(1) data on the thickness of the fluff layer from Noordhuis (2008) and Vijverberg (2008);
(2) data on former Zuiderzee sea shell distributions on the bed, used as a proxy for bed roughness (Noordhuis (2008); see also Fig. 17.5);
(3) data on the subsurface bed composition from the Geological Atlas of Lake Markermeer (Lenselink and Menke, 1995).

Figure 17.3 suggests maximal SPM-concentrations of about 400 mg/L during storm, though close to the bed, a few g/L has been measured, as mentioned before. At a water depth of 4 m, 400 mg/L is equivalent to about

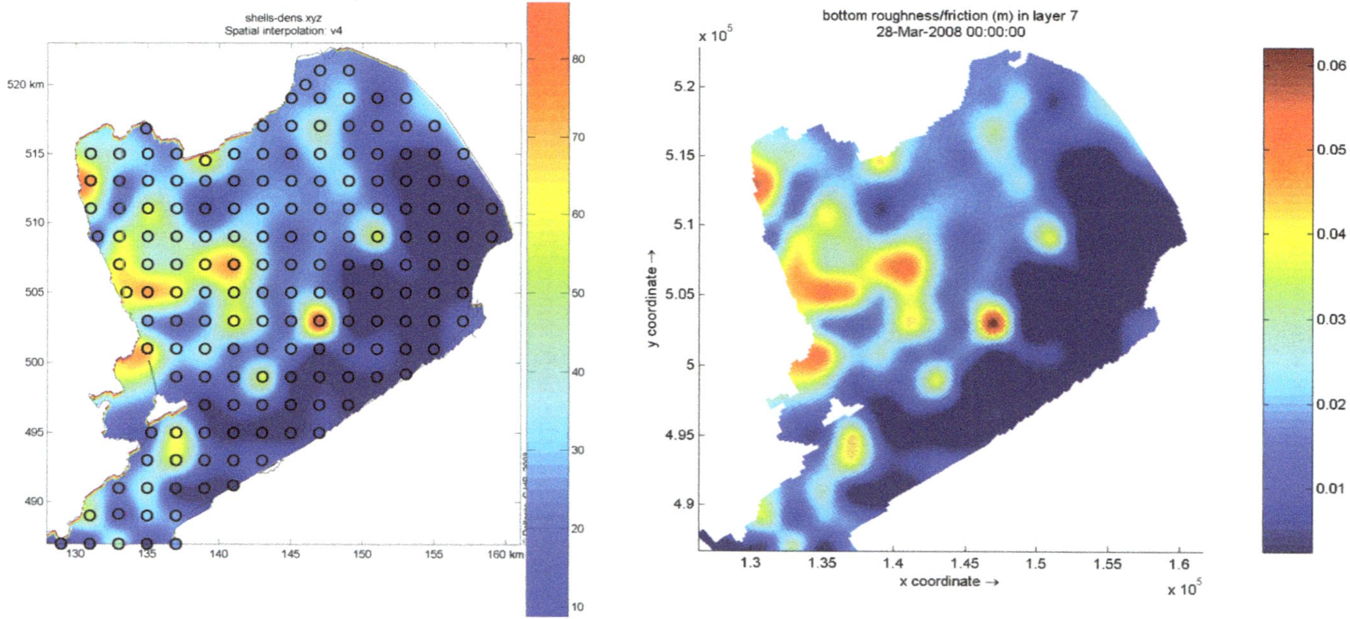

Fig. 17.5. Left panel: Historical shell distribution (%); right panel: Nikuradse roughness height (m) derived from shell distribution.

8 mm of the anoxic fluff layer, assuming $\rho_{dry} = 200\,\text{kg/m}^3$. This is about 4% of the anoxic layer thickness. In other words, under average conditions, less than 1% of the fluff layer is mobilized and reformed, i.e. active. We discuss below (Sec. 17.4.1) how this information is used.

17.2.4 *Model requirements*

A fine sediment transport model of Lake Markermeer should cover not only the entire lake, but also the narrow, long branch, the "Eemmeer" (Fig. 17.1) to properly simulate circulation patterns induced by wind and wind-induced setup.

The water movement in the lake is not uniform, with considerable currents closer to the bed opposite to the main wind-induced flow, in particular during periods of alternating winds. To resolve these complicated current patterns, a 3D model is required with at least 5–10 vertical layers.

Generally, SPM-concentrations do not exceed 100 mg/L, thus the turbulence field is not much affected by SPM (Sec. 7.1). A decoupled approach is therefore feasible. Only during storm conditions, the near-bed SPM-concentrations of a few g/L may affect the flow field, f.i. in the form of SPM-induced density currents. These can therefore not be simulated in a decoupled approach.

The organic fractions of the sediment in the lake is high. Therefore, at least two fine sediment fractions have to be accounted for, one "coarse" fraction, mainly representing the more mineral fraction, and one "fine" fraction, the organic material. Note that both fractions may contain minerals and organic matter, owing to flocculation. Further, as a result of flocculation and sorting, particle size increases under more dynamic conditions, which should be accounted for.

SPM dynamics in the lake are characterized by small SPM-concentrations with finer sediment during calm weather conditions, and larger SPM-concentrations, containing coarser sediment during storm. Modeling the bed response for these conditions over longer periods requires the use of at least the two-layer buffer model (Sec. 12.3). This approach allows for simulating seasonal variations in SPM values as well.

The literature describes a number of examples of suspended sediment modeling in lakes. The approach followed here is in many respects similar to the one presented by Luettich *et al.* (1990), which is further discussed in the next section. More recent examples of modeling fine sediment in shallow lakes are discussed by Bailey and Hamilton (1997); Chao *et al.* (2008); Chung *et al.* (2009); James *et al.* (1997); Li *et al.* (2008); Malmaeus and Håkanson (2003). These approaches range from completely process-based formulations (with different levels of sophistication) to empirical data-driven models (Aalderink *et al.*, 1985). An early example of comparison with remote sensing data is presented by Dekker *et al.* (2001). A concise overview of modeling the sediment dynamics in lakes can be found in Ji (2008).

Thus, modeling wind-induced fine sediment dynamics in a shallow lake is well-established. In the present application we introduce three new features:

(1) Quality of observations: both long-term high-frequency observations to assess temporal variability and remote-sensing images to assess spatial variability are used.
(2) An objective, analytical calibration procedure beyond standard curve-fitting of parameters is discussed.
(3) A (dynamic) equilibrium bed sediment distribution is established from the model simulations, instead of prescribed by the user. Starting with a uniform distribution, the simulation period is set sufficiently long (typically 20 years for Lake Markermeer) to attain a dynamic equilibrium.

As Lake Markermeer is a closed system, the empty-bed approach (Sec. 12.3) is not feasible, unless the production of fines by bioturbation is accounted for, together with its response to anoxic conditions, as described in Sec. 8.3. This is however beyond the state-of-the-art. As the total amount of fines in the lake is well known, and its distribution is fairly even, deployment of the full-bed approach will not induce the problems described in Sec. 12.3.

17.3 Hydrodynamic model

17.3.1 *Model domain and grid resolution*

For the hydrodynamic model, the Delft3D-flow module was used (Lesser *et al.*, 2004; Stelling and Van Kester, 1994), with a curvi-linear grid and seven equidistant sigma layers, accounting for the three-dimensional flow patterns, and using the k–ϵ turbulence closure model. The Lake Markermeer model grid resolution is set at a 200 ± 50 m, with a time step of 150 s.

The narrow Eemmeer branch has a resolution of about 500 m, sufficient for its role as water storage. The computational grid and bathymetry are shown in Fig. 17.1. Water can enter or leave the model through two boundary, i.e. the sluices at Krabbersgat and Houtrib (Fig. 17.1). The water fluxes through these sluices are prescribed further to the water balance for the entire lake, rainfall and evaporation time series, and seasonal variations in water level, with a summer level of NAP −0.2 m, and a winter level of NAP −0.4 m, with modulations of about 0.2 m due to rainfall (NAP is Dutch datum level).

17.3.2 *Calibration and validation*

Subsequently, the model was calibrated for the high-pass water-level response to wind at five stations in the lake by tuning the wind friction coefficient. Variable wind forcing was applied at an hourly interval. In response, water level variations of about 3 cm are predicted at station FL42 in the center of the lake, increasing with distance, up to 20 cm at Schellingwouderbrug (Fig. 17.6) in the SW corner. This response is a consequence of the setup in the lake. The results for station Krabbersgat are show in Fig. 17.6. Note that the simulation period covers the period for which high-frequency SPM data is available, i.e. from 1/8/2007 to 1/9/2008.

17.3.3 *Analysis of hydrodynamics*

The flow patterns in a lake depend on wind conditions. In periods with persistent wind, the currents near the water surface are almost in the direction of the wind, whereas the near-bed currents are often in reverse direction. This so-called conveyor belt mechanism is a typical feature of deep lakes with a flat bed. In lakes with variations in water depth, also 2D horizontal eddies are generated. Lake Markermeer flow patterns bear characteristics of both. Thus to simulate advection and dispersion of SPM in Lake Markermeer, modeling fundamentally requires a 3D approach, as argued before.

Waves

Wave-induced bed shear stresses steer the exchange of sediment between bed and water column, and have therefore to be accounted for. However, wave simulations with a process-based numerical wave model, such as SWAN, are computationally expensive, and therefore only feasible at three hour time interval. As the wind and wave forcing may change significantly within this time, a semi-analytical approach was adopted. Wind-induced waves are

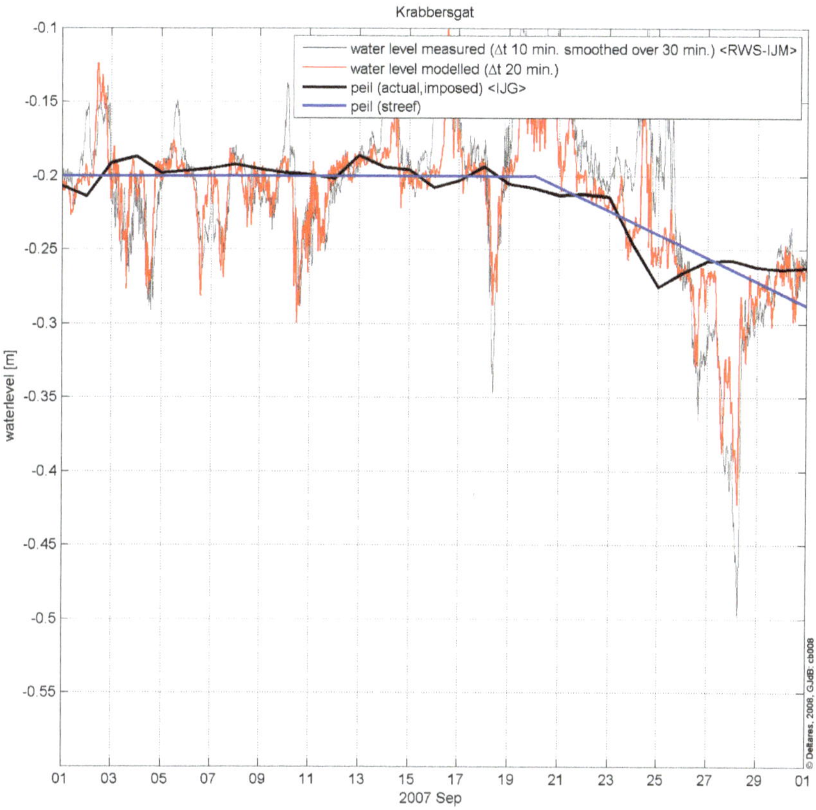

Fig. 17.6. Water levels in September 2007 at location Krabbersgat. The blue and gray line represent the spatially averaged target water level, and the actual measured water level, respectively. The black line represents the low-pass filtered water level at Krabbersgat sluice, imposed on the model, and the red line is the computed high-pass filtered response due to wind.

computed dynamically at every computational time step of the hydrodynamic model with an analytical fetch-length model (Bretschneider, 1954). Thus all high frequency variations due to wind are resolved within the model, but at lower accuracy as could be achieved with SWAN. This high-frequency response was deemed more important than the accuracy in wave height and period.

Figure 17.7 shows that this fetch-length wave model can predict wave height variations satisfactorily at station FL42. Because of small variations in water depth, and regular geometry, the fetch-length approach works well

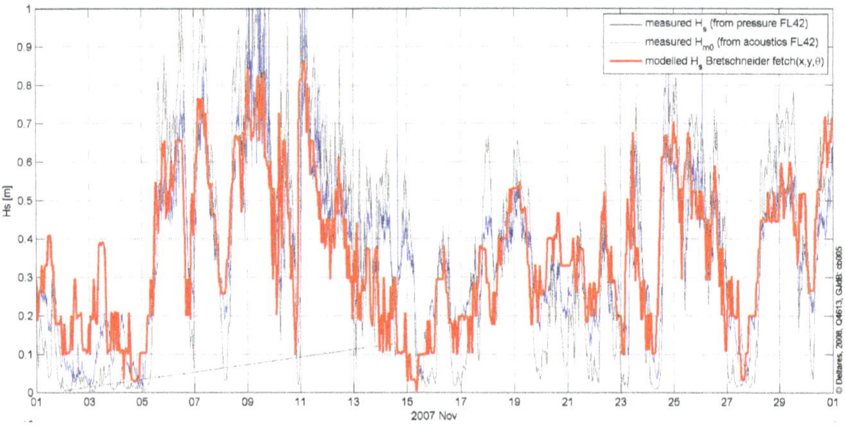

Fig. 17.7. Measured and computed wave heights in November 2007 at Station FL42.

for the entire lake, except in the narrow Eemmeer branch, but this is outside the domain of interest.

17.4 Sediment transport model

As argued above, the fine sediment transport model is run decoupled from the hydrodynamics. Thus the water movement parameters computed with the hydrodynamic model are stored on file and read as input to the sediment transport model (Sec. 13.3.1).

17.4.1 *Model setup*

The buffer model is designed to simulate seasonal variations in SPM-concentrations. However, the function of some model parameters is modified in the present application. Contrary to the description in Sec. 12.3, the thickness of the lower buffer layer δ_2 is kept constant, and variations in sediment mass m_2 in that layer are accounted for through the mud fraction ξ_2^m.

The sediment model consists of the seven σ-layers in the water column with the same distribution as in the hydrodynamic model, and three bed layers (S1, S2 and S3). Layer S1 represents the oxygenated, thin upper part of the layer of fluffy mud (the active layer), in equilibrium with SPM in the water column during periods of calm weather. Layer S2 represents the bulk of the (anoxic) fluff layer (the buffer layer), whereas layer S3 represents the

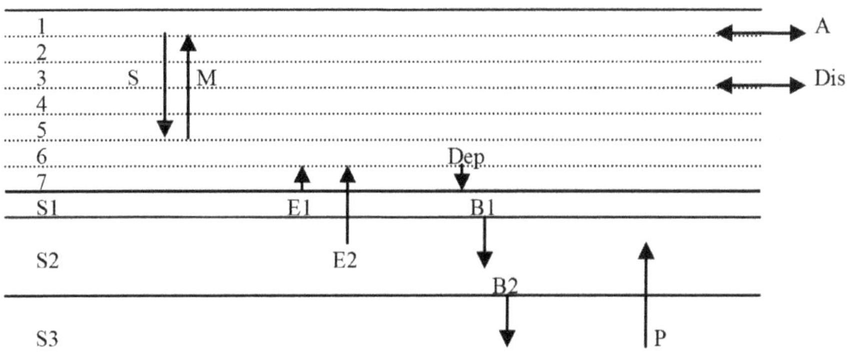

Fig. 17.8. Schematic representation of sediment fluxes in model (all fractions); S1 = active layer, S2 = buffer layer, S3 = parent bed.

stiff Holocene deposits underneath, i.e. the parent bed. The layer distribution is illustrated in Fig. 17.8. The arrows represent the sediment fluxes discussed below.

Two non-interacting sediment fractions, denoted with subscript $\cdot_1$ and $\cdot_2$, respectively, are introduced to simulate observed vertical SPM-concentration gradients, in conjunction with observed temporal SPM-concentration variations during calm weather conditions, following a period of rough weather. Further to Van Duin (1992), these fractions have the following properties: $w_{s01} = 2.3 \cdot 10^{-5}$ m/s and $w_{s02} = 4.6 \cdot 10^{-6}$ m/s.[1] However, larger particles are formed by flocculation and/or picked up from the bed during more dynamic conditions, as was found from the 1DV-analysis of the data (see below). This can be accounted for with a ero-order flocculation model through a concentration-dependent effective settling velocity:

$$w_{s,i} = w_{s0i} c / C_0 \tag{17.1}$$

where w_{s0i} is the settling velocity of fraction i at $c = C_0$, with the reference concentration $C_0 = 25$ mg/L, while the linear parameterization in Sec. 4.3 is used.

The erosion parameters for both fractions have to be identical, as explained in Sec. 12.2. Hence, the properties of the fractions differ only with respect to their settling velocity.

Owing to its shallowness, a dynamic balance in the water column exists between settling S and mixing M of the sediment, yielding a Rouse profile

[1]These low settling velocities reflect very fine sediment, with a large impact on light conditions/turbidity.

at equilibrium (Sec. 4.1). The vertical mixing coefficient is derived from the hydrodynamic computations, using the k–ϵ turbulence closure model. The settling flux S is computed from $S = w_s c$, where w_s is settling velocity.

The deposition flux from the near-bed water layer 7 to bed layer S1 is expressed by $D(= w_{s,1}c_{1,7} + w_{s,2}c_{2,7})$ (Sec. 4.2), while neither a critical shear stress for deposition ($\tau_d = \infty$), nor hindered deposition ($p_d = 1$) is applied. At equilibrium, the deposition flux is balanced by the erosion fluxes E_1 and E_2 from bed layers S1 and S2 respectively:

$$\begin{aligned} E_{1,i} &= \min\left(1; \frac{m_1}{m_{cr}}\right) M_{E,1}\left(\frac{\tau_b}{\tau_{e,1}} - 1\right) \\ &= \min\left(1; \frac{m_1}{m_{cr}}\right) M_{E,1}\Theta_{e1} \end{aligned} \tag{17.2}$$

$$E_{2,i} = \xi^m_{2,i} M_2 \left(\frac{\tau_b}{\tau_{e,2}} - 1\right)^{1.5} = \xi^m_{2,i} M_2 \Theta_{e2} \tag{17.3}$$

where $\tau_{e,j}$ is the critical shear stress for erosion of layer Sj, $\xi^m_{j,i}$ its mud fraction, and M the erosion flux parameter. The bed shear stress τ_b consists of a flow- and a wave-induced component, while the dimensionless excess bed shear stress is given by Θ_{ej}. The power "1.5" is introduced as part of the lake bed is sandy; if purely muddy, this power would be unity (Sec. 6.6). The results are not very sensitive to this power, however, hence one value was set for the entire domain, whether the bed is sandy or muddy. These equations apply to the sum of both sediment fractions. As layer S1 represents the fluff oxic layer atop S2, representing the anoxic fluff sediment or sandy bed, in general: $\tau_{e,1} < \tau_{e,2}$.

The mud fraction ξ^m_2 (both fractions) is computed from m_2, the total mass of mud per unit area in layer S2, according to $\xi^m_2 = m_2/\delta_2(1 - n_2)\rho_s$, where n_2 is the bed porosity, δ_2 the thickness of layer S2 and ρ_s the solids specific density. The maximum mud fraction ξ^m_2 is unity, whereas ξ^m_1 (both fractions) is always unity owing to the formulations. When ξ^m_2 exceeds unity, excess material is buried into the parent bed layer S3 (term B2 in Fig. 17.8), as δ_2 is kept constant (Sec. 12.2).

The erosion flux parameter M_1 depends on the thickness δ_1 of fluff layer S1 (in fact its mass). If its thickness is smaller than the local roughness height of the bed, no spatially uniform fluff layer can develop. Then a first-order resuspension flux is required (see Sec. 4.2), and the erosion flux scales linearly with m_1, the mass of sediment per unit area in layer S1. The transition between first- and zeroth-order erosion is set at $m_{cr} = 4\,\mathrm{kg/m^2}$, at an erosion flux parameter $M_{E,1} = 10^{-6}\,\mathrm{kg/m^2/s}$. This is equivalent to a 0.2 m thick fluff layer, assuming a fluff layer concentration of $200\,\mathrm{kg/m^3}$.

Both layers may erode concurrently. However, the fluff layer generally erodes faster than the bed layer underneath. When the fluff layer becomes depleted (for instance during a storm), erosion can occur from the second bed layer S2 only.

Transfer of sediment from layer S1 towards S2 is modeled with a burial term, expressed as: $B_1 = m_1 V_b$, i.e. the burial rate scales linearly with the sediment mass m_1. Upward sediment transport from layer S2 is accounted for by erosion (term E_2). This burial term parameterizes processes such as consolidation and bioturbation.

Finally, a production (source) term P is introduced, transferring sediment from the parent bed S3 towards S2. This term can be used to take into account the slow erosion of older deposits (such as Zuiderzee clay or exposed peat layers along the shoreline), as explained in Sec. 8.3. On short time scales, sediment production plays a minor role, but at longer times, it may become significant. Here, term P is set at a constant value of $1\,\mathrm{kg/m^2/year}$, yielding a total gross production of 720 kton/yr over the lake's $720\,\mathrm{km^2}$ area. The term B_2, representing the flux from layer S2 towards the parent bed S3, is neglected in the present study, although in some conditions it may become important as discussed later.

Apart from vertical fluxes, also horizontal SPM fluxes are important, governing the dispersion of sediment over the lake. Horizontal fluxes are either caused by advection $\mathrm{A} = \overline{U}c$ or by dispersion Dis $(= -D_x \partial C/\partial x - D_y \partial C/\partial y)$ (Chapter 11 and Sec. 12.1). Advective transport is determined using the flow velocity computed with the hydrodynamic model, thus varies in time and space. Diffusive transport follows from the constant user-defined horizontal dispersion coefficients D_x, D_y. In most cases, advective transport is dominant.

The formulations described above capture the most essential processes involved with fine sediment transport in Lake Markermeer. Some processes, however, are not included. For example, no sediment-density coupling is applied, which may play a role during storm periods when very high SPM-concentrations have been observed (Vijverberg, 2008; Vijverberg *et al.*, 2011). If the fluff layer is thick enough, at low strength, it may flow under gravity (density current — Sec. 7.5). Flocculation is only taken into account through a linear relationship between total suspended sediment concentration and settling velocity, whereas the number of sediment fractions is limited to two. No production and decay of organic matter is accounted for either, as neither the effects of bioturbation are implemented (apart from the constant production term P). The effects of temperature, such as ice formation or temperature-induced density currents are also not accounted for.

17.4.2 *Calibration and validation*

The high-frequency data of Sec. 17.2.3 are available for the period 1/8/2007–1/9/2008. Because of this long period of detailed SPM data, it is feasible to explicitly split the tuning of the sediment transport model in a calibration and validation phase, for which the data and forcing of 1/8/2007–1/4/2008 and of 1/4/2008–1/9/2008 are used, respectively. Also the hydrodynamic model was run for these periods.

17.4.3 *Calibration procedure*

In the calibration procedure, the following features of the SPM-dynamics in the lake should be predicted correctly:

(1) average SPM-concentrations, and typical values during storm and calm weather,
(2) rates of increase and reduction in SPM-concentration during conditions of increasing and decreasing wind speed, respectively,
(3) characteristic vertical gradients in SPM-concentration,
(4) characteristic spatial SPM-concentration distributions at various wind directions,
(5) spatial distribution of (equilibrium) bed composition,
(6) sedimentation rates in deep pits and navigation channels,
(7) location and extent of areas exhibiting net sedimentation, and
(8) sediment budget of the lake (net sink and source terms e.g. by consolidation or bioturbation).

The numerical model contains the following parameters to be optimized for calibration:

(1) w_{s0i}: settling velocity of fraction i (in this case the measured values were used — see above),
(2) $M_{E,1}$: erosion flux parameter for bed layer S1,
(3) $\tau_{e,1}$: critical shear stress for erosion of layer S1,
(4) M_2: erosion flux parameter for bed layer S2,

(5) $\tau_{e,2}$: critical shear stress for erosion of layer S2,
(6) V_b: burial rate from layer S1 towards S2 (the burial flux thus becomes $m_1 V_b$ [kg/m^2/s]),
(7) δ_2: constant thickness of layer S2,
(8) P: production, contributing to the sediment buffer in layer S2,
(9) k_s: bed roughness (see below).

As Lake Markermeer is very shallow, sediment processes in the vertical are expected to dominate local SPM-concentrations (Chapters 17 and 13). The calibration procedure therefore consists of a 1DV-phase to attain a first estimate of the model parameters, followed by fine-tuning with the full 3D model.

We advocate the empty-bed approach, where feasible (Sec. 12.3). This approach, however, is unpractical for semi-closed systems as Lake Markermeer, as spin-up times would become 5–10 years (De Lucas Pardo, 2014; De Lucas Pardo *et al.*, 2013). Hence, the initial amount of sediment in the lake is assessed from data and our system understanding.

17.4.3.1 *Calibration in 1DV mode*

In view of the large number of parameters, a trial-and-error calibration procedure is cumbersome and time consuming. Therefore, an alternative step-by-step procedure is introduced, starting with a 1DV-calibration phase. In this approach, horizontal fluxes are ignored. By neglecting horizontal advection (A) and dispersion (Dis), production (P) and deep burial (B2), i.e. the total sediment mass per unit area remains constant. However, the vertical distribution of SPM over the water column and bed layers S1 and S2 will vary in response to variations in bed shear stress. With these assumptions, analytical expressions can be derived for these vertical distributions.

Mass balances

Three balance equations, summing over both fractions, form the basis for these analytical expressions. The first is the overall mass balance:

$$m_{tot} = m_1 + m_2 + Ch \tag{17.4}$$

where h is the local water depth, C is the (a characteristic) time-average total SPM-concentration and m_1 and m_2 are the total mass (i.e. the sum of both fractions) per unit area in bed layers S1 and S2, respectively. The second balance equation is based on the assumption of long-term equilibrium between sedimentation and erosion per fraction i:

$$w_{si}C_i = \min\left(1; \frac{m_1}{m_{cr}}\right) M_{E,1}\bar{\Theta}_{e,1} + \xi_{2i}^m M_2\bar{\Theta}_{e,2} \tag{17.5}$$

where $\bar{\Theta}_{ej}$ is the long-term time-averaged excess bed shear stress for bed layer Sj, defined as

$$\bar{\Theta}_{e,1} = \frac{1}{T^*}\int^{T^*}\left(\frac{\tau_b}{\tau_{e,1}} - 1\right)\mathrm{d}t \quad \text{and} \quad \bar{\Theta}_{e,2} = \frac{1}{T^*}\int^{T^*}\left(\frac{\tau_b}{\tau_{e,2}} - 1\right)\mathrm{d}t \tag{17.6}$$

where T^* is the averaging time. Note that (17.5) is invalid for areas with net sedimentation, such as deep pits. In such areas deep burial (B2) will occur (i.e. within the current modeling framework), which is ignored in the 1DV calibration phase.

The third balance equation is based on the assumption of long-term equilibrium between burial towards layer S2 and erosion from layer S2:

$$m_1 V_b = \xi_2^m M_2\bar{\Theta}_{e,2} \tag{17.7}$$

The relationship between mud fraction ξ_2^m and sediment mass per unit area m_2 reads:

$$m_2 = \xi_2^m \rho_s(1 - n_2)\delta_2 \tag{17.8}$$

where δ_2 is layer S2 thickness and n_2 is the bed porosity. The residence time T_{r2} in layer S2 is expressed as

$$T_{r2} = m_2/(\xi_2^m M_2\bar{\Theta}_{e,2}) = \rho_s(1 - n_2)\delta_2/(M_2\bar{\Theta}_{e,2}) \tag{17.9}$$

During storm, layer S1 and S2 will be eroded/resuspended. For a storm with duration T_{st} and average excess bed shear stress $\Theta_{e,1,s}$ and $\Theta_{e,2,s}$, SPM-concentrations increase by

$$\Delta C \approx (T_{st}/h)\left(\xi_2^m M_2\Theta_{e,2,s} + \min\left(1; \frac{m_1}{m_{cr}}\right) M_{E,1}\Theta_{e,1,s}\right) \tag{17.10}$$

The maximum contribution of layer S1 to a storm-induced SPM peak is limited to $m_1/h\,\mathrm{kg/m^3}$, as layer S1 then becomes depleted. Layer S2 contains so much sediment that it cannot become depleted during a storm event (see also last para in Sec. 17.2.3). The excess bed shear stress depends on the values of $\tau_{e,1}$ and $\tau_{e,2}$. The lower these are, the higher the excess bed shear stress and the more the erosion flux and SPM-concentration increase.

Parameter assessment

From these equations, parameter settings can be assessed from the characteristic values for Lake Markermeer for C, ΔC, T_{r2} and m_1 (or T_{r1}), assuming realistic values for n_2, ξ_2^m and δ_2 (or m_2). It is advised to set ξ_2^m to a value which represents the observed amount of mud in the lake, and δ_2 at a value representative for the typical mixing depth observed (about 0.1 m, see also Chapter 8).

Target values for C and ΔC are derived from SPM observations. The value of m_1 is proportional to the fluff layer thickness $\delta_1(\approx m_1/c_{gel}$, with c_{gel} the gelling concentration (Sec. 6.5). Typically, $50 < c_{gel} < 200$ g/L for inorganic mud, while for organic rich mud, such as peat $c_{gel} < 50$ g/L. The residence time T_{r2} determines the model response time to variations in forcing (wind climate), sediment supply and/or geometry. Unfortunately, average residence times and their spatial distribution are difficult to establish. For Lake Markermeer, we assume a mean residence time of one year, representing extreme events with a one year recurrence, based on our system understanding. For $T_{r2} = 1$ year, a spin-up time of the model of three years is required to attain 95% equilibrium. Note, T_{r2} has no influence on the final equilibrium conditions, while its influence is small on short-term mud dynamics. Therefore, the setting of T_{r2} is of minor importance in the present study.

From the balance equations (17.4)–(17.7) and target settings for C, ΔC, T_{r2}, ξ_2^m and m_1, the parameters settings are established as follows:

$$m_2 = \xi_2^m \rho_s (1 - n_2)\delta_2 \tag{17.11}$$

$$m_{tot} = m_1 + m_2 + Ch \tag{17.12}$$

$$V_b = m_2/(m_1 T_{r2}) \tag{17.13}$$

$$m_{cr} = M_{E,1}(\bar{\Theta}_{e,1} m_1)/(w_s C - m_1 V_b) \tag{17.14}$$

$$M_2 = \rho_s (1 - n_2)\delta_2/(T_{r2}\bar{\Theta}_{e,2}) \tag{17.15}$$

The resulting values are given in Table 17.1.

Settling velocity and deposition flux

In general, the settling velocity $w_{s,i}$ and reference concentration C_0 of Eq. (17.1) should be tuned such that a model reproduces observed vertical SPM-concentration gradients and the rate of concentration decrease after a storm event. Vertical SPM-concentration gradients as a function of settling velocity can be determined from the Rouse profile (Sec. 3.4, Eq. (3.63) or (3.64)), assuming equilibrium conditions.

However, here w_s values from literature are used (Van Duin, 1992) as a basis. These values were corroborated with 1DV simulations, while C_0 followed from fine tuning of the model calibration — see below. The resulting values are given in Table 17.1. Application of the Rouse profile equation (3.63), assuming $U = 0.1\,\text{m/s}$, $n = 0.024$, and $w_s = 0.02\,\text{mm/s}$, gives $\beta = 0.006$ and $C_s/C_b \approx 1.02$. Thus the water column is well-mixed under calm conditions, as observed.

The time scale for SPM decay after a storm is given by $T_d = h/w_s$, assuming a vertically mixed system. For $h = 4\,\text{m}$ and $w_s = 0.02\,\text{mm/s}$, $T_d \approx 55\,\text{hrs}$, which agrees with the high-frequency SPM observations from the measuring rigs FL41 and FL42. Of course, the finer fraction yields larger T_d. This implies that the fine sediment dynamics in Lake Markermeer exhibit profound memory effects.

The deposition flux is determined by the near-bed SPM-concentration. In areas where resuspension is negligible, such as in deep pits, the gross deposition flux equals the net deposition flux, i.e. the sedimentation flux (Sec. 4.2). In the following, both fractions are included (summed). The gross flux follows from $D = w_s C_b$. As an example, for $w_s = 0.02\,\text{mm/s}$ and $C_b = 100\,\text{mg/L}$, $D \approx 2 \cdot 10^{-6}\,\text{kg/m}^2\text{/s}$ (i.e. $0.17\,\text{kg/m}^2\text{/day}$, or 0.08–0.16 m/year, assuming $\rho_{dry} = 800$–$400\,\text{kg/m}^3$ De Lucas Pardo, 2014). This deposition rate is much smaller than observed in these pits, which is explained by larger near-bed SPM-concentrations during storm, and possible sediment-driven density currents (Vijverberg *et al.*, 2011). In the latter case, the current model without sediment–flow interactions (sediment-induced density currents) is not suitable to assess such siltation rates. Note that the contribution of the coarser sediment fractions tends to dominate the total deposition rate.

During storm, larger particles, with *same* settling velocity, are picked up from the fluff anoxic layer. However, these cannot be mixed over the entire water column, owing to the small flow velocities. Therefore, the SPM-concentration profile becomes stratified in the lower part of the water column. This coarsening of SPM can be modeled with Eq. (17.1), tuning C_0, which thus simulates grading merely than flocculation. However, properly simulating the large SPM-concentration gradients reported by Vijverberg *et al.* (2011) requires accounting for sediment-flow interactions, as argued above.

Further to our system understanding, the 1DV-simulations showed also that at least two fractions are needed to attain a satisfactory reproduction of SPM observations during both calm weather periods and storm events.

These simulations suggest the following values: $w_{s01} = 2.3 \cdot 10^{-5}$ m/s; $w_{s02} = 4.6 \cdot 10^{-6}$ m/s and $C_0 = 25$ mg/L (in line with observations). The coarser fraction is resuspended during periods with moderate to strong wind, whereas the finer fraction remains partly in suspension during calm weather as well.

Of course, more fractions can be used, which may yield better accuracy, but at the cost of additional calibration parameters. For the present model, two fractions provides a fair balance between simplicity and accuracy. From the calibration of the 1DV-model, a mass ratio of 3:1 between the fractions was established, i.e. the coarser fraction is three times more abundant in the bed than the finer fraction. In the water column and in the fluff layer this ratio varies strongly with hydrodynamic conditions. During calm whether the coarse fraction hardly contributes to the SPM-concentrations, whereas during storms it is dominant.

It is stressed that using measured w_s values as input to a numerical sediment transport model may highly be misleading, and may yield unrealistic results, as discussed in Sec. 4.1. This was not the case here, owing to the small settling velocities.

Thickness of bed layers

Under calm weather conditions, typical SPM-concentrations are about 10 mg/L, which is equivalent to a sediment mass of 0.04 kg/m^2 at a water depth of 4 m. A 1 cm thick fluff layer with $\rho_{dry} = 400$ kg/m^3 contains 4 kg/m^2 ($\equiv m_1$).

A 5 cm thick layer S2 with $\xi_2^m = 0.2$ contains $m_2 = 15.6$ kg/m^2. Most sediment is therefore stored in layer S2. Under storm conditions, the fluff layer is resuspended completely ($m_1 = 0$), while SPM-concentrations may be up to 500 mg/L. This is equivalent with an eroded mass of 2 kg/m^2 for a water depth of 4 m. Thus sediment from layer S2 will be eroded as well contributing to the peak concentrations during storm. However, the majority of the sediment in layer S2 will not be mobilized, even during storms. Integrated over the lake area (about 720 km^2), the active fine sediment mass is thus about 10 Mton.

Critical shear stress

The next parameter to consider is the critical shear stress for erosion for layers S1 and S2, $\tau_{e,1}$ and $\tau_{e,2}$, respectively. Parameter $\tau_{e,1}$ is set at a value to achieve onset of resuspension of the fluff layer at a wind speed of 4 m/s (Bft 3). For this wind speed, the computed wave-induced bed shear stress is about 0.02 Pa. Parameter $\tau_{e,2}$ is chosen such that net sedimentation is predicted in known sinks, such as deep pits and channels, but not in areas where the bed is not predominantly muddy.

Depending on its settings in relation to the local bed shear stress, either of two conditions occurs for layer S2:

(1) $M_2\bar{\Theta}_{e,2} > m_1 V_b$: no net accumulation occurs, layer S2 will reach an equilibrium mud fraction $\xi_2^m = m_1 V_b / M_2\bar{\Theta}_{e,2}$, or
(2) $M_2\bar{\Theta}_{e,2} < m_1 V_b$: ξ_2^m may become larger than unity; excess sediment is buried towards bed layer S3 and net sediment accumulation occurs ("deep burial").[2]

Condition 1 occurs in areas with moderate to large bed shear stresses, whereas condition 2 occurs in areas where bed shear stresses are small, such as in deep pits. These are therefore net sediment sinks. In contrast, areas with condition 1 may be in dynamic equilibrium, or provide sediment by gradual erosion of old deposits, thus become a net sediment source. In the Lake Markermeer model, such a source is implemented as a constant, spatially uniform term ($P = 1\,\text{kg/m}^2/\text{year}$). If sink and source terms are initially not in equilibrium, the total amount of active sediment mass increases or decreases, in conjunction with the SPM-concentration. The time scale T_e involved is at least T_{r2}. At the scale of the whole lake, $T_e \approx M_{eq}/P^*$, which may be of the order of 10 years or more. Herein, M_{eq} is defined as the active sediment mass in the whole lake at equilibrium (estimated at about 10 Mton) and P^* the source term integrated over the lake (order 1 Mton/year).

Therefore, the model has to be run for 5–10 years to attain equilibrium conditions, which implies that sink and source terms, integrated over the lake, become identical. Locally, non-equilibrium conditions may occur, but these are not important for the present study, as they hardly change the active sediment mass at the short time scale considered (order of 1 year).

17.4.3.2 *Calibration in 3D mode*

Because of horizontal advection and dispersion, and spatial variations in bathymetry, geometry and (wind and wave) forcing, the parameter settings from the 1DV calibration phase need to be further tuned in the 3D SPM transport model, while also a small source term (P) is added to account

[2]Deep burial simulates the immobilization of the deposited sediment, rather than merely burial into the parent bed.

for ongoing erosion of the Holocene parent bed (Sec. 8.3), represented by layer S3. Layers S1 and S2 should not deliver or store any sediment over longer time scales, i.e. they should not be a net source or sink. Of course, temporal variations in the mass of sediment per unit area in these layers will occur in response to variations in forcing. In the 3D calibrated model, these layers should therefore be in a dynamic equilibrium, their mass fluctuating around a long-term constant value, for layer S1 at time scales of hours to a few days, and for layer S2 weeks to a few months. The overall time scale to reach this equilibrium amounts to about $3T_{r2}$, i.e. about 10 year, as discussed above.

The parameter settings from the 1DV calibration phase are used as initial, spatially uniform settings for the 3D calibration phase. Of course, bathymetry, geometry and forcing, however, are no longer uniform. With these settings, the computed bed composition does not match observations, with too low mud content in layer S2 in the deepest south-east part of the lake near Lelystad, whereas too high mud contents along the north-west coast near Hoornse Hop are predicted. The bed composition is mainly driven by the time-integrated bed shear stress distribution, affected by (1) the wind climate and (2) the bed roughness.

Calibration in 1DV-mode was based on an eight month wind forcing (1/8/2007–1/4/2008), which is short compared to $3T_{r2}$. In 3D mode, calibration is therefore based on a fivefold repetition of this period, i.e. 40 months of forcing. Moreover, wind direction and fetch play an important role, contrary to simulations in 1DV-mode, and are therefore included.

These effects were established by constructing an equilibrium bed composition based on the results of 16 stationary model simulations at two different wind speeds from eight different directions — the hydrodynamic model was run a few days, sufficient to attain hydrodynamic equilibrium. The computational results were weighted with the frequency of occurrence of wind speed and direction, obtained from the long-term wind climate (results not presented here). The resulting bed composition is similar to the one obtained with the actual, thus varying wind forcing in the period 1/8/2007–1/4/2008. This implies that this period is sufficiently representative for the long-term average wind climate.

The simulation time required to approach an equilibrium bed composition is at least five times the calibration period. The wind forcing in the calibration period is therefore also repeated at least five times (> 3 years).

Note that this time scale differs from the time scale to attain equilibrium between sinks and sources (the sediment balance), which was estimated at 10 year.

The left panel of Fig. 17.5 presents a map of the historic shell distribution in Lake Markermeer. The effective hydraulic roughness over shell banks can be large, and was parameterized by Van Leeuwen (2008) as $k_s = \max(0.003,\ 0.07\xi^{sh})$, where ξ^{sh} is the shell fraction, varying from 0 (no shells) to 1 (fully covered). Adopting these values, k_s in Lake Markermeer model varies between 3 mm and 7 cm, as a function of shell cover. These k_s-values are used for computing the wave-induced bed shear stresses. Flow-induced bed shear stresses are computed in the hydrodynamic model with a constant Manning coefficient ($n = 0.024\,\mathrm{s/m^{1/3}}$), which was also used in the 1DV-simulations. The total bed shear stress is obtained using Soulsby's approach, combining the wave- and flow-stress vectors (Sec. 3.3.2). Neglecting the shell effects on the flow computations does not affect the hydrodynamic results, as the wave-induced stresses are much larger than the flow-induced bed shear stress, whereas the flow patterns and wave field themselves are not affected by these local variations in bed roughness.

By applying this spatially varying bed roughness, the model now reproduces the observed bed composition pattern quite well. As a local reduction in τ_b also reduces resuspension, SPM-concentrations above smooth areas should decrease. The model captures this well, as shown in a comparison of model results with remote sensing data, see Sec. 17.2.3.

The best agreement between the 3D model and observations, particularly those of the high frequency data at stations FL41 and FL42, is obtained with the parameter settings of Table 17.1, the majority of which are identical to those found from the 1DV calibration phase, while also production P is accounted for. However, the mass in the fluff layer m_1 becomes too small using these settings. Therefore, in the 3D model, m_{cr}, M_2 and V_b are modified by a factor 2–5.

Apparently, vertical fluxes dominate the horizontal ones, which of course explains why the 1DV approach is so fertile. Note that this finding is consistent with the shallow water approach described in Chapter 1.

17.4.3.3 *Validation*

After calibration against the FL41 and FL42 high-frequency data for the period 1/8/2007–1/4/2008, the model is validated against the following data:

- Station FL41 and FL42 high-frequency turbidity data for the period 1 April–1 September 2008.
- Remote sensing data for the year 2006.

For the analysis against the remote sensing data, the hydrodynamic and SPM model were re-run with 2006 wind forcing and water balance. The model settings from the 3D calibration phase were used (Table 17.1). Also a comparison with data of bed composition and accretion is made. However, this is not an independent validation, as these data were also used in the calibration phase.

Figures 17.9 and 17.10 show the computed and observed SPM-concentrations at locations FL41 and FL42, respectively. In general, the model performance is similar to that of the calibration phase. Trends and concentration peaks are overall well reproduced, though some peaks are over- or underestimated, and sometimes a bias (see Table 17.2 for formal definition) in the computed average concentration with respect to the observations is found. However, also the OBS data may contain errors because of sensor fouling and the inherently difficult calibration of turbidity to

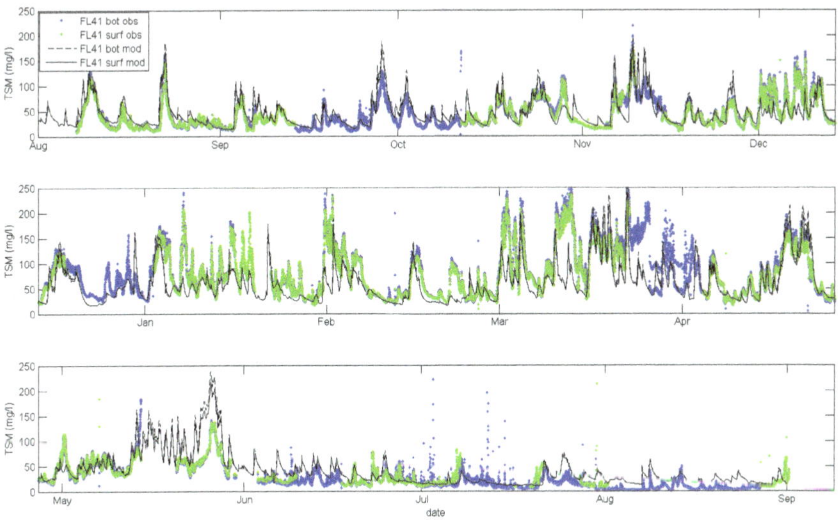

Fig. 17.9. Observed and computed near-surface and near-bed SPM-concentration (mg/L) at location FL41 for the period 01/08/2007–01/09/2008.

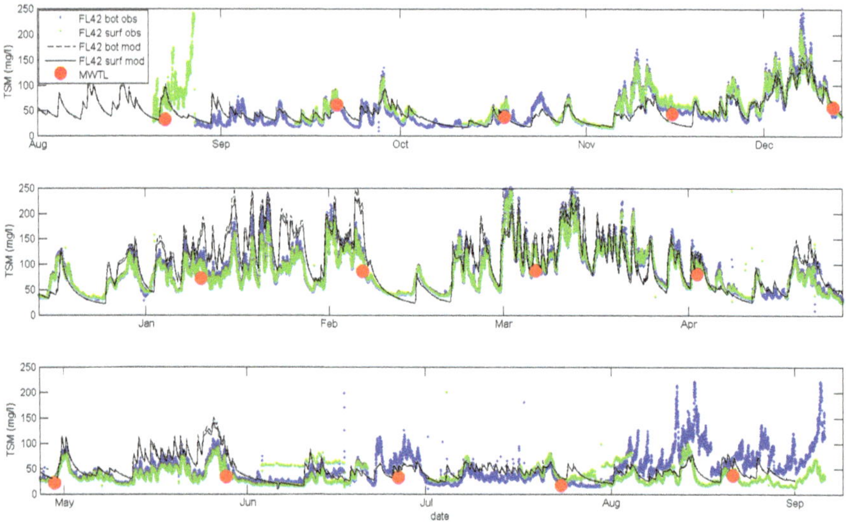

Fig. 17.10. Observed and computed near-surface and near-bed SPM-concentration (mg/L) at location FL42 for the period 01/08/2007–01/09/2008.

solids concentration. For example, data after August 1, 2008 at FL42 are less reliable, as the SPM levels of the top and bottom sensor diverge much more than during the one year period prior to this date.

Figures 17.11 and 17.12 show a comparison between SPM distributions obtained from satellite images and computational results (upper layer) for four days in 2006. The computed and observed distributions compare well. Only at very low wind speeds (Fig. 17.12), the comparison is less favorable: the model predicts SPM-concentrations between 10 and 20 mg/L, whereas the satellite data suggests levels below 10 mg/L. The likely cause for this is that the model was not calibrated against these low SPM-concentrations as the OBS data at FL41 and FL42 in the calibration period hardly contain 10 mg/L values.

Evaluation of the calibration/validation

In the above, a phenomenological calibration of the model (Sec. 13.5.2) is implicitly included. However, it is possible to compare model results and observations more explicitly at a quantitative level.

An important conclusion from both remote sensing data and model simulations is that the turbidity in Lake Markermeer is far from homogeneous: suspended sediment concentrations vary strongly in space and time.

Table 17.2. Statistical values at FL41 and FL42 (surface) for observed SPM concentration (in mg/L) (O) and model prediction (P) (Lynch *et al.*, 2009).

	FL41	FL42
R	0.71	0.80
$\overline{O}$	54.6	60.1
$\overline{P}$	51.7	70.8
std(O)	41.2	39.1
std(P)	35.5	46.1
std* = std(P)/std(O)	0.861	1.18
bias = $\overline{P} - \overline{O}$	−2.89	10.78
Bias* =bias/std(O)	−0.070	0.276
RMSE = RMS($P - O$)	29.7	29.7
RMSD = $\sqrt{(\text{mean}(\text{res}(P) - \text{res}(O))^2)}$	29.6	27.7
RMSD* = $\sqrt{(1 + \text{std}^{*2} - 2\text{std}^{*} R)}$	0.718	0.709
RMSD′* = sgn(std(O)-std(P))× RMSD*	−0.718	0.709
$c_f = \frac{1}{n}\frac{\sum \lvert P-O \rvert}{\text{std}(O)}$	0.471	0.545
Average absolute error $E_{aa} = \frac{1}{n}\sum \lvert P - O \rvert$	19.4	21.3
Reliability index $I_r = \exp\left\{\sqrt{\frac{1}{n}\sum [\log(O/P)]^2}\right\}$	1.24	1.21
Modeling efficiency $M_e = \frac{\sum (O-\bar{O})^2 - \sum (P-\bar{P})^2}{\sum (O-\bar{O})^2}$	0.48	0.42

Also for strong winds, low turbidity areas may persist in some parts or corners of the lake, depending on wind direction.

However, no quantitative criteria exist to measure the accuracy of the spatial patterns in predicted SPM-concentrations. A one-to-one comparison at grid-level of instantaneous observations and predictions is generally very poor, as affected by small spatial or temporal shifts in data points, see Lynch *et al.* (2009) for further discussion. Assessment of model performance with respect to spatial distributions can therefore be done only qualitatively, by comparing patterns visually (Figs. 17.11 and 17.12).

Auto-correlation

Next, the auto-correlation r of the SPM-concentrations from the FL41 10-minute near-surface OBS-sensor and the model computations at that point and level have been established for the entire measuring period of 1/8/2007–1/9/2008, i.e. more than 37,000 data points. The auto-correlation

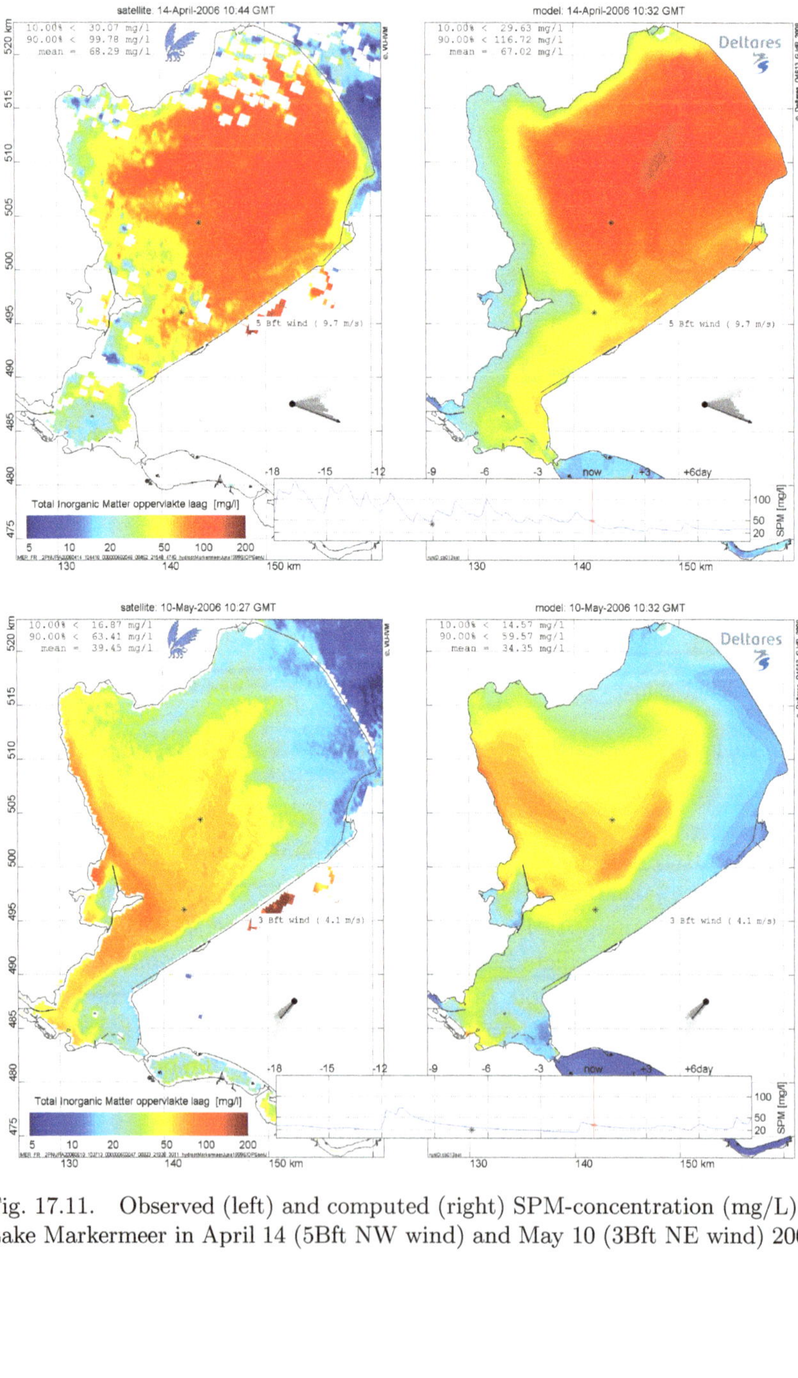

Fig. 17.11. Observed (left) and computed (right) SPM-concentration (mg/L) at Lake Markermeer in April 14 (5Bft NW wind) and May 10 (3Bft NE wind) 2006.

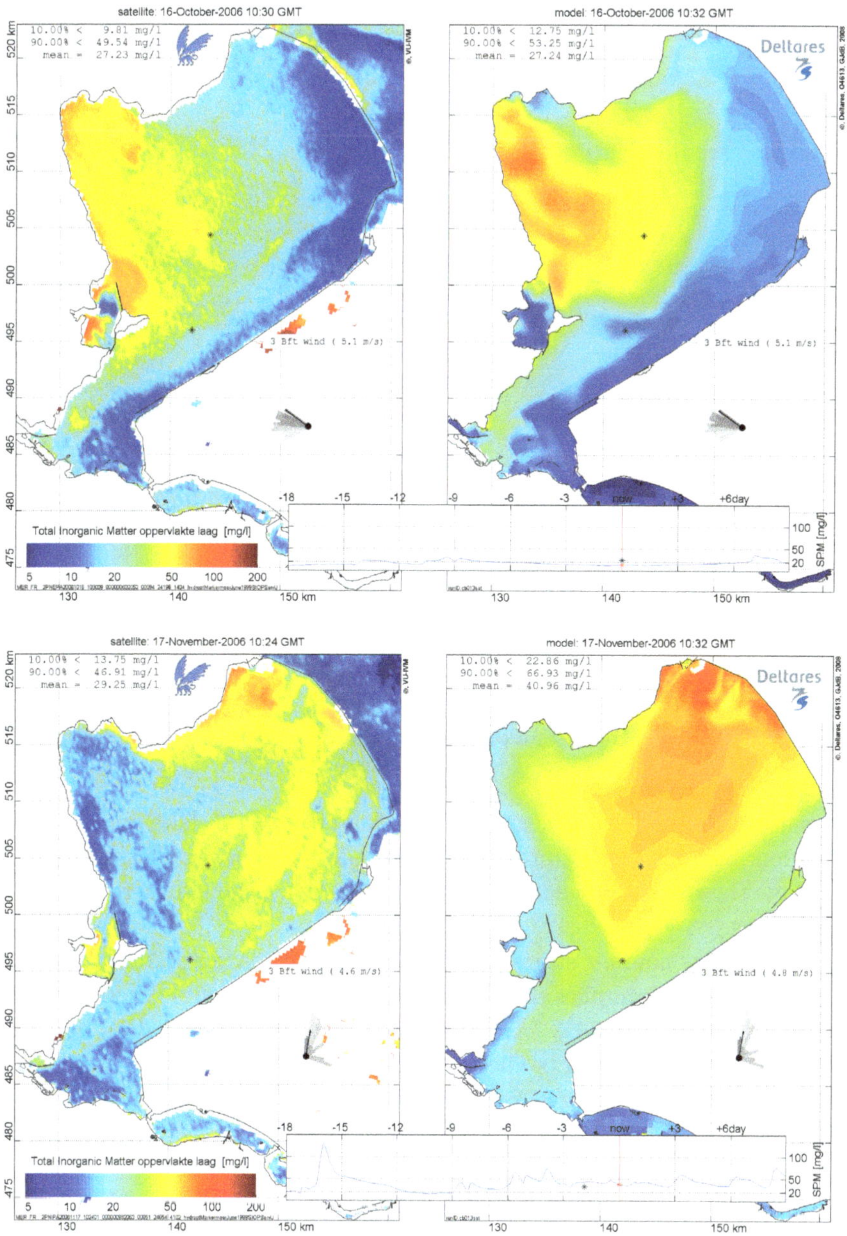

Fig. 17.12. Observed (left) and computed (right) SPM-concentration (mg/L) at Lake Markermeer in October 16 (3 Bft SE wind) and November 17 (3 Bft SSW wind) 2006.

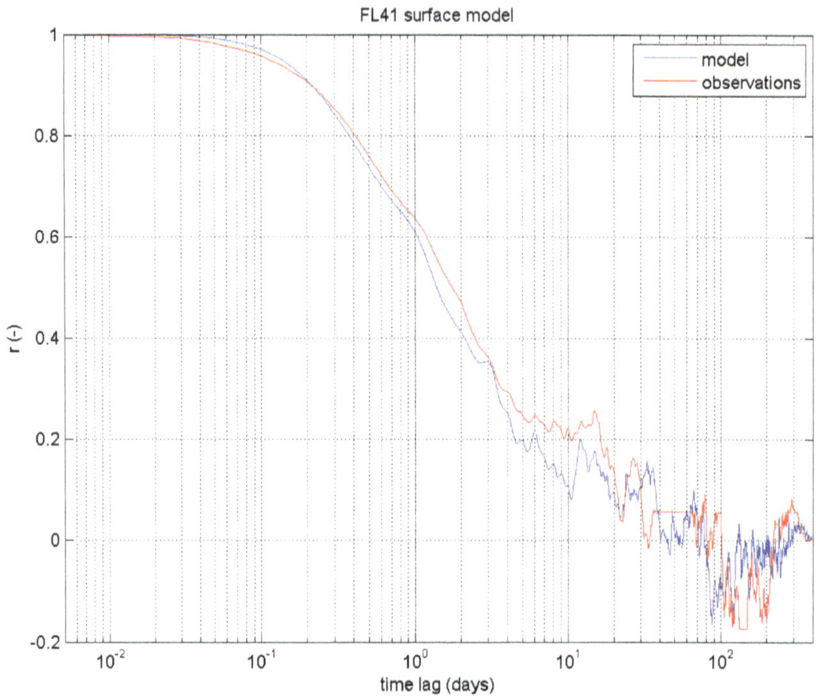

Fig. 17.13. Auto-correlation of observed and predicted SPM-concentrations.

can be interpreted as a measure of the memory of the lake at a specific location. The results are presented in Fig. 17.13, showing that the model predicts these memory effects in a satisfactory way, though there are some deviations.

After a few hours, r starts to deviate from unity. After one day, r is still about 60%, which implies that local SPM-values are for 60% determined by the conditions at the day before. After a week, model and data deviate: after ten days the model predicts r-values of about 10%, whereas the data suggest 20%. Integrating r over ten days yields $R(= \int_T r\mathrm{d}t) \approx 2$ days for the observations and 2.5 days for the model results. This can be interpreted as an overall 20% error in the model results.

After about one month, r becomes virtually zero, both for the FL42 data and model results. This implies that local SPM data are no longer correlated to the conditions one month earlier. One may conclude that, depending on the definition, the memory time of the lake amounts to a few days

($r = 0.5$) to one month ($r = 0$). This is consistent with the estimate of T_d above.

The auto-correlation remains almost zero for longer times, but becomes negative after about 100–200 days. This reflects seasonal effects, i.e. low SPM values in summer negatively correlate with high SPM values in winter. During spring and autumn, comparable SPM values are expected, which explains the irregular behavior of r between 80 and 200 days. After one year, r returns to zero, as expected.

Statistical analysis

All observations and predictions are summarized in the scatter plot of Figure 17.14, which shows that the majority of predictions are within a factor two off the observations, with a few outliers, though.

The statistical properties are given in Table 17.2 and Fig. 17.15. For a detailed discussion of these parameters, the reader is referred to Chung *et al.* (2009); Los and Blaas (2010); Lynch *et al.* (2009). However, the more more important statistical parameters are Bias* and RMSD′*, which score between "reasonable" and "good".

Based on the model validation, it is concluded that the Lake Markermeer model is well capable of capturing the essentials of the observed SPM dynamics. Computed SPM-concentration levels are globally in agreement with typical observed levels, while the observed temporal variability forced by wind is also reproduced satisfactorily. The computed equilibrium bed composition resembles the observed one, but only when a spatially varying bed roughness is used, based on the observed shell distribution. Apparently, the presence of mussels and old shell deposits influences erosion and deposition fluxes in the Lake Markermeer significantly.

These results have been obtained with a transparent calibration method linking parameter settings directly to target values for mean concentration, concentration peaks, etc. In this way, the huge parameter space of the model can be largely confined.

The model compares favorably against remote sensing images of 2006. For most images available, the model reproduces characteristic features of the SPM-concentration distribution. However, the computed concentration levels at low wind speed are substantially higher than suggested by remote sensing data. This is caused by limitations of the calibration data set, as such low concentrations do not occur in the 2007/2008 OBS calibration data at locations FL41 and FL42.

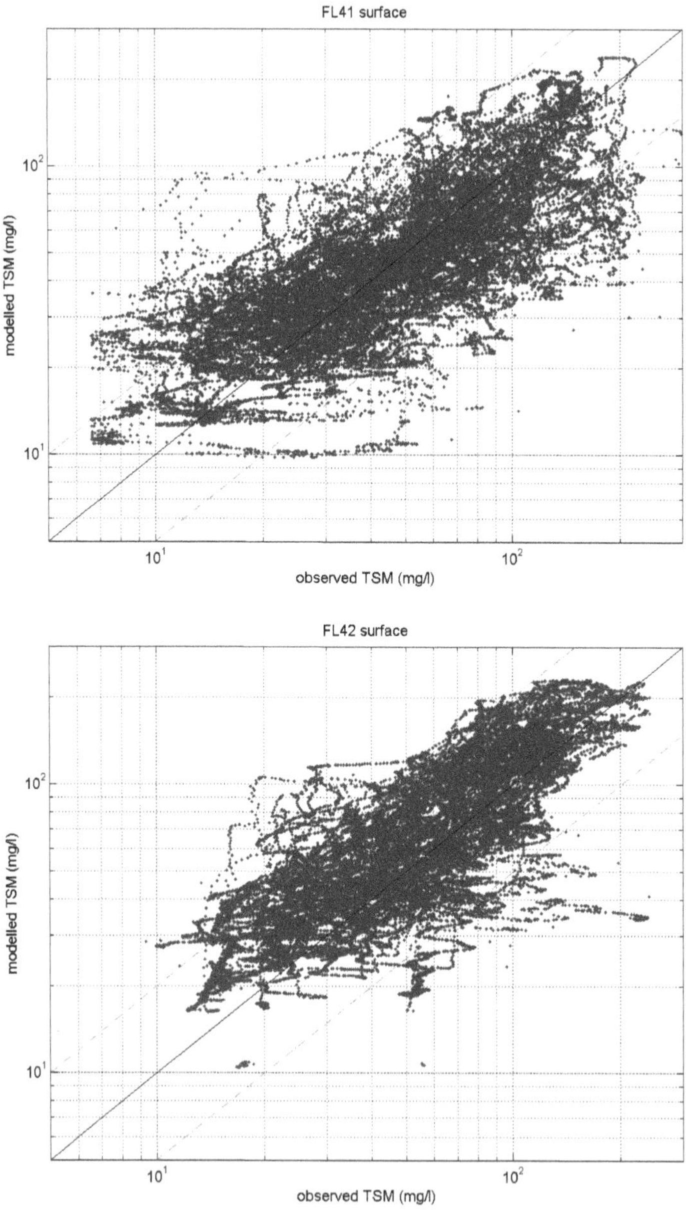

Fig. 17.14. Scatter plot of predicted versus observed SPM-concentrations. Top panel: station FL41 (37,571 points); bottom panel: station FL42 (45,385 points). For statistical values see Table 17.2. Perfect agreement is indicated by the solid diagonal, while dashed lines represent a factor two error.

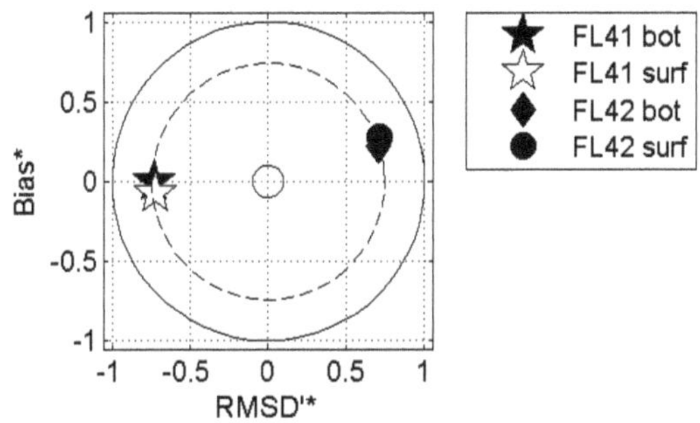

Fig. 17.15. Target diagram according to Los and Blaas (2010). The model performance on the dashed circle represents the boundary between "reasonable" and "good" performance. Performance outside the solid circle should be judged as poor.

17.5 Model application

The model described above has been applied to evaluate the response of the SPM-concentrations in the lake to measures/interventions, though at short time scales only, i.e. up to a few years (Boderie and Vijverberg, 2008; Noordhuis *et al.*, 2014).

However, at present the model is not calibrated to properly reproduce long-term trends in SPM. On longer term the active sediment mass or sediment properties is expected to change significantly, affecting model performance in a yet unknown way. Re-calibration would then be necessary, including additional processes which may influence sediment availability and properties (e.g. bioturbation of old deposits changing bed strength or algae composition changing floc properties such as settling velocity).

The present model is neither suitable to describe near-bed processes such as sediment-driven flow which may enhance sediment transport during energetic conditions, as discussed above.

An example of a short-term application is the analysis of the impact of an island archipelago in the eastern part of the lake. Such an archipelago blocks currents locally, and, more important, creates shadow areas for waves, resulting is less resuspension and more deposition at its lee side. As wind and waves may be from any direction, the year-average SPM-concentration is expected to decrease everywhere around the archipelago. This is quantified through simulation with the model, the results of which are presented in Fig. 17.16, showing the relative change in SPM-concentration by the archipelago.

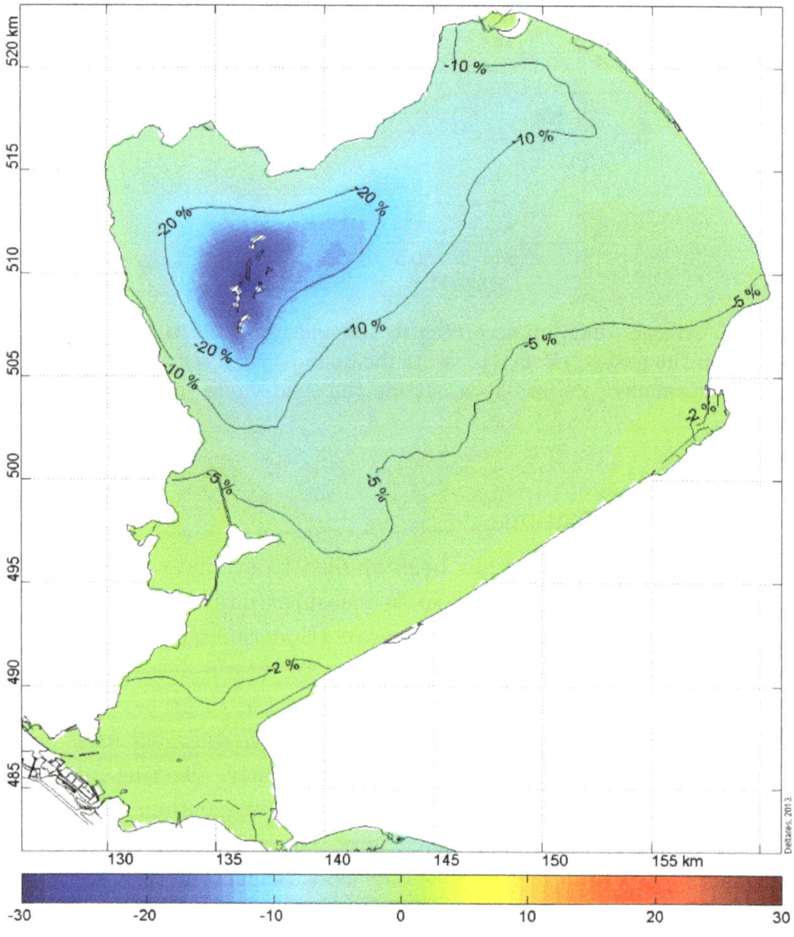

Fig. 17.16. Year-averaged relative change in SPM-concentration by an island archipelago in the eastern part of Lake Markermeer.

Figure 17.17 shows the concurrent change of net sedimentation/erosion around the archipelago. At most locations sedimentation occurs. However, near both edges of the archipelago, the bed erodes locally, owing to local flow contraction/acceleration, whereas wave sheltering is limited here. The overall effect on the SPM-concentration field is more smeared out than the bed response, because of the diffusive character of horizontal transport.

These simulations have contributed to the design of the "Markerwadden", a new man-made wetland in the Markermeer. Its main purpose is to

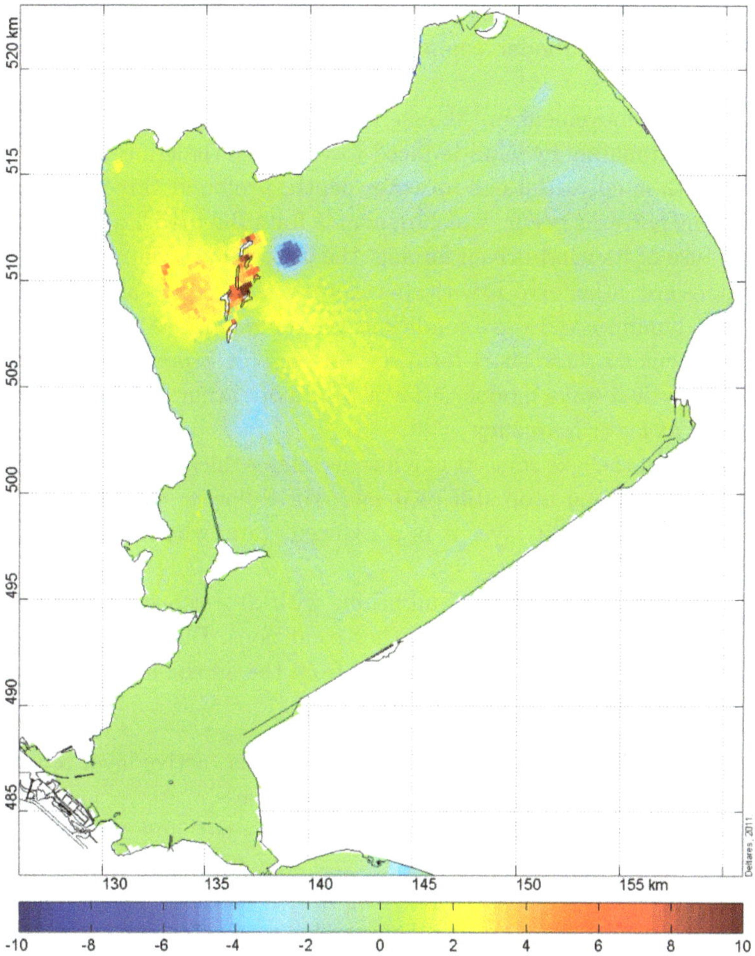

Fig. 17.17. Change (in kg/m^2) of net sedimentation (+) and erosion (−) over a year by an island archipelago in the eastern part of Lake Markermeer.

create more habitats for bird life, and reduce the high turbidity in the lake locally, improving the aquatic ecosystem as well.

17.5.1 *Crucial system understanding aspects*

This chapter describes the setup, calibration and application of an engineering model to address management questions related to the transport and fate of fine sediment in Lake Markermeer at larger spatial and temporal scales. The most crucial elements in our understanding of the lake system, accounted for in the model, comprise:

(1) The water movement in the lake is wind-driven, while bed sediment is mobilized mainly by wind-induced waves. Even though the lake is shallow with small variations in water depth, profound three-dimensional current patterns occur, which demands for a 3D model approach.
(2) Because of its wind-driven forcing, the water movement and wave conditions are rather erratic with short term variations — calibration of the hydrodynamics and wave conditions therefore requires ample attention, with focus on these short-term responses. This is possible by applying an analytical wave model, in favor of a more accurate numerical wave model at lower frequency.
(3) The bed of the lake consists of a few decimeter thick soft layer of mineral and organic mud atop stiff peat and stiff Holocene clay deposits. The majority of the soft layer is in a reduced state, whereas its upper few mm is oxidized.
(4) The oxidized layer is in equilibrium with SPM-concentrations in the water column during calm weather conditions. During a storm, more and coarser sediment is remobilized from the anoxic layer underneath. This behavior can be modeled with:

- The two-layer buffer model, where the upper, active layer represents the oxidized fluff layer of a few mm thickness,
- The use of two sediment fractions, one "coarse" and one "fine" fraction, and
- A simple "flocculation" model in which settling velocity increases with SPM-concentration, merely modeling grading than flocculation, though.

(5) From a hydro-sedimentological point of view, the lake should be regarded as a closed system. The empty bed approach is therefore not practical, as erosion rates of the stiff parent (peat and clay) bed are so slow that equilibrium requires at least about 10 year simulation time. The initial amount of mud in the full-bed approach is obtained from measurements. The small erosion of the parent bed is modeled with a user-defined source term.
(6) A 1DV calibration step speeds up convergence in the large parameter space, as the lake is shallow, thus on short term, SPM-concentrations are dominated by water-bed exchange processes.

(7) Observed spatial inhomogeneities can be predicted by increasing the bed friction coefficient locally, where relic shell banks occur on the lake bed.

(8) The relevant time scales and memory effects of the lake can be quantified with an auto-correlation coefficient for model and field data, which also provides metrics for the accuracy of the model.

Chapter 18

Modeling SPM in Harbor Basins: Port of Antwerp

18.1 Introduction

In response to ongoing growth in transoceanic trade, ships become ever larger, increasing in draft as well. This requires further deepening of fairways, and larger and deeper harbor basins. Deepening of navigation channels within estuaries often leads to tidal amplification and increasing tidal asymmetry, as well as enhanced salinity intrusion and salinity-driven gravitational circulation. These mechanisms bring fine and coarser sediment into the fairways and basins, requiring regular maintenance dredging to safeguard navigation. The costs and environmental impact of dredging motivate quantitative analyses of siltation rates, ranging from feasibility studies for projected ports to mitigating measures and sediment management strategies for existing ports.

Fairways in sheltered estuaries and harbor basins in coastal and estuarine waters generally suffer from siltation of fine sediments (Sec. 9.5), whereas channels exposed at open sea often suffer from siltation with coarser, sandy sediment (Sec. 10.5). This section focuses on siltation with fines in harbor basins, exemplified by the Deurganckdok in the port of Antwerp. This dock is situated along the left bank of the Lower Sea Scheldt River, was opened in 2005 and completed in 2008. It is the largest fully open container dock in the world. The theoretical background of port siltation (mechanisms driving exchange of water and sediment between a dock and adjacent basin) is described in Sec. 9.5, and not repeated here. The majority of the work in this section was carried out when the Deurganckdok was just opened, with the aim of designing measures to mitigate previsioned large siltation rates. As a result, hydrodynamic validation data for exchange flows was available, whereas information on siltation rates was not. The majority of the presented model results are based on Van Maren *et al.* (2011, 2009a), supplemented with siltation information IMDC (2013) which became available after completion of the modeling work.

Note that Chapter 15 also addresses the siltation rates in harbor basins along the Scheldt estuary. However, owing to the coarse model resolution, required to simulate the large spatial domain of the entire estuary, siltation rates could be predicted correctly only by the parameterization of some physical processes. However, that model was setup to simulate the large-scale impact of harbor maintenance and not the siltation processes themselves, treated in this chapter. Hence Chapter 15 and the present one support our arguments of Chapter 13 that model setup is strongly controlled by the questions to be addressed.

18.2 Conceptual model

18.2.1 *Purpose of the study*

The Deurganckdok was constructed to increase the capacity for container handling of the Port of Antwerp, Belgium. In contrast to existing docks, accessible through locks only, the entrance to the Deurganckdok has open access to the Lower Sea Scheldt. This is of course efficient for navigation, but unfavorable with respect to siltation. Likely, sedimentation rates in the dock are high, yielding high maintenance costs, as the dock is located at the river's Estuarine Turbidity Maximum (ETM). Near-bed SPM-concentrations in the ETM vary from several hundred mg/L up to 1 g/L (Chen *et al.*, 2005).

In order to minimize siltation rates, the port authorities investigated measures to reduce the influx of SPM into the dock. These measures included the layout of quay walls, a sill at the entrance of the dock, and the construction of a Current Deflecting Wall (CDW) (see Hofland *et al.*, 2001; Kuijper *et al.*, 2005; Winterwerp, 2005). A CDW aims to minimize sediment influx into a harbor basin by (1) deflecting sediment-laden near-bed water away from the dock entrance, and (2) reduce the strength of exchange flows driven by horizontal eddies and density-driven currents. The purpose of this study is to evaluate the effectiveness of a CDW, including several design alternatives, using a numerical model. This model needs to resolve the sediment dynamics on the Lower Sea Scheldt, and needs to be sufficiently refined to evaluate the effect of a CDW on siltation rates.

18.2.2 *Aims and research questions*

The aim of the study is to quantify the effect of CDW alternatives on siltation rates in the Deurganckdok. Research questions to be addressed are as follows:

(1) What are the sediment transport processes in the Lower Sea Scheldt, in particular in relation to water exchange mechanisms with the Deurganckdok (check differences with Chapter 15)?
(2) What are water and sediment exchange mechanisms between the Scheldt and the Deurganckdok?
(3) What is the effect of a CDW on siltation rates in the Deurganckdok, and how can its design be optimized with respect to these rates?

18.2.3 *System understanding*

Siltation in harbors is the result of the exchange of sediment-laden water between a harbor basis and its ambient waters, in combination with trapping within the basin. In tidal environments, water exchange is driven by horizontal entrainment, density-driven flow, and tidal filling and emptying (see Sec. 9.5). The relative importance of these mechanisms differs per harbor, and need to be identified for each individual site. Even more site-specific is the relation between these exchange mechanisms and sediment supply: water exchange mechanisms that are important during periods of large sediment supply dominate harbor siltation. The supply of sediment is driven by estuarine sediment transport mechanisms (Secs. 9.1 and 9.2) responsible for the formation and maintenance of the Estuarine Turbidity Maximum (ETM) in the Scheldt River.

The Scheldt River has a catchment area around 22,000 km^2, flowing through the north of France and Belgium before discharging into the North Sea through the Western Scheldt Estuary (see Fig. 15.1). The average freshwater discharge is around 100 m^3/s, with minimum and maximum values between 20 and 600 m^3/s (Fettweis *et al.*, 1998). The tides in the Western Scheldt are meso-tidal and semi-diurnal, and amplify up-estuary to a tidal range exceeding 5 m near Antwerp. Sediment near the Deurganckdok is both of fluvial and marine origin (Verlaan, 2000). About 0.75–2.2 Mton is annually supplied by the Scheldt River (Fettweis *et al.*, 1998). An additional amount of sediment originates from the North Sea and is transported up-estuary, mostly through tidal asymmetry. Most sediment accumulates in the turbidity maximum, located somewhere between the Deurganckdok and the city of Antwerp. The location of this turbidity maximum varies over the seasons. Close to the Deurganckdok, approximately one third of the sediment is of marine origin and two thirds of fluvial origin (Verlaan, 2000). However, marine suspended matter, imported from the North Sea, can reach much further upstream during low river discharge periods (Verlaan *et al.*, 1998).

The bed of the Lower Sea Scheldt channel is primarily composed of solid substrate (strongly consolidated tertiary clay, shell fragments and gravel, IMDC, 2013). Muddy sand and sand deposits are found in the inner bends and near sills; (sandy) mud deposits in lock basins. The ETM of the Scheldt hardly interacts with the bed (see also Chapter 15). This is illustrated by SPM observations collected in the Scheldt close to the Deurganckdok (Fig. 18.1) in conjunction with the flow velocities computed with a numerical model (Fig. 18.5); see also Chapter 15. During a flood, SPM-concentrations are highest along the left bank of the Scheldt (station Ka), whereas SPM-concentrations are highest along the opposite bank during ebb (station Kd). Peak flow velocities though, show an opposite pattern: at station Ka the highest flow velocities occur during ebb, and at station Kd during flood, especially close to the bed (see Fig. 18.5).

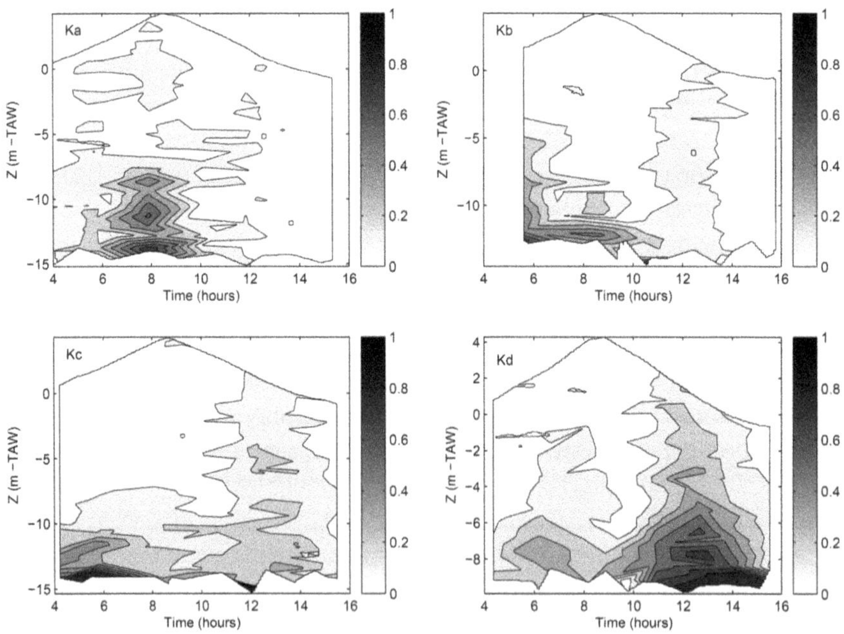

Fig. 18.1. SPM concentration [g/L] measured with the so-called siltprofiler at positions a–d of the K-transect locations near the Deurganckdok, three days after spring tide, on 16 February 2005. Location Ka is on the left bank (close to Deurganckdok) — see Fig. 18.2.

Understanding these sediment dynamics is crucial for predicting siltation rates and mitigating measures to minimize siltation in the Deurganckdok, which is located close to station Ka.

These observations imply that SPM transport in the Lower Sea Scheldt is mainly driven by horizontal advection, rather than by local resuspension. The spatial asymmetry in SPM and velocity is induced by differential advection (Sec. 9.1.2) in the river bend. Temporal variations in SPM are thus governed by the migration of the ETM over the tidal cycle, while lateral gradients are induced by differential advection (Winterwerp *et al.*, 2006a).

18.2.4 *Model requirements*

The setup of the numerical model is governed by the research questions to be addressed and our system understanding of the processes relevant for answering these questions (check difference with Chapter 15). This results in the following requirements:

(1) High vertical resolution is needed to resolve the effect of a CDW on vertical stratification and density currents by salinity and sediment. Turbulence needs to be modeled with a closure scheme resolving effects of density differences on vertical mixing, such as a k–ϵ model.
(2) The horizontal resolution in the river (see also the discussion in Chapter 15) and around the entrance of the Deurganckdok needs to be sufficiently detailed to (i) evaluate the effects of local measures to reduce siltation, (ii) model exchange of water and sediment through horizontal eddies and (iii) to simulate the patterns described in Fig. 18.1. The effects may be evaluated over relatively short periods, covering two to four spring-neap cycles.
(3) The sediment transport model should capture the ETM dynamics in the Lower Sea Scheldt. In particular, the dependency of SPM on tidal flow velocities is important as this controls the sediment availability at the entrance of the dock. This ETM is not in equilibrium with the bed (Chapters 1 and 15) but depends on convergence of sediment pathways. Such a model setup requires simulation periods of months to years to attain dynamic equilibrium.

Attaining dynamic equilibrium in SPM conditions in the Lower Sea Scheldt (requirement 3) requires long simulation times (typically several months to a year) which conflicts with requirements on vertical (requirement 1) and horizontal (requirement 2) resolution. These contradicting requirements can be resolved through the setup of a cascade of models at different resolution (nesting), as elaborated in the next section (see also Chapter 15).

18.3 Hydrodynamic model

18.3.1 *Grid design and boundary conditions*

Accurately resolving salinity-driven exchange flows requires a high vertical model resolution with 20 or more (equidistant) σ-layers. To simulate the expected sediment-induced density currents, an even higher resolution is required (*ca* 40 layers). Also the horizontal resolution needs to be high. The effects of relatively small variations in the geometry of quay walls and CDW on the exchange flows between dock and river have to be studied, requiring a horizontal grid size down to 20 m near the mouth of the dock.

The model domain covers approximately 60 km of the Western Scheldt estuary and Lower Sea Scheldt (Fig. 18.2). This ensures that, when migrating, salinity intrusion and ETM remain within the model domain (i.e. two tidal excursion). Simulations with the vertical and horizontal resolution mentioned above can be executed for a maximum of two to four spring-neap cycles (resulting in computational times of 1–2 weeks).

Long-term sediment dynamics in the Lower Sea Scheldt are simulated with a second model that is three times coarser in horizontal resolution (Fig. 18.2), reducing computational times by a factor 27 (3×3 for horizontal resolution, and a third factor three from the larger time step). Thus sufficiently long simulation periods to attain a dynamic equilibrium of SPM-concentrations become feasible. The computed SPM and bed-sediment distribution are subsequently transferred to the high resolution model to quantify the exchange of sediment between the Lower Sea Scheldt and the Deurganckdok with a hot start of the fine-mesh model.

18.3.2 *Calibration and validation*

The coarse grid Lower Sea Scheldt model is first calibrated against water levels, which was straightforward and therefore not shown here. Exchange flows between dock and river are sensitive to the velocity distribution on the river. The main focus during model setup was therefore on the velocity distribution between the Lower Sea Scheldt and the Deurganckdok. For this purpose ADCP-measurements were carried out in the entrance of the dock (Fig. 18.2), the results of which are used for model validation.

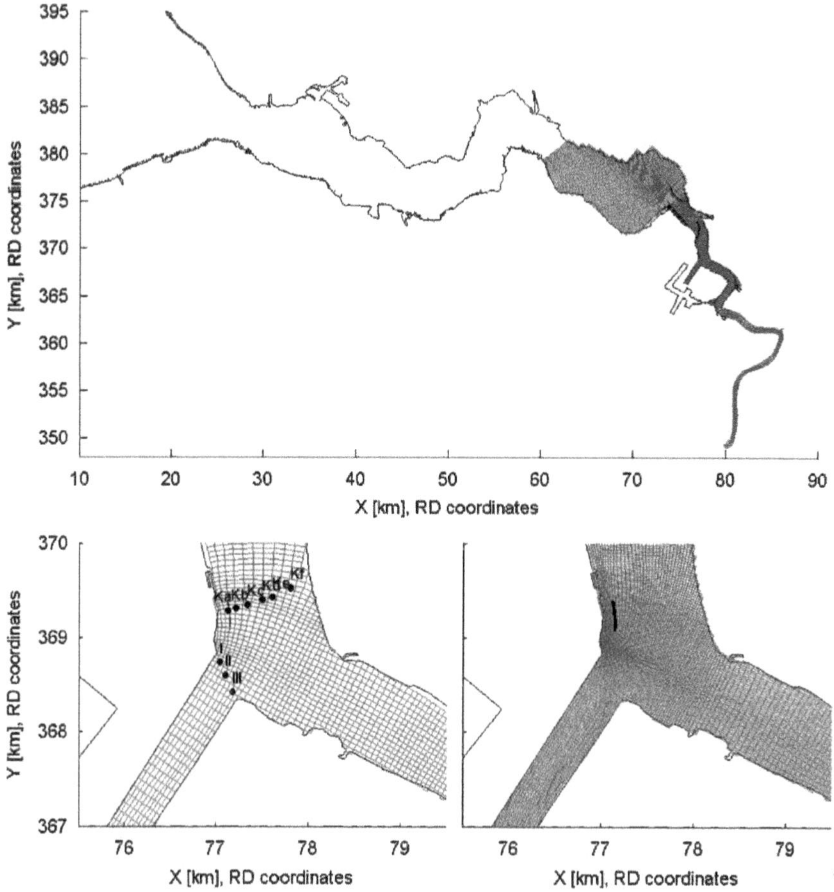

Fig. 18.2. Model layout with overall numerical grid (top panel), a detail of the coarse model grid (lower left panel, with locations and names of output stations) and the fine grid (lower right panel) near the Deurganckdok. The CDW is drawn in the lower right panel as a thick black line near Ka-station.

The combination of horizontal and vertical circulations governing the exchange of water between dock and estuary, conceptually visualized in Fig. 9.30, results in a complex distribution of inflow and outflow in the entrance of the dock. The ADCP-measurements (Fig. 18.3) reveal that for a period of 4 hours ($t = 14$–18 hrs), the near-surface and near-bed currents are in opposite directions. Phasing as well as the magnitude of this two-layer flow structure is well reproduced by the model.

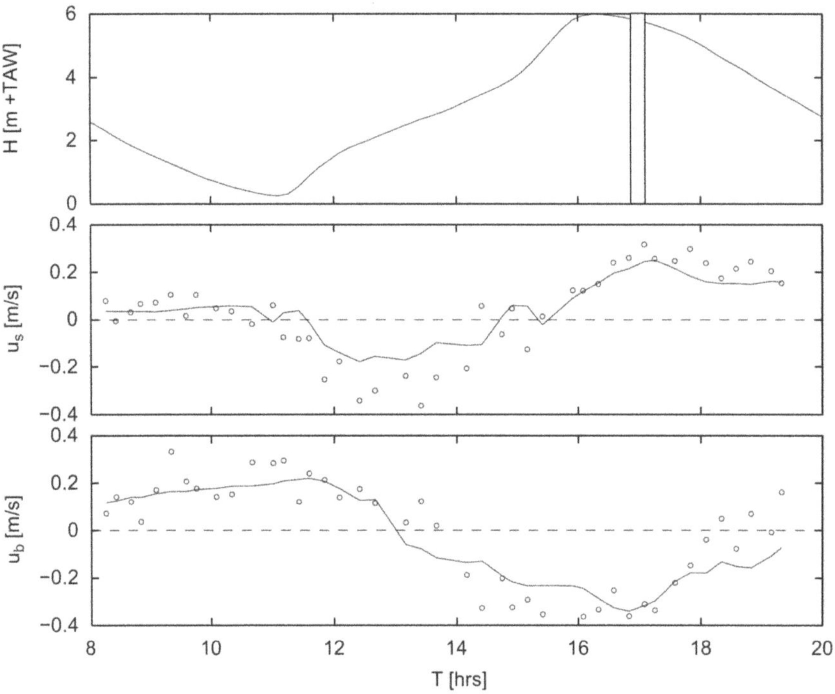

Fig. 18.3. Water level (top panel, with the gray box indicating the time of Fig. 18.4). Lower two panels: measured (circles) and computed (solid line) current velocity [m/s] into (negative) and out of (positive) the Deurganckdok, near-surface (u_s, middle panel) and near-bed currents (u_b, lower panel), at location III (see Fig. 18.2 for location and Fig. 18.4 for vertical position), on 17 November 2005.

The vertical flow structure during a period of maximum density differences ($t = 17\,\text{hrs}$) is examined in more detail in Fig. 18.4. Water flows into the dock (negative velocities) over the larger part of the water column at the SE side of the dock ($x = 500\,\text{m}$) whereas outflow prevails at the NW side of the dock ($x = 0\,\text{m}$). This is characteristic for horizontal circulation cells (Fig. 9.30(a)) driven by the flood current. Superimposed on these horizontal exchange currents are density driven flows, resulting in near-bed inflow and near-surface outflow. Halfway the cross-section inflow and outflow are evenly distributed over the vertical, but at its NW and SE edges strongly influenced by the horizontal circulation cells described above. This complex pattern is well reproduced with the model.

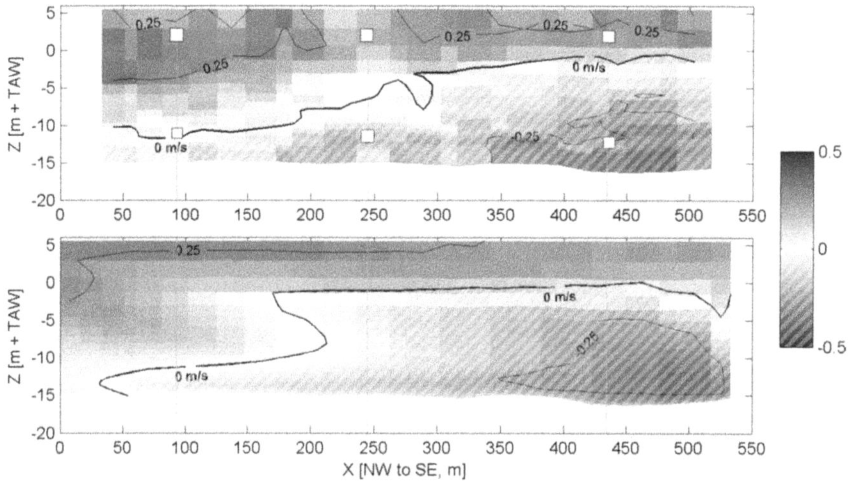

Fig. 18.4. Measured (top panel) and computed (lower panel) current velocity (m/s) into (negative, shading with diagonal lines) and out of (positive) the Deurganckdok, on 17 November 2005. $x = 0$ corresponds to the NW and $x = 500$ to the SE part of the entrance. Location III (Fig. 18.3) is the white square at $x = 440$. See the gray shaded area in Fig. 18.3 for the phasing within the tidal cycle.

18.3.3 *Analysis hydrodynamics*

The computed flow velocities during the SPM observations in a cross-section just down-estuary of the dock (Fig. 18.2) are visualized in Fig. 18.5. The flow velocities in front of the Deurganckdok (station Ka) are strongly asymmetric, much larger during ebb than during flood. In contrast, the sediment concentrations are highest during flood (Fig. 18.1). This is induced by the migration of the ETM over the tidal cycle, with lateral gradients induced by differential advection (see Sec. 18.2.3).

Highest SPM-concentrations at the entrance of the dock occur from 2 hr before to 2 hrs after HW. This occurs when the two-layer exchange flow is maximal (Fig. 18.3), while horizontal circulations are also pronounced (Fig. 18.4). Both mechanisms promote sediment fluxes into the dock, as turbid near-bottom water enters the dock through density currents and horizontal circulation. The period around HW is therefore a crucial period, in which (1) the sediment influx is high, and (2) the effect of a CDW might be large (see Sec. 18.2.1).

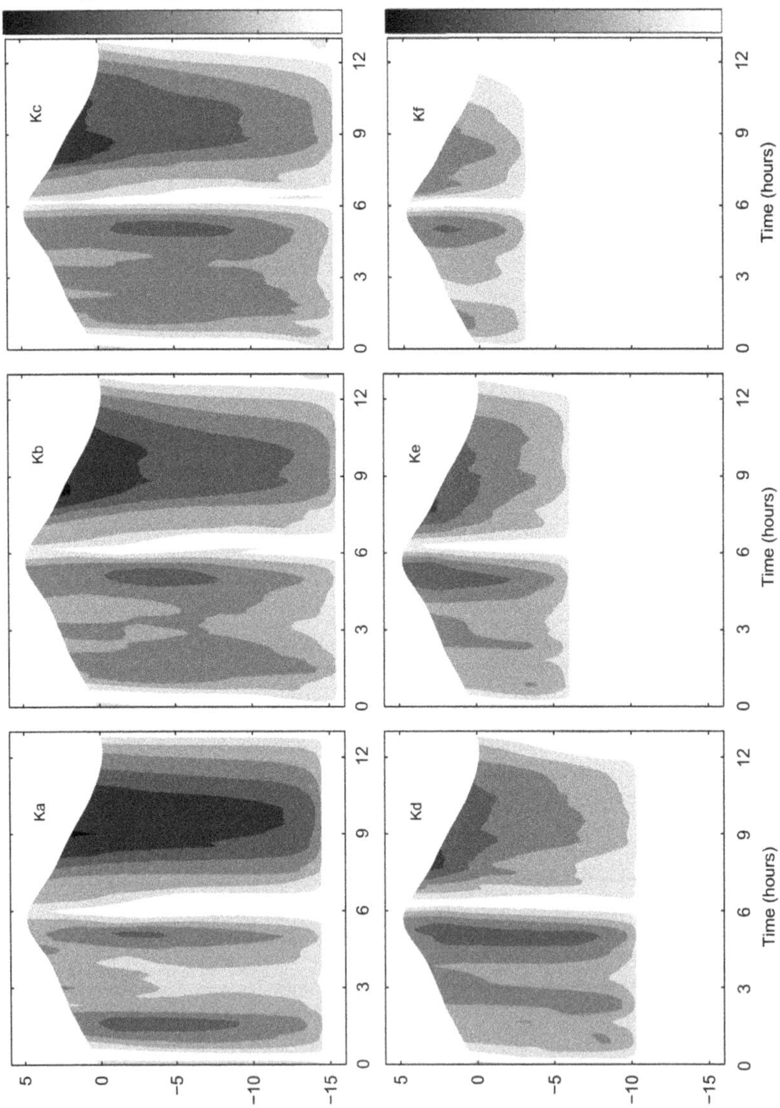

Fig. 18.5. Flow velocity (m/s) computed with the hydrodynamic model at positions a–f along the K-transect near Deurganckdok, on 16 February 2005. Location Ka is near the left bank (close to Deurganckdok) and Kf near the right bank (Fig. 18.2).

18.4 Sediment transport model

18.4.1 *Sediment fractions, properties and boundary conditions*

The sediment transport model is setup for non-equilibrium bed and supply-limited conditions (Chapter 1), further to the system understanding above. This implies an empty-bed approach (Sec. 12.3), where SPM enters the model domain through its boundaries while no sediment is prescribed on the bed, as in Chapter 15. Computed SPM-concentrations attain a dynamic equilibrium when the net SPM fluxes through the model boundaries equal siltation rates in low-energy environments, such as the Deurganckdok. SPM-concentrations then fluctuate around a mean value over a (spring-neap) tidal cycle. These dynamic equilibrium conditions are established with the coarse-grid model.

At the down-estuary model boundary (Fig. 18.2), a constant SPM-concentration of 100 mg/L is prescribed, whereas at the up-estuary, riverine boundary, 210 mg/L is prescribed. To account for flocculation processes in the river, the parameterized flocculation model (4.31) by Winterwerp *et al.* (2006a) is used, yielding variations in w_s as a function of SPM-concentrations and turbulence levels (flow velocity) — see Fig. 18.6. This algebraic model accounts for higher settling velocities at increasing sediment concentration, particularly at intermediate turbulence levels, which is in agreement with observations.

The deposition flux D is modeled as the product of the near-bed SPM-concentration c_b and w_s (Sec. 4.2). The erosion flux E is modeled with $E = M_E(\tau_b - \tau_e)$. Note that because of the short simulation times, we do not need to use the two-layer buffer model as in Chapter 15. As elaborated in Sec. 18.2.3, sediment in the ETM interacts little with the bed. Therefore the critical shear stress for erosion τ_e is set to a low value (0.05 Pa). The calibration therefore consisted of varying M_E until settings are obtained for which (1) sediment converges in the ETM zone, and (2) the sediment concentration levels correspond to observations. Note that this approach parameterizes the fluff layer dynamics described in Sec. 6.5.

18.4.2 *Phenomenological calibration*

Next, the sediment transport model is calibrated qualitatively against observed sediment deposition patterns and fluxes (both not shown here) and SPM-concentrations measured at the locations Ka–Kd. The main calibration criterium hereby is the lateral and temporal variability of the

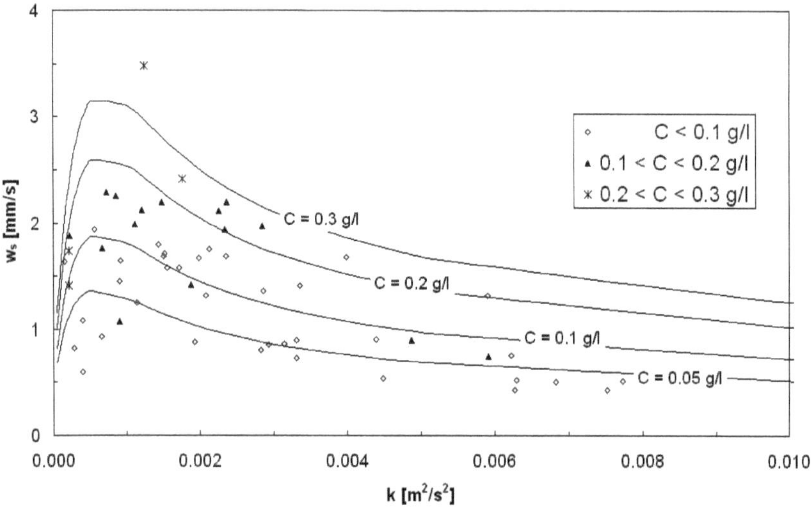

Fig. 18.6. Settling velocity w_s as a function of turbulent kinetic energy k for a variety of SPM-concentrations (after Winterwerp *et al.*, 2006a). The settling velocity is highest at (1) intermediate levels of turbulent energy, and (2) at high SPM-concentrations (see Van Maren *et al.*, 2011, for details).

SPM-concentration. The key parameter is the SPM distribution at station Ka around HWS, as then near-bed exchange flows and horizontal circulations promote sediment transport into the Deurganckdok (see Sec. 18.3.3). Comparing Fig. 18.7 with Fig. 18.1 shows that the model captures the following features:

(1) The tidal asymmetry in SPM-concentrations along the left bank of the Scheldt (station Ka), with the higher concentrations during flood, despite the larger peak ebb velocity, see Fig. 18.5.
(2) The large vertical sediment concentration gradient with near-bed SPM-concentrations of about 1 g/L to very low concentrations halfway the water column.
(3) Well-mixed SPM-concentration profiles at stations Kb–Kd.
(4) A pronounced tidal asymmetry in SPM-concentrations (maximal during ebb) at station Kd, despite symmetric ebb and flood currents, see Fig. 18.5.

This agreement shows that the complex dynamics of the ETM, including the lateral variability in SPM, driven by complex cross-sectional residual

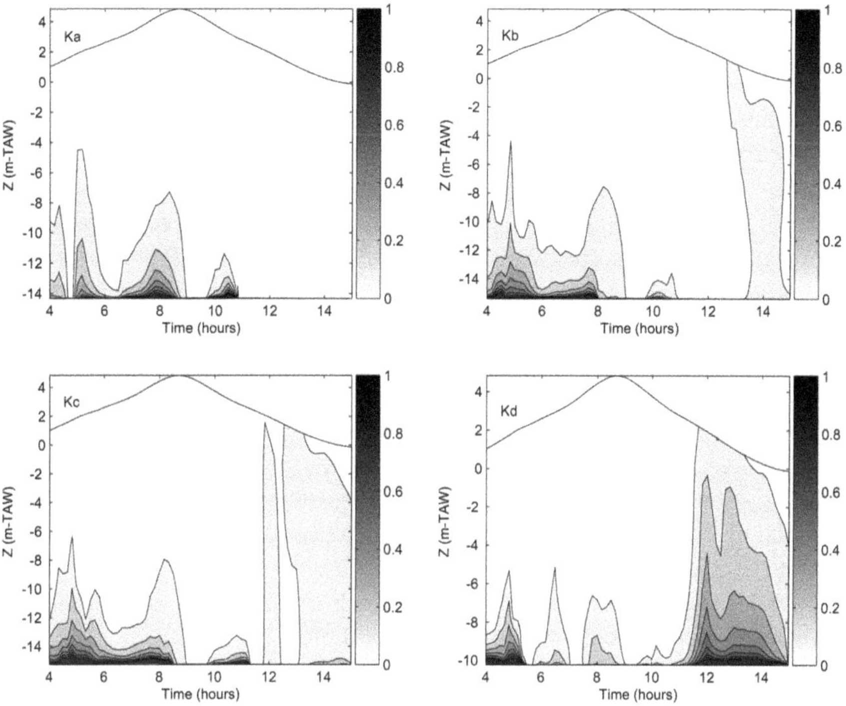

Fig. 18.7. SPM-concentration [g/L] computed at locations Ka–Kd (see Fig. 18.2).

flows, can be represented by a model in which (1) the vertical and horizontal structure of salinity and velocity is resolved at high resolution, and (2) sediment interacts little with the bed. The developed model can predict the exchange flows between the Scheldt River and the Deurganckdok, and the phasing of sediment availability near the entrance of the Deurganckdok. In a next step, the model is used to assess the effects of mitigating measures.

18.5 Model application

The observation that the phasing between water exchange and sediment availability is crucial for predicting harbor siltation in fresh-salt water mixing regions agrees with earlier findings by De Nijs *et al.* (2009) and Van Maren *et al.* (2009a). An important implication of this conclusion is that only complex 3D sediment transport models can be applied to analyze mitigating measures. Here we use the high resolution 3D model (using the coarse model output as initial conditions) to evaluate several CDW configurations. For this purpose, a CDW is implemented in the numerical model.

As explained in Sec. 18.2.1, a CDW is designed to deflect near-bed sediment-laden currents away from the dock and near-surface sediment-poor currents into the dock (to maintain the water balance). In the current model study, a CDW is schematized through local modification of the bathymetry, a guidance wall, and an additional energy loss term in the momentum equation (see Van Maren *et al.*, 2011 for details). The CDW is constructed down-estuary of the Deurganckdok (see Fig. 18.8 for location).

The reference model predicts siltation rates of 0.94 Mton/yr (Table 18.1). This compares favorably with the observed siltation rates in the period 2008 (completion of the Deurganckdok) through 2009 (1.0 Mton/yr — see Table 18.1 and IMDC, 2013). Note that these observations were not yet available during the setup and analysis of the models — the predicted siltation rates were solely based on a detailed reproduction of hydrodynamics and thorough analysis of sedimentary processes, i.e. our system understanding.

When designing mitigating measures, it is important to understand the mechanism responsible for siltation. The sediment concentration cannot be decomposed into a contribution by density-driven flow, horizontal eddies, and tidal filling as is common for the exchange flow. Therefore, the sediment flux cannot be decomposed in a straightforward way, as the exchange flow itself. As an alternative, the combined flux resulting from horizontal eddies and tidal filling/emptying $F_{t,h}$ is defined as the tide-integrated product of the depth-mean SPM-concentration $\overline{c}$ and velocity $\overline{u}$; the contribution by density currents F_d is defined as $F - F_d$ (Van Maren *et al.*, 2009a). Using this simplified approach, the estimated contribution of density-driven flows to the total sediment flux is about 70%.

The effect of various CDW designs was subsequently assessed with the model. The so-called design CDW was a construction optimized in a physical scale model (Van Kessel and Cornelisse, 2003). The numerical model suggests that this design CDW leads to an 18% reduction. Fig. 18.8 presents snapshots of near-bed SPM-concentrations 1 hour before HW (the period of maximum sediment availability), showing that the CDW deflects turbid near-bed water away from the dock. The effect of two CDW alternatives were additionally evaluated using the numerical model: one with a wider channel in between the CDW and river bank, and one with a lower sill underneath the CDW. Both yield lower efficiency, i.e. reductions in siltation rates by

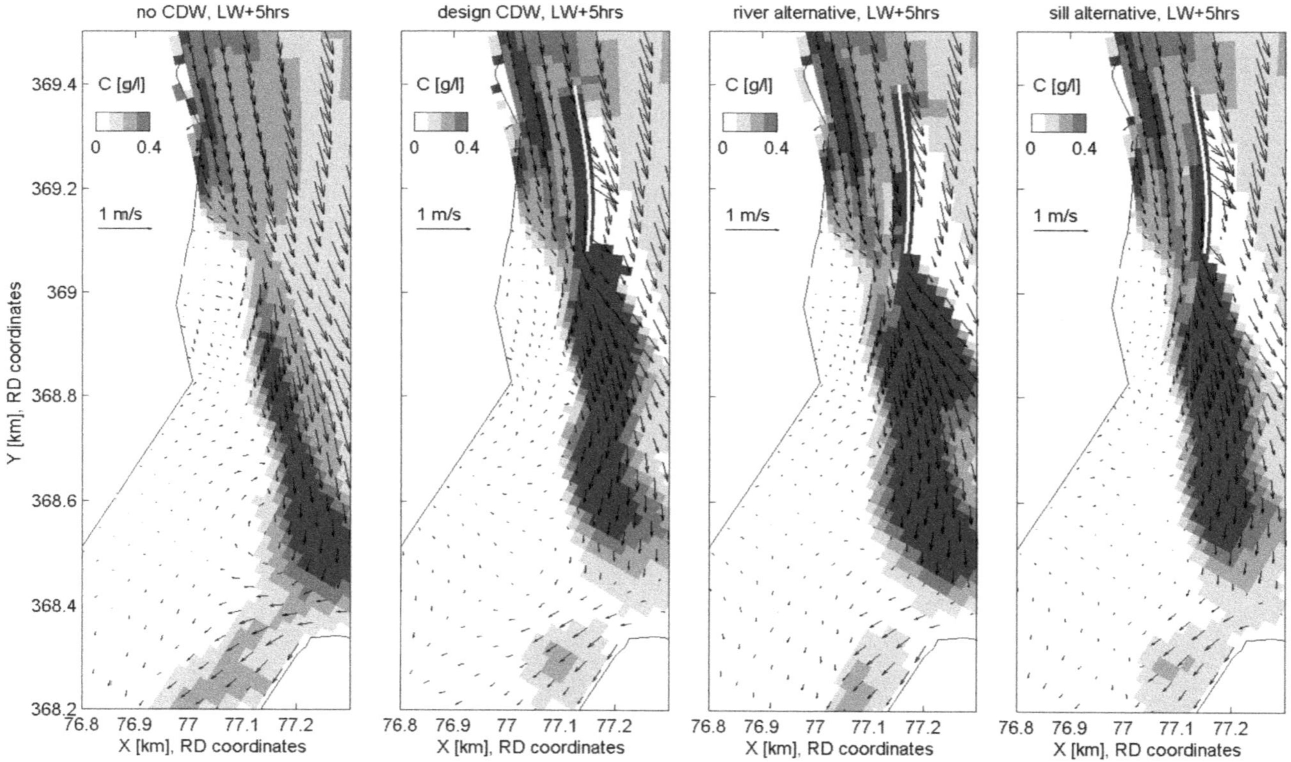

Fig. 18.8. Near-bed SPM-concentration (gray tones) and velocity vectors around the entrance of the Deurganckdok, 5 hrs after LW, without CDW and with 3 CDW alternatives.

Table 18.1. Sedimentation in the Deurganckdok (absolute sedimentation in million ton/year): observed (2008–2009), from IMDC (2013), predicted without CDW and for three scenarios with CDW (the design CDW, alternative 1 (widened CDW channel), and alternative 2 (lowered CDW channel). The predicted change in siltation relative to scenarios without CDW "effect CDW") and relative to the design CDW (effect CDW alternative) are given in %.

	Sedimentation [10^6 ton/year]	Effect CDW [%]	Effect CDW alternative [%]
Observed 2008–2009	1.0	–	–
Predicted, no CDW	0.94	–	–
Predicted, design CDW	0.77	−18	–
Predicted, CDW alt. 1	0.80	−15	+4
Predicted, CDW alt. 2	0.79	−16	+2

15% and 16%, respectively, implying 4% and 2% more sedimentation than the reference design CDW.

Partly based on these numerical model results, the design CDW was constructed in 2010 down-estuary of the DGD. Unfortunately, a direct comparison between observed siltation before and after construction of the CDW is obscured the high natural variability in siltation rates. Hence a two-year period for assessing baseline conditions without CDW is short. Moreover, an increase in maintenance depth of the fairway was implemented as well. Observations before and after construction of the CDW (IMDC, 2013) reveal a substantial decrease in the strength of the horizontal eddy and in the near-bed sediment concentration of (0–30%), despite comparable SPM-concentrations on the Scheldt River itself.

Summarizing, this study provided the following information to answer the managerial questions formulated in Sec. 18.2.2:

(1) Sediment transport in the Lower Sea Scheldt is supply-limited: SPM-concentrations are not in equilibrium with local flow velocities, but determined by their availability in the water column. This implies that SPM dynamics are controlled by advection, lateral flows and residuals, rather than by peak flow velocities.
(2) Exchange of water between the Lower Sea Scheldt and the Deurganckdok is primarily driven by horizontal entrainment (eddies) and density-driven vertical exchange flows. The highest SPM-concentrations on the Lower Sea Scheldt synchronize with near-bed and horizontal entrainment directed into the dock.

(3) It is possible to simulate the effects of a Current Deflecting Wall with a numerical model, as shown by comparison with physical scale model results.
(4) The designed CDW can deflect near-bed flows away from the Deurganckdok, resulting in reductions in siltation rates between 15% and 18%. Model results suggest that the efficiency of the CDW was not very sensitive to small changes in the designs studied.

18.5.1 *Crucial system understanding aspects*

This chapter describes the setup, calibration and application of an engineering model to establish the siltation rate with fine sediment in the Deurganckdok, a harbor basin along the Scheldt estuary, and to analyze possible mitigating measures to reduce maintenance of the dock. The most crucial elements in our understanding of the local sediment dynamics, accounted for in the model, comprise:

(1) In this chapter, fine sediment dynamics around a harbor basin are studied, at scales substantially smaller than the overall Scheldt model presented in Chapter 15. This sustains our arguments that the setup of a model strongly depends on the questions to be addressed, as outlined in Chapter 13.
(2) The siltation rates in a harbor basin along an estuary is governed by four parameters, their relative contribution being site-specific:
    - water exchange between river and dock by tidal filling,
    - water exchange between river and dock by horizontal circulations,
    - water exchange between river and dock by density currents (salt-fresh water- and/or sediment-driven),
    - the availability of sediment, and its phasing with the water exchange mechanisms.
(3) SPM-concentrations in the Scheldt River adjacent to the dock reveal profound asymmetric patterns, largely driven by differential advection in the sinusoidal river. During flood SPM-concentrations are highest along the left bank of the river (thus near the mouth of the dock), whereas during ebb, highest SPM-concentrations are found along the opposite, right bank.
(4) For the Deurganckdok system, salt-fresh water induced density currents in conjunction with the asymmetry in SPM-concentrations are responsible for about 70% of the siltation rates in the dock.

(5) Deflection of the sediment-laden near-bed flood current would be an efficient measure to reduce these siltation rates.
(6) Resolving the asymmetry in SPM-concentrations, the salt-fresh water induced density currents, and deflection of near-bed currents requires a high-resolution model, which prohibits deployment of a large spatial model and large simulation times. The fine-grid dock-model is therefore nested within a larger-scale model, which provides initial and boundary conditions for the dock-model.
(7) This model application illustrates the arguments of Chapter 13 that a model setup is strongly governed by the questions to be addressed, leading to rather large differences with the Scheldt model presented in Chapter 15.

Appendix A

Measuring Fine Sediment Properties

In this appendix, we describe procedures for determining the mud properties necessary for building system understanding (e.g. Chapter 13) and input to numerical models (e.g. Chapters 14–18), with emphasis on fine sediment dynamics in the marine environment. The reader is also referred to Winterwerp and Van Kesteren (2004), Appendix C.

Standards for measuring sediment properties have been developed in many countries, based on work in soil mechanics, sedimentology, etc. Unfortunately, these standards are not identical everywhere, as f.i. in defining particle size (see Winterwerp and Van Kesteren, 2004). In the following we will refer to the origin of these standards where relevant. However, in a number of cases the procedures proposed in this appendix have not been standardized, or international standards are not adequate for our purpose, as discussed below.

A.1 Sampling and storage

The first step in measuring sediment properties in the laboratory, of course, is collecting the samples. We distinguish between water and bed sampling, the latter disturbed or undisturbed, depending on the type of analyses to be be performed:

(1) Bed/soil samples
   (a) undisturbed for studying stratigraphy, sedimentology, etc.,
   (b) disturbed for analyses in this appendix.
(2) Water samples for determining SPM values and bed sample dilution for laboratory analyses.

Bed sampling can be done under water, or above water on e.g. intertidal mud flats. The latter is often easy, but not always, in particular when such flats are large and soft. Moreover, these are generally subject to overconsolidation. Undisturbed bed samples are taken through piston coring, box corers, etc. (Van Rijn, 1986). Water samples should be taken isokinetic, or in still water. Isokinetic sampling implies siphoning sediment-laden water from the water column at the local water velocity so that inertia cannot result in over- or undersampling, as explained in detail by Van Rijn (1986).

Note that the procedures below require mixing of the sediments, sometimes diluting with (local) water — these procedures of course do not require undisturbed sampling. Disturbed samples can be taken with a Van Veen grab (Van Rijn, 1986), provided no pore water is lost — which is straight forward if the mud is not too watery. In general, grabs of a few liter volume are preferable, though smaller grabs may be more practical. While sampling, other physical parameters have to be established as well, such as flow velocity, wave height, salinity and water temperature, as well as the weather conditions. Often, such conditions prior to sampling have also to be known.

Chemical pore water properties have to be measured as soon as the samples are on board, i.e. pore water salinity S, acidity pH, redox potential Eh and pore water temperature T. Note that Eh may change rapidly if the sample oxidizes above water. Often it is advisable to bring local water as well for diluting the bed samples in the laboratory.

The water and sediment samples should be stored away from sunlight, preferably in dark containers at low temperature minimizing algae/bacterial growth. When stored properly, analysis of the samples is not urgent, and may be postponed a few weeks if conditions require. However, it is always better to carry out the analyses as soon as possible after *in-situ* sampling. This is in particular the case when organic matter analysis is required. The samples then have to stored at low temperature as well (preferably at about 4°C).

A.2 Bulk and dry density, organic and carbonate content

The first step in assessing sediment properties is measuring the bulk density ρ_b of the soil samples. The samples are mixed with the pore water that may have egressed during storage. The procedure is then straightforward: the wet sample is placed in a container/beaker of known volume, and from the container's weight, the bulk density of the soil follows.

Next, a sub-sample is dried at 105°C for at least 12 hours. This subsample is again placed in a container of known volume and weighed, from which the soil's dry density ρ_{dry} follows. Note that this dry density may have to be corrected for salts, if the salinity of the pore water is significant.

Archimedes' method may be used for instance in the field if no lab facilities are available. A transparent measuring cup/column is partly filled with local water, and its weight is measured (cup + water). Then a certain amount of wet sediment is added to the cup, the difference in volume and the new weight (cup + water + sample) is measured. Some simple algebra gives a good estimate of the sediment bulk density.

In soil mechanics, also water content W and, void ratio e and porosity n are used, which all relate to ρ_{dry} and ρ_b, following the definitions in Chapter 2. For determining these soil mechanical parameters, the specific density ρ_s of the minerals has to be known as well.

The specific density ρ_s of quartz sediment is about 2650 kg/m^3. Yet, clay and silt minerals, and some sands as well, may have different specific densities, and it is recommended assessing the specific density of the minerals at hand, in particular when the site is new to an engineer/scientist. The specific density ρ_s of granular sediments is measured with a pycnometer. The procedure, and the instrument itself are standardized (e.g. ASTM D854-14, ISO 11508, NEN 5111, DIN 18124 and BS1377-Part2-8.3).

The total organic content TOM consists of an organic and an inorganic part. For the analyses in this book, one does not need to distinguish, and we discuss assessing total TOM only. Basically there exists two methods:

(1) removal of TOM by heating,
(2) removal of TOM by oxidation.

Removal of TOM by heating is standardized in e.g. ASTM 2000, NEN 5756. The samples is oven-dried at 350–440°C; not higher as then also carbonates will be lost. Weighing the sample before and after treatment yields the TOM. A disadvantage of this method is that clay minerals may dissociate at these high temperatures (thus we recommend the lower value of 350°C, minimizing dissociation).

To the knowledge of the authors of this book, the second method (oxidation) is not standardized. However, it is widely used, as it is simple and does not dissociate clay minerals. A sample is treated with hydro peroxide (H_2O_2). H_2O_2 is continually added until frothing stops. This method is less accurate than heating a sample, but often sufficient for our analyses.

Calcium carbonates refer to carbonate minerals, such as calcite ($CaCO_3$), magnesite ($MgCO_3$) and dolomite ($CaMg_2CO_3$). Determination of the carbonate content is done through oxidation of an oven-dried soil sample (105°C) with hydrochloric acid HCl. The volume of escaping CO_2 is measured. This method is described in detail in ASTM D4373. As the method is quite laborious, various commercial instruments are available on the market.

A.3 Particle size distribution

Particle size classifies sediment type, such as clay, silt and sand. However, different classifications are used in the various standards across the world. For instance, the transition between silt and sand is set at 76 μm in the USA (ASTM D422), whereas in The Netherlands (NEN5104) and Great Britain (BS1377) the transition is set at 63 and 60 μm, respectively. The latter two standards define clays as particles with a diameter $< 2 \mu$m, whereas ASTM D422 distinguishes between clays ($< 5\,\mu$m) and colloids ($< 1\,\mu$m). In the following, we follow Dutch definitions, but this is not crucial for our discussions.

In this appendix, we focus on the finer fraction ($< 64\,\mu$m) only. This fraction is obtained by wet sieving, using standardized sieves, splitting a sample in a finer and a coarser fraction. The particle size distribution (psd) of the coarser fraction is often determined by dry sieving of the dried coarser fraction of the sample (consult various standards). The finer fraction is never to be dried for determining its psd.

Note that this classification of sediments refers to size only, whereas many clay minerals are sized up to the silt range, while quartz may be in the clay size (see Chapter 2 and Winterwerp and Van Kesteren, 2004).

Various standards have been developed, e.g. ASTM D4822, and the literature on measuring particle size distribution (psd) is overwhelming as well, while a large number of commercial psd-instruments is also available. However, it is to be appreciated that a unique psd does not exist, in particular for the smaller sediment classes ($< 60\,\mu$m), because of a number of reasons:

(1) different instruments yield different psd of the same material (e.g. Fig. A.1, Jacobs, 2011 and Van Rijn and Barth, 2018), and
(2) sample preparation is critical.

We stress that the psd of (bed) sediments is one of the more important sediment properties. Yet, prior to determining this psd, one should contemplate on how that psd fits into the studies to be carried out. In the modeling

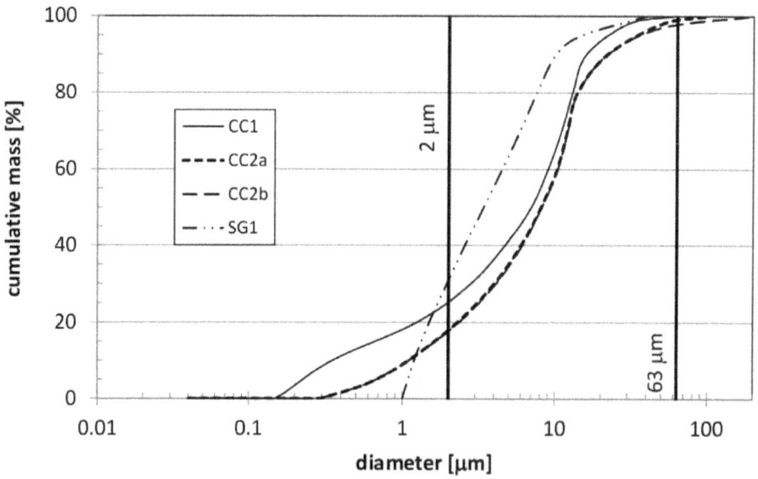

Fig. A.1. Four different particle size distributions of one industrial kaolinite powder, measured with a Coulter Counter by two different laboratories (CC1, and two twin measurements CC2a and CC2b), and measured with a SediGraph SG1.

approach discussed in this book, the four major psd-related processes to be resolved are:

(1) transition between cohesive and non-cohesive soil behavior,
(2) erosion of a bed by turbulent flow/waves,
(3) floc formation and settling, and
(4) consolidation of mud deposits.

The critical clay content at the transition between cohesive and non-cohesive soil behavior follows from a critical plasticity index of about $\text{PI}_c \approx 7\%$ (see Sec. 2.2, the activity plot). As PI_c is based on empirical soil mechanical data, the Atterberg limits (see Sec. 10.4) and clay content have to be determined following soil mechanical procedures, which mainly implies that the sediment has to be deflocculated (see below).

For the other three psd-related processes, information on the floc structure is required. This is not an easy task, as the in-situ floc structure is easily affected/destroyed upon sampling and storage.

Deflocculation (peptization) of soil samples can be done in two ways, i.e. with sodium pyrophosphate ($Na_4P_2O_7$; NEN 5753) or with hydrochloric acid (HCl) and hydrogen peroxide (H_2O_2). Both treatments take about 12 hours. Samples are gently stirred, not in an ultrasonic bath (Veld *et al.*, 2016).

The psd of this deflocculated sample can be established with a variety of instruments:

(1) hydrometer,
(2) pipette method,
(3) SediGraph,
(4) Coulter counter, and/or
(5) Malvern particle sizer.

Next to determining the psd of the deflocculated sediment, it is strongly recommended establishing the psd of a non-treated sample. The soil has to be diluted, preferably with local (*in-situ*) water, and mixed carefully. It is to be appreciated that the floc size distribution thus obtained will differ from the *in-situ* floc size distribution. Yet this procedure gives a first idea on the floc sizes one can expect in nature. We advocate the use of one of the following instruments:

(1) Malvern particle sizer,
(2) sedimentation balance.

The hydrometer test is based on differential settling of a sediment-water mixture in a column (this is basically a settling column). During differential settling, the buoyancy of the mixture reduces. This reduction is measured with the so-called hydrometer (a float, generally made of glass) with a scale for measuring the immersion of the hydrometer with time. From the 1DV sediment balance, the distribution of the settling velocity of the sediment mixture is obtained.

Its simplicity is also the hydrometer's pitfall, and experience is required to carry out the measurements correctly, obtaining accurate results. The hydrometer has to be calibrated properly, ambient and equipment temperature have to controlled, and the mixture has to be homogenized securely.

The pipette method is also based on differential settling of a sediment-water mixture in a column (one to a few liters). Small samples of the settling suspension are withdrawn with a pipette from the column at a specific height at predetermined times. The sediment content of these samples is determined by standard procedures (filter paper, oven). From the times series of the subtracted mass and the 1DV mass balance equation, the settling velocity distribution of the mixture is obtained.

The pipette method requires more preparation than the hydrometer test, as proper sub-sampling demands a sophisticated experimental setup.

Commercial setups are therefore widely available. Also for the pipette method, experience is required for obtaining accurate results, while the same care has to be taken with respect to temperature, and homogenizing the mixture as with the hydrometer.

A SediGraph consists of a (small) settling column/container in which the sediment-water mixture is homogeneously mixed. The settling velocity of the particles and the sediment concentration in the container are measured from the deflection and adsorption of a light beam, using the Beer–Lambert model. Commercial instruments are available, and the executing the measurements is fairly straightforwards.

In a Coulter counter a (small) sediment sample is put in an electrolyte (a fluid with ions — see also Sec. 2.1). An (alternating) electric current drives the particles (one by one) through an orifice, while the local change in impedance is measured. This change is a function of the size of the particles, while the count yields the number of particles. This instrument is commercially available, and fairly easy to use.

The Malvern particle sizer is based on the diffraction of a laser beam by the particles suspended in a small container. The diffraction is measured with concentric diodes, and from the Lorenz–Mie method (and some corrections implemented by the manufacturer), the particle size distribution is obtained.

Though developed for very small particles ($<2\,\mu$m), the method is applicable for a wide range of sizes. However, in ill-sorted mixtures, smaller particles may be obscured by larger ones, hence not "seen" by the laser. Therefore, Ibanez Sanz (2018) developed a procedure for measuring the entire spectrum of the psd. The basic idea behind this method is to split the original sample in two or more sub-samples through differential settling, and than measuring the psd of each sub-sample.

The Malvern particle sizer is commercially available, and fairly easy to use. An advantage is that a very wide range of particle sizes can be measured, provided that a procedure to capture the smaller fraction as well is used. After some experience, these measurements are not difficult and do not require much sample preparation. However, the instruments are quite expensive to buy.

The sedimentation balance consists of a settling column (about 20 cm height) and a pan. The pan is suspended to a scale near the bottom of the column, collecting the settling sediment. The sediment is brought into the column and gently mixed homogeneously. From the increase in mass on the pan (the measured weight), and using a 1DV mass balance model, the distribution of the sediment settling velocity is obtained.

This instrument is commercially available, but the setup is simple and it can be in-house made as well. However, it is quite sensitive to vibrations and temperature variations (gusts), thus the sedimentation balance needs to be placed carefully on an isolated place.

Other methods can be used as well. For instance, particles settling in a column can be registered with camera, using PIV or other techniques obtaining the settling velocity distribution. This technique is frequently used for *in-situ* measurements of the settling velocity (Manning *et al.*, 2010).

Note that four instruments measure the settling velocity distribution from the sediment balance. The particle size distribution is then obtained using Stokes' law. Stokes' law, assuming spherical particles, can theoretically be corrected for non-spherical particles, provided that the particle's shape is known. As this is generally not the case, the use of Stokes' law is one of the reasons that the resulting psd of these instruments always deviates from the two direct particle size measurements obtained with the other two instruments.

A.4 Atterberg limits

The two Atterberg limits relevant for this book are the plastic limit (PL) and the liquid limit (LL). Procedures obtaining these limits are described in ASTM D4318, which is also followed in The Netherlands. The basic procedure for determining the LL is wetting a dried sample until a groove in the sample closes upon distortion of the soil. Further to this procedure, the PL follows from rolling a thread of soil until it crumbles.

Soil mechanical procedures demand that the soil sample is oven-dried at 105°C. The dry sample is then pulverized and rewetted with distilled water. Here we propose a slightly different procedure, air-drying a sample (with a hair-drier) sufficiently to carry out the tests. Moreover, we advocate the use of natural (*in-situ*) water for wetting a sample, if possible. With the latter procedure the effects of non-minerals on the mechanical properties of the soil are best maintained.

The LL is determined with the Casagrande instrument. The sample is placed in the Casagrande cup, and a groove of 13.5 mm is made in the soil. Next, the cup is dropped over a distance of 1 cm repeatedly. This procedure is repeated at a variety of moisture contents of the soil. At the liquid limit, the groove closes upon 25 drops.

The PL is determined from rolling a thread of soil. While rolling, the roll looses water, thus dries. At the PL, a 3.2 mm diameter thread breaks within two minutes.

It is obvious that these procedures can be quite subjective, and that two measurements may yield (quite) different results. It is strongly recommended that these tests are carried out by experienced staff, following a strict protocol for minimizing experimental variations.

Currently, the Atterberg limits are also determined with more sophisticated tests, such as the Swedish fall-cone test for the liquid limit (Wood, 1985), which yields better controlled and repeatable conditions. The cone consists of stainless steel, with a weight of 80 g, and cone angle of 30°. The cone is place on the surface of the soil, which is placed in 40 mm high container of 55 mm diameter. The cone is then released for 5 s. At the liquid limit, the cone penetrates 20 mm into the soil. This 20 mm is obtained through interpolation of a series of cone tests at various water contents of the soil sample.

A.5 Zeta potential

The zeta potential is an indicator of the stability of colloidal dispersions. Therefore, this parameter characterize the flocculation potential of cohesive sediment. High zeta potential (negative or positive) indicates that the particles are electrically stabilized, while low zeta potentials reflect the tendency of sediment to coagulate or flocculate.

The zeta potential cannot be measured directly, but can be derived from the experimentally determined electrophoretic mobility during an electrophoresis process with a commercially available instrument. A small sediment sample is diluted with distilled water, and put into the measuring container of the instrument. An electric current is applied to the mixture, accelerating the particles. This process is referred to as electrophoresis. The velocity of the particles is measured with laser-doppler velocimetry (the particle's equilibrium velocity does not depend on its mass; its acceleration, of course, does).

In practice, alternating electrical fields are used to prevent (1) contamination and deterioration of the electrodes and (2) electro-osmosis in the cell. The measurements give the electrophoretic mobility of the particles. From this mobility, the zeta potential (electrokinetic surface potential) can be deduced. The simplest method is the Smoluchowski relation, which gives a linear relation between the mobility and the zeta potential at high salinity. More advanced theories are available for a large range of salinity. In that case the relation between mobility and zeta potential is not linear, and depends on additional parameters, such as particle size and salinity. The reader is referred to for instance https://en.wikipedia.org/wiki/Zeta_potential.

After some experience, these measurements are not difficult and do not require much sample preparation. However, the instruments are quite expensive to buy.

A.6 Mechanical properties — Rheology

Rheology studies the mechanical response of a material to deformations, either by stress or by shear. Rheology is a science by itself with an overwhelming amount of literature, (standard) text books, and commercial and non-commercial instruments available. Rheological studies are carried out in many industries (oil, food, metal, ...). Here we focus on the rheological properties of fine sediment soils at small shear rates, as generally occurring in nature. Fine sediment generally depicts shear-thinning behavior, i.e. the soil viscosity decreases with increasing shear. Owing to the reversible nature of the soil floc structure, the rheology of fine sediment soils shows time/history-dependent behavior, known as thixotropy.

The two working horses for determining the rheological properties of cohesive soils are the vane test and the roto-viscometer. We do not discuss other methods. Commercially, fairly simple to highly advanced instruments are offered on the market, and such purchases need therefore to be evaluated carefully. The authors are not aware of any (inter)national standards for the tests described below.

A vane test is used to determine the soil undrained shear strength c_u. The vane is rotated at a specific angular speed, which is kept constant during a test. A typical result of a vane test is shown in Fig. 2.19 of Chapter 2. The following procedures have to be followed:

(1) *Sample preparation*: The vane test is generally carried out on remolded samples, homogenized carefully (minimize floc rupture), whereupon the sample bulk (dry) density is measured. The soil sample is to be placed in a container large enough to prevent wall effects — generally a few vane diameters is sufficient.

    Vane tests can of course also be carried out on (stratified) undisturbed samples, aiming to measure the vertical variation in soil strength.

(2) *Vane size and depth*: A vane of proper size has to be selected in relation to the strength of the soil sample. Possibly some trial and error is required. The vane is lowered slowly and carefully into the soil, minimizing disruption of the soil structure. This is done preferably with a mechanical device, ensuring vertical insertion.

The top of the vane should be deep enough with respect to the sample's surface; generally one vane length is sufficient.

(3) *Rotational speed*: The rotational speed of the vane is selected (generally a few rpm) and kept constant during a test. Again some trial and error may be required. A typical output of a vane test is given in Fig. 2.19 of Chapter 2. This output is characterized by a peak, the so-called peak strength τ_p and a shoulder, which represents the c_u. A good experiment is characterized by a peak strength about 10–50% larger than c_u. Larger τ_p indicate too large rotational speed, inducing either pore water underpressures, or disrupture of too much of the soil structure. Both affect the accuracy of the measurements. At too low rotational speeds, soil failure takes place at drained conditions, resulting in an under-estimation of c_u.

(4) *Reproducibility*: It is advisable to carry out a few measurements per sample checking procedure and reproducibility — of course this is only possible if the soil sample is large enough.

With a roto-viscometer, the shear stress — shear strain relation of a sediment-water mixture is measured. Various configurations can be used, depending on the consistency of the sample, such as beaker-cone, cone-plate, etc. One element (the cone) is connected to the central shaft of the roto-viscometer, while the second element is placed on the base-plate of the instrument. The shaft drives the deformation of the sample, while the soils response is measured at the base plate. Consult the manufacturer's manual for selecting the proper configuration.

An experiment can be carried out under shear stress controlled conditions or under shear rate controlled conditions — in advanced roto-viscometers, the user can choose. In the stress-controlled mode, the shafts torque (stress) is set, while the shear rate is measured. Under shear rate controlled conditions, the rotational speed is set, while the torque is measured.

Roto-viscometer tests yield the so-called flow curve of a soil (e.g. Fig. 2.20), the soil response in a stress–strain diagram. This flow curve can be established in two ways:

(1) applying a constant stress or strain rate,
(2) applying an accelerating stress or strain rate.

By applying a constant stress or strain on the sample, the equilibrium flow curve is found by repeating the test at steps of increasing stress or strain rate. Each sub-test provides one point on the equilibrium flow curve.

Stress or strain-rate can also be increased at a constant rate — this is a very common procedure. However, the resulting flow rate may be a function of the selected acceleration. Therefore, one test in general covers an accelerating phase, directly followed by a decelerating phase, up to zero stress or strain-rate. This procedure allows studying the thixotropic behavior of the soil as well. Often the acceleration–deceleration cycle is repeated a number of times until the two flow curves (one accelerating and one decelerating branch) do not change anymore.

The slope of the flow curve represents the soil viscosity, while the crossing with the vertical y-axis yields the soil yield strength. Various models exist for analyzing these flow curves, i.e. the Bingham model, the Herschel–Bulkley model, just to name two (see also Sec. 7.3). The parameters of these models are obtained through curve-fitting the experimental data through these models.

Interpretation of roto-viscometers results should be done carefully. Often, the shear rates deployed are very high ($100\,\mathrm{s}^{-1}$), well beyond shear rates encountered in nature. At real-world shear rates, the roto-viscometer results are quite sensitive to the instrument settings (acceleration/deceleration), while viscosities at these small shear rates can be very high (Fig. 2.20). Stress-controlled tests are therefore often carried out, enabling high resolution near the onset of soil failure (yield strength).

The roto-viscometer can also be used in oscillating mode, a mode frequently used for studying the soil response to wave forcing. This mode provides the (complex) storage and loss moduli of a soil, and their phase lag. Frequently used models for analyzing the test results are the visco-elastic Kelvin–Voigt model and the visco-plastic Maxwell model. For further information the reader is referred to Mehta (2013).

Using a roto-viscometer requires attention for:

(1) The gap between beaker and cone, cone and plate and other configurations. This gap should be small enough to guarantee constant shear over this gap. However, the gap size should measure at least ten times the largest particle diameter in the sediment-water mixture (Coussot, 1997). An additional point of attention is that flocs may (re)form during an experiment, which process should not be affected by the gap size.

(2) In particular for more dilute suspensions, (differential) settling of particles is a problem, changing the mixtures composition over time. Apparatus with counter flow have been developed, but have not proved very successful.

A.7 Mechanical properties — Slump test

The slump test is a simple means for a first assessment of the bearing capacity of a soil and its yield strength. It is often used to test the workability of concrete, as described in the British and American standards, BS 1991 (1970) and ASTM C 143-90. A soil sample is put into a straight cylinder with a diameter of about 0.1 m, and length varying between 0.08 m and 0.2 m (Baudez *et al.*, 2002; Pashias *et al.*, 1996). Cylinder material is not too critical, and it can be made of steel, pvc or perspex, as long as mud does not adhere to the cylinder wall. The cylinder is placed on a smooth horizontal plate, and slowly lifted. The initial cylindrically shaped soil sample will collapse under its own weight. After equilibrium, the height of the yielded sample $H - \Delta h$ is measured, where H is the initial height of the cylindrically shaped sample and Δh the distance over which the sample yielded, referred to as the slump height in literature.

Figure A.2 shows the schematic of a slump test, as described by Pashias *et al.* (1996). The vertical stress σ_z at height z in the undisturbed sample reads:

$$\sigma_z = \alpha \rho g z \tag{A.1}$$

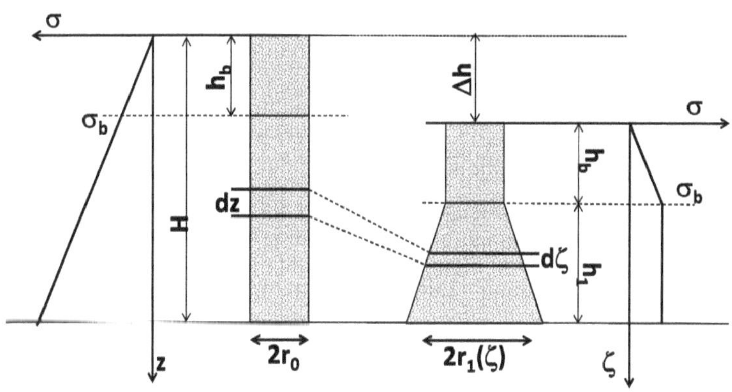

Fig. A.2. Schematic of slump test with initial sample (left panel) and yielded sample at equilibrium (right panel).

where ρ is the bulk density of the soil. For homogeneous samples, $\alpha = 1/2$. However, such samples are seldom fully homogeneous, and Coussot (2005) advises $\alpha = 1/\sqrt{3}$. At $z = h_b$, this stress equals the bearing strength of the soil σ_b. When left unsupported by the cylinder, the soil will yield. However, the upper part of the soil with thickness h_b will not yield, as it is strong enough. The bearing strength can thus be derived directly by measuring h_b. As this is not very accurate, Pashias *et al.* (1996) proposed another method.

It is assumed that yielding induces horizontal flow only, i.e. all horizontal planes remain horizontal. Pashias *et al.* (1996) introduced a material coordinate ζ. Because of continuity, a relation exists between the thickness and radius of a disk in the initial and yielded sample, i.e. dz and dζ:

$$\pi r_0^2 \rho \mathrm{d}z = \pi r_1^2 \rho \mathrm{d}\zeta \tag{A.2}$$

where r_0 is the inner radius of the cylinder, $r_1(\zeta)$ is the radius of the yielded sample at height ζ (assuming undrained deformation).

The thickness of the yielded sample is

$$h_1 = \int_{h_b}^{H} \mathrm{d}\zeta = \int_{h_b}^{H} \left(\frac{r_0}{r_1}\right)^2 \mathrm{d}z \tag{A.3}$$

in which Eq. (A.2) is substituted. Further, the total load above the levels z and ζ should be equal, thus:

$$\pi r_0^2 \sigma_z = \pi r_1^2 \sigma_b \tag{A.4}$$

Substitution of Eqs. (A.1), (A.2) and (A.4) into Eq. (A.3) gives:

$$h_1 = \int_{h_b}^{H} \frac{\sqrt{3}\sigma_b}{\rho g z} \mathrm{d}z = \frac{\sqrt{3}\sigma_b}{\rho g} \ln \frac{H}{h_b} \tag{A.5}$$

Thus the slump height Δh follows from:

$$\Delta h = H - h_b - h_1 = H - \frac{\sqrt{3}\sigma_b}{\rho g}\left(1 - \ln\left(\frac{\sqrt{3}\sigma_b}{\rho g H}\right)\right) \tag{A.6}$$

which is a parameter much easier and more accurate to measure than h_b. However, Eq. (A.6) is implicit, and needs to be solved iteratively or graphically.

The bearing strength σ_b can be related to the yield stress τ_y with the Mohr–Coulomb failure envelope, Sec. 2.2, Eq. (2.2). For cohesive soils, the angle of repose ϕ_d amounts to 10° and 30°. Thus $\tau_y = (0.2\text{–}0.6)\sigma_b$.

The length of the cylinder should be chosen such that h_b measures at least a few cm to obtain accurate results.

A.8 Settling and consolidation

While the tests described above have been elaborated upon widely in literature, and a number of them have been standardized in (inter)national norms, the tests and analyses described in this and the next section have not been standardized. We therefore provide a detailed protocol, and provide the relevant equations for analyzing the results.

The procedure described below is partly based on the work by Merckelbach and Kranenburg (2004a) in analyzing settling/consolidation experiments in a settling column. At the end of this section we summarize a protocol and the minimum requirements for such experiments. A schematic settling/consolidation curve is depicted in Fig. A.3, showing the lowering of the sediment-water mixtures interface with time. In the first phase (HS), settling occurs in the hindered settling regime, whereas beyond $t_{c=c_{gel}}$, consolidation takes place (phase C_a and C_b).

The measured settling curve is fit to parts of the settling/consolidation equation, tuning the relevant coefficients. The settling equation describes the effects of hindered settling, while the consolidation equation yields Gibson's equation (Merckelbach and Kranenburg 2004a; Winterwerp and Van Kesteren, 2004; Dankers and Winterwerp, 2007). The consolidation equation in terms of the solids content ϕ_s and in Eulerian

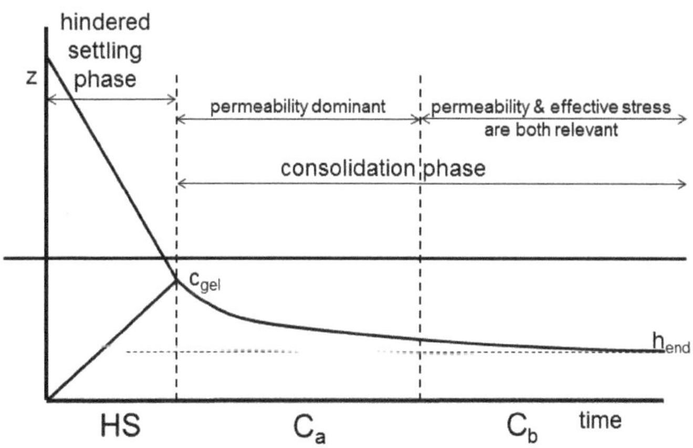

Fig. A.3. Schematic settling/consolidation curve with time.

coordinates reads:

$$\underbrace{\frac{\partial \phi_s}{\partial t}}_{\text{I}} - \frac{\partial}{\partial z}\left[\underbrace{w_{s,0}\mathrm{f}\left(\phi_s, c_{gel}\right)}_{\text{II}} + \underbrace{\Delta_\rho k \phi_s^2}_{\text{III}}\right] - \underbrace{\frac{\partial}{\partial z}\Gamma_c \frac{\partial \phi_s}{\partial z}}_{\text{IV}} = 0 \tag{A.7}$$

in which t and z are time and vertical coordinate (positive in upward direction), $w_{s,0}$ = settling velocity of individual mud floc, f = hindered settling function (Kynch parameter — see Sec. 4.4), Δ_ρ = relative sediment density ($\Delta_\rho = (\rho_s - \rho_w)/\rho_w$), ρ_s and ρ_w are the specific density of the solids and the pore water.

Further, self-similarity is assumed and the consolidation parameters are modeled with a fractal approach (Sec. 5.1 and Merckelbach and Kranenburg, 2004a). The hindered settling function f follows from Dankers and Winterwerp (2007) — see also Sec. 4.4:

$$\mathrm{f} = \frac{\left(1-\phi_f\right)^2\left(1-\phi_s\right)}{1+2.5\phi_f} \tag{A.8}$$

in which ϕ_f = volumetric floc concentration ($\phi_f = c/c_{gel}$), c = solids mass concentration, c_{gel} = gelling concentration (mass concentration at which the flocs form a space-filling mud skeleton, also known as the structural density), and ϕ_s = volumetric concentration of solids ($\phi_s = c/\rho_s$).

Further to the fractal approach, the soils hydraulic conductivity k, vertical effective stress σ_{zz}^{sk} and consolidation coefficient Γ_c (see also Eq. (5.12) are given by

$$k = K_k \phi_s^{-2/3-n_f}, \quad \sigma_{zz}^{sk} = K_p \phi_s^{2/3-n_f}, \quad \Gamma_c = \frac{2}{3-n_f}\frac{K_k K_p}{g\rho_w} \tag{A.9}$$

The hindered settling phase is described by the terms I and II from Eq. (A.7). Assuming $\mathrm{d}h/\mathrm{d}t = w_{s,eff} \equiv \mathrm{f}w_{s,0}$ provides one equation (i.e. Eq. (4.37)) with the two unknowns $w_{s,0}$ and c_{gel}. In case the transition between the hindered settling and consolidation phase is very sharp, the settling curve in Fig. A.3 depicts a sharp kink at c_{gel}. Note however, that one cannot read c_{gel} directly from that kink, as part of the deposited bed is already in consolidating mode.

Therefore a second settling test has to be carried out at a different initial concentration. Thus two equations are obtained with the two unknowns, which can be solved to get the two unknowns. At the end of this section, some comments on a protocol to be followed are given.

Merckelbach and Kranenburg (2004a) realized that during the initial phases of consolidation, effective stresses are small, and that the consolidation process is mainly governed by the soil hydraulic conductivity, i.e. phase C_a in Fig. A.3. This consolidation phase is thus described with a simple-wave equation, i.e. the terms I and III of Eq. (A.7). This equation can be solved with the method of characteristics, given by:

$$\frac{\mathrm{d}z}{\mathrm{d}t} = K_k \Delta_\rho (2-n) \phi_s^{1-n} \tag{A.10}$$

in which the parameter $n(=2/(3-n_f))$ is introduced for convenience. Integration along the characteristics (e.g. Merckelbach and Kranenburg, 2004a) yields:

$$h(t) = \left(\frac{2-n}{1-n}\zeta_m\right)^{\frac{1-n}{2-n}} \left((n-2)\Delta_\rho K_k\right)^{\frac{1}{2-n}} t^{\frac{1}{2-n}} \tag{A.11}$$

with the Gibson height for the total amount of mud in the column:

$$\zeta_m = \int_h \phi_s \mathrm{d}z \tag{A.12}$$

Hence, from plotting the measured mud interface h against time t and fitting equation (A.11) to these data, the n (thus n_f) and K_k are found (see Merckelbach and Kranenburg (2004a) for example).

After some time, effective stresses build up, and the consolidation rate can no longer be described by the terms I and III of Eq. (A.7) alone. At equilibrium ($t = t_{h=h_{end}}$), the mud balance is described by the terms III and IV of Eq. (A.7). Integrating once over depth yields the vertical dry density profile at equilibrium, whereas a second integration yields the final thickness (h_{end}), e.g. Merckelbach and Kranenburg (2004a):

$$\phi_s(z) = \left[\frac{n-1}{n}\frac{(\rho_s-\rho_w)g}{K_p}\right]^{\frac{1}{n-1}} (h-z)^{\frac{1}{n-1}} \tag{A.13}$$

$$h_{end} = \frac{nK_p}{(n-1)(\rho_s-\rho_w)g}\left[\frac{(\rho_s-\rho_w)g}{K_p}\zeta_m\right]^{\frac{n-1}{n}} \tag{A.14}$$

From a fit of Eq. (A.13) against the measured equilibrium dry density profile, K_p is found, using the n_f value obtained from phase C_a analysis. These values can also be obtained from the final thickness, i.e. Eq. (A.14) (again, see Merckelbach and Kranenburg (2004a) for example).

The method proposed by Merckelbach and Kranenburg (2004a) is very versatile, and easy to execute. However, a firm protocol should be set, which should contain the following elements:

(1) *Sample preparation*: Merckelbach and Kranenburg (2004a) elaborate on the presence of sand, which is treated as a passive material in the consolidation process. Hence, it is possible carrying out these tests with a sandy mud. However, it is recommended to sieve the soil, removing the sand fraction. The reason is that the tests have to be carried out from below-structural densities (see below and Sec. 5.3) — then sand will settle first, as it cannot be carried by a mud skeleton. Yet, not all sand will settle, and the amount of sand in a test, affecting the soils material parameters is unknown.

Further, it is strongly recommended to dilute the soil with local water (same salinity and pH) for preparing suspensions of lower concentration.

(2) *Column dimensions*: From an accuracy point of view, a tall, wide column is better than a smaller one. Wall effects affect consolidation because of (i) adhesion and/or (ii) the formation of drainage channels along the wall. Obviously, these effects reduce in wider columns. Also the reading of the interface in general, and of the final dry density distribution of consolidated mud in particular is more accurate in a large column. However, mixing of a sediment–water mixture attaining initial homogeneous conditions is easier in a small column, and virtually impossible in large columns. A 0.5 m tall, 0.15 m wide column may prove optimal.

(3) *Initial mixing/conditions*: The settling tests need to be carried out at below-structural density (i.e. $c_0 < c_{gel}$). If not, part of the column is in an over-consolidated state initially, and swelling needs to be accounted for (Sec. 5.3).

The method above assumes homogeneous initial conditions. Hence, the sediment–water mixture has to be mixed thoroughly, yet preventing floc breakup. Gently hand-mixing with a disk, slightly smaller than the column diameter, perforated with large holes, works well. If flocs are broken during mixing, reflocculation in the column is expected. In that case, the suspension will only start settling after some time (this may even take several tens of minutes).

(4) *Initial concentration*: In the hindered settling phase two unknowns have to be determined. Hence, tests with at least two initial concentrations have to be carried out. With three tests, confidence in the test results

and their accuracy increase. Also, the initial concentration cannot be too small, as then reading of the interface becomes cumbersome. Initial values for c_0 of 20, 40 and 60 g/L have proven to work well.

(5) *Particle distribution*: The entire procedure assumes a single particle size (one settling velocity, no differential settling). This is generally obtained with flocculated material. If not, the settling curve becomes very gradual, and distinction between the hindered settling and consolidation regime becomes difficult (Te Slaa, 2015).

Further it is to be noted that the structure of a soil sample will be affected by sampling, transport, storage and sample preparation. Thus these tests give at best a qualitative description of the soil behavior in the natural environment. More confidence can be attained by measuring strength distribution *in-situ*, though spatial variability may be large. Yet, better procedures are not available at present.

A.9 Capillary suction time

The capillary suction time (CST) is measured with the CST-instrument, which is commercially available. The instrument is cheap, easy to use, and requires small samples (*ca* 9 cm^3). The mud is placed in a small cylinder, open on both sides. The sample container is placed on filter paper, generating a suction pressure on the samples pore water. The filtration of water in the filter paper is measured with two electrodes, and the filter time between these electrodes defines the capillary suction time t_{CST}. A sketch of the CST-apparatus is shown in Fig. A.4, where t_{CST} is the travel time of filter water between the rings D_1 and D_2, which is measured by the apparatus.

Huisman and Van Kesteren (1998) describe CST application for assessing the soil consolidation coefficient, resolving Terzhagi's equation as a function of void ratio e. However, transforming e into volumetric concentration ϕ_s is not advisable as accuracy is lost largely. Therefore, the relevant

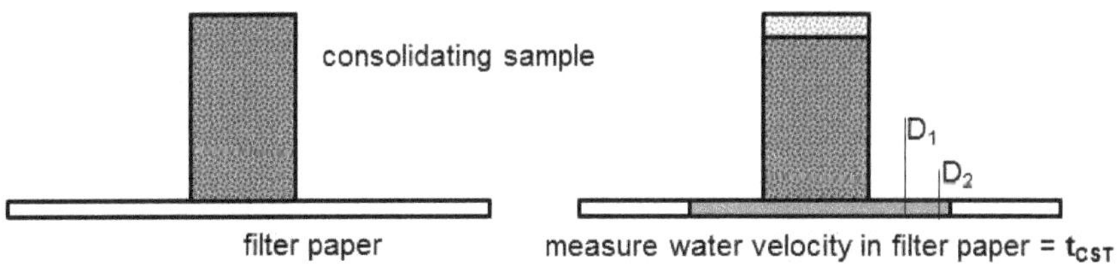

Fig. A.4. Diagram of CST-apparatus with initial (left panel) and pore water level after some time (right panel).

equations for assessing hydraulic conductivity and the consolidation coefficient are given in terms of ϕ_s below.

The hydraulic conductivity can be obtained from the terms I and III of Eq. (A.7), Sec. A.8 above. The consolidation equation can then be rewritten as:

$$\frac{\partial \phi_s}{\partial t} - v_s(\phi)\frac{\partial \phi_s}{\partial z} = 0 \tag{A.15}$$

in which the advective velocity v_s reads:

$$v_s(\phi_s) = \Delta_\rho K_k \frac{\mathrm{d}}{\mathrm{d}\phi_s}\phi_s^{\frac{4-2n_f}{3-n_f}} = \Delta_\rho K_k \frac{4-2n_f}{3-n_f}\phi_s^{\frac{1-n_f}{3-n_f}} \tag{A.16}$$

The simple wave equation (A.15) can be solved, yielding a relation between ϕ_0 and t_{CST},assuming $\phi_s \approx \phi_0$:

$$\phi_0^{\frac{1-n_f}{3-n_f}} = \frac{3-n_f}{4-2n_f}\frac{1}{\Delta_\rho K_k}\frac{(D_2^2 - D_1^2)}{D_{\mathrm{CST}}^2} d_f p_f s_f \frac{1}{t_{\mathrm{CST}}} \tag{A.17}$$

in which ϕ_0 is the initial volumetric sediment concentration of the sample when put into the sample container. With $s_f p_f = 0.22$ (Huisman and Van Kesteren, 1998), the filter paper thickness $d_f = 1.10$ mm, diameter of the sample container $D_{\mathrm{CST}} = 18$ mm, $D_2 = 45$ mm and $D_1 = 32$ mm, one finds:

$$\phi_0 = \left(1.20.10^{-6}\frac{3-n_f}{2n_f-4}\frac{1}{K_k}\frac{1}{t_{\mathrm{CST}}}\right)^{\frac{3-n_f}{1-n_f}} \tag{A.18}$$

A CST-test is carried out for a number of initial concentrations ϕ_0, diluting the original sample preferably with *in-situ* water. Plotting these ϕ_0-values against the corresponding t_{CST} yields n_f and K_k — see further thoughts on a proper protocol below.

The consolidation coefficient Γ_c is obtained through CST-tests on pre-consolidated samples. Equation (A.7) is reduced to Terzaghi's equation, which is a diffusion equation (terms I and IV of Eq. (A.7)), see also Sec. 5.1. With the initial condition $\phi_{s,t=0} = \phi_0$ and the boundary conditions $\phi_{s,z=0} = \phi_{eq}$ and $\phi_{s,z=\infty} = \phi_0$, the following solution is obtained:

$$\frac{\phi_s(z) - \phi_{s,eq}}{\phi_{s,0} - \phi_{s,eq}} = \mathrm{erf}\left\{\frac{z}{\sqrt{4\Gamma_c t}}\right\} \tag{A.19}$$

Next, we integrate Eq. (A.19) over the height of the CST instrument h_{CST}:

$$\begin{aligned}\int_0^{h_{\mathrm{CST}}} &\operatorname{erf}\left\{\frac{z}{\sqrt{4\Gamma_c t}}\right\}\mathrm{d}z \\ &= h_{\mathrm{CST}}\cdot\operatorname{erf}\left\{\frac{h_{\mathrm{CST}}}{\sqrt{4\Gamma_c t}}\right\} \\ &\quad + \frac{1}{\sqrt{\pi}}\left(\sqrt{4\Gamma_c t}\cdot\exp\left\{-\left(\frac{h_{\mathrm{CST}}}{\sqrt{4\Gamma_c t}}\right)^2\right\} - \sqrt{4\Gamma_c t}\right)\end{aligned} \tag{A.20}$$

The consolidation coefficient Γ_c is typically of the order of 10^{-10} m/s. Hence, we can expand Eq. (A.20) for small denominator, e.g. $\Gamma_c t$:

$$\begin{aligned}\lim_{\Gamma_c t\to 0} &\left\{h_{\mathrm{CST}}\cdot\operatorname{erf}\left\{\frac{h_{\mathrm{CST}}}{\sqrt{4\Gamma_c t}}\right\}\right. \\ &\left.+ \frac{1}{\sqrt{\pi}}\left(\sqrt{4\Gamma_c t}\cdot\exp\left\{-\left(\frac{h_{\mathrm{CST}}}{\sqrt{4\Gamma_c t}}\right)^2\right\} - \sqrt{4\Gamma_c t}\right)\right\} \\ \approx h_{\mathrm{CST}} &- \sqrt{\frac{4\Gamma_c t}{\pi}}\end{aligned} \tag{A.21}$$

Substitution of Eq. (A.21) into (A.19) and some rearrangement, we find the equation for the capillary suction time t_{CST}:

$$t_{\mathrm{CST}} = \frac{\pi}{4\Gamma_c}\left(\frac{d_f s_f p_f}{(\phi_{eq}-\phi_0)\,D_{\mathrm{CST}}^2}\right)^2\left(D_2^4 - D_1^2\right) \tag{A.22}$$

where d_f, s_f, p_f, D_{CST}, D_1 and D_2 are given above. Substitution of these CST-apparatus dimensions, and those of the filter paper, one finds:

$$\phi_0 = \phi_{eq} - \frac{1.16\cdot 10^{-3}}{\sqrt{\Gamma_c}}\frac{1}{\sqrt{t_{\mathrm{CST}}}} \tag{A.23}$$

The consolidation coefficient Γ_c is found from plotting ϕ_0 from a series of consolidation tests against $1/\sqrt{t_{\mathrm{CST}}}$.

A convenient way to verify the accuracy of this procedure is plotting ϕ_0 also against t_{CST} on double-log scale. The slope of this curve should not deviate too much from 0.5. If not, ϕ_0 does not scale with $1/\sqrt{t_{\mathrm{CST}}}$, and the CST-test will not yield accurate predictions of the consolidation coefficient Γ_c.

The derivation of the CST-formulas assumes a homogeneous sample in the samples container. Because of its small size, this is not too difficult a job.

A permeability test is carried out by diluting a mother sample a number of times, preferably with local water (same salinity and pH) attaining five or more sub-samples with different ϕ_0. For each initial condition, t_{CST} is determined, and the results are plotted in a ϕ_0–t_{CST} diagram. As the procedure assumes small effective stresses (neglecting term IV in Eq. (A.7)), the samples should be quite dilute — some trial-and-error may be required.

Note that the procedure presumes $2 < n_f < 3$, which is a minimum check on the reliability of the procedure.

A consolidation test is to be carried out on pre-consolidated samples. This is to be done by placing a sample in a consolidation device (the so-called API-apparatus). This API consists of a sealed cylinder, with a small outflow opening at its bottom and a small inflow opening at its top. Pressurized air is injected into the API-cylinder through the inflow opening. The over-pressure induces a pore water flow through the lower outlet. At equilibrium, i.e. when the pore water flow stops, the sample is pre-consolidated. Prior to a CST-test, the sample is remolded, obtaining homogeneous conditions. A sub-sample is taken to determine the actual dry density of the sample.

This procedure is repeated a number of times obtaining at least five samples with different ϕ_0. The procedure presumes that the effects of permeability are small (neglecting term III of Eq. (A.7)), so the samples should be dense enough — some trial-and-error may be required.

As discussed in Sec. 5.1, the consolidation coefficients Γ_c and the coefficients by Terzaghi c_v and Gibson c_F are related (almost identical): $\Gamma_c = c_v \approx c_F$.

Appendix B

Terminology

Asymmetry	
Peak velocity	Difference between peak flood and peak ebb velocity
Slack water	Difference between slack water period during flood and during ebb
Bed	More or less firm sedimentary substrate (sand, consolidated mud, clay)
Capacity	Maximal sediment load carried by a turbulent water flow
Below-capacity load	Sediment transport flux at below-capacity conditions
Capacity load	Sediment transport flux at capacity conditions
Clay	Minerals with $D < 2\mu$m
Coarse sediment	Large sediment particles, may contain fines in its pores
Cohesion	Attraction between clay particles and/or organic polymers
Cohesive sediment	Sediments with intrinsic cohesive properties
Concentration	
Mass	Mass of solids per unit volume of soil or water
Volumetric	Volume of solids per unit volume of soil or water
Consolidation	Compaction of soil by expelling pore water
Critical state	Soil packing at which deformation induce no volume change
Deposition	Particles falling/depositing on a sediment bed
Density	
Bulk	Mass of wet soil per volume wet soil
Dry	= Mass concentration
Specific	Mass of particles per particle volume
Drained	Pore water can flow away upon sample deformation; no or little excess pore water pressures are induced
Equilibrium	
Dynamic	Sedimentological state where bed mud content and SPM-concentrations vary around a mean value at system scale
Transport	= Capacity load
Erosion	Removal of particles from a firm sediment bed
First-order	Erosion parameter depends on mass in/on the bed
Zeroth-order	Erosion parameter is constant
Entrainment	Removal of particles from fluid mud or soft bed

(*Continued*)

(Continued)

Fines	Small sediment particles, easily brought into suspension — whether or not cohesive is not relevant
Floc	Porous aggregate with clay, silt, organic material and water
Fluff layer	Thin layer of loose fines atop a firm bed
Fluid mud	Fluid-like mud with non-Newtonian properties, well below LL, and transient — will consolidate
Gelling	Formation of fluid mud at gelling concentration
Morphodynamic	Feedback between bed level changes and hydrodynamics
Morphostatic	Bed level changes do not affect hydrodynamics
Mud	Soil with cohesive properties ($<63\,\mu$m); mud = clay + silt + organic material + water
Non-cohesive sediment	Inert sediment, such as sand
Over-consolidation	Soil stronger than own-weight to be carried
Particle	Irrelevant whether floc or solid
Pseudo-cohesion	Apparent attraction between particles induced by pore-water under-pressures
Sand	Minerals with $D > 64\,\mu$m
Saturation	
Auto-	Formation of beyond-capacity conditions by flow- and/or wave induced erosion
SPM	Beyond capacity conditions: start of fluid mud formation
Soil	No air in pores
Sedimentation	Net changes in bed level — bed formation
Settling	Particles falling through the water column
Silt	Minerals with $2 < D < 64\,\mu$m
Soft mud	Non-specified mud deposit, easily brought into suspension
SPM	Suspended particulate matter, cohesive or not
Stratification	Vertical gradients in density, damping of vertical mixing
Self-	Generation of stratification by settling particles
Swelling	
Microscopic	Sedimentology: inherent properties of clay, adsorbing water
Macroscopic	Soil mechanics: swelling or elastic rebound in response to pore-water under-pressure
Transport modes	
Erosion limited	Too few fines can be eroded from the bed to reach capacity
Settling limited	Part of SPM remain in water column during slack tide
Supply limited	Too few fines are available within the bed to reach capacity
Undrained	Pore water cannot flow away upon sample deformation; buildup of excess pore water pressures
Virgin consolidation	Bed stresses in equilibrium with own-weight to be carried

Nomenclature

Latin letters

A	activity
a	Rouse reference height
a_b	wave orbital excursion at the bed
a_μ	material parameter for viscosity
C	Chézy coefficient
c	mass concentration
c_a	Rouse reference concentration
C_c	compression coefficient
c_D	drag coefficient
c_d	drained strength
c_F	Gibson consolidation coefficient
C_s	swelling coefficient
c_u	undrained shear strength
c_v	Terzaghi consolidation coefficient
C_{cap}	capacity concentration
c_{gel}	gelling concentration
c_{ps}	pseudo cohesion
D	deposition rate
D	particle size
D_f	floc size
D_p	size primary particles
D_s	molecular diffusion coefficient
D_{10}	10th-percentile of the grain size
D_{50}	median grain size
D_{90}	90th-percentile of the grain size
e	void ratio
E_k	kinetic energy of main flow

f	Darcy–Weisbach friction coefficient
f_T	frequency turbulent eddies
f_w	wave friction coefficient
G	shear rate at the Kolmogorov length scale
g	acceleration of gravity
h	water depth
H_s	significant wave height
I	irradiance
K	consistency index Herschel–Bulkley model
k	(turbulence) wave number
k	hydraulic conductivity
k	wave number (surface waves)
K_0	coefficient of earth pressure at rest
k_0	wave number for deep water
k_A	aggregation coefficient in floc model
k_B	breakup coefficient in floc model
K_d	total attenuation coefficient
K_k	material parameter for hydraulic conductivity
K_p	material parameter for effective stress
k_s	Nikuradse roughness height — equivalent sand roughness
k_T	turbulent kinetic energy
K_y	material parameter for yield strength
K_{im}	attenuation coefficient by SPM
m_v	coefficient of volumetric compressibility
n	Manning friction coefficient
n	porosity
n_f	fractal dimension
p	isotropic stress
p_d	coefficient of reduced deposition
p_e	excess pore pressure
p_f	total water pressure
p_h	hydrostatic pressure
q	deviatoric stress
S	Strickler friction coefficient
S	salinity
S	sedimentation rate
s_u	undrained shear strength
$S_{T,ij}$	turbulent strain tensor
T	temperature

T	tidal period
T	wave period
T_b	time scale for floc breakup
T_D	time scale for deposition
T_f	time scale for flocculation
T_m	time scale for vertical mixing
T_e	time scale for erosion
T_s	time scale for settling
T_T	time scale for turbulence dissipation
U	characteristic velocity
u	velocity
u_*	shear velocity
U_w	near-bed wave orbital velocity
W	(gravimetric) water content
W_s	characteristic settling velocity
w_s	settling velocity
$w_{s,\mu}$	settling velocity micro flocs
$w_{s,M}$	settling velocity macro flocs
z_0	roughness height

Greek letters

$\Delta\rho_f$	excess density flocs
δ	characteristic thickness
δ_S	Stokes boundary layer thickness
δ_T	turbulent wave boundary layer thickness
δ_ℓ	thickness laminar sublayer
$\dot{\gamma}$	shear rate
η	empirical settling parameter
Γ_c	consolidation coefficient
Γ_T	eddy diffusivity
κ	Von Kármàn constant
λ_K	Kolmogorov length scale
μ	dynamic viscosity
μ_a	apparent viscosity
μ_T	turbulent viscosity
ν_T	eddy viscosity
ω	wave frequency
ϕ	angle of internal friction

ϕ_f	volumetric concentration flocs
ϕ_s	volumetric solids concentration
ρ_b	bulk density
ρ_f	floc density
ρ_s	specific solids density
ρ_w	specific water density
ρ_{dry}	dry density ($\equiv c$)
σ_{zz}	total vertical stress
σ_{zz}^{sk}	vertical effective stress
σ	normal stress
σ_T	turbulent Prandtl–Schmidt number
$\sigma_{T,ij}$	turbulent stress tensor
τ	shear stress
τ_B	Bingham strength
τ_d	critical shear stress for deposition
τ_T	turbulent time scale
τ_y	yield strength
τ_{xz}	(turbulent) shear stress in vertical x–z plane
ε	volumetric strain
ε_T	turbulence dissipation rate per unit mass
ξ^i	solids content fraction i
ξ_0	critical clay content for cohesive behavior
Ξ_s	settling function
ζ_m	Gibson height for mud
σ_i^{sk}	effective stress

Scaling numbers

β	Rouse number
$\mathrm{Pe_w}$	Péclet number
Re	overall Reynolds number
Re_*	roughness Reynolds number
$\mathrm{Re_p}$	particle Reynolds number
$\mathrm{Re_T}$	turbulence Reynolds number
$\mathrm{Re_w}$	wave Reynolds number
Ri	gradient Richardson number
Ri_*	bulk Richardson number

Superscripts

cl	clay
m	mud
sa	sand
si	silt
sk	skeleton related parameter

Subscripts

0	reference/initial conditions
d	drained conditions
e	equilibrium conditions
f	flocculated phase
f	flow related parameters
s	sediment/solid phase
u	undrained conditions
w	water/fluid phase
w	wave related parameters
fw	flow and wave related parameters

Other

$\bar{\cdot}$	time of spatial mean
$\cdot'$	time of spatial deviation from mean
ℓ	length scale

Abbreviations

K_0-line	normal consolidation line
CD	consolidated drained
CSL	critical state line
CSM	critical state model
CSR	constant strain rate instrument
CU	consolidated undrained
LI	liquidity index
LL	liquid limit
OCR	over-consolidation ratio
PI	plasticity index
PI*	normalized plasticity index
PI_c	critical plasticity index for cohesive behavior

PL	plastic limit
SIC	seepage induced consolidation instrument
SL	shrinkage limit
SL	swelling line
SPM	suspended particulate matter
UU	unconsolidated undrained
VCL	virgin compression line
ZP	zeta potential
L	liter

Bibliography

Aalderink, R. H., Lijklema, L., and Breukelman, J. (1985). Quantification of wind induced resuspension in a shallow lake, *Water Science and Technology* **17**, 6–7, pt 3, pp. 903–914, doi:10.2166/wst.1985.0189.

Abraham, G. and Vreugdenhil, C. B. (1971). Discontinuities in stratified flows, *Journal of Hydraulic Research* **9**, 3, pp. 293–308, doi: 10.1080/00221687109500356.

Abraham, G., Karelse, M., and Van Os, A. G. (1979). On the magnitude of interfacial shear of subcritical stratified flows in relation with interfacial stability, *Journal of Hydraulic Research* **17**, 4, p. 273.

Adachi, Y. (2019). Aspects of colloid and interface in the engineering science of soil and water with emphasis on the flocculation behavior of model particles, *Paddy and Water Environment* doi:10.1007/s10333-019-00712-7.

Adriaens, R., Zeelmaekers, E., Fettweis, M., Vanlierde, E., Vanlede, J., Stassen, P., Elsen, J., Środoń, J., and Vandenberghe, N. (2018). Quantitative clay mineralogy as provenance indicator for recent muds in the southern North Sea, *Marine Geology* **398**, pp. 48–58, doi:10.1016/j.margeo.2017.12.011.

Agrawal, Y. C. and Pottsmith, H. C. (2000). Instruments for particle size and settling velocity observations in sediment transport, *Marine Geology* **168**, 1–4, pp. 89–114, doi:10.1016/S0025-3227(00)00044-X.

Allersma, E. (1971). Mud on the oceanic shelf off Guiana, in *Symposium on Investigations and Resources of the Caribbean Sea and Adjacent Regions, UNESCO, Paris*, pp. 193–203.

Allison, M. A. and Lee, M. T. (2004). Sediment exchange between Amazon mudbanks and shore-fringing mangroves in French Guiana, *Marine Geology* **208**, 2–4, pp. 169–190, doi:10.1016/j.margeo.2004.04.026.

Amos, C. L., Daborn, G. R., Christian, H. A., Atkinson, A., and Robertson, A. (1992a). In situ erosion measurements on fine-grained sediments from the Bay of Fundy, *Marine Geology* **108**, 2, pp. 175–196, doi:10.1016/0025-3227(92)90171-D.

Amos, C. L., Grant, J., Daborn, G. R., and Black, K. (1992b). Sea Carousel A benthic, annular flume, *Estuarine, Coastal and Shelf Science* **34**, 6, pp. 557–577, doi:10.1016/S0272-7714(05)80062-9.

Andersen, T. J., Jensen, K. T., Lund-Hansen, L., Mouritsen, K. N., and Pejrup, M. (2002). Enhanced erodibility of fine-grained marine sediments by Hydrobia ulvae, *Journal of Sea Research* **48**, 1, pp. 51–58, doi:10.1016/S1385-1101(02)00130-2.

Anthony, E. J., Gardel, A., Gratiot, N., Proisy, C., Allison, M. A., Dolique, F., and Fromard, F. (2010). The Amazon-influenced muddy coast of South America: A review of mud–bank–shoreline interactions, doi: 10.1016/j.earscirev.2010.09.008.

Ariathurai, C. (1974). *A Finite Element Model for Sediment Transport in Estuaries*, Ph.D. thesis, University of California, Davis.

Ariathurai, R. and Arulanandan, K. (1978). Erosion rates of cohesive soils, *Journal of the Hydraulics Division of the American Society of Civil Engineers* **104**, 2.

Armanini, A. (1995). Non-uniform sediment transport: Dynamics of the active layer, *Journal of Hydraulic Research* **33**, 5, pp. 611–622, doi: 10.1080/00221689509498560.

Augustinus, P. (1988). Suriname, in *Artificial Structures and Shorelines* (Springer), pp. 689–693.

Augustinus, P. G. (1978). *The Changing Shoreline of Surinam (South America).*, Ph.D. thesis, University Utrecht.

Augustinus, P. G. E. F. (2004). The influence of the trade winds on the coastal development of the Guianas at various scale levels: A synthesis, *Marine Geology* **208**, 2–4, pp. 145–151.

Austen, I., Andersen, T. J., and Edelvang, K. (1999). The influence of benthic diatoms and invertebrates on the erodibility of an intertidal mudflat, the Danish Wadden Sea, *Estuarine, Coastal and Shelf Science* **49**, 1, doi: 10.1006/ecss.1998.0491.

Bagnold, R. A. (1954). Experiments on a gravity-free dispersion of large solid spheres in a Newtonian fluid under shear, *Proceedings of the Royal Society*, London, Vol. A225, pp. 49–63.

Bailey, M. C. and Hamilton, D. P. (1997). Wind induced sediment resuspension: A lake-wide model, *Ecological Modelling* **99**, 2–3, pp. 217–228, doi: 10.1016/S0304-3800(97)01955-8.

Baldock, T. E., Tomkins, M. R., Nielsen, P., and Hughes, M. G. (2004). Settling velocity of sediments at high concentrations, *Coastal Engineering* **51**, 1, pp. 91–100, doi:10.1016/j.coastaleng.2003.12.004.

Balke, T., Webb, E. L., Van den Elzen, E., Galli, D., Herman, P. M., and Bouma, T. J. (2013). Seedling establishment in a dynamic sedimentary environment: A conceptual framework using mangroves, *Journal of Applied Ecology* **50**, 3, pp. 740–747, doi:10.1111/1365-2664.12067.

Baptist, M. J., Babovic, V., Rodriguez Uthurburu, J., Keijzer, M., Uittenbogaard, R. E., Mynett, A., and Verwey, A. (2009). On inducing equations for vegetation resistance, *Journal of Hydraulic Research Research* **47**, 2, pp. 281–282.

Barciela Rial, M. (2019). *Consolidation and Drying of Slurries: A Building with Nature Study for the Marker Wadden*, Ph.D. thesis, TU Delft University.

Barciela-Rial, M., Van Paassen, L. A., Griffioen, J., Van Kessel, T., and Winterwerp, J. C. (2020). The effect of solid-phase composition on the drying behavior of Markermeer sediment, *Vadose Zone Journal* doi: 10.1002/vzj2.20028.

Barenblatt, G. I. (1953). On the motion of suspended particles in a turbulent flow. *Prikladnaya Matematika i Mekhanika (Applied Mathematics and Mechanics)* **17**, 3, pp. 261–274 (in Russian).

Barr, D. I. H. (1967). Densimetric exchange flow in rectangular channels III. Large scale experiments, *La Houille Blanche* **22**, pp. 619–632.

Baudez, J. C., Chabot, F., and Coussot, P. (2002). Rheological interpretation of the slump test, *Applied Rheology* **12**, 3, pp. 133–141, doi:10.1515/arh-2002-0008.

Beardsley, R., Candela, J., Limeburner, R., Geyer, W., Lentz, S., Castro, B., Cacchione, D., and Carneiro, N. (1995). The M2 tide on the Amazon shelf, *Journal of Geophysical Research* **100**, pp. 2283–2319.

Bedford, K. W., Wai, O., Libicki, C. M., and Van Evra, R. (1987). Sediment entrainment and deposition measurements in long Island sound, *Journal of Hydraulic Engineering* **113**, 10, pp. 1325–1342, doi:10.1061/(asce)0733-9429(1987)113:10(1325).

Been, K. and Sills, G. C. (1981). Self-weight consolidation of soft soils: An experimental and theoretical study, *Geotechnique* **31**, 4, pp. 519–535, doi: 10.1680/geot.1981.31.4.519.

Benson, T. and French, J. R. (2007). InSiPID: A new low-cost instrument for in situ particle size measurements in estuarine and coastal waters, *Journal of Sea Research* **58**, 3, pp. 167–188, doi:10.1016/j.seares.2007.04.003.

Bergaya, F. and Lagaly, G. (2013). *Handbook of Clay Science*, 2nd edn. (Elsevier).

Biggs, R. B. and Howell, B. A. (1984). The estuary as a sediment trap: Alternate approaches to estimating its filtering efficiency, in *The Estuary As a Filter* (Elsevier), pp. 107–129, doi:10.1016/b978-0-12-405070-9.50012-8.

Bijker, W. E. (1980). Sedimentation in channels and trenches. *In Proc. Seventeenth Coastal Engng. Conf.,* **2**, (Sydney, Australia: Mar. 23-28, 1980), pp. 1708–1718, doi:10.1061/9780872622647.104.

Bird, R. B., Stewart, W. E., and Lightfoot, E. N. (1960). *Transport Phenomena* (Wiley, New York).

Blaas, M. and Van den Boogaard, H. (2006). Statistical methods to assess the impact of MV2 on SPM along the Dutch coast. Results of WP 3 and 5. Tech. Rep., WL—Delft Hydraulics Report Z4046 (RKZ-1661), Delft.

Blanchard, G. F., Paterson, D. M., Stal, L. J., Richard, P., Galois, R., Huet, V., Kelly, J., Honeywill, C., De Brouwer, J., Dyer, K., Christie, M., and Seguignes, M. (2000). The effect of geomorphological structures on potential biostabilisation by microphytobenthos on intertidal mudflats, *Continental Shelf Research* **20**, 10–11, pp. 1243–1256, doi:10.1016/S0278-4343(00)00021-2.

Blok, B. (2008). Monitoring Markermeer. Data rapport on measurement program FL41/FL42 at Markermeer, August–September 2007, Tech. Rep., Deltares Report no. G0200.10, Delft.

Bockelmann, F. D., Puls, W., Kleeberg, U., Müller, D., and Emeis, K. C. (2018). Mapping mud content and median grain-size of North Sea sediments: A geostatistical approach, *Marine Geology* **397**, pp. 60–71, doi: 10.1016/j.margeo.2017.11.003.

Boderie, P. and Vijverberg, T. (2008). Scenario computations Markermeer (in Dutch). Tech. Rep., Deltares report Q4613, Delft.

Booij, N., Ris, R. C., and Holthuijsen, L. H. (1999). A third-generation wave model for coastal regions: 1. Model description and validation, *Journal of Geophysical Research: Oceans* **104**, C4, pp. 7649–7666.

Bosboom, J. and Stive, M. J. F. (2015). Coastal Dynamic I: Lecture Notes CIE4305, *Delft, The Netherlands: Delft University of Technology,* p. 537.

Bouma, T. J., De Vries, M. B., and Herman, P. M. (2010). Comparing ecosystem engineering efficiency of two plant species with contrasting growth strategies, *Ecology* **91**, 9, doi:10.1890/09-0690.1.

Box, G. E. (1976). Science and statistics, *Journal of the American Statistical Association* **71**, 356, pp. 791–799, doi:10.1080/01621459.1976.10480949.

Braat, L., Van Kessel, T., Leuven, J. R., and Kleinhans, M. G. (2017). Effects of mud supply on large-scale estuary morphology and development over centuries to millennia, *Earth Surface Dynamics* **5**, 4, doi:10.5194/esurf-5-617-2017.

Brakenhoff, L. B., Smit, Y., Donker, J. J., and Ruessink, G. (2019). Tide-induced variability in beach surface moisture: Observations and modelling, *Earth Surface Processes and Landforms* doi:10.1002/esp.4493.

Bretschneider, C. (1954). Generation of wind waves over a shallow bottom, Tech. Rep., Texas A & M Research Foundation College Station.

Brière, C. and Van Kessel, T. (2006). Modelling of seasonal mud supply in the Dutch coastal zone. Tech. Rep., WL — Delft Hydraulics Report Z4150, Delft.

Bronick, C. J. and Lal, R. (2005). Soil structure and management: A review, doi:10.1016/j.geoderma.2004.03.005.

Bruens, A., Van Kessel, T., De Kok, J., and De Grave, P. (2007). Results R&D mud 2007, Tech. Rep., WL — Delft Hydraulics report Z4374, Delft.

Bruens, A. W., Winterwerp, J. C., and Kranenburg, C. (2012). Physical and numerical modeling of the entrainment by a high-concentration mud suspension, *Journal of Hydraulic Engineering* **138**, 6, pp. 479–490.

Burchard, H., Schuttelaars, H. M., and Ralston, D. K. (2018). Sediment trapping in estuaries, *Annual Review of Marine Science* **10**, pp. 371–395, doi: 10.1146/annurev-marine-010816-060535.

Buschmann, M. H. and Gad-El-Hak, M. (2010). Normal and cross-flow Reynolds stresses: Differences between confined and semi-confined flows, *Experiments in Fluids* **49**, 1, pp. 213–223, doi:10.1007/s00348-010-0834-z.

Calliari, L. J., Winterwerp, J. C., Fernandes, E., Cuchiara, D., Vinzon, S. B., Sperle, M., and Holland, K. T. (2009). Fine grain sediment transport and deposition in the Patos Lagoon-Cassino beach sedimentary system, *Continental Shelf Research* **29**, 3, doi:10.1016/j.csr.2008.09.019.

Camenen, B. and Van Bang, D. P. (2011). Modelling the settling of suspended sediments for concentrations close to the gelling concentration, *Continental Shelf Research* **31**, 10 Suppl., pp. S106–S161, doi:10.1016/j.csr.2010.07.003.

Chao, X., Jia, Y., Shields Jr, F. D., Wang, S. S. Y., and Cooper, C. M. (2008). Three-dimensional numerical modeling of cohesive sediment transport and wind wave impact in a shallow oxbow lake, *Advances in Water Resources* **31**, 7, pp. 1004–1014.

Chassagne, C. and Bedeaux, D. (2008). The dielectric response of a colloidal spheroid, *Journal of Colloid and Interface Science* **326**, 1, pp. 240–253, doi: 10.1016/j.jcis.2008.06.055.

Chassagne, C., Mietta, F., and Winterwerp, J. C. (2009). Electrokinetic study of kaolinite suspensions, *Journal of Colloid and Interface Science* **336**, 1, pp. 352–359, doi:10.1016/j.jcis.2009.02.052.

Chen, M. S., Wartel, S., Van Eck, B., and Van Maldegem, D. (2005). Suspended matter in the Scheldt estuary, *Hydrobiologia* **540**, 1–3, pp. 79–104, doi:10.1007/s10750-004-7122-y.

Chernetsky, A. S., Schuttelaars, H. M., and Talke, S. A. (2010). The effect of tidal asymmetry and temporal settling lag on sediment trapping in tidal estuaries, *Ocean Dynamics* **60**, 5, doi:10.1007/s10236-010-0329-8.

Chien, K.-Y. (1982). Predictions of channel and boundary-layer flows with a low-Reynolds-number turbulence model, *AIAA Journal* **20**, 1, pp. 33–38.

Chou, H.-T., Foda, M. A., and Hunt, J. R. (1993). Rheological response of cohesive sediments to oscillatory forcing, in *Coastal and Estuarine Studies*, Vol. 42 (Wiley Online Library), pp. 126–147.

Christie, M. C., Dyer, K. R., Blanchard, G., Cramp, A., Mitchener, H. J., and Paterson, D. M. (2000). Temporal and spatial distributions of moisture and organic contents across a macro-tidal mudflat, *Continental Shelf Research* **20**, 10–11, pp. 1219–1241, doi:10.1016/S0278-4343(00)00020-0.

Chung, E. G., Bombardelli, F. A., and Schladow, S. G. (2009). Modeling linkages between sediment resuspension and water quality in a shallow, eutrophic, wind-exposed lake, *Ecological Modelling* **220**, 9–10, pp. 1251–1265, doi: 10.1016/j.ecolmodel.2009.01.038.

Cleveringa J. (2013). Grootschalige sedimentbalans Van de Westerschelde. Tech. Rep., LTV V&T-Rapport K-17. C04021.002666.0100.

Coleman, N. L. (1981). Velocity profiles with suspended sediment, *Journal of Hydraulic Research* **19**, 3, pp. 211–229, doi:10.1080/00221688109499516.

Coles, D. (1956). The law of the wake in the turbulent boundary layer, *Journal of Fluid Mechanics*, **1**, 2, pp. 191–226.

Colosimo, I., De Vet, P. L. M., Van Maren, D. S., Reniers, A. J. H. M., Winterwerp, J. C., and Van Prooijen, B. C. (2020). The impact of wind on flow and sediment transport over intertidal flats, *J. Mar. Sci. Eng.* **8**, 11, p. 910; https://doi.org/10.3390/jmse8110910.

Corey, A. (1949). *Influence of Shape on the* 1 *Fall Velocity of Sand Grains.*, Ph.D. thesis, Colorado A and M College, Fort Collins, Dankers, P. (2006). *On the Hindered Settling of Suspensions of Mud and Mud-Sand Mixtures*, Ph.D. thesis, Delft University of Technology, Delft, The Netherlands.

Cournane, S., León Vintró, L., and Mitchell, P. I. (2010). Modelling the reworking effects of bioturbation on the incorporation of radionuclides into the sediment column: Implications for the fate of particle-reactive radionuclides in Irish Sea sediments, *Journal of Environmental Radioactivity* **101**, 11, pp. 985–991, doi: 10.1016/j.jenvrad.2010.07.006.

Coussot, P. (1997). *Mudflow Rheology and Dynamics* (Routledge), doi: 10.1061/(asce)0733-9429(2001)127:4(331).

Coussot, P. (2005). *Rheometry of Pastes, Suspensions, and Granular Materials: Applications in Industry and Environment* (John Wiley & Sons), doi: 10.1002/0471720577.

Cox, T. J., Maris, T., Van Engeland, T., Soetaert, K., and Meire, P. (2019). Critical transitions in suspended sediment dynamics in a temperate meso-tidal estuary, *Scientific Reports* **9**, 1, doi:10.1038/s41598-019-48978-5.

Croad, R. (1981). Physics of erosion of cohesive soils, Tech. Rep., University of Auckland, Auckland.

Cronin, K. (2013). MoS2-II: Deterministic model calibration: updates of the ZUNO-DD hydrodynamic and SPM model. Tech. Rep., Deltares Report 1204561, Delft.

Cronin, K. and Blaas, M. (2015). Maasvlakte 2 and fine sediment fluxes towards the Wadden Sea. A model analysis of pre and post construction phases. Tech. Rep., Deltares Report 1210059, Delft.

Cross, M. M. (1965). Rheology of non-Newtonian fluids: A new flow equation for pseudoplastic systems, *Journal of Colloid Science* **20**, 5, pp. 417–437, doi:10.1016/0095-8522(65)90022-X.

Cuong, C. V. and Brown, S. (2012). Coastal rehabilitation and mangrove 32 restoration using Melaleuca Fences: Practical experience form Kien Giang, Province. Agricultural Publishing House, Ho Chi Minh City, Vietnam.

Cuthbertson, A. J., Ibikunle, O., McCarter, W. J., and Starrs, G. (2016). Monitoring and characterisation of sand-mud sedimentation processes, *Ocean Dynamics* **66**, 6–7, pp. 867–891, doi:10.1007/s10236-016-0958-7.

Dalrymple, R. A. and Liu, F. P. (1978). Waves over soft muds: A two-layer fluid model. *Journal of Physical Oceanography* **8**, 6, Nov.1978, pp. 1121–1131. doi:10.1175/1520-0485(1978)008<1121:wosmat>2.0.co;2.

Damen, J. M., Van Dijk, T. A., and Hulscher, S. J. (2018). Spatially varying environmental properties controlling observed sand wave morphology, *Journal of Geophysical Research: Earth Surface* **123**, 2, pp. 262–280, doi: 10.1002/2017JF004322.

Dankers, N. and Koelemaij, K. (1989). Variations in the mussel population of the Dutch Wadden Sea in relation to monitoring of other ecological parameters, *Helgoländer Meeresuntersuchungen* **43**, 3–4, pp. 529–535, doi: 10.1007/BF02365909.

Dankers, P. (2006). *On the Hindered Settling of Suspensions of Mud and Mud–Sand Mixtures*, Ph.D. thesis, Delft University of Technology, Delft.

Dankers, P. J. and Winterwerp, J. C. (2007). Hindered settling of mud flocs: Theory and validation, *Continental Shelf Research* **27**, 14, pp. 1893–1907, doi:10.1016/j.csr.2007.03.005.

Das, B. (2008). *Advanced Soil Mechanics*, 3rd edn. (Taylor and Francis, New York).

De Boer, G. J. (2009). *On the Interaction Between Tides and Stratification in the Rhine Region of Freshwater Influence*, Ph.D. thesis, Delft University of Technology, Delft, The Netherlands.

De Boer, G. J., Pietrzak, J. D., and Winterwerp, J. C. (2006). On the vertical structure of the Rhine region of freshwater influence, *Ocean Dynamics* **56**, 3–4, pp. 198–216, doi:10.1007/s10236-005-0042-1.

De Brouwer, J. F., Bjelic, S., De Deckere, E. M., and Stal, L. J. (2000). Interplay between biology and sedimentology in a mudflat (Biezelingse Ham, Westerschelde, The Netherlands), *Continental Shelf Research* **20**, 10–11, pp. 1159–1177, doi:10.1016/S0278-4343(00)00017-0.

De Deckere, E. M., Kornman, B. A., Staats, N., Termaat, G. R., Winder, B., Stal, L. J., and Heip, C. H. (2002). The seasonal dynamics of benthic (micro) organisms and extracellular carbohydrates in an intertidal mudflat and their effect on the concentration of suspended sediment, in *Proceedings in Marine Science*, Vol. 5, pp. 429–440, doi:10.1016/S1568-2692(02)80031-6.

De Jonge, V. N., Schuttelaars, H. M., Van Beusekom, J. E., Talke, S. A., and De Swart, H. E. (2014). The influence of channel deepening on estuarine turbidity levels and dynamics, as exemplified by the Ems estuary, *Estuarine, Coastal and Shelf Science* **139**, pp. 46–59, doi:10.1016/j.ecss.2013.12.030.

De Lucas Pardo, M. A. (2014). *Effect of biota on fine sediment transport processes. A study of Lake Markermeer*, Ph.D. thesis, Delft University of Technology, Delft, The Netherlands.

De Lucas Pardo, M. A., Bakker, M., Van Kessel, T., Cozzoli, F., and Winterwerp, J. C. (2013). Erodibility of soft freshwater sediments in Markermeer: The role of bioturbation by meiobenthic fauna Topical Collection on the 11th International Conference on Cohesive Sediment Transport, *Ocean Dynamics* **63**, 9–10, pp. 1137–1150, doi:10.1007/s10236-013-0650-0.

De Nijs, M. A., Winterwerp, J. C., and Pietrzak, J. D. (2009). On harbour siltation in the fresh-salt water mixing region, *Continental Shelf Research* **29**, 1, pp. 175–193, doi:10.1016/j.csr.2008.01.019.

De Swart, H. E. and Zimmerman, J. T. (2009). Morphodynamics of tidal inlet systems, doi:10.1146/annurev.fluid.010908.165159.

De Vet, P. L. M., Van Prooijen, B. C., Schrijvershof, R. A., Van der Werf, J. J., Ysebaert, T., Schrijver, M. C., and Wang, Z. B. (2018). The Importance of Combined Tidal and Meteorological Forces for the Flow and Sediment Transport on Intertidal Shoals, *J. Geophys. Res. Earth Surf.* **123**, pp. 2464–2480.

De Vriend, H. J. (1981). Velocity redistribution in curved rectangular channels, *Journal of Fluid Mechanics* **107**, HY10, pp. 423–439, doi: 10.1017/S0022112081001833.

De Vriend, H. J. (1991a). Mathematical modelling and large-scale coastal behavior — Part 2: Predictive models, *Journal of Hydraulic Research* **29**, 6, pp. 741–753.

De Vriend, H. J. (1991b). Mathematical modelling and large-scale coastal behaviour — Part 1: Physical processes, *Journal of Hydraulic Research* **29**, 6, pp. 727–740.

De Wit, P. J. and Kranenburg, C. (1993). Liquefaction and erosion of china clay due to waves and current, in *Proceedings of the Coastal Engineering Conference*, Vol. 3, ISBN 0872629333, pp. 2937–2948, doi: 10.1061/9780872629332.224.

De Wit, P. J. and Kranenburg, C. (1997). On the liquefaction and erosion of mud due to waves and current, in *Cohesive Sediments* (John Wiley & Sons), pp. 331–340.

Dean, R. G. and Dalrymple, R. A. (2001). *Coastal Processes with Engineering Applications* (Cambridge University Press), doi:10.1017/cbo9780511754500.

Decho, A. W. (2000). Microbial biofilms in intertidal systems: An overview, *Continental Shelf Research* **20**, 10–11, pp. 1257–1273, doi:10.1016/S0278-4343(00)00022-4.

Dekker, A. G., Vos, R. J., and Peters, S. W. (2001). Comparison of remote sensing data, model results and in situ data for total suspended matter (TSM) in the southern Frisian lakes, *Science of the Total Environment* **268**, 1–3, pp. 197–214, doi:10.1016/S0048-9697(00)00679-3.

Delefortrie, G. and Vantorre, M. (2006). Effects of a muddy bottom on the straight-line stability, in *7th IFAC Conference on Manoeuvring and Control of Marine Craft 2006: proceedings.*

Delefortrie, G. and Vantorre, M. (2016). Ship manoeuvring behaviour in muddy navigation areas: State of the art, in *Proceedings of the 4th International Conference on Ship Manoeuvring in Shallow and Confined Water with Special Focus on Ship Bottom Interaction, Hamburg, Germany, 23 to 25 May 2016 (4th MASHCON)*, pp. 26–36.

Delgado, A. V., González-Caballero, F., Hunter, R. J., Koopal, L. K., and Lyklema, J. (2005). Measurement and interpretation of electrokinetic phenomena (IUPAC technical report), *Pure and Applied Chemistry* **77**, 10, pp. 1753–1805.

Deltares (2018). Delft3D-FLOW User Manual. Version: 3.15. SVN Revision: 54312. February 7, 2018. Tech. Rep., Deltares, Delft.

Dijkstra, Y. M., Schuttelaars, H. M., and Schramkowski, G. P. (2019a). A regime shift from low to high sediment concentrations in a tide-dominated estuary, *Geophysical Research Letters* **46**, 8, pp. 4338–4345, doi: 10.1029/2019GL082302.

Dijkstra, Y. M., Schuttelaars, H. M., and Schramkowski, G. P. (2019b). Can the Scheldt River Estuary become hyperturbid? *Ocean Dynamics* **69**, 7, pp. 809–827, doi:10.1007/s10236-019-01277-z.

Dijkstra, Y. M., Schuttelaars, H. M., Schramkowski, G. P., and Brouwer, R. L. (2019c). Modeling the transition to high sediment concentrations as a response to channel deepening in the Ems River Estuary, *Journal of Geophysical Research: Oceans* **124**, 3, pp. 1578–1594, doi:10.1029/2018JC014367.

Dijkstra, Y. M., Schuttelaars, H. M., and Winterwerp, J. C. (2018). The hyperturbid state of the water column in estuaries and rivers: The importance of hindered settling, *Ocean Dynamics* **68**, 3, pp. 377–389.

Dingemans, M. W. (1997). *Water Wave Propagation Over Uneven Bottoms*, Vol. 13 (World Scientific).

Dong, L., Wolanski, E., and Li, Y. (1997). Field and modeling studies of fine sediment dynamics in the extremely turbid Jiaojiang River Estuary, China, *Journal of Coastal Research* **13**, 4, pp. 995–1003.

Dronkers, J. (2016). *Dynamics of Coastal Systems: Second edition*, Vol. 41 (World Scientific), doi:10.1142/9818.

Duran-Matute, M., Gerkema, T., and Sassi, M. G. (2016). Quantifying the residual volume transport through a multiple-inlet system in response to wind forcing: The case of the western Dutch Wadden Sea, *Journal of Geophysical Research: Oceans* **121**, 12, pp. 8888–8903, doi:10.1002/2016JC011807.

Dyer, K. (1997). *Estuaries: A Physical Introduction.*, 2nd edn. (John Wiley & Sons, Chichester).

Dyer, K. R. and Manning, A. J. (1999). Observation of the size, settling velocity and effective density of flocs, and their fractal dimensions, *Journal of Sea Research* **41**, 1–2, pp. 87–95, doi:10.1016/S1385-1101(98)00036-7.

Eckart, C. (1948). The approximate solution of one-dimensional wave equation, *Review of Modern Physics* **20**, pp. 399–417, doi: 10.1103/RevModPhys.20.399.

Edmonds, D. A. and Slingerland, R. L. (2007). Mechanics of river mouth bar formation: Implications for the morphodynamics of delta distributary networks, *Journal of Geophysical Research: Earth Surface* **112**, 2, doi: 10.1029/2006JF000574.

Einstein, H. A. (1950). The bed-load function for sediment transportation in open channel flows, Tech. Rep. 1026, US Dept. of Agriculture.

Einstein, H. A. and Qian, N. (1955). Effects of heavy sediment concentration near the bed on velocity and sediment distribution. *University of California, Institute of Engineering Research, Berkeley, California*, pp. 1–98.

Eisma, D. (1981). Supply and deposition of suspended matter in the North Sea, *Holocene Marine Sedimentation in the North Sea Basin*, pp. 415–428.

Eisma, D. (2009). Supply and deposition of suspended matter in the North Sea, in *Holocene Marine Sedimentation in the North Sea Basin* (Wiley), pp. 415–428, doi:10.1002/9781444303759.ch29.

Ekman, V. W. (1904). On dead water, *Norwegian North Polar Expedition*, 1893–1896, pp. 1–150.

Elder, J. (1959). The dispersion of marked fluid in turbulent shear flow, *Journal of Fluid Mechanics* **5**, 4, pp. 544–560.

El Ganaoui, O., Schaaff, E., Boyer, P., Amielh, M., Anselmetc, F., and Grenz, C. (2004). The deposition and erosion of cohesive sediments determined by a multi-class model, *Estuar Coast Shelf S* **60**, pp. 457–475.

Ettema, R. and Mutel, C. (2014). *Hans Albert Einstein: His Life as a Pioneering Engineer* (ASCE Press), https://ascelibrary.org/doi/abs/10.1061/9780784413302.

Eysink, W. D. (1989). Sedimentation in harbour basins. Small density differences may cause serious effects, in *9th International Harbour Congress*, 417, p. 13.

Eysink, W. D. and Vermaas, H. (1983). Computational methods to estimate the sedimentation in dredged channels and harbour basins in estuarine environments, in *International Conference on Coastal and Port Engineering in*

Developing Countries, 1st International Association for Hydraulic Research, p. 12.

Fernandes, E. H. L., Dyer, K. R., Moller, O. O., and Niencheski, L. F. H. (2002). The Patos lagoon hydrodynamics during an El Nino event (1998), *Continental Shelf Research* **22**, 11–13, pp. 1699–1713.

Fettweis, M., Sas, M., and Monbaliu, J. (1998). Seasonal, neap-spring and tidal variation of cohesive sediment concentration in the Scheldt Estuary, Belgium, *Estuarine, Coastal and Shelf Science* **47**, 1, pp. 21–36, doi: 10.1006/ecss.1998.0338.

Fettweis, M. and Van den Eynde, D. (2003). The mud deposits and the high turbidity in the Belgian-Dutch coastal zone, southern bight of the North Sea, *Continental Shelf Research* **23**, 7, pp. 669–691, doi:10.1016/S0278-4343(03)00027-X.

Fischer, H. B. (1976). Mixing and Dispersion in Estuaries, *Annual Review of Fluid Mechanics* **8**, 1, pp. 107–133, doi:10.1146/annurev.fl.08.010176.000543.

Flemming, B. W. (2000). A revised textural classification of gravel-free muddy sediments on the basis of ternary diagrams, *Continental Shelf Research* **20**, 10–11, pp. 1125–1137, doi:10.1016/S0278-4343(00)00015-7.

Flemming, B. W. and Delafontaine, M. T. (1994). Biodeposition in a juvenile mussel bed of the east Frisian Wadden Sea (Southern North Sea), *Netherlands Journal of Aquatic Ecology* **28**, 3–4, pp. 289–297, doi:10.1007/BF02334197.

Foda, M. A. and Tzang, S. Y. (1994). Resonant fluidization of silty soil by water waves, *Journal of Geophysical Research* **99**, C10, doi:10.1029/94jc02040.

Fontein, W. F. and Byrd, R. W. (2007). The nautical depth approach: A review for implementation. WODCON XVIII Annual Dredging Seminar.

Fredlund, M. D., Wilson, G. W., and Fredlund, D. G. (2002). Representation and estimation of the shrinkage curve.

Fredsøe, J. and Deigaard, R. (1992). *Mechanics of Coastal Sediment Transport, Advanced Series on Ocean Engineering*, Vol. 3 (World Scientific), doi: 10. 1142/1546.

Friedrichs, C. T. (2010). Barotropic tides in channelized estuaries, in *Contemporary Issues in Estuarine Physics* (Cambridge University Press), pp. 27–61, doi:10.1017/CBO9780511676567.004.

Friedrichs, C. T. (2012). Tidal flat morphodynamics: A synthesis, in *Treatise on Estuarine and Coastal Science*, Vol. 3 (Academic Press, Cambridge), pp. 137–170, doi:10.1016/B978-0-12-374711-2.00307-7.

Friedrichs, C. T. and Aubrey, D. G. (1988). Non-linear tidal distortion in shallow well-mixed estuaries: A synthesis, *Estuarine, Coastal and Shelf Science* **27**, 5, pp. 521–545.

Fugate, D. C. and Friedrichs, C. T. (2003). Controls on suspended aggregate size in partially mixed estuaries, *Estuarine, Coastal and Shelf Science* **58**, 2, pp. 389–404, doi:10.1016/S0272-7714(03)00107-0.

Gabioux, M., Vinzon, S. B., and Paiva, A. M. (2005). Tidal propagation over fluid 18 mud layers on the Amazon shelf, *Continental Shelf Research* **25**, 1, pp. 113–125, doi:10.1016/j.csr.2004.09.001.

Gade, H. G. (1958). Effects of a nonrigid, impermeable bottom on plane surface waves in shallow water, *Journal of Marine Research* **16**, pp. 61–81.

Galappatti, G. and Vreugdenhil, C. B. (1985). A depth-integrated model for suspended sediment transport, *Journal of Hydraulic Research* **23**, 4, pp. 359–377.

Garvine, R. W. (1987). Estuary plumes and fronts in shelf waters: A layer model, *Journal of Physical Oceanography* **17**, 11, pp. 1877–1896.

Gatto, V. M., Van Prooijen, B. C., and Wang, Z. B. (2017). Net sediment transport in tidal basins: Quantifying the tidal barotropic mechanisms in a unified framework, *Ocean Dynamics* **67**, 11, pp. 1385–1406.

Gedan, K. B., Kirwan, M. L., Wolanski, E., Barbier, E. B., and Silliman, B. R. (2011). The present and future role of coastal wetland vegetation in protecting shorelines: Answering recent challenges to the paradigm, doi: 10.1007/s10584-010-0003-7.

Georgiadis, N. J., Chitsomboon, T., and Zhu, G. (1994). *Modification of the Two-Equation Turbulence Model in NPARC to a Chien Low Reynolds Number Ke Formulation*, Vol. 106710 (Lewis Research Center).

Gerkema, T. (2019). *An Introduction to Tides* (Cambridge University Press), doi:10.1017/9781316998793.

Geyer, W. R. (2010). Estuarine salinity structure and circulation, in *Contemporary Issues in Estuarine Physics* (Cambridge University Press), pp. 12–26, doi:10.1017/CBO9780511676567.003.

Gibson, R. E., England, G. L., and Hussey, M. J. (1967). The theory of one-dimensional consolidation of saturated clays, *Geotechnique* **17**, 3, pp. 261–273, doi:10.1680/geot.1967.17.3.261.

Goldberg, U. and Apsley, D. (1997). A wall-distance-free low Re k - epsilon turbulence model, *Computer Methods in Applied Mechanics and Engineering* **145**, 3–4, doi:10.1016/S0045-7825(96)01202-9.

Gowland, J. E., Mehta, A. J., Stuck, J. D., John, C. V., and Parchure, T. M. (2007). Organic-rich fine sediments in Florida — Part II: Resuspension in a lake, in *Proceedings in Marine Science*, Vol. 8 (Elsevier), pp. 167–188.

Graf, G. and Rosenberg, R. (1997). Bioresuspension and biodeposition: A review, *Journal of Marine Systems* **11**, 3–4, pp. 269–278, doi:10.1016/S0924-7963(96)00126-1.

Grant, W. D. and Madsen, O. S. (1979). Combined wave and current interaction with a rough bottom, *Journal of Geophysical Research: Oceans* **84**, C4, pp. 1797–1808, doi:10.1029/jc084ic04p01797.

Grasso, F., Le Hir, P., and Bassoullet, P. (2015). Numerical modelling of mixed-sediment consolidation, *Ocean Dynamics* **65**, 4, pp. 607–616, doi: 10.1007/s10236-015-0818-x.

Groen, P. (1967). On the residual transport of suspended matter by an alternating tidal current, *Netherlands Journal of Sea Research* **3**, 4, pp. 564–574, doi: 10.1016/0077-7579(67)90004-X.

Gualtieri, C., Angeloudis, A., Bombardelli, F., Jha, S., and Stoesser, T. (2017). On the values for the turbulent schmidt number in environmental flows, doi: 10.3390/fluids2020017.

Guo, J. (2002). Logarithmic matching and is applications in computational hydraulics and sediment transport, *Journal of Hydraulic Research* **40**, 5, pp. 555–565.

Gust, G. and Müeller, V. (1997). Interfacial hydrodynamics and entrainment functions of currently used erosion devices, in *Cohesive Sediments*, 1 (John Wiley & Sons ltd, New York), pp. 149–174.

Hansen, D. V. and Rattray, M. (1965). Gravitational circulation in straits and estuaries, *Journal of Marine Research* **23**, pp. 104–122.

Harbeck, G. E. and Meyers, J. S. (1970). Present day evaporation measurement techniques, *Journal of the Hydraulics Division* **96**, 7, pp. 1381–1390.

Harrison, A. J. M. and Owen, M. W. (1971). Siltation of fine sediments in estuaries, in *Proceedings of the XIV IAHR Congress*, Paper D1.

Harrison, W. D., Musgrave, D., and Reeburgh, W. S. (1983). A wave-induced transport process in marine sediments. *Journal of Geophysical Research* **88**, C12, pp. 7617–7622, doi:10.1029/JC088iC12p07617.

Hayter, E. J. (1986). XVI. Estuarial sediment bed model. *Estuarine Cohesive Sediment Dynamics. Proc. Workshop*, Tampa, 1984, pp. 326–359.

Heaps, N. S. (1972). Estimation of density currents in the Liverpool Bay area of the Irish Sea, *Geophysical Journal of the Royal Astronomical Society* **30**, 4, pp. 415–432, doi:10.1111/j.1365-246X.1972.tb05825.x.

Hendriks, E. and Schuurman, F. (2017). Modellering alternatieve loswallocaties. Tech. Rep., Deltares and Royal Haskoning DHV Report 1230860, Delft.

Herrling, G. and Niemeyer, H. D. (2008). Comparison of the hydrodynamic regime of 1937 and 2005 in the Ems-Dollard estuary by applying mathematical modeling, *NLWKN Coastal Research Station*.

Herschel, V. W. H. (1926). Consistency of measurements rubber-benzene solutions, *Kolloid-Zeitschrift* **39**, pp. 291–300 (in German).

Hibma, A., Schuttelaars, H. M., and de Vriend, H. J. (2004). Initial formation and long-term evolution of channel-shoal patterns, *Continental Shelf Research* **24**, 15, pp. 1637–1650, doi:10.1016/j.csr.2004.05.003.

Hinze, J. O. (1975). *Turbulence* (McGraw-Hill, New York).

Hofland, B. and Battjes, J. A. (2006). Probability density function of instantaneous drag forces and shear stresses on a bed, *Journal of Hydraulic Engineering* **132**, 11, pp. 1169–1175, doi:10.1061/(asce)0733-9429(2006)132:11(1169).

Hofland, B., Christiansen, H., Crowder, R., Kirby, R., Leeuwen, C. V., and Winterwerp, J. (2001). The current deflecting wall in an estuarine harbour, *Proceedings of the XXIV IAHR Congress*, Beijing, pp. 242–250.

Hoitink, A. J., Hoekstra, P., and Van Maren, D. S. (2003). Flow asymmetry associated with astronomical tides: Implications for the residual transport of sediment, *Journal of Geophysical Research: Oceans* **108**, 10, doi: 10.1029/2002jc001539.

Hoitink, A. J. F. and Jay, D. A. (2016). Tidal river dynamics: Implications for deltas, *Reviews of Geophysics* **54**, 1, pp. 240–272.

Holthuijsen, L. (2010). *Waves in Oceanic and Coastal Waters* (Cambridge University Press).

Horstman, E. M., Bryan, K. R., Mullarney, J. C., Pilditch, C. A., and Eager, C. A. (2018). Are flow-vegetation interactions well represented by mimics? A case study of mangrove pneumatophores, *Advances in Water Resources* **111**, pp. 360–371.

Houwing, E.-J. and Van Rijn, L. C. (1998). In situ erosion flume (ISEF): Determination of bed-shear stress and erosion of a kaolinite bed, *Journal of Sea Research* **39**, 3–4, pp. 243–253.

Hsu, T. J., Traykovski, P. A., and Kineke, G. C. (2007). On modeling boundary layer and gravity-driven fluid mud transport, *Journal of Geophysical Research: Oceans* **112**, 4, doi:10.1029/2006JC003719.

Huettel, M. and Rusch, A. (2000). Transport and degradation of phytoplankton in permeable sediment, *Limnology and Oceanography* **45**, 3, pp. 534–549, doi:10.4319/lo.2000.45.3.0534.

Huettel, M., Ziebis, W., and Forster, S. (1996). Flow-induced uptake of particulate matter in permeable sediments, *Limnology and Oceanography* **41**, 2, pp. 309–322, doi:10.4319/lo.1996.41.2.0309.

Huisman, M. and Van Kesteren, W. (1998). Consolidation theory applied to the capillary suction time (CST) apparatus, *Water Science and Technology* **37**, 6–7, doi:10.2166/wst.1998.0743.

Hwang, C. B. and Lin, C. A. (1998). Improved low-reynolds number k-ϵ model based on direct numerical simulation data, *AIAA Journal* **36**, 1, pp. 38–43.

Ibanez Sanz, M. E. (2018). *Flocculation and Consolidation of Cohesive Sediments Under the Influence of Coagulant and Flocculant*, Ph.D. thesis, Delft University of Technology.

IMDC (2013). LTV O & M Veiligheid en Toegankelijkheid: Harde lagen Beneden-Zeeschelde. Tech. Rep., IMDC, Report I/RA/11387/12.112/GVH.

Inglis, C. and Allen, F. (1957). Maritime paper no. 38: The regimen of the thames estuary as affected by currents, salinities, and river flow. *Proc. Institution of Civil Engineers* doi:10.1680/iicep.1957.2705.

Ippen, A. T. and Harleman, D. R. (1961). One-dimensional analysis of salinity intrusion in estuaries, Tech. Rep., Army Engineer Waterways Experiment Station Vicksburg Ms, Technical Bulletin No. 5, Committee on Tidal Hydraulics.

Jacobs, W. (2011). *Sand–Mud Erosion from a Soil Mechanical Perspective*, Ph.D. thesis, Delft University of Technology, Delft, The Netherlands.

Jain, M. and Mehta, A. J. (2009). Role of basic rheological models in determination of wave attenuation over muddy seabeds, *Continental Shelf Research* **29**, 3, pp. 642–651.

Jalón-Rojas, I., Schmidt, S., and Sottolichio, A. (2017). Comparison of environmental forcings affecting suspended sediments variability in two macrotidal, highly-turbid estuaries, *Estuarine, Coastal and Shelf Science* **198**, pp. 529–541, doi:10.1016/j.ecss.2017.02.017.

James, R. T., Martin, J., Wool, T., and Wang, P. F. (1997). A sediment resuspension and water quality model of lake Okeechobee 1, *JAWRA Journal of the American Water Resources Association* **33**, 3, pp. 661–678.

Jaramillo, S., Sheremet, A., Allison, M. A., Reed, A. H., and Holland, K. T. (2009). Wave-mud interactions over the muddy Atchafalaya subaqueous clinoform, Louisiana, United States: Wave-supported sediment transport, *Journal of Geophysical Research: Oceans* **114**, C4.

Jay, D. A. and Musiak, J. D. (1994). Particle trapping in estuarine tidal flows, *Journal of Geophysical Research: Oceans* **99**, C10, pp. 20445–20461.

Jay, D. A. and Musiak, J. D. (1996). Internal tidal asymmetry in channel flows: Origins and consequences, *Coastal and Estuarine Studies*, pp. 211–249.

Jennes, M. and Duineveld, G. (1985). Effects of tidal currents on chlorophyll *a* content of sandy sediments in the southern North Sea, *Marine Ecology Progress Series* **21**, pp. 283–287, doi:10.3354/meps021283.

Jensen, J. H., Madsen, E. S., and Fredsøe, J. (1999). Oblique flow over dredged channels. II: Sediment transport and morphology, *Journal of Hydraulic Engineering* **125**, 11, doi:10.1061/(asce)0733-9429(1999)125:11(1190).

Jepsen, R., Roberts, J., and Lick, W. (1997). Effects of bulk density on sediment erosion rates, *Water, Air, and Soil Pollution* **99**, 1–4, pp. 21–31.

Jeuken, M. and Wang, Z. B. (2010). Impact of dredging and dumping on the stability of ebb-flood channel systems, *Coastal Engineering* **57**, 6, pp. 553–566.

Ji, Z. G. (2008). Water quality and eutrophication. Hydrodynamic and Water 6 Quality: Modeling Rivers, Lakes, and Estuaries. Wiley-Interscience, Hoboken, New Jersey.

Jiang, B. (2012). *Assessing the Influence of Suspended Sediment on Effective Hydraulic Roughness from Tidal Damping Aspect a Study Based on Yangtze River Estuary*, Ph.D. thesis, Delft University of Technology.

Jiang, L. and Zhao, Z. (1989). Viscous damping of solitary waves over fluid-mud seabeds, *Journal of Waterway, Port, Coastal, and Ocean Engineering* **115**, 3, pp. 345–362.

Jolliff, J. K., Kindle, J. C., Shulman, I., Penta, B., Friedrichs, M. A., Helber, R., and Arnone, R. A. (2009). Summary diagrams for coupled hydrodynamic-ecosystem model skill assessment, *Journal of Marine Systems* **76**, 1–2, doi: 10.1016/j.jmarsys.2008.05.014.

Jones, C. G., Lawton, J. H., and Shachak, M. (1994). Organisms as ecosystem engineers, in *Ecosystem Management* (Springer), pp. 130–147.

Kamps, R., Koster, E., Eleveld, M., Kuijper, M., and Laanen, M. (2007). Zwevend stof atlas Markermeer 2006, *RWS IJG Report* **3**.

Kandiah, A. and Arulanandan, K. (1974). Hydraulic erosion of cohesive soils, *Transportation Research Record* **497**, pp. 60–68.

Keetels, G., Harezlak, V., Van Kessel, T., Van Rooijen, A., Friocourt, Y., Van der Kaaij, T., and Los, H. (2012). Sand mining North Sea for coastal nourishments 2013-2017. Validation report. Tech. Rep., Deltares Report 1204963, Delft.

Kessel, T. V. (1997). *Generation and Transport of Subaqueous Fluid Mud Layers*, Ph.D. thesis, Technische Universiteit Delft.

Keulegan, G. H. (1966). The mechanism of an arrested saline wedge, *Estuary and Coastline Hydrodynamics* **546**, p. 574.

Khelifa, A. and Hill, P. S. (2006). Models for effective density and settling velocity of flocs, *Journal of Hydraulic Research* **44**, 3, pp. 390–401, doi: 10.1080/00221686.2006.9521690.

Kirby, R. and Parker, W. R. (1974). Seabed density measurements related to echo sounder records. *Dock and Harbour Authority* **54**, 641, pp. 423–424.

Kirby, R. and Parker, W. R. (1983). Distribution and behavior of fine sediment in the Severn Estuary and inner Bristol Channel, U.K. *Canadian Journal of Fisheries and Aquatic Sciences* doi:10.1139/f83-271.

Kirby, R., Parker, W. R., and Van Oostrum, W. H. A. (1980). Definition of the 40 seabed in navigation routes through mud areas, *Proceedings of the International Hydrographic Review*, Monaco LVII, (1) pp. 107–117.

Kirichek, A., Chassagne, C., Winterwerp, H., and Vellinga, T. (2018). How navigable are fluid mud layers, *Terra et Aqua* **151**.

Kirk, J. T. (2010). *Light and Photosynthesis in Aquatic Ecosystems, Third Edition* (Cambridge University Press), ISBN 9781139168212, doi: 10.1017/CBO9781139168212.

Kleinhans, M., Montfort, O., Dankers, P. J. T., Van Rijn, L. C., and Bonne, W. (2005). Mud dynamics on the shoreface and upper shelf (Noordwijk, The Netherlands), in *Sandpit Project*, Vol. 5 (Aqua Publications), pp. Q1–Q16.

Knoch, D. and Malcherek, A. (2011). A numerical model for simulation of fluid mud with different rheological behaviors, *Ocean Dynamics* **61**, 2–3, doi: 10.1007/s10236-010-0327-x.

Komar, P. D. (1998). *Beach Processes and Sedimentation* (Prentice-Hall, New York).

Kranenburg, C. (1992). Hindered settling and consolidation of mud — analytical results, Tech. Rep., Delft University of Technology, Delft.

Kranenburg, C. (1994). The fractal structure of cohesive sediment aggregates, *Estuarine, Coastal and Shelf Science* **39**, pp. 451–460.

Kranenburg, C. and Winterwerp, J. C. (1997). Erosion of fluid mud layers. I: Entrainment model, *Journal of Hydraulic Engineering* **123**, 6, pp. 504–511.

Kranenburg, W. M. (2008). *Modelling Wave Damping by Fluid Mud, Derivation of a Dispersion Equation and an Energy Dissipation Term and Implementation into SWAN*, Ph.D. thesis, Delft University of Technology.

Kranenburg, W. M., Winterwerp, J. C., De Boer, G. J., Cornelisse, J. M., and Zijlema, M. (2011). SWAN-Mud: Engineering model for mud-induced wave damping, *Journal of Hydraulic Engineering* **137**, 9, pp. 959–975.

Krebs, M. and Weilbeer, H. (2008). Ems-dollart estuary, in *Die Küste, 74 ICCE*, 74, pp. 252–262.

Kristensen, E., Penha-Lopes, G., Delefosse, M., Valdemarsen, T., Quintana, C. O., and Banta, G. T. (2012). What is bioturbation? The need for a precise definition for fauna in aquatic sciences, *Marine Ecology Progress Series* **446**, pp. 285–302.

Krone, R. (1962). *Flume Studies of the Transport of Sediment in Estuarial Shoaling Processes* (Hydraulic Engineering Laboratory and Sanitary Engineering Research Laboratory, Berkeley).

Krone, R. (1993). Sedimentation revisited, Elsevier, *Coastal and Estuarine Studies*, pp. 108–125.

Krone, R. B. (1986). The significance of aggregate properties to transport processes, in *Estuarine Cohesive Sediment Dynamics*, pp. 66–84, doi: 10.1007/978-1-4612-4936-8_4.

Kuijper, C., Christiansen, H., Cornelisse, J. M., and Winterwerp, J. C. (2005). Reducing Harbor Siltation. II: Case Study of Parkhafen in Hamburg, *Journal of Waterway, Port, Coastal, and Ocean Engineering* **131**, 6, pp. 267–276, doi: 10.1061/(asce)0733-950x(2005)131:6(267).

Kuijper, K. and Van Rijn, L. C. (2011). Analytical and numerical analysis of tides and salinities in estuaries; Part II: Salinity distributions in prismatic and convergent tidal channels, *Ocean Dynamics* **61**, 11, pp. 1743–1765, doi: 10.1007/s10236-011-0454-z.

Kurup, P. G. (1972). Littoral currents in relation to the mud bank formation along the coast of Kerala, *Mahasagar* **5**, 3, p. 158.

Kusuda, T., Umita, T., Koga, K., Futawatari, T., and Awaya, Y. (1984). Erosional processes of cohesive sediment, *Water Science Engineering* **17**, pp. 891–901.

Kynch, G. (1952). A theory of sedimentation, *Transactions of the Faraday Society* **58**, pp. 166–176.

Lagaly, G. and Jasmund, K. (1993). Tonminerale und tone, *Struktur, Eigenschaften, Anwendungen und Einsatz in Industrie und Umwelt ed K Jasmund and G Lagaly (Darmstadt: Steinkopff Verlag).*

Lamb, H. (1916). XLIV. On waves due to a travelling disturbance, with an application to waves in superposed fluids, *The London, Edinburgh, and Dublin Philosophical Magazine and Journal of Science* **31**, 184, pp. 386–399, doi: 10.1080/14786440408635511.

Lamb, H. (1932). *Hydrodynamics* (Cambridge University Press, Cambridge).

Lambe, T. W. and Whitman, R. V. (1979). *Soil Mechanics, SI Version* (John Wiley & Sons, New York).

Launder, B. E. and Spalding, D. B. (1972). *Mathematical Models of Turbulence* (Academic Press, London).

Le Hir, P., Bassoullet, P., and Jestin, H. (2000). Application of the continuous modeling concept to simulate high-concentration suspended sediment in a macrotidal estuary, *Proceedings in Marine Science* **3**, C, pp. 229–247, doi: 10.1016/S1568-2692(00)80124-2.

Le Hir, P., Cann, P., Jestin, H., and Bassoullet, P. (2006). Instrumentation légère pour la mesure de lérodabilité des sédiments vaseux ou sablo-vaseux, in *IXèmes Journées Nationales Génie Civil-Génie Côtier, Brest*, pp. 539–546, doi:10.5150/jngcgc.2006.052-l.

Le Hir, P., Cann, P., Waeles, B., Jestin, H., and Bassoullet, P. (2008). Erodibility of natural sediments: Experiments on sand/mud mixtures from laboratory and field erosion tests, in *Proceedings in Marine Science*, Vol. 9 (Elsevier), pp. 137–153.

Le Hir, P. and Cayocca, F. (2002). 3D application of the continuous modeling concept to mud slides in open seas, in *Proceedings in Marine Science*, pp. 545–562.

Le Hir, P., Cayocca, F., and Waeles, B. (2011). Dynamics of sand and mud mixtures: A multiprocess-based modelling strategy, *Continental Shelf Research* **31**, 10 Suppl., pp. S135–149, doi:10.1016/j.csr.2010.12.009.

Le Hir, P., Monbet, Y., and Orvain, F. (2007). Sediment erodability in sediment transport modelling: Can we account for biota effects? *Continental Shelf Research* **27**, 8, pp. 1116–1142.

Lecroart, P., Schmidt, S., Anschutz, P., and Jouanneau, J.-M. (2007). Modeling sensitivity of biodiffusion coefficient to seasonal bioturbation, *Journal of Marine Research* **65**, pp. 417–444.

Lee, B. J., Toorman, E., Molz, F. J., and Wang, J. (2011). A two-class population balance equation yielding bimodal flocculation of marine or estuarine sediments, *Water Research* **45**, 5, pp. 2131–2145, doi: 10.1016/j.watres.2010.12.028.

Lenselink, G. and Menke, U. (1995). *Geologische en bodemkundige atlas van het Markermeer* (Rijkswaterstaat, Directie IJsselmeergebied, Lelystad.).

Lesser, G. R., Roelvink, J. A., Van Kester, J. A., and Stelling, G. S. (2004). Development and validation of a three-dimensional morphological model, *Coastal Engineering* **51**, 8–9, pp. 883–915, doi:10.1016/j.coastaleng.2004.07.014.

Levich, V. G. and Tobias, C. W. (1963). Physicochemical hydrodynamics, *Journal of the Electrochemical Society*, doi:10.1149/1.2425619.

Li, C., Walker, N., Hou, A., Georgiou, I., Roberts, H., Laws, E., McCorquodale, J. A., Weeks, E., Li, X., and Crochet, J. (2008). Circular plumes in Lake Pontchartrain estuary under wind straining, *Estuarine, Coastal and Shelf Science* **80**, 1, pp. 161–172, doi:10.1016/j.ecss.2008.07.020.

Lick, W. (1982). The transport of contaminants in the Great Lakes. *Annual Review of Earth & Planetary Sciences* **10**, pp. 327–353, doi: 10.1146/annurev.ea.10.050182.001551.

Lie-Yauw Oey and Mellor, G. L. (1993). Subtidal variability of estuarine outflow, plume, and coastal current: A model study, *Journal of Physical Oceanography* **23**, 1, pp. 164–171.

Lindenberg, J., Van Rijn, L. C., and Winterwerp, J. C. (1989). Some experiments on wave-induced liquefaction of soft cohesive soils, *Journal of Coastal Research*, pp. 127–137.

Liu, J. C. and Znidarčić, D. (1991). Modeling one-dimensional compression characteristics of soils, *Journal of Geotechnical Engineering*, doi: 10.1061/(ASCE)0733-9410(1991)117:1(162).

Liu, K. and Mei, C. C. (1989). Effects of wave-induced friction on a muddy seabed modelled as a Bingham-plastic fluid, *Journal of Coastal Research* **5**, 4, pp. 777–789.

Liu, P. L. (1973). Damping of water waves over porous bed. *ASCE Journal of Hydraulic Division* **99**, HY12, pp. 2263–2271.

Locat, J. and Demers, D. (1988). Viscosity, remolded strength and liquidity index relationship for high water content soils, *Canadian Geotechnical Journal* **25**, pp. 799–806.

Lofquist, K. (1960). Flow and stress near an interface between stratified liquids, *Physics of Fluids* **3**, 2, pp. 158–175, doi:10.1063/1.1706013.

Longuet-Higgins, M. S. (1953). Mass transport in water waves, *Philosophical Transactions of the Royal Society of London. Series A, Mathematical and Physical Sciences* **245**, 903, pp. 535–581.

Longuet-Higgins, M. S. (1958). The mechanics of the boundary layer near the bottom in a progressive wave, in *Proceedings from 6th International Conference Coastal Engineering, Berkeley (CA), ASCE*, Vol. 184, p. 193.

Los, F. J. and Blaas, M. (2010). Complexity, accuracy and practical applicability of different biogeochemical model versions, *Journal of Marine Systems* **81**, 1–2, pp. 44–74, doi:10.1016/j.jmarsys.2009.12.011.

Luettich, R. A., Harleman, D. R., and SomlyÓdy, L. (1990). Dynamic behavior of suspended sediment concentrations in a shallow lake perturbed by episodic wind events, *Limnology and Oceanography* **35**, 5, pp. 1050–1067, doi:10.4319/lo.1990.35.5.1050.

Lynch, D. R., McGillicuddy, D. J., and Werner, F. E. (2009). Skill assessment for coupled biological/physical models of marine systems, doi: 10.1016/j.jmarsys.2008.05.002.

Maa, J. P. and Mehta, A. J. (1990). Soft mud response to water waves, *Journal of Waterway, Port, Coastal, and Ocean Engineering* **116**, 5, pp. 634–650, doi:10.1061/(asce)0733-950x(1990)116:5(634).

Maa, J. P., Wright, L. D., Lee, C. H., and Shannon, T. W. (1993). VIMS sea carousel: A field instrument for studying sediment transport, *Marine Geology* **115**, 3–4, pp. 271–287, doi:10.1016/0025-3227(93)90056-2.

MacCready, P. and Geyer, W. R. (2010). Advances in estuarine physics, *Annual Review of Marine Science* **2**, 1, pp. 35–58, doi:10.1146/annurev-marine-120308-081015.

Macpherson, H. (1980). The attenuation of water waves over a non-rigid bed, *Journal of Fluid Mechanics* **97**, 4, pp. 721–742, doi: 10.1017/S002211 2080002777.

Maggi, F. (2005). *Flocculation Dynamics of Cohesive Sediment*, Ph.D. thesis, Delft University of Technology, Delft, The Netherlands.

Maggi, F., Mietta, F., and Winterwerp, J. (2007). Effect of variable fractal dimension on the floc size distribution of suspended cohesive sediment, *Journal of Hydrology* **343**, pp. 43–55.

Malmaeus, J. M. and Håkanson, L. (2003). A dynamic model to predict suspended particulate matter in lakes, *Ecological Modelling* **167**, 3, pp. 247–262, doi: 10.1016/S0304-3800(03)00166-2.

Malvern, L. (1969). *Introduction to the Mechanics of a Continuous Medium* (Prentice-Hall).

Mango, A. J., Schmeeckle, M. W., and Furbish, D. J. (2004). Tidally induced groundwater circulation in an unconfined coastal aquifer modeled with a Hele–Shaw cell, *Geology* **32**, 3, pp. 233–236, doi:10.1130/G19922.1.

Manning, A. J. and Dyer, K. R. (2002). The use of optics for the in situ determination of flocculated mud characteristics, *Journal of Optics A: Pure and Applied Optics*, doi:10.1088/1464-4258/4/4/366.

Manning, A. J., Schoellhamer, D. H., Mehta, A. J., Nover, D., and Schladow, S. G. (2010). Video measurements of flocculated sediment in lakes and estuaries in the USA, in *Proc. Joint Federal Interagency Conference Sedimentation and Hydrologic Modeling*, Riviera Hotel, Las Vegas, Nevada, USA, 27th June–1st July.

Martinius, A. and Van den Berg, J. (2011). *Atlas of Sedimentary Structures in Estuarine and Tidally-Influenced River Deposits of the Rhine-Meuse-Scheldt System* (EAGE Publications), ISBN 978-90-73834-11-8.

Mastbergen, D. R. and Van den Berg, J. H. (2003). Breaching in fine sands and the generation of sustained turbidity currents in submarine canyons, *Sedimentology* **50**, 4, pp. 625–637, doi:10.1046/j.1365-3091.2003.00554.x.

Mathew, R. and Winterwerp, J. C. (2017). Surficial sediment erodibility from time-series measurements of suspended sediment concentrations: Development and validation, *Ocean Dynamics* **67**, 6, pp. 691–712, doi: 10.1007/s10236-017-1055-2.

Mathew, R. and Winterwerp, J. C. (2020). Sediment dynamics and transport regimes in a narrow microtidal estuary, *Ocean Dynamics* **70**, 4, pp. 435–462, doi:10.1007/s10236-020-01345-9.

Mayor-Mora, R., Mortensen, P., and Fredsoe, J. (1977). Sedimentation studies on the Niger River Delta, in *Fifteenth Coastal Engineering Conference* (Hawaii, USA, 1976), Vol. 2, pp. 2151–2169, doi:10.1061/9780872620834.126.

McAnally, W. H., Friedrichs, C., Hamilton, D., Hayter, E., Shrestha, P., Rodriguez, H., Sheremet, A., and Teeter, A. (2007). Management of fluid mud in estuaries, bays, and lakes. I: present state of understanding on character and behavior, *Journal of Hydraulic Engineering* **133**, 1, pp. 9–22, doi: 10.1061/(asce)0733-9429(2007)133:1(9).

McBride, M., Boll, M., Briggs, M., *et al.* (2014). Harbour approach channels — Design guidelines, Tech. Rep., PIANC.

McIvor, A. L., Möller, I., Spencer, T., and Spalding, M. (2012). Reduction of wind and swell waves by mangroves, Tech. Rep., The Nature Conservancy and Wetlands International.

McIvor, A. L., Möller, I., Spencer, T., and Spalding, M. (2013). Mangroves as a sustainable coastal defence, in *Proceedings of international Conference Asian and Pacific Coasts, APAC 2013*, pp. 24–26.

McLaren (1994). Sediment transport in the Western Scheldt between Baarland and Rupelmonde. Tech. Rep., Report prepared for the Antwerp Harbour Authorities.

McNeil, J., Taylor, C., and Lick, W. (1996). Measurements of erosion of undisturbed bottom sediments with depth, *Journal of Hydraulic Engineering* **122**, 6, pp. 316–324, doi:10.1061/(asce)0733-9429(1996)122:6(316).

Meakin, P. (1986). Computer simulation of growth and aggregation processes, in H. Stanley and N. Ostrowsky (eds.), *On Growth and Form, Fractal and Non-fractal Patterns in Physics* (Martinus Nijhoff Publishers), pp. 111–135, doi:10.1007/978-94-009-5165-57.

Meakin, P. (1998). *Fractals, Scaling and Growth far from Equilibrium*, Vol. 5 (Cambridge University Press, Cambridge), doi:10.5860/choice.36-0383.

Mehta, A. (2013). *An Introduction to Hydraulics of fine Sediment Transport* (World Scientific Publishing Company), ISBN 978-981-4449-48-9.

Mehta, A. J. (1991). Review notes on cohesive sediment erosion, in *Coastal Sediments'91*, pp. 40–53.

Mehta, A. J., Hwang, K. N., and Khare, Y. P. (2015). Critical shear stress for mass erosion of organic-rich fine sediments, *Estuarine, Coastal and Shelf Science* **165**, pp. 97–103, doi:10.1016/j.ecss.2015.08.020.

Mehta, A. J. and Partheniades, E. (1982). Resuspension of deposited cohesive sediment beds, in *Proceedings of the Coastal Engineering Conference, 1982*, Vol. 2 (Cape Town, South Africa), pp. 1569–1588, doi:10.9753/icce.v18.95.

Mehta, A. J. and Srinivas, R. (1993). Observations on the entrainment of fluid mud by shear flow, *Nearshore and Estuarine Cohesive Sediment Transport* **42**, pp. 224–246.

Mei, C. C. (1989). *The Applied Dynamics of Ocean Surface Waves*, Vol. 1 (World Scientific), ISBN 0471064076, doi:10.1016/0029-8018(84)90033-7.

Mei, C. C. and Ko-Fei Liu (1987). A Bingham-plastic model for a muddy seabed under long waves, *Journal of Geophysical Research* **92**, C13, doi: 10.1029/jc092ic13p14581.

Meirelles, S. and Vinzon, S. B. (2016). Field observation of wave damping by fluid mud, *Marine Geology* **376**, pp. 194–201, doi:10.1016/j.margeo.2016.03.006.

Mendez, F. J. and Losada, I. J. (2004). An empirical model to estimate the propagation of random breaking and nonbreaking waves over vegetation fields, *Coastal Engineering* **51**, 2, pp. 103–118, doi:10.1016/j.coastaleng. 2003.11.003.

Merckelbach, L. M. and Kranenburg, C. (2004a). Determining effective stress and permeability equations for soft mud from simple laboratory experiments, *Geotechnique* **54**, 9, pp. 581–591, doi:10.1680/geot.2004.54.9.581.

Merckelbach, L. M. and Kranenburg, C. (2004b). Equations for effective stress and permeability of soft mud-sand mixtures, *Geotechnique* **54**, 4, pp. 235–243, doi:10.1680/geot.2004.54.4.235.

Meysman, F. J., Boudreau, B. P., and Middelburg, J. J. (2003). Relations between local, nonlocal, discrete and continuous models of bioturbation, *Journal of Marine Research* **61**, 3, pp. 391–410, doi:10.1357/002224003322201241.

Meysman, F. J., Galaktionov, O. S., Gribsholt, B., and Middelburg, J. J. (2006). Bioirrigation in permeable sediments: Advective pore-water transport induced by burrow ventilation, *Limnology and Oceanography* **51**, 1 I, pp. 142–156, doi:10.4319/lo.2006.51.1.0142.

Meysman, F. J., Malyuga, V. S., Boudreau, B. P., and Middelburg, J. J. (2008). A generalized stochastic approach to particle dispersal in soils and sediments, *Geochimica et Cosmochimica Acta* **72**, 14, pp. 3460–3478, doi: 10.1016/j.gca.2008.04.023.

Michalowski, R. L. (1997). Stability of uniformly reinforced slopes, *Journal of Geotechnical and Geoenvironmental Engineering* **123**, 6, pp. 546–556, doi: 10.1061/(asce)1090-0241(1997)123:6(546).

Middelburg, J. J., Soetaert, K., and Herman, P. M. (1997). Empirical relationships for use in global diagenetic models, *Deep-Sea Research Part I: Oceanographic Research Papers* **44**, 2, pp. 327–344, doi:10.1016/S0967-0637(96)00101-X.

Mietta, F., Chassagne, C., and Winterwerp, J. C. (2009). Shear-induced flocculation of a suspension of kaolinite as function of pH and salt concentration, *Journal of Colloid and Interface Science* **336**, 1, pp. 134–141, doi: 10.1016/j.jcis.2009.03.044.

Migniot, C. (1968). Étude des propriétés physiques de différents sédiments très fins et de leur comportement sous des actions hydrodynamiques, *La Houille Blanche*, 7, pp. 591–620, doi:10.1051/lhb/1968041.

Migniot, C. (1989a). Tassement et rhéologie des vases. Deuxième partie, *La Houille Blanche*, 2, pp. 95–112, doi:10.1051/lhb/1989006.

Migniot, C. (1989b). Tassement et rhéologie des vases. Première partie, *La Houille Blanche*, 1, pp. 11–29, doi:10.1051/lhb/1989001.

Millero, F. J. and Poisson, A. (1981). International one-atmosphere equation of state of seawater, *Deep Sea Research Part A, Oceanographic Research Papers* **28**, 6, pp. 625–629, doi:10.1016/0198-0149(81)90122-9.

Mitchell, J. (1976). *Fundamentals of Soil Behavior* (John Wiley & Sons, New York).

Möller, O. O., Castaing, P., Salomon, J. C., and Lazure, P. (2001). The influence of local and non-local forcing effects on the subtidal circulation of Patos Lagoon, *Estuaries* **24**, 2, pp. 297–311, doi:10.2307/1352953.

Montserrat, F., Suykerbuyk, W., Al-Busaidi, R., Bouma, T. J., Van der Wal, D., and Herman, P. M. (2011). Effects of mud sedimentation on lugworm ecosystem engineering, *Journal of Sea Research* **65**, 1, pp. 170–181, doi: 10.1016/j.seares.2010.09.003.

Moore, F. (1959). The rheology of ceramic slip and bodies, *Trans. Brit. Ceram. Soc.* **58**, pp. 470–492.

Munk, W. and Anderson, E. (1948). Notes on a theory of the thermocline, *J. Marine Res.*, 7, p. 276.

Muste, M. and Patel, V. C. (1997). Velocity profiles for particles and liquid in open-channel flow with suspended sediment, *Journal of Hydraulic Engineering* **123**, 9, pp. 742–751.

Nauw, J. J., Merckelbach, L. M., Ridderinkhof, H., and Van Aken, H. M. (2014). Long-term ferry-based observations of the suspended sediment fluxes through the Marsdiep inlet using acoustic Doppler current profilers, *Journal of Sea Research* **87**, pp. 17–29, doi:10.1016/j.seares.2013.11.013.

Nedeco (1968). *Surinam Transportation Study: Report on Hydraulic Investigation* (Delft Hydraulics, The Hague, The Netherlands), p. 293.

Nederlof, L. (1978). "Varen boven slib" in het Rotterdamse haven- en rivierengebied, Tech. Rep., Gementewerken Rotterdam. 98.21-R7332., Rotterdam.

Nepf, H. M. (2011). Flow and transport in regions with aquatic vegetation, *Annual Review of Fluid Mechanics* **44**, pp. 123–142, doi:10.1146/annurev-fluid-120710-101048.

Nepf, H. M. (2012). Hydrodynamics of vegetated channels, *Journal of Hydraulic Research* **50**, 3, pp. 262–279, doi:10.1080/00221686.2012.696559.

Nezu, I. and Nakagawa, N. (1993). Turbulence in open channel flow, *International Association for Hydraulic Research, Monograph Series.*

Nezu, I. and Rodi, W. (1986). Open-channel flow measurements with a laser doppler anemometer, *Journal of Hydraulic Engineering* **112**, 5, pp. 335–355, doi:10.1061/(asce)0733-9429(1986)112:5(335).

Ng, C. O. (2000). Water waves over a muddy bed: A two-layer Stokes' boundary layer model, *Coastal Engineering* **40**, 3, pp. 221–242, doi:10.1016/S0378-3839(00)00012-0.

Noordhuis, R. (2008). *Data overview of Markermeer.* Waterdienst, Lelystad, The Netherlands.

Noordhuis, R., Genseberger, M., and Thiange, C. (2014). Luwtemaatregelen Hoornse Hop — Bijdrage Deltares aan MIRT verkenning Hoornse Hop waterbeweging, slib en ecologie. Tech. Rep., Deltares report 1207128, Delft.

Noordhuis, R. and Houwing, E.-J. (2003). Decrease in zebra mussels in Markermeer, Tech. Rep., RWS RIZA report 2003.016.

Obi, S., Inoue, K., Furukawa, T., and Masuda, S. (1996). Experimental study on the statistics of wall shear stress in turbulent channel flows, *International Journal of Heat and Fluid Flow* **17**, 3, pp. 187–192, doi:10.1016/0142-727X(96)00041-0.

Odd, N. and Cooper, A. (1988). A two-dimensional model of the movement of fluid mud in a high energy turbid estuary. Technical Report. Hydraulics Research Wallingford, *Journal of Coastal Research*, 5, pp. 185–193.

Oreskes, N., Shrader-Frechette, K., and Belitz, K. (1994). Verification, validation, and confirmation of numerical models in the earth sciences, *Science* **263**, 5147, pp. 641–646, doi:10.1126/science.263.5147.641.

Orvain, F. and Sauriau, P. G. (2002). Environmental and behavioural factors affecting activity in the intertidal gastropod Hydrobia ulvae, *Journal of Experimental Marine Biology and Ecology* **272**, 2, pp. 191–216, doi: 10.1016/S0022-0981(02)00130-2.

Orvain, F., Sauriau, P. G., Sygut, A., Joassard, L., and Le Hir, P. (2004). Interacting effects of Hydrobia ulvae bioturbation and microphytobenthos on the erodibility of mudflat sediments, *Marine Ecology Progress Series* **278**, pp. 205–223, doi:10.3354/meps278205.

Papanastasiou, T. C. and Alexandrou, A. N. (1987). Isothermal extrusion of non-dilute fiber suspensions, *Journal of Non-Newtonian Fluid Mechanics* **25**, 3, pp. 313–328, doi:10.1016/0377-0257(87)85032-2.

Papenmeier, S., Schrottke, K., Bartholomä, A., and Flemming, B. W. (2013). Sedimentological and rheological properties of the water-solid bed interface in the weser and ems estuaries, North Sea, Germany: Implications for fluid mud classification, *Journal of Coastal Research* **29**, 4, pp. 797–808, doi: 10.2112/JCOASTRES-D-11-00144.1.

Parchure, T. M. and Mehta, A. J. (1985). Erosion of Soft Cohesive Sediment Deposits, *Journal of Hydraulic Engineering* **111**, 10, pp. 1308–1326, doi: 10.1061/(asce)0733-9429(1985)111:10(1308).

Parker, G. (2006). Gary Parker's Morphodynamics Web Page: http://hydrolab.illinois.edu/people/parkerg/default.asp.

Parker, G., Fukushima, Y., and Pantin, H. M. (1986). Self-accelerating turbidity currents, *Journal of Fluid Mechanics* **171**, pp. 145–181.

Parker, W., Abbott, M., Chapter, W. P., and 1994, U. (1994). Determining depth and navigability in fine sediment areas, *Coastal, Estuarial and Harbor Engineer's Reference Book*, pp. 611–614.

Partheniades, E. (1962). *A study of erosion and deposition of cohesive soils in salt water*, Ph.D. thesis, University of California, Berkeley.

Partheniades, E. (1965). Erosion and deposition of cohesive soils, *Journal of the Hydraulics Division* **91**, pp. 105–139.

Partheniades, E. (1986). A fundamental framework for cohesive sediment dynamics, in *Estuarine Cohesive Sediment Dynamics* (Springer), pp. 219–250.

Partheniades, E. (2009). *Cohesive Sediments in Open Channels* (Butterworth-Heinemann), doi:10.1016/B978-1-85617-556-2.X0001-7.

Pashias, N., Boger, D. V., Summers, J., and Glenister, D. J. (1996). A fifty cent rheometer for yield stress measurement, *Journal of Rheology* **40**, 6, pp. 1179–1189, doi:10.1122/1.550780.

Patel, V. C., Rodi, W., and Scheuerer, G. (1985). Turbulence models for near-wall and low reynolds number flows — A review, *AIAA Journal* **23**, 9, pp. 1308–1319, doi:10.2514/3.9086.

Paterson, D. M. (1989). Short-term changes in the erodibility of intertidal cohesive sediments related to the migratory behavior of epipelic diatoms, *Limnology and Oceanography* **34**, pp. 223–234, doi:10.4319/lo.1989.34.1.0223.

Pemberton, G. S., Risk, M. J., and Buckley, D. E. (1976). Supershrimp: Deep bioturbation in the Strait of Canso, Nova Scotia, *Science* **192**, 4241, pp. 790–791, doi:10.1126/science.192.4241.790.

PIANC (2014). Harbour Approach Channels — Design Guidelines. Report MarComWG 121.

Poggi, D., Porporato, A., Ridolfi, L., Albertson, J. D., and Katul, G. G. (2004). The effect of vegetation density on canopy sub-layer turbulence, *Boundary-Layer Meteorology* **111**, pp. 565–587.

Pollock, L. W. and Hummon, W. D. (1971). Cyclic changes in interstitial water content, atmospheric exposure, and temperature in a Marine Beach, *Limnology and Oceanography*, doi:10.4319/lo.1971.16.3.0522.

Pope, S. (2001). *Turbulent flows* (Cambridge University Press).

Postma, H. (1961). Transport and accumulation of suspended matter in the Dutch Wadden Sea, *Netherlands Journal of Sea Research* **1**, 1–2, pp. 148–190.

Prandle, D. (1981). Salinity intrusion in estuaries. *Journal of Physical Oceanography* **11**, 10, pp. 1311–1324.

Prandle, D. (2009). *Estuaries: Dynamics, Mixing, Sedimentation and Morphology* (Cambridge University Press).

Proudman, J. and Doodson, A. T. (1924). V. The principal constituent of the tides of the North Sea, *Philosophical Transactions of the Royal Society of London. Series A, Containing Papers of a Mathematical or Physical Character* **224**, 616-625, pp. 185–219, doi:10.1098/rsta.1924.0005.

Rakotonimaro, T. V., Neculita, C. M., Bussière, B., Benzaazoua, M., and Zagury, G. J. (2017). Recovery and reuse of sludge from active and passive treatment of mine drainage-impacted waters: A review, *Environmental Science and Pollution Research*, doi:10.1007/s11356-016-7733-7.

Ralston, D. K., Geyer, W. R., and Lerczak, J. A. (2008). Subtidal salinity and velocity in the Hudson River estuary: Observations and modeling, *Journal of Physical Oceanography* **38**, 4, pp. 753–770, doi:10.1175/2007JPO3808.1.

Raudkivi, A. J. and Witte, H. (1990). Development of bed features, *Journal of Hydraulic Engineering* **116**, 9, pp. 1063–1079, doi:10.1061/(asce)0733-9429(1990)116:9(1063).

Refsgaard, J. C., Van der Sluijs, J. P., Brown, J., and Van der Keur, P. (2006). A framework for dealing with uncertainty due to model structure error, *Advances in Water Resources* **29**, 11, pp. 1586–1597, doi: 10.1016/j.advwatres.2005.11.013.

Richardson, J. and Zaki, W. (1954). Sedimentation and fluidisation: Part I, *Transactions of the Institution of Chemical Engineers* **32**, pp. 35–53.

Riedel, T., Lettmann, K., Beck, M., and Brumsack, H. J. (2010). Tidal variations in groundwater storage and associated discharge from an intertidal coastal aquifer, *Journal of Geophysical Research: Oceans*, doi: 10.1029/2009JC005544.

Ris, R. C., Holthuijsen, L. H., and Booij, N. (1999). A third-generation wave model for coastal regions 2. Verification, *Journal of Geophysical Research: Oceans* **104**, C4, doi:10.1029/1998jc900123.

Rodi, W. (1984). *Turbulence Models and Their Applications in Hydraulics-a State-of-the-art Review* (IAHR), doi:10.1016/0045-7825(81)90171-7.

Rodriguez, H. N. (2000). *Mud Bottom Evolution at Open Coasts*, Ph.D. thesis, University of Florida.

Rodriguez, H. N., Mehta, A. J., Beach, W. P., Rodriguezt, H. N., and Mehta, A. J. (2001). Modeling muddy coast response to waves, *Journal of Coastal Research* **27**, pp. 137–148.

Roelvink, D. and Reniers, A. (2012). *A Guide to Modeling Coastal Morphology*, Vol. 12 (World Scientific).

Rogers, W. E. and Holland, K. T. (2009). A study of dissipation of wind-waves by mud at Cassino Beach, Brazil: Prediction and inversion, *Continental Shelf Research* **29**, 3, pp. 676–690, doi:10.1016/j.csr.2008.09.013.

Roscoe, K. H., Schofield, A. N., and Wroth, C. P. (1958). On the yielding of soils, *Géotechnique* **8**, 1, pp. 22–53, doi:10.1680/geot.1958.8.1.22.

Rouse, H. (1939). An analysis of sediment transportation in the light of fluid turbulence, Tech. Rep., United States Department of Agriculture.

Ruddy, G., Turley, C. M., and Jones, T. E. R. (1998). Ecological interaction and sediment transport on an intertidal mudflat I. Evidence for a biologically mediated sediment-water interface, *Geological Society, London, Special Publications* **139**, 1, pp. 135–148.

Russi, D., ten Brink, P., Farmer, A., Badura, T., Coates, D., Förster, J., Kumar, R., and Davidson, N. (2013). The economics of ecosystems and biodiversity for water and wetlands, *IEEP, London and Brussels* **78**.

Sakakiyama, T. and Byker, E. W. (1989). Mass transport velocity in mud layer due to progressive waves, *Journal of Waterway, Port, Coastal, and Ocean Engineering* **115**, 5, pp. 614–633, doi:10.1061/(asce)0733-950x(1989)115:5(614).

Sanford, L. P. (2008). Modeling a dynamically varying mixed sediment bed with erosion, deposition, bioturbation, consolidation, and armoring, *Computers and Geosciences* **34**, 10, pp. 1263–1283, doi:10.1016/j.cageo.2008.02.011.

Sanford, L. P. and Maa, J. P. (2001). A unified erosion formulation for fine sediments, *Marine Geology* **179**, 1–2, pp. 9–23, doi:10.1016/S0025-3227(01)00201-8.

Sanford, L. P., Panageotou, W., and Halka, J. P. (1991). Tidal resuspension of sediments in northern Chesapeake Bay, *Marine Geology* **97**, 1–2, pp. 87–103, doi:10.1016/0025-3227(91)90020-5.

Sassi, M., Duran-Matute, M., Van Kessel, T., and Gerkema, T. (2015). Variability of residual fluxes of suspended sediment in a multiple tidal-inlet system: The Dutch Wadden Sea, *Ocean Dynamics* **65**, 9–10, pp. 1321–1333, doi: 10.1007/s10236-015-0866-2.

Savenije, H. H. (2005). *Salinity and Tides in Alluvial Estuaries* (Elsevier), doi: 10.1016/B978-0-444-52107-1.X5000-X.

Schofield, A. N. and Wroth, P. (1968). *Critical State Soil Mechanics*, Vol. 310 (McGraw-Hill, London).

Sellmeijer, R. and Oortmerssen, G. (1984). The effect of mud on tanker manoeuvres, *Royal Institution of Naval Architects Transactions* **126**.

Sheng, Y. P. (1986). XVII. Modeling bottom boundary layer and cohesive sediment dynamics in estuarine and coastal waters, *Estuarine Cohesive Sediment Dynamics. Proc. Workshop, Tampa, 1984*, pp. 360–400, doi:10.1007/978-1-4612-4936-8_17.

Sheremet, A., Jaramillo, S., Su, S. F., Allison, M. A., and Holland, K. T. (2011). Wave-mud interaction over the muddy Atchafalaya subaqueous clinoform, Louisiana, United States: Wave processes, *Journal of Geophysical Research: Oceans* **116**, 6, doi:10.1029/2010JC006644.

Simmons, H. (1955). Some Effects of Upload Discharge on Estuarine Hydraulics, *Proceedings of the American Society of Civil Engineers* **81**, 9, pp. 1–20.

Simpson, J. H., Brown, J., Matthews, J., and Allen, G. (1990). Tidal straining, density currents, and stirring in the control of estuarine stratification, *Estuaries* **13**, 2, pp. 125–132.

Skelland, A. H. P. (1967). *Non-Newtonian Flow and Heat Transfer* (John Wiley & Sons, New York).

Skempton, A. W. (1961). Effective stress in soils, concrete and rocks, pore pressure and suction in soils, in *Conference Organized by the British National Society of International Society of Soil Mechanics and Foundation Eng* (Butterworths, London).

Skempton, A. W. (1985). Residual strength of clays in landslides, folded strata and the laboratory, *Geotechnique* **35**, 1, pp. 3–18, doi:10.1680/geot.1985.35.1.3.

Sloff, C. J., Jagers, H. R., Kobayashi, E., and Kitamura, Y. (2006). Enhanced flushing of sediment deposits in the backwater reach of reservoirs, in *River, Coastal and Estuarine Morphodynamics: RCEM 2005 — Proceedings 4th IAHR Symptoms River, Coastal and Estuarine Morphodynamics*, Vol. 1, ISBN 0415393752, pp. 483–491, doi:10.1201/9781439833896.ch53.

Smagorinsky, J. (1963). General circulation experiments with the primitive equations, *Monthly Weather Review* **91**, 3, pp. 99–164.

Smerdon, E. T., Beasley, R. P., *et al.* (1959). *The tractive force theory applied to stability of open channels in cohesive soils* (University of Missouri, College of Agriculture, Agricultural Experiment Station).

Snarski, S. and Lueptow, R. (2006). Wall pressure and coherent structures in a turbulent boundary layer on a cylinder in axial flow, *Journal of Fluid Mechanics* **137**.

Soltanpour, M., Shibayama, T., and Noma, T. (2003). Cross-shore mud transport and beach deformation model, *Coastal Engineering Journal* **45**, 3, pp. 363–386, doi:10.1142/S0578563403000804.

Song, D., Wang, X. H., Kiss, A. E., and Bao, X. (2011). The contribution to tidal asymmetry by different combinations of tidal constituents, *Journal of Geophysical Research: Oceans* **116**, 12, doi:10.1029/2011JC007270.

Soulsby, R. L. and Clarke, S. (2005). Bed shear-stress under combined waves and currents on smooth and rough beds (TR 137), Tech. Rep., HR Wallingford.

Soulsby, R. L., Hamm, L., Klopman, G., Myrhaug, D., Simons, R. R., and Thomas, G. P. (1993). Wave-current interaction within and outside the bottom boundary layer, *Coastal Engineering* **21**, 1–3, pp. 41–69, doi: 10.1016/0378-3839(93)90045-A.

Soulsby, R. L., Manning, A. J., Spearman, J., and Whitehouse, R. J. (2013). Settling velocity and mass settling flux of flocculated estuarine sediments, *Marine Geology* **339**, pp. 1–12, doi:10.1016/j.margeo.2013.04.006.

Spalding, M. (2010). *World Atlas of Mangroves* (Routledge), doi: 10.4324/9781849776608.

Spierenburg, S. E. J. (1987). *Seabed Response to Water Waves*, Ph.D. thesis, Delft University of Technology.

Spingat, F. and Oumeraci, H. (2000). Schwebstoffdynamik in der Trübungszone des Ems-Ästuars-Anwendung eines Analysekonzeptes für hoch aufgelöste und dauerhaft betriebene Gewässergütemessungen, *Die Küste, 62* **62**, pp. 159–219.

Sposito, G. (2008). *The Chemistry of Soils* (Oxford University Press, Oxford).

Stelling, G. S. and Van Kester, J. A. T. M. (1994). On the approximation of horizontal gradients in sigma co-ordinates for bathymetry with steep bottom slopes, *International Journal for Numerical Methods in Fluids* **18**, 10, pp. 915–935.

Stowa (1999). *Good Modelling Practice Handbook.* (WL — Delft Hydraulics, TNO, Stowa, Wageningen University, Alterra and Rijkswaterstaat RIZA, STOWA report 99-05, Dutch Dept. of Public Works, Institute for Inland Water Management and Waste Water Treatment, report 99.036), ISBN ISBN 90-5773-056-1.

Suijlen, J. and Duin, R. (2002). Atlas of near-surface total suspended matter concentrations in the Dutch coastal zone of the North Sea, Tech. Rep., Rijkswaterstaat, RIKZ.

Sumer, B. and Fredsøe, J. (2002). *The Mechanics of Scour in the Marine Environment* (World Scientific Publishing Company).

Swart, D. H. (1977). Predictive equations regarding coastal transports, *Coastal Engineering 1976* **2**, pp. 1113–1132, doi:10.1061/9780872620834.066.

Talke, S. A., De Swart, H. E., and Schuttelaars, H. M. (2008). An analytical model of the equilibrium distribution of suspended sediment in an estuary, in *River, Coastal and Estuarine Morphodynamics: RCEM 2007 — Proc. 5th IAHR Symposium on River, Coastal and Estuarine Morphodynamics*, Vol. 1, pp. 403–411, doi:10.1201/noe0415453639-c51.

Tanino, Y. and Nepf, H. M. (2008). Laboratory investigation of mean drag in a random array of rigid, emergent cylinders, *Journal of Hydraulic Engineering* **134**, 1, pp. 34–41, doi:10.1061/(asce)0733-9429(2008)134:1(34).

Taylor, D., Hall, K., and MacDonald, N. (2007). Investigating Ship Induced Scour in a Confined Shipping Channel, in *Coastal Sediments'07*, pp. 2213–2228.

Taylor, G. I. (1954). The dispersion of matter in turbulent flow through a pipe, *Proceedings of the Royal Society of London. Series A. Mathematical and Physical Sciences* **223**, 1155, pp. 446–468.

Te Slaa, S. (2020). *Deposition and Erosion of Silt–Rich Sediment–Water Mixtures*, Ph.D. thesis, Delft University of Technology.

Temmerman, S., Bouma, T. J., Van de Koppel, J., Van der Wal, D., De Vries, M. B., and Herman, P. M. (2007). Vegetation causes channel erosion in a tidal landscape, *Geology* **35**, 7, pp. 631–634, doi:10.1130/G23502A.1.

Tennekes, H., Lumley, J. L., Lumley, J. L. (1972). *A First Course in Turbulence* (MIT press).

Terzaghi, K. (1943). *Theoretical Soil Mechanics* (John Wiley & Sons, New York), doi:10.1002/9780470172766.

Thatcher, M. L. and Harleman, D. R. F. (1972). A mathematical model for the prediction of unsteady salinity intrusion in estuaries, Tech. Rep., MIT School of Engineering Massachusetts Institute of Technology, Department of Civil Engineering.

Tolhursf, T. J., Gust, G., and Paterson, D. M. (2002). The influence of an extracellular polymeric substance (EPS) on cohesive sediment stability, in *Proceedings in Marine Science*, Vol. 5 (Elsevier), pp. 409–425, doi: 10.1016/S1568-2692(02)80030-4.

Tolhurst, T. J., Black, K. S., Shayler, S. A., Mather, S., Black, I., Baker, K., and Paterson, D. M. (1999). Measuring the *in situ* erosion shear stress of intertidal sediments with the cohesive strength meter (CSM), *Estuarine, Coastal and Shelf Science* **49**, 2, pp. 281–294, doi:10.1006/ecss.1999.0512.

Toorman, E. A. (1994). An analytical solution for the velocity and shear rate distribution of non-ideal Bingham fluids in concentric cylinder viscometers, *Rheologica Acta* **33**, 3, pp. 193–202, doi:10.1007/BF00437304.

Toorman, E. A. (1996). Sedimentation and self-weight consolidation: General unifying theory, *Geotechnique* **46**, 1, pp. 103–113, doi:10.1680/geot.1996.46.1.103.

Toorman, E. A. (2002). Modelling of turbulent flow with suspended cohesive sediment, in *Proceedings in Marine Science*, Vol. 5 (Elsevier), pp. 155–169, doi:10.1016/S1568-2692(02)80014-6.

Toorman, E. A. (2008). Vertical Mixing in the Fully Developed Turbulent Layer of Sediment-Laden Open-Channel Flow, *Journal of Hydraulic Engineering* **134**, 9, pp. 1225–1235, doi:10.1061/(asce)0733-9429(2008)134:9(1225).

Torfs, H. (1995). *Erosion of mud/sand mixtures*, Ph.D. thesis, Katholieke Universiteit Leuven, Belgium, https://core.ac.uk/download/pdf/35111687.pdf.

Traykovski, P., Trowbridge, J., and Kineke, G. (2015). Mechanisms of surface wave energy dissipation over a high-concentration sediment suspension, *Journal of Geophysical Research: Oceans* **120**, 3, pp. 1638–1681, doi: 10.1002/2014JC010245.

Trowbridge, J. H. and Kineke, G. C. (1994). Structure and dynamics of fluid muds on the Amazon continental shelf, *Journal of Geophysical Research* **99**, C1, pp. 865–874, doi:10.1029/93JC02860.

Tsiampousi, A., Zdravković, L., and Potts, D. M. (2013). A new Hvorslev surface for critical state type unsaturated and saturated constitutive models, *Computers and Geotechnics* **48**, pp. 156–166, doi:10.1016/j.compgeo.2012.09.010.

Tubman, M. W. and Suhayda, J. N. (1976). Wave Action And Bottom Movements In Fine Sediments, *Coastal Engineering Proceedings* **1**, 15, p. 68, doi: 10.9753/icce.v15.68.

Turner, J. S. (1973). *Buoyancy Effects in Fluids* (Cambridge University Press), doi:10.1017/cbo9780511608827.

Uiterwijk Winkel, A. (1975). Micro-biological aspects and the sedimentation behaviour of riverine cohesive sediment, Tech. Rep., Rijkswaterstaat, Directie Waterhuishouding en Waterbeweging, District Zuid-West, Report 44.006.01.

Uittenbogaard, R. E. (1994). Physics of turbulence: Technical report on sub-task 5.2, Tech. Rep., Delft Hydraulics Delft, The Netherlands.

Uncles, R. J., Stephens, J. A., and Smith, R. E. (2002). The dependence of estuarine turbidity on tidal intrusion length, tidal range and residence time, *Continental Shelf Research* **22**, 11-13, pp. 1835–1856, doi:10.1016/S0278-4343(02)00041-9.

Valle-Levinson, A. (2010). *Contemporary Issues in Estuarine Physics* (Cambridge University Press), doi:10.1017/CBO9780511676567.

Van Alphen, J. (1990). A mud balance for Belgian-Dutch coastal waters between 1969 and 1986, *Netherlands Journal of Sea Research* **25**, 1–2, pp. 19–30.

Van Craenenbroeck, K. and Vantorre, M. (1991). Navigation in Muddy Areas: Establishing the Navigable Depth in the Port of Zeebrugge, in *Proceedings CEDA PIANC Conference, Accessible Harbours*, Vol. 57, pp. 21–28.

Van den Bosch, B. A. P. (2016). *The Effect of Initial Concentration on the Consolidation Behaviour of Mud: A Study on lake Markermeer Sediment.*, Ph.D. thesis, Delft University of Technology.

Van der Kaaij, T., Boon, J., and Los, F. (2003). Kust- en Zeestudies ONL, Effectketen Noordzee: Voortgangsverslag fase 2. Tech. Rep., WL — Delft Hydraulics report Z3376. Delft, The Netherlands (in Dutch).

Van der Kaaij, T., Van Kessel, T., Troost, T., Herman, P., Van Duren, L., and Villars, M. (2017). Modelondersteuning MER winning suppletie- en ophoogzand Noordzee 2018–2027: Modelvalidatie. Tech. Rep., Deltares. Rapport no. 1230888, Delft, The Netherlands (in Dutch).

Van der Loeff, M. M. (1981). Wave effects on sediment water exchange in a submerged sand bed, *Netherlands Journal of Sea Research* **15**, 1, pp. 100–112, doi:10.1016/0077-7579(81)90009-0.

Van der Wegen, M. and Jaffe, B. E. (2013). Towards a probabilistic assessment of process-based, morphodynamic models, *Coastal Engineering* **75**, pp. 52–63, doi:10.1016/j.coastaleng.2013.01.009.

Van der Wegen, M. and Roelvink, J. A. (2008). Long-term morphodynamic evolution of a tidal embayment using a two-dimensional, process-based model, *Journal of Geophysical Research: Oceans* **113**, C3.

Van Duin, E. H. S. (1992). *Sediment Transport, Light and Algal Growth in the Markermeer*, Ph.D. thesis, Agricultural University, Wageningen.

Van Genuchten, M. T. (1980). A closed-form equation for predicting the hydraulic conductivity of unsaturated soils, *Soil Science Society of America Journal* **44**, 5, pp. 892–898, doi:10.2136/sssaj1980.03615995004400050002x.

Van Gils, J. and Tatman, S. (2003). Light penetration in the water column, Tech. Rep., WL — Delft Hydraulics, Delft, The Netherlands, MARE Report, WL2003001 Z3379.

Van Heteren, S., Kiden, P., and Hettelaar, J. (2007). Inventarisatie slibgehalte Noordzeebodem tussen Cadzand en Den Helder. Tech. Rep., TNO-rapport 2007-U-R0584/C.

Van Kessel, T. and Cornelisse, J. M. (2003). Physical Scale Model Zeeschelde, *Delft Hydraulics, Report Z* **2516**.

Van Kessel, T., De Goede, E. D., and Van den Boogaard, H. (2006). Year simulations with the ZUNO-DD model on the impact of Maasvlakte-2 on large-scale silt transport. A comparison with 14-day simulations. Tech. Rep., WL — Delft Hydraulics report Z4160., Delft.

Van Kessel, T. and Kranenburg, C. (1998). Wave-induced liquefaction and flow of subaqueous mud layers, *Coastal Engineering* **34**, 1–2, pp. 109–127.

Van Kessel, T. and Van Maren, D. S. (2013). Far-field and long-term dispersion of released dredged material, in *Proceedings of XXth WODCON Conference, 9p.*

Van Kessel, T. and Vanlede, J. (2010). Impact of harbour basins on mud dynamics Scheldt estuary. Tech. Rep., Deltares report 1200253., Delft.

Van Kessel, T., Vanlede, J., and De Kok, J. (2011a). Development of a mud transport model for the Scheldt estuary, *Continental Shelf Research* **31**, 10 Suppl., pp. S165–S181, doi:10.1016/j.csr.2010.12.006.

Van Kessel, T., Vanlede, J., Eleveld, M., Van der Wal, D., and B., D. M. (2011b). Validation and application of mud model Scheldt estuary. Tech. Rep., Deltares report 1202021, Delft.

Van Kessel, T., Winterwerp, H., Van Prooijen, B., Van Ledden, M., and Borst, W. (2011c). Modelling the seasonal dynamics of SPM with a simple algorithm for the buffering of fines in a sandy seabed, *Continental Shelf Research* **31**, 10 Suppl., pp. S124–S134, doi:10.1016/j.csr.2010.04.008.

Van Kessel, T., Van Oeveren-Theeuwes, M. C., and Van Rooijen, A. A. (2012). Kalibratieslibtransportmodel voor de Hollandse kustzone aan de hand van cadmium-metingen, Tech. Rep., Deltares.

Van Kesteren, W. and Bakker, W. (1985). Near bottom velocities in waves with a current: Analytical and numerical computations, in *Coastal Engineering 1984* (American Society of Civil Engineers, New York, NY), pp. 1161–1177, doi:10.1061/9780872624382.080.

Van Ledden, M. (2002). A process-based sand-mud model, *Proceedings in Marine Science*, pp. 577–594.

Van Ledden, M. (2003). *Sand-Mud Segregation in Estuaries and Tidal Basins*, Ph.D. thesis, Delft University of Technology, Delft.

Van Ledden, M., Wang, Z. B., Winterwerp, H., and De Vriend, H. (2004). Sand-mud morphodynamics in a short tidal basin, in *Ocean Dynamics*, Vol. 54, pp. 385–391, doi:10.1007/s10236-003-0050-y.

Van Leeuwen, B. (2008). Modeling mussel bed influence on fine sediment dynamics on a Wadden Sea intertidal flat, *Mastersthesis, University of Twente.*

Van Leussen, W. (1988). Aggregation of particles, settling velocity of mud flocs a review, in *Physical Processes in Estuaries* (Springer), pp. 347–403.

Van Leussen, W. (1994). *Estuarine Macroflocs and Their Role in Fine-Grained Sediment Transport*, Ph.D. thesis, University of Utrecht, Utrecht.

Van Maldegem, D. (2004). Overzicht gegevens voor LTV-slibmodel Schelde-estuarium. Tech. Rep., RIKZ, Werkdocument RIKZ/AB/2004.821x.

Van Maren, D. S. (2007). Grain size and sediment concentration effects on channel patterns of silt-laden rivers, *Sedimentary Geology* **202**, 1–2, pp. 297–316.

Van Maren, D. S. and Cronin, K. (2016). Uncertainty in complex three-dimensional sediment transport models: Equifinality in a model application of the Ems Estuary, The Netherlands, *Ocean Dynamics* **66**, 12, pp. 1665–1679.

Van Maren, D. S. and Gerritsen, H. (2012). Residual flow and tidal asymmetry in the Singapore Strait, with implications for resuspension and residual transport of sediment, *Journal of Geophysical Research: Oceans* **117**, C4.

Van Maren, D. S., Liew, S. C., and Hasan, G. M. J. (2014). The role of terrestrial sediment on turbidity near Singapore's coral reefs, *Continental Shelf Research* **76**, pp. 75–88.

Van Maren, D. S., Van Kessel, T., Cronin, K., and Sittoni, L. (2015a). The impact of channel deepening and dredging on estuarine sediment concentration, *Continental Shelf Research* **95**, pp. 1–14.

Van Maren, D. S., Vroom, J., Fettweis, M., and Vanlede, J. (2020). Formation of the Zeebrugge coastal turbidity maximum: The role of uncertainty in near-bed exchange processes, *Marine Geology*, doi:10.1016/j.margeo.2020.106186.

Van Maren, D. S. and Winterwerp, J. C. (2013). The role of flow asymmetry and mud properties on tidal flat sedimentation, *Continental Shelf Research* **60**, pp. S71–S84.

Van Maren, D. S., Winterwerp, J. C., Decrop, B., Wang, Z. B., and Vanlede, J. (2011). Predicting the effect of a current deflecting wall on harbour siltation, *Continental Shelf Research* **31**, 10, p. S182–S198.

Van Maren, D. S., Winterwerp, J. C., Sas, M., and Vanlede, J. (2009a). The effect of dock length on harbour siltation, *Continental Shelf Research* **29**, 11–12, pp. 1410–1425.

Van Maren, D. S., Winterwerp, J. C., and Vroom, J. (2015b). Fine sediment transport into the hyper-turbid lower Ems River: the role of channel deepening and sediment-induced drag reduction, *Ocean Dynamics* **65**, 4, pp. 589–605.

Van Maren, D. S., Winterwerp, J. C., Wu, B. S., and Zhou, J. J. (2009b). Modelling hyperconcentrated flow in the Yellow River, *Earth Surface Processes and Landforms* **34**, 4, pp. 596–612.

Van Maren, D. S. and Cronin, K. (2016). Uncertainty in complex three-dimensional sediment transport models: Equifinality in a model application of the Ems Estuary, the Netherlands, *Ocean Dynamics* **66**, 12, pp. 1665–1679.

Van Prooijen, B. C., Montserrat, F., and Herman, P. M. (2011). A process-based model for erosion of Macoma balthica-affected mud beds, *Continental Shelf Research* **31**, 6, pp. 527–538, doi:10.1016/j.csr.2010.12.008.

Van Prooijen, B. C. and Wang, Z. B. (2013). A 1D model for tides waves and fine sediment in short tidal basins: Application to the Wadden Sea, *Ocean Dynamics* **63**, 11–12, pp. 1233–1248, doi:10.1007/s10236-013-0648-7.

Van Prooijen, B. C. and Winterwerp, J. C. (2010). A stochastic formulation for erosion of cohesive sediments, *Journal of Geophysical Research: Oceans* **115**, 1, doi:10.1029/2008JC005189.

Van Rijn, L. (1993). Principles of Sediment Transport in Rivers, Estuaries and 39 Coastal Seas, *Principles of Sediment Transport in Rivers, Estuaries and 40 Coastal Seas* Aqua Publications.

Van Rijn, L. C. (1984). Sediment Pick-Up Functions, *Journal of Hydraulic Engineering* **110**, 10, doi:10.1061/(asce)0733-9429(1984)110:10(1494).

Van Rijn, L. C. (1986). *Manual Sediment Transport Measurements* (Delft Hydraulics Laboratory).

Van Rijn, L. C. (2005). *Principles of Sedimentation and Erosion Engineering n Rivers, Estuaries and Coastal Seas* (Aqua Publications).

Van Rijn, L. C. and Barth, R. (2018). Settling 1 and consolidation of soft mud–sand layers, *Journal of Waterway, Port, Coastal and Ocean Engineering* **145**, 1, doi:10.1061/(ASCE)WW.1943-5460.0000483.

Van Straaten, L. M. J. U. and Kuenen, P. H. (1957). Accumulation of fine grained sediments in the Dutch Wadden Sea, *Geologie en Mijnbouw* **19**, 1954, pp. 329–354.

Van Vuren, B. G. (2005). *Stochastic Modelling of River Morphodynamics*, Ph.D. thesis, Delft University of Technology.

Vanlede, J., Delecluyse, K., Primo, B., Verheyen, B., Leyssen, G., Plancke, Y., Verwaest, T., and Mostaert, F. (2015). Verbetering randvoorwaardenmodel: Subreport 7 — Calibration of NEVLA 3D. Tech. Rep., Flanders Hydraulics Research and IMDC: Antwerp, Belgium.

Vanlede, J., Dujardin, A., Fettweis, M., Van Hoestenberghe, T., and Martens, C. (2019). Mud dynamics in the Port of Zeebrugge, *Ocean Dynamics* **69**, 9, doi:10.1007/s10236-019-01273-3.

Vanoni, V. A. (1940). Experiments on the transportation of suspended sediment by water. Dissertation (Ph.D.), California Institute of Technology. doi: 10.7907/CYT2-W932

Vantorre, M. (1991). Ship behaviour and control at low speed in layered fluids, in *Hadmar'91, Int. Symp. Hydro- and Aerodynamics in Marine engineering*, pp. 1–9.

Vantorre, M. (2001). Nautical bottom approach. Application to the access to the harbour of Zeebrugge, *HANSA, Schiffahrt, Schiffbau, Hafen, 138 Jahrgang, Nr.* **6**, pp. 93–97.

Vardon, P. J., Yao, Y., Van Paassen, L. A., and Van Tol, A. F. (2015). The use of a large-strain consolidation model to optimise multilift tailing deposits, in *Proceedings Tailings and Mine Waste Management for the 21st Century, Sydney (Australia), 27-28 July, 2015* (The AusIMM), pp. 1–8.

Veld, H., De Weert, J., Van Kessel, T., Verheul, M., De Kleine, M., Vonhögen, L. (2016). Protocol for sampling and measuring for geological studies for sand mining in the North Sea. Deltares report 1230624-000, Delft, the Netherlands (in Dutch).

Verbeek, H., Wang, Z. B., and Thoolen, P. M. C. (1999). Secondary currents in estuarine morphodynamic modelling: A case-study of the Western Scheldt, *River, Coastal and Estuarine Morphodynamics*, pp. 649–659.

Verlaan, P. (1998). *Mixing of Marine and Fluvial Particles in the Scheldt Estuary*, Ph.D. thesis, Delft University.

Verlaan, P. A. (2000). Marine vs fluvial bottom mud in the Scheldt Estuary, *Estuarine, Coastal and Shelf Science* **50**, 5, pp. 627–638, doi: 10.1006/ecss.1999.0599.

Verlaan, P. A., Donze, M., and Kuik, P. (1998). Marine vs fluvial suspended matter in the Scheldt Estuary, *Estuarine, Coastal and Shelf Science* **46**, 6, pp. 873–883, doi:10.1006/ecss.1997.0323.

Verney, R., Lafite, R., Brun-Cottan, J. C., and Le Hir, P. (2011). Behaviour of a floc population during a tidal cycle: Laboratory experiments and numerical modelling, *Continental Shelf Research* **31**, 10 Suppl., pp. S64–S83, doi: 10.1016/j.csr.2010.02.005.

Verwey, E. J. W. and Overbeek, J. T. G. (1999). *Theory of the Stability of Lyophobic Colloids 1948* (Elsevier, Amsterdam).

Verwey, J. (1954). On the ecology of distribution of cockle and mussel in the Dutch Waddensea, their role in sedimentation and the source of their food supply, *Archives néerlandaises de Zoologie* **10**, 2, pp. 171–239.

Vicsek, T. (1992). *Fractal Growth Phenomena* (World Scientific), doi: 10.1142/1407.

Vijverberg, T. (2008). Mud dynamics in the Markermeer, *Technische Universiteit Delft, Delft* .

Vijverberg, T., Winterwerp, J. C., Aarninkhof, S. G. J., and Drost, H. (2011). Fine sediment dynamics in a shallow lake and implication for design of hydraulic works, *Ocean Dynamics* **61**, 2-3, pp. 187–202.

Vinzon, S. B. and Mehta, A. J. (1998). Mechanism for formation of lutoclines by waves, *Journal of Waterway, Port, Coastal, and Ocean Engineering* **124**, 3, pp. 147–149.

Vinzon, S. B., Winterwerp, J. C., Nogueira, R., and De Boer, G. J. (2009). Mud deposit formation on the open coast of the larger Patos Lagoon-Cassino Beach system, *Continental Shelf Research* **29**, 3, pp. 572–588, doi: 10.1016/j.csr.2008.09.021.

Visser, M. (1993). *On the Transport of Fine Marine Sediment in the Netherlands Coastal Zone*, Ph.D. thesis, University of Twente.

Volkenborn, N., Hedtkamp, S. I. C., Van Beusekom, J. E. E., and Reise, K. (2007). Effects of bioturbation and bioirrigation by lugworms (Arenicola marina) on physical and chemical sediment properties and implications for intertidal habitat succession, *Estuarine, Coastal and Shelf Science* **74**, 1–2, pp. 331–343.

Vreugdenhil, C. B. (1994). *Numerical Methods for Shallow-Water Flow*, Vol. 13 (Springer Science & Business Media).

Vroom, J., Van Maren, D. S., Van der Werf, J., and Van Rooijen, A. (2017). Zand-slib modellering voor het mondingsgebied van het Schelde estuarium. Deltares report 1210301-002-ZKS-0002, p. 117.

Walker, W., Harremoës, P., Rotmans, J., Van Der Sluijs, J., Van Asselt, M., Janssen, P., and Krayer von Krauss, M. (2003). Defining uncertainty: a conceptual basis for uncertainty management in model-based decision support, *Integrated Assessment* **4**, 1, doi:10.1076/iaij.4.1.5.16466.

Wall, S. (2010). The history of electrokinetic phenomena, *Current Opinion in Colloid & Interface Science* **15**, 3, pp. 119–124.

Walles, B., Salvador de Paiva, J., Van Prooijen, B., Ysebaert, T., and Smaal, A. (2015). The ecosystem engineer crassostrea gigas affects tidal flat morphology beyond the boundary of their reef structures, *Estuaries and Coasts* **38**, 3, doi: 10.1007/s12237-014-9860-z.

Wang, L. (2010). *Tide Driven Dynamics of Subaqueous Fluid Mud Layers in Turbidity Maximum Zones of German Estuaries*, Ph.D. thesis, University of Bremen, Germany.

Wang, Y., Bohlen, W. F., and O'Donnell, J. (2000). Storm enhanced bottom shear stress and associated sediment entrainment in a moderate energetic estuary, *Journal of Oceanography* **56**, 3, doi:10.1023/A:1011155717622.

Wang, Y. H. (2003). The intertidal erosion rate of cohesive sediment: A case study from Long Island Sound, *Estuarine, Coastal and Shelf Science* **56**, 5–6, doi: 10.1016/S0272-7714(02)00215-9.

Wang, Z. B., Jeuken, C., and De Vriend, H. J. (1999). Tidal asymmetry and residual sediment transport in estuaries, *Delft Hydraulics report Z2749.*

Wang, Z. B., Jeuken, M., Gerritsen, H., De Vriend, H. J., and Kornman, B. A. (2002). Morphology and asymmetry of the vertical tide in the Westerschelde estuary, *Continental Shelf Research* **22**, 17, pp. 2599–2609.

Wang, Z. Y., Larsen, P., Nestmann, F., and Dittrich, A. (1998). Resistance and drag reduction of flows of clay suspensions, *Journal of Hydraulic Engineering* **124**, 1, pp. 41–49, doi:10.1061/(ASCE)0733-9429(1998)124:1(41).

Warwick, R. M. and Clarke, K. R. (1984). Species size distribution in marine benthic communities, *Oecologia*, **61**, 1, pp. 32–41, doi:10.1007/BF00379085.

Guan, W. B. E. Wolanski, and L. X. Dong (1998). Cohesive sediment transport in the Jiaojiang River estuary, China, *Estuarine, Coastal and Shelf Science* **46**, pp. 861–871.

Weaver, C. E. (1989). *Clays, Muds, and Shales* (Elsevier).

Webb, J. and Theodor, J. (1972). Wave-induced circulation in submerged sands, *Journal of the Marine Biological Association of the United Kingdom* **52**, pp. 903–914.

Weerman, E. J., Herman, P. M., and Van De Koppel, J. (2011). Top-down control inhibits spatial self-organization of a patterned landscape, *Ecology* **92**, 2, doi: 10.1890/10-0270.1.

Wehr, D. and Malcherek, A. (2012). Numerical simulation of fluid mud dynamics–The isopycnal model MudSim, *Die Küste, 79* , 79, pp. 1–52.

Weilbeer, H. (2008). Numerical simulation and analyses of sediment transport processes in the Ems-Dollard estuary with a three-dimensional model, *Proceedings in Marine Science*, pp. 447–462.

Wells, J. (1987). Entrapment of shelf-destined mud, Cape Lookout Bight, NC, in *South-eastern Section of the Geological Society of America, Abstracts with Progress*, p. 135.

Wells, J. T. and Kemp, G. P. (1986). III. Interaction of surface waves and cohesive sediments: field observations and geologic significance (Louisiana and Surinam). *Estuarine Cohesive Sediment Dynamics. Proceedings Workshop, Tampa, 1984*.

Whitehouse, R. J., Bassoullet, P., Dyer, K. R., Mitchener, H. J., and Roberts, W. (2000). The influence of bedforms on flow and sediment transport over intertidal mudflats, *Continental Shelf Research* **20**, 10–11, pp. 1099–1124, doi:10.1016/S0278-4343(00)00014-5.

Whitlow, R. (1990). *Basic Soil Mechanics* (Harlow, Essex).

Whitman, G. B. (1974). *Linear and Nonlinear Waves.*, Vol. 42 (John Wiley & Sons), doi:10.4249/scholarpedia.4308.

Widdows, J., Brinsley, M. D., Bowley, N., and Barrett, C. (1998). A benthic annular flume for in situ measurement of suspension feeding/biodeposition rates and erosion potential of intertidal cohesive sediments, *Estuarine, Coastal and Shelf Science* **46**, 1, pp. 27–38.

Widdows, J., Friend, P. L., Bale, A. J., Brinsley, M. D., Pope, N. D., and Thompson, C. E. (2007). Inter-comparison between five devices for determining erodability of intertidal sediments, *Continental Shelf Research* **27**, 8, doi: 10.1016/j.csr.2005.10.006.

Widdows, J., Lucas, J. S., Brinsley, M. D., Salkeld, P. N., and Staff, F. J. (2002). Investigation of the effects of current velocity on mussel feeding and mussel bed stability using an annular flume, *Helgoland Marine Research* **56**, 1, p. 3.

Williamson, H. and Ockenden, M. (1996). ISIS: An instrument for measuring erosion shear stress in situ, *Estuarine, Coastal and Shelf Science* **42**, 1, pp. 1–18.

Willmott, C. J. (1981). On the validation of models, *Physical Geography* **2**, 2, pp. 184–194.

Winterwerp, J. (2007). On the deposition flux of cohesive sediment, *Proceedings in Marine Science* **8**, pp. 209–226.

Winterwerp, J. C. (1998). A simple model for turbulence induced flocculation of cohesive sediment, *Journal of Hydraulic Research* **36**, 3, pp. 309–326, doi: 10.1080/00221689809498621.

Winterwerp, J. C. (1999). *On the Dynamics of High-concentrated Mud Suspensions*, Ph.D. thesis, Delft University of Technology.

Winterwerp, J. C. (2001). Stratification effects by cohesive and noncohesive sediment, *Journal of Geophysical Research: Oceans* **106**, C10, pp. 22559–22574, doi:10.1029/2000jc000435.

Winterwerp, J. C. (2002). On the flocculation and settling velocity of estuarine mud, *Continental Shelf Research* **22**, pp. 1339–1360.

Winterwerp, J. C. (2005). Reducing harbor siltation. I: Methodology, *Journal of Waterway, Port, Coastal, and Ocean Engineering* **131**, 6, pp. 258–266, doi:10.1061/(asce)0733-950x(2005)131:6(258).

Winterwerp, J. C. (2011). Fine sediment transport by tidal asymmetry in the high-concentrated Ems River: Indications for a regime shift in response to channel deepening, *Ocean Dynamics* **61**, 2–3, pp. 203–215.

Winterwerp, J. C. and Kranenburg, C. (1997). Erosion of fluid mud layers. II: Experiments and model validation, *Journal of Hydraulic Engineering* **123**, 6, pp. 512–519.

Winterwerp, J. C., Lely, M., and He, Q. (2009). Sediment-induced buoyancy destruction and drag reduction in estuaries, *Ocean Dynamics* **59**, 5, pp. 781–791.

Winterwerp, J. C., Manning, A. J., Martens, C., De Mulder, T., and Vanlede, J. (2006a). A heuristic formula for turbulence-induced flocculation of cohesive sediment, *Estuarine, Coastal and Shelf Science* **68**, pp. 195–207, doi: 10.1016/j.ecss.2006.02.003.

Winterwerp, J. C. and Van Kessel, T. (2003). Siltation by sediment-induced density currents, *Ocean Dynamics* **53**, 3, pp. 186–196.

Winterwerp, J. C. and Van Kesteren, W. (2004). *Introduction to the Physics of Cohesive Sediment in the Marine Environment* (Elsevier, New York).

Winterwerp, J. C., Van Kesteren, W. G. M., Van Prooijen, B., and Jacobs, W. (2012). A conceptual framework for shear flow-induced erosion of soft cohesive sediment beds, *Journal of Geophysical Research: Oceans* **117**, C10, doi:10.1029/2012jc008072.

Winterwerp, J. C., Vroom, J., Wang, Z.-B., Krebs, M., Hendriks, E. C. M., Van Maren, D. S., Schrottke, K., Borgsmüller, C., and Schöl, A. (2017). SPM response to tide and river flow in the hyper-turbid Ems River, *Ocean Dynamics* **67**, 5, pp. 559–583.

Winterwerp, J. C. and Wang, Z. B. (2013). Man-induced regime shifts in small estuaries — I: Theory, *Ocean Dynamics* **63**, 11–12, pp. 1279–1292, doi: 10.1007/s10236-013-0662-9.

Winterwerp, J. C., Wang, Z. B., der Kaaij, T., Verelst, K., Bijlsma, A., Meersschaut, Y., and Sas, M. (2006b). Flow velocity profiles in the Lower Scheldt estuary, *Ocean Dynamics* **56**, 3–4, pp. 284–294.

Winterwerp, J. C., Wang, Z. B., Van Braeckel, A., Van Holland, G., and Kösters, F. (2013). Man-induced regime shifts in small estuaries — II: A comparison of rivers, *Ocean Dynamics* **63**, 11–12, pp. 1293–1306, doi:10.1007/s10236-013-0663-8.

Winterwerp, J. C., Wang, Z. B., Van Der Kaaij, T., Verelst, K., Bijsma, A. C., Meersschaut, Y., and Sas, M. (2006c). Secondary currents by weak salinity gradients in estuaries, *Ocean Dynamics* **56**, pp. 284–294.

Winterwerp, J. C., Zhou, Z., Battista, G., Van Kessel, T., Jagers, H. R. A., Van Maren, D. S., and Van der Wegen, M. (2018). Efficient consolidation model for morphodynamic simulations in low-SPM environments, *Journal of Hydraulic Engineering* **144**, 8, p. 4018055.

Wolanski, E. and McLusky, D. S. (2011). *Treatise on Estuarine and Coastal Science* (Academic Press).

Wood, D. J. (1997). *Pressure-Impulse Impact Problems and Plunging Wave Jet Impact*, Ph.D. thesis, University of Bristol.

Wood, D. M. (1985). Some fall-cone tests, *Geotechnique* **35**, 1, doi: 10.1680/geot.1985.35.1.64.

Worrall, W. E. and Tuliani, S. (1964). Viscosity changes during the ageing of clay-water suspensions, *Transactions British Ceramic Society* **63**, pp. 167–185.

Wrytki, K. (1961). Physical Oceanography of The Southeast Asian Water, Naga Report, Volume 2, Scripps Institute of Oceanography, University of California.

Wurpts, R. and Torn, P. (2005). 15 years experience with fluid mud: Definition of the nautical bottom with rheological parameters, *Terra et Aqua* **99**, pp. 22–32.

Yamamoto, T., Koning, H. L., Sellmeijer, H., and Van Hijum, E. V. (1978). On the response of a poro-elastic bed to water waves, *Journal of Fluid Mechanics* **87**, 1, doi:10.1017/S0022112078003006.

Yamamoto, T. and Takahashi, S. (1985). Wave Damping by Soil Motion, *Journal of Waterway, Port, Coastal, and Ocean Engineering* **111**, 1, doi: 10.1061/(asce)0733-950x(1985)111:1(62).

Yang, J. Q., Kerger, F., and Nepf, H. M. (2015). Estimation of the bed shear stress in vegetated and bare channels with smooth beds, *Water Resources Research* **51**, 5, doi:10.1002/2014WR016042.

Yusuf, M. and Yanagi, T. (2013). Numerical modeling of tidal dynamics in the Java Sea, *Coastal Marine Science* **36**, 1, pp. 1–12.

Zhang, Q., Gong, Z., Zhang, C., Lacy, J. R., Jaffe, B. E., and Xu, B. (2018). Bed shear stress estimation under wave conditions using near-bottom measurements: comparison of methods, *Journal of Coastal Research*, **85**, pp. 241–245, doi:10.2112/SI85-049.1.

Zhou, Z., Van der Wegen, M., Jagers, B., and Coco, G. (2016). Modelling the role of self-weight consolidation on the morphodynamics of accretional mudflats, *Environmental Modelling & Software* **76**, pp. 167–181.

Zhu, Q., Van Prooijen, B., Maan, D., Wang, Z., and Yao, P. (2019). The heterogeneity of mudflat erodibility, *Geomorphology* **345**.

Zhu, Q., Van Prooijen, B., Wang, Z., Ma, Y., and Yang, S. (2016). Bed shear stress estimation on an open intertidal flat using in situ measurements, *Estuarine, Coastal and Shelf Science* **182**, pp. 190–201, doi:10.1016/j.ecss.2016.08.028.

Ziebis, W., Forster, S., Huettel, M., and Jørgensen, B. B. (1996). Complex burrows of the mud shrimp Callianassa truncata and their geochemical impact in the sea bed, *Nature* **382**, 6592, pp. 619–622.

Index

C

F

G

Lightning Source UK Ltd.
Milton Keynes UK
UKHW020614191121
394242UK00001B/33

9 789811 243615